Forensic DNA Analysis

Current Practices and Emerging Technologies

Forensic DNA Analysis

Current Practices and Emerging Technologies

Editor
Jaiprakash G. Shewale, Ph.D.
Technical Director
Life Technologies
Foster City, California, USA

Coeditor
Ray H. Liu
Professor Emeritus
University of Alabama at Birmingham
Birmingham, Alabama, USA
and
Editor-in-Chief
Forensic Science Review
Vancouver, Washington, USA

CRC Press
Taylor & Francis Group
Boca Raton London New York

CRC Press is an imprint of the
Taylor & Francis Group, an **informa** business

CRC Press
Taylor & Francis Group
6000 Broken Sound Parkway NW, Suite 300
Boca Raton, FL 33487-2742

Printed on acid-free paper
Version Date: 20130709

International Standard Book Number-13: 978-1-4665-7126-6 (Hardback)

Library of Congress Cataloging-in-Publication Data

Forensic DNA analysis : current practices and emerging technologies / editors, Jaiprakash G. Shewale, Ray H. Liu.
p. ; cm.
Includes bibliographical references and index.
ISBN 978-1-4665-7126-6 (hardcover : alk. paper)
I. Shewale, Jaiprakash G., editor of compilation. II. Liu, Ray H., editor of compilation.
[DNLM: 1. Forensic Genetics--methods. 2. DNA--analysis. 3. Genotyping Techniques--methods. W 750]

RA1057
614'.12--dc23 2013025792

Visit the Taylor & Francis Web site at
http://www.taylorandfrancis.com

and the CRC Press Web site at
http://www.crcpress.com

Dedicated to my wife Manisha J. Shewale
who loves me more than herself and has supported me during ups
and downs in my scientific career as well as in family life.

Jaiprakash G. Shewale

Contents

Foreword xi
Preface xiii
Contributors xv

Section I

SAMPLE COLLECTION, SAMPLE STORAGE, AND DNA EXTRACTION

1 **Forensic DNA Evidence Collection at a Crime Scene: An Investigator's Commentary** 3
JOSEPH BLOZIS

2 **Optimizing Storage and Handling of DNA Extracts** 19
STEVEN B. LEE, CECELIA A. CROUSE, AND MARGARET C. KLINE

3 **Extraction of DNA from Forensic Biological Samples for Genotyping** 39
JAMES E. STRAY, JASON Y. LIU, MAXIM G. BREVNOV, AND JAIPRAKASH G. SHEWALE

4 **Extraction of DNA from Human Remains** 65
JAMES E. STRAY AND JAIPRAKASH G. SHEWALE

Section II

SAMPLE ASSESSMENT

5 **RNA Profiling for the Identification of the Tissue Origin of Dried Stains in Forensic Biology** 81
ERIN K. HANSON AND JACK BALLANTYNE

6 **Assessment of DNA Extracted from Forensic Samples Prior to Genotyping** 101
MAURA BARBISIN AND JAIPRAKASH G. SHEWALE

Section III

STRs—PROVEN GENOTYPING MARKERS

7 **Principles, Practice, and Evolution of Capillary Electrophoresis as a Tool for Forensic DNA Analysis** 131

JAIPRAKASH G. SHEWALE, LIWEI QI, AND LISA M. CALANDRO

8 **Next-Generation STR Genotyping Kits for Forensic Applications** 163

JULIO J. MULERO AND LORI K. HENNESSY

9 **Biology and Genetics of New Autosomal STR Loci Useful for Forensic DNA Analysis** 181

JOHN M. BUTLER AND CAROLYN R. HILL

10 **Hidden Variation in Microsatellite Loci: Utility and Implications for Forensic DNA** 199

JOHN V. PLANZ AND THOMAS A. HALL

11 **Additional Y-STRs in Forensics: Why, Which, and When** 221

KAYE N. BALLANTYNE AND MANFRED KAYSER

Section IV

EXPANDING THE GENOTYPING CAPABILITIES

12 **Forensic Mitochondrial DNA Analysis: Current Practice and Future Potential** 249

MITCHELL HOLLAND, TERRY MELTON, AND CHARITY HOLLAND

13 **Applications of Autosomal SNPs and Indels in Forensic Analysis** 279

CHRISTOPHER PHILLIPS

14 **Deep-Sequencing Technologies and Potential Applications in Forensic DNA Testing** 311

ROXANNE R. ZASCAVAGE, SHANTANU J. SHEWALE, AND JOHN V. PLANZ

15 **Sample-to-Result STR Genotyping Systems: Potential and Status** **349**

JENNY A. LOUNSBURY, JOAN M. BIENVENUE, AND JAMES P. LANDERS

Section V

TRAINING

16 **Training of Forensic DNA Scientists—A Commentary** **381**

MEREDITH A. TURNBOUGH, ARTHUR J. EISENBERG, LISA SCHADE, AND JAIPRAKASH G. SHEWALE

Index **391**

Foreword

The world population is expected to reach over 9 billion people by 2050, with the majority of this growth occurring in the developing regions of the world. This unparalleled expansion in population density will challenge our existing social infrastructure as governments deal with complex and evolving issues in public health, food production and distribution, environmental protection, and public safety.

Within any society, the need to identify and convict those individuals who commit harm to others and to exonerate those who are falsely accused is fundamental to a modern criminal justice system. No field of science has contributed more to this goal than DNA forensics, and today it is difficult to imagine a system of public safety not supported by the scientific methods developed, tested, and validated for genetic identification and human individualization. The fields of forensics and criminalistics have captured the imagination of the public, as evidenced by the proliferation of books, television shows, and media attention given to the topic. However, beyond mass media appeal, the need exists to provide the most up-to-date and scientifically accurate road map for forensic DNA analysis for those engaged in performing, analyzing, and explaining this global and rapidly evolving field. It has been estimated* that by 2015, 60% of the world's population will live in countries that have either passed DNA database legislation or operate under governmental policies requiring DNA collection from individuals suspected of specific criminal acts. Today, over 40 million profiles populate forensic DNA databases around the globe.

In the early days of DNA forensics, the technology for analyzing biological crime scene evidence was based on radioactive labeling and the detection of DNA, and it could take weeks or months to acquire and analyze the samples. In the 1990s, an alternative to radioactivity was developed: enzyme reactions coupled with chemical reagents that emit light reduced the time to complete a case to a week or so. Today, using the latest technology of DNA amplification (PCR), a complete analysis can often be performed in less than a day, depending on the specifics of the case. However, there are still difficult scientific issues and unsolved needs for the forensic DNA scientist.

Criminal justice systems are challenging the limits of forensic DNA technology, as DNA test results are requested from an expanding array of identity management situations, sample types, and criminal cases. Challenges of forensic DNA investigation include the application to biological samples that yield smaller quantities of DNA, resolution of DNA mixtures (this is of particular importance as DNA testing is applied to more cases involving trace evidence, such as guns and fingerprints), and the application of next-generation sequencing methods to increase the information from crime scene evidence. These new technologies will challenge the scientific community to provide reliable, reproducible, and validated forensic protocols, while legal and ethical considerations will determine the

* Gordon Thomas Honeywell Governmental Affairs.

extent to which these technologies are employed. This volume provides insight into the latest advances as they evolve into 21st century DNA forensics.

Forensic scientists collecting DNA reference samples or performing casework and analysis are presented with an array of possible choices to determine the best operational methods and policies for their laboratories. The latest technologies for collection, storage, and extraction of forensic samples, choice of DNA analysis methods, automation, training, and a path to next-generation advances in DNA forensics are presented in this book. The editors have assembled a comprehensive overview of DNA forensics methodologies that should be of interest to researchers, students, and forensic scientists, both as practitioners and as visionaries of the future.

Leonard Klevan, Ph.D.
Orinda, California

Preface

The first recorded application of medical knowledge to the solution of criminal cases was reported in a 1248 A.D. Chinese book, *Hsi Duan Yu* ("the washing away of wrongs"), which contains a description of how to distinguish drowning from strangulation, and which became an official textbook for coroners (http://www.forensicdna.com/Timeline020702.pdf and http://www.crimezzz.net/forensic_history/index.htm). Nearly seven centuries later, the discovery of ABO blood groups in 1902 by Karl Landsteiner helped to solve crime and paternity cases by a simple immunological technique. In less than another century, the discovery of restriction fragment length polymorphism (RFLP) in 1980 by Ray White, David Botstein, and colleagues, and the generation of individual-specific DNA "fingerprints" using multilocus variable numbers of tandem repeats (VNTR) in 1985 by Sir Alec Jeffreys laid the foundation for current DNA-based human identification methodologies. The invention of polymerase chain reaction (PCR) and multiplex capabilities further enhanced genotyping capabilities. Since then, forensic DNA analysis is probably the fastest growing method for crime investigation. This is evident from the fact that the number of countries adopting a national DNA database has more than tripled, from 16 in 1999 to 54 in 2008, according to a survey by INTERPOL. Many more countries are expected to pass legislation to establish their own national DNA databases. It is estimated that 60% of the world's population will soon live in a country with a DNA database program. The exponential growth in forensic DNA analysis can be attributed to the high power of discrimination provided by these genetic markers, acceptance of DNA results by court systems, legislation passed by government agencies, increased funding, advancements in DNA analysis technologies, and continued success in worldwide case resolution. Genotyping of biological samples is now routinely performed in human identification (HID) laboratories for applications including paternity, forensic casework, DNA databasing, the hunt for missing persons, family lineage studies, identification of human remains, mass disasters, and more. It is important to note that milestone contributions in several other areas played key roles in shaping the currently used genotyping methods in forensic DNA analysis. Some of these milestones are related to the discovery and optimization of restriction endonucleases, Southern blotting, the polymerase chain reaction, multiplex PCR, genetically engineered *Taq* polymerases, spectrally resolvable fluorescence dyes, capillary electrophoresis and automated DNA sequencers, liquid-handling robots and automated systems, and software capabilities.

Forensic DNA analysis in casework encompasses activities ranging from sample collection to testimony in court. The whole process includes multiple steps such as sample collection, sample preservation, evidence examination, body fluid identification, extraction of DNA, assessment of DNA recovered, amplification of target loci, detection of amplified products, data analysis, results interpretation, and report generation. Automation and workflow integration streamlines the entire process.

The types and quality of samples received in forensic laboratories vary to a great extent. These can be grouped on different platforms like body fluids/tissues, source, nature of substrate on which the biological sample is deposited, age, quantity of biological sample, and so forth. Needless to say, attempts have been made to develop "tailor-made" workflows, protocols, and/or genotyping systems for the different sample types. Some examples of such dedicated workflows or systems are direct amplification workflows for reference/single-source samples, miniSTRs for degraded samples, robust short tandem repeat (STR) genotyping systems for inhibited samples, differential extraction for sexual assault samples, and cell separation methods for separation of cell types, and so forth.

It is evident from the literature that STRs received a great deal of attention from forensic scientists over the course of the past two decades. Nevertheless, the potential for applications of single nucleotide polymorphism (SNP) genotyping was not forgotten. Technological challenges and the high cost of analysis have so far prohibited the utilization of these approaches. However, innovations in multiplex PCR design, microarrays, next-generation sequencing, automation, and analytical software provide a promise that new methods for SNPs that are amenable to forensic scientists for routine analysis may be developed in the near future. In recent years, attempts have been made to expand the capabilities of forensic DNA analysis, for example, obtaining genotypes from samples containing minimal quantities of DNA (trace/touch evidence), mixture resolution, obtaining genotypes for investigation tools, DNA profiling at the collection site, and so forth. The potential of next-gen sequencing in forensic investigations is just now beginning to be explored.

The topics covered in this book encompass almost all aspects of forensic DNA analysis, from sample collection at a crime scene to generation of genotypes as well as the utility of new technologies such as next-gen sequencing and sample-to-answer systems.

All chapters are either revised versions or adapted from select review articles published in *Forensic Science Review* volumes 22 (2), 24 (1), 24 (2), and 25 (1) and the images therein are used with permission from the journal's publisher.

Jaiprakash G. Shewale
Life Technologies Corporation
Foster City, California

Contributors

Jack Ballantyne
National Center for Forensic Science
Department of Chemistry
University of Central Florida
Orlando, Florida

Kaye N. Ballantyne
Office of the Chief Forensic Scientist
Forensic Services Department
Victoria Police
Macleod, Victoria
Australia

Maura Barbisin
Life Technologies Corporation
Foster City, California

Joan M. Bienvenue
University of Virginia
Charlottesville, Virginia

Joseph Blozis
Retired NYPD Detective Sergeant
Forensic Investigations Division
Crime Scene Unit
New York, New York

Maxim G. Brevnov
Life Technologies Corporation
Foster City, California

John M. Butler
National Institute of Standards and Technology
Applied Genetics Group
Gaithersburg, Maryland

Lisa M. Calandro
Life Technologies Corporation
Foster City, California

Cecelia A. Crouse
Forensic Biology Unit
Palm Beach County Sheriff's Office Crime Laboratory
West Palm Beach, Florida

Arthur J. Eisenberg
Forensic and Investigative Genetics
University of North Texas Health Science Center
Fort Worth, Texas

Thomas A. Hall
Ibis Biosciences
Carlsbad, California

Erin K. Hanson
National Center for Forensic Science
University of Central Florida
Orlando, Florida

Lori K. Hennessy
Life Technologies Corporation
Foster City, California

Carolyn R. Hill
National Institute of Standards and Technology
Applied Genetics Group
Gaithersburg, Maryland

Charity Holland
Mitotyping Technologies, an AIBioTech Company
State College, Pennsylvania

Mitchell Holland
The Pennsylvania State University
State College, Pennsylvania

Manfred Kayser
Department of Forensic Molecular Biology
Erasmus MC University Medical Center Rotterdam
Rotterdam, South Holland
The Netherlands

Margaret C. Kline
Human Identity Project Team
National Institute of Standards and Technology
Gaithersburg, Maryland

James P. Landers
University of Virginia
Charlottesville, Virginia

Steven B. Lee
Forensic Science Program
Justice Studies Department
San Jose State University
San Jose, California

Jason Y. Liu
Life Technologies Corporation
Foster City, California

Ray H. Liu
Forensic Science Review
Vancouver, Washington

Jenny A. Lounsbury
University of Virginia
Charlottesville, Virginia

Terry Melton
Mitotyping Technologies, an AIBioTech Company
State College, Pennsylvania

Julio J. Mulero
Life Technologies Corporation
Foster City, California

Christopher Phillips
Forensic Genetics Unit
Institute of Legal Medicine
University of Santiago de Compostela
Spain

John V. Planz
Department of Forensic and Investigative Genetics
University of North Texas Health Science Center
Fort Worth, Texas

Liwei Qi
Life Technologies Corporation
Foster City, California

Lisa Schade
Life Technologies Corporation
Foster City, California

Jaiprakash G. Shewale
Life Technologies Corporation
Foster City, California

Shantanu J. Shewale
Department of Forensic and Investigative Genetics
University of North Texas Health Science Center
Fort Worth, Texas

James E. Stray
Life Technologies Corporation
Foster City, California

Meredith A. Turnbough
Forensic and Investigative Genetics
University of North Texas Health Science Center
Fort Worth, Texas

Roxanne R. Zascavage
Department of Forensic and Investigative Genetics
University of North Texas Health Science Center
Fort Worth, Texas

Sample Collection, Sample Storage, and DNA Extraction

I

Forensic DNA Evidence Collection at a Crime Scene
An Investigator's Commentary

1

JOSEPH BLOZIS

Contents

1.1 Introduction 4
1.2 Personal Protective Equipment 5
1.3 Establishing a Forensic Technical Plan and Documentation of the Crime Scene 6
1.4 Recovery of Biological DNA Evidence 6
1.4.1 Sources of DNA Evidence 7
1.4.2 Touch DNA Sample Sizes 7
1.4.3 Probative and Nonprobative Evidence 8
1.4.4 Alternative Light Source 8
1.4.5 Chemical Enhancements 8
1.4.6 Fingerprints and DNA 9
1.4.7 DNA Recovery Supplies 9
1.4.8 Swabbing Techniques for Touch DNA 11
1.4.9 Recovery of Biological DNA Evidence from Various Substrates 11
1.4.10 Types of DNA Samples 15
1.4.11 Packaging DNA Evidence 15
1.4.12 Transporting and Storing DNA Evidence 16
1.5 Conclusions 17
Acknowledgments 17

Abstract: The purpose of this chapter is twofold. The first is to present a law enforcement perspective of the importance of a crime scene, the value of probative evidence, and how to properly recognize, document, and collect evidence. The second purpose is to provide forensic scientists who primarily work in laboratories with insight on how law enforcement personnel process a crime scene. Among all the technological advances in the various disciplines associated with forensic science, none have been more spectacular than those in the field of DNA. The development of sophisticated and sensitive instrumentation has enabled forensic scientists to detect DNA profiles from minute samples of evidence in a much timelier manner. In forensic laboratories, safeguards and protocols associated with American Society of Crime Laboratory Directors/Laboratory Accreditation Board (ASCLD/LAB) International, Forensic Quality Services, and/or ISO/IEC 17020:1998 accreditation have been established and implemented to ensure proper case analysis. But no scientist, no instrumentation, and no laboratory could come to a successful conclusion about evidence if that evidence had been compromised or simply missed at a crime scene. Evidence collectors must be trained thoroughly to process a crime scene and to be able to distinguish between probative evidence and nonprobative evidence. I am a firm believer in the well-known phrase "garbage in, garbage out." The evidence collector's goal is to recover sufficient DNA so that an eligible Combined DNA Index System (CODIS) profile can be generated to not only identify an offender but also, more importantly, to exonerate the innocent.

1.1 Introduction

As a former detective sergeant with 28 years of service, including 20 years assigned to the Forensic Investigations Division of the New York Police Department (NYPD), I believe that the most important factor pertaining to an investigation is the value of the crime scene. It is imperative that, from the outset, the crime scene be handled properly by first responders. The scene must be properly safeguarded and preserved to best maintain its integrity, since it is the crucial source of probative evidence. Proper handling of the crime scene ultimately will ensure a successful conclusion to an investigation and prosecution.

The investigation begins with the initial 911 call. It then is followed by the first responders pivotal actions at the crime scene. First responders have many responsibilities and some are described as follows. First responders must be cognizant of their own safety, including on-scene arrival. They quickly must assess and evaluate what actions must be taken. These actions may include suspect confrontation, victim aid, searching for a suspect, a complainant interview, the determination of potential witnesses and, subsequently, provisions to separate them. First responders also will request additional assistance, determine if there are multiple crime scenes, make command notifications, take copious notes, prepare detailed reports, and safeguard and preserve the scene. All of this must be done in an expeditious manner while maintaining scene integrity.

To successfully preserve a scene, the first responder or responsible party must keep unauthorized persons out, often including police personnel. From firsthand experience, many police officers, including high-ranking supervisors, have a morbid curiosity for the dead. At times, we are our own worst enemies. Once first responders deem the scene safe and victims are tended to, there is no reason to reenter the scene until arrival of the crime scene unit. After the crime scene has been established, the only official personnel permitted into the area are crime scene detectives, medical personnel, medical examiners, the district attorney, the detective supervisor, and the assigned detective. A safeguarding officer should maintain a log of all responding officers at the scene, including those who previously departed. DNA elimination samples of those individuals at the scene are as important as the unknown DNA samples recovered.

Crime scene investigators should establish a single path in and use the same path out to minimize contamination. Edmond Locard's theory of evidence transference states that whenever you enter a crime scene, you leave something in that scene, and whenever you leave the scene, you take a part of that scene out with you. Crime scene investigators should escort the personnel listed above into the scene only when the crime scene investigator deems it safe to do so and in such a way that forensic evidence is not jeopardized. Investigators, by necessity, conduct walkthroughs of the scene to make an assessment and evaluate what occurred; however, these should be conducted only when it is forensically safe to do so.

Once the scene is safeguarded and preserved, detectives and the crime scene unit are notified to respond. The assigned detective and his or her supervisor are in charge of the investigation. The crime scene unit is a support unit to the detectives and forensically processes the scene. However, it is important that both the detectives and the evidence collectors work as a team with open communication and respect for each other's professional experience. The primary functions of a crime scene investigator are to document, process, reconstruct, and collect evidence. The following sections are a brief summary of the equipment, procedures, and materials involved in the important role of the crime scene investigator.

1.2 Personal Protective Equipment

Personal protective equipment (PPE) is used to both prevent contamination and protect the wearer. It consists of a full Tyvek® suit including a hood, booties, face mask, and gloves. DNA evidence becomes contaminated when DNA from another source is mixed in with DNA relevant to the case. There are potentially serious health and safety concerns associated with touching biological evidence. For these reasons, crime scene investigators and laboratory personnel should always wear PPE, use clean instruments, and avoid touching other objects (including their own bodies) when handling evidence or items used to collect evidence. To prevent sample contamination, it is imperative that crime scene investigators change their disposable gloves after the recovery of *each* sample. Contamination of a sample could jeopardize the investigation and subsequent prosecution of suspects. In addition, the investigator must always remember that allowing biological evidence to contact his or her skin may be hazardous to his or her health. PPE protects investigators from hazards such as bloodborne pathogens, prevents contamination of the evidence DNA samples with the collector's DNA, and eliminates chances of cross-contamination of samples collected at the same site.

Not only are frequent glove changes critical to preventing contamination, but equipment must also be free of any DNA contaminants. Equipment should be cleaned prior to use and after each sample is collected. To clean the equipment:

i. Dip the instruments in a 10% chlorine bleach and water solution and swish them around.
ii. Remove the instruments, dip them in a 70% ethanol and water solution, and swish them around.
iii. Rinse the instruments with plain water and allow them to air-dry.

The three-step method above is necessary to ensure that the instrument is not only free of any DNA residue from the previous sample but also safe for use with the next sample. The bleach solution sterilizes the instrument, but if left on the instrument, bleach residue will destroy any future DNA samples. The ethanol solution removes any bleach residue, and the plain water removes any ethanol. Bleach degrades quickly, so it is important to prepare a new bleach solution weekly, or more frequently if practical.

Since it is easier to change gloves than to sterilize instruments, consider collecting items such as cigarette butts and clothing with a gloved hand instead of using tweezers or forceps. Using the tweezers and forceps only when necessary saves time and effort. Equipment brought into the crime scene, such as flashlights, tripods, alternative light sources, and so forth, also can contaminate the scene by introducing trace evidence from other crime scenes. Consider using clean laboratory mats beneath any equipment transported from another scene. Laboratory mats are laminated sheets with a plastic material on the bottom layer and an absorbent paper layer on top. This prevents any equipment contamination from being deposited on a substrate containing DNA evidence.

1.3 Establishing a Forensic Technical Plan and Documentation of the Crime Scene

Upon arrival at the crime scene, a crime scene investigator should confer with first responders, investigators, and victims to ascertain what has happened and what each person's role was at the scene. The investigator then should establish a singular entry-and-exit path within the scene to minimize contamination and disruption of the scene. Perform a scene walkthrough to determine a sequence of events for forensic scene processing.

Before an item with possible DNA evidence is recovered, it must be documented in place. The item must be photographed, documented in the investigator's notes, measured, sketched, and logged into each of the crime scene logs (crime scene photography, evidence log, etc.) before recovery is attempted. When documenting an item with DNA evidence:

i. *Photograph the item in place*, showing how it looked before collection (Figure 1.1).
ii. *Take written notes*, describing the condition of the evidence, what was collected, and how it was collected.
iii. *Take measurements*, showing the location of the evidence and its position relative to other objects.
iv. *Sketch the location* of the evidence in the crime scene sketchbook.

1.4 Recovery of Biological DNA Evidence

Investigators do not always consider DNA as a factor unless there is blood at the scene. However, an investigator can expect to find DNA evidence anywhere within a crime scene with which a suspect has had contact; this may be referred to as contact DNA, transfer DNA, or touch DNA. Almost every component of the human anatomy is a potential source for DNA. Any personal contact between individuals, or between an individual and an object, has the

Figure 1.1 A crime scene investigator photographically documents the scene. Photographs consist of overalls, midrange, close-ups, and/or macros. (Photograph from an unknown source; text was inserted.)

possibility of transferring DNA, even where the possibility of recovering a fingerprint itself is remote. DNA may be extracted from many kinds of biological evidence, including blood, semen, saliva, perspiration, hair (root), skin cells, bone marrow, tooth root/pulp, urine/feces (which may contain epithelial cells), and vomit (which may contain cells from the throat). DNA evidence is readily transferred from the human body and deposited nearly anywhere.

1.4.1 Sources of DNA Evidence

DNA deposit sources are not limited to obvious biological material, such as tissue, stains, or fluids, but also include items such as drinking containers, clothing, surface areas, and any other touched items, and items that may have "received" bodily fluids or skin cells, such as perspiration or saliva. Personal items handled regularly often contain skin cell deposits that are an excellent DNA source. If an item is touched or handled repeatedly, it is likely that skin cells have been deposited onto the item. Items in close daily contact with an individual are potentially rich sources for DNA. Items such as clothing, bedding, and eyeglasses potentially hold thousands of skin cells, hair, perspiration, and oils that have been transferred from the body. Items discarded by suspects at crime scenes, such as drink containers, gum, hats, masks, and bandanas, should be documented, collected, packaged, and sent to the laboratory for DNA analysis. Any physical evidence repeatedly or forcibly handled should be considered a prime source to recover DNA. For example, ligatures often contain skin cells and perspiration from being handled and pulled on by the assailant.

Vehicle surfaces, both interior and exterior, are also prime sources of DNA evidence. For example, a vehicle that strikes an individual and was in close proximity to the detonation of a bomb, or when the vehicle itself was used as a vehicle-borne improvised explosive device (VBIED), also called a "car bomb," is certain to contain DNA evidence. Often the vehicle's grill, fender, and undercarriage are prime locations from which to recover physical evidence with DNA. A vehicle's interior also can contain numerous sources for DNA recovery, including skin cells or perspiration deposited on the arm rests, blood or other trace evidence on the floor mats, and hair stuck to the headliner. Discarded trash in a vehicle also is an important source for DNA. Look for cigarette butts, drink containers, candy, napkins, tissues, and partially eaten food.

The point of entry or exit to/from a crime scene is another excellent place for DNA evidence. It is not uncommon for a perpetrator to be injured during a forced entry, so look for blood, tissue, and hair near windows and doors. Biological evidence recovered at the point of entry is compelling evidence in a judicial proceeding.

1.4.2 Touch DNA Sample Sizes

DNA technological advances of the last 25 years have reduced the size of a usable biological sample. In the 1980s, a 1–2 cm sample drop of blood had a reasonable probability of yielding a DNA profile. In the 1990s, the size was reduced to 1 cm or less. Presently, the recovery of 10 to 20 skin cells has the probability of yielding a DNA profile. In other words, DNA can be extracted from the miniscule amount of skin cells that the human body naturally sheds when it comes in contact with an object. One of the challenges investigators face when recovering touch DNA is that it is rarely visible to the human eye. The investigator must determine where touch DNA is likely to be located, swab the area, and hope that the forensic scientist has enough skin cells to obtain a CODIS-eligible DNA profile.

1.4.3 Probative and Nonprobative Evidence

Probative evidence is recovered from a crime scene and would provide the case investigator with probable cause to make an arrest. It is evidence that would prove or disprove an alleged fact relevant to the investigation. Nonprobative evidence is evidence recovered from a crime scene that would not provide the case investigator probable cause to make an arrest. However, it may provide the case investigator with an investigatory lead or it may be evidence that, at a later date, would prove significant to the investigation. Crime scene investigators must be trained in recognition and collection of probative evidence and the ability to differentiate probative from nonprobative evidence. Evidence collectors must be trained not to burden their crime laboratories with too much nonprobative evidence. Evidence that may yield an investigatory lead should be noted on laboratory requests. Nonprobative evidence at times may be recovered; however, it should be forwarded to an approved storage facility with proper notification to the assigned case investigator.

1.4.4 Alternative Light Source

An alternative light source (ALS) is one of the many commercially available pieces of equipment that will facilitate the search for DNA evidence at a scene. The ALS uses a variety of wavelengths to detect trace evidence normally invisible to the naked eye. Various wavelengths cause certain types of trace evidence to fluoresce and become visible when viewed with filtered goggles, indicating their precise location on a substrate. Examples of trace biological evidence that will fluoresce include fingerprints on both porous and nonporous surfaces, body fluids, skin damage resulting from bitemarks and bruising, bone fragments, and hair. All of these have the potential to yield a DNA sample.

1.4.5 Chemical Enhancements

Chemical enhancements such as BLUESTAR®, luminol, and leuco crystal violet are used at crime scenes to detect the presence of blood. BLUESTAR is a latent bloodstain reagent used to reveal bloodstains that have been washed out, wiped off, or are otherwise invisible to the naked eye. Luminol is a chemoluminescent reaction in the presence of an oxidizing agent on contact with blood; it is visible without the use of an ALS. Leuco crystal violet and hydrogen peroxide in contact with blood triggers a chemical reaction that turns the solution to a purple/violet color (Figure 1.2).

Figure 1.2 (See color insert.) The luminescent reaction that occurs when BLUESTAR®, luminol, or leuco crystal violet are applied to a substrate. (Photograph from BLUESTAR Forensic Web site, http://www.bluestar-forensic.com/, accessed June 9, 2010.)

1.4.6 Fingerprints and DNA

Touch DNA exists where a suspect has touched a surface at the crime scene, possibly leaving a fingerprint as well as DNA. Therefore, before the scene can be processed, it must be decided whether to process for DNA, fingerprints, or both. Areas that are smooth, hard, and nonporous should be processed for fingerprints. Areas that have irregular surfaces should be processed for DNA. Before swabbing an area for DNA, make sure the area is not conducive for fingerprints, since one may be swabbing through a latent fingerprint. As a critical piece of evidence in criminal proceedings, fingerprints always should be a priority at any crime scene for several reasons:

i. Fingerprint processing is more cost-effective than DNA analysis.
ii. Databases containing fingerprints are significantly larger than DNA databases, resulting in a greater opportunity for an identification to be made.
iii. Fingerprint identifications can be made in hours, whereas DNA results can take weeks.

When evaluating a fingerprint, use a magnifying glass to determine if there is a sufficient amount of ridge detail to recover the print. When there is an insufficient amount of ridge detail or the fingerprint is clearly a smudge, consider processing it for DNA. It is possible to recover DNA from areas that have already been processed for fingerprints by swabbing the area after the fingerprint was lifted. Be mindful of contamination issues concerning the fingerprint brush: a fingerprint brush used at other crime scenes has the potential to contaminate DNA from those scenes and be transferred to surfaces at the present crime scene. Substrates, such as drinking glasses, can be processed for both DNA and fingerprints. First, process the rim area for DNA, then process the entire glass for fingerprints. Other small substrates with minute surface areas that will not provide sufficient ridge detail for comparison should be processed for DNA.

1.4.7 DNA Recovery Supplies

Proper supplies at the scene, and the knowledge to use them, are critical to collection of usable DNA samples. A crime scene investigator's toolkit should include basic and necessary supplies for DNA evidence recovery. The supplies normally used for processing and recovering DNA samples include:

i. Sterile cotton-tipped applicators (swabs) to collect samples. Some examples of swabs are presented in Figures 1.3, 1.4, and 1.5.
ii. Hydration to moisten the swab.
   a. Distilled water is acceptable.
   b. Sterile water is better.
   c. Sterile phosphate-buffered saline (PBS) solution is the best.
iii. Plastic pipettes for transferring the distilled water to the swabs.
iv. Paper envelopes used for packaging.

Figure 1.3 (See color insert.) Swabs, distilled water, plastic pipettes, and paper envelopes are components of a basic DNA recovery kit. (Photograph by the author.)

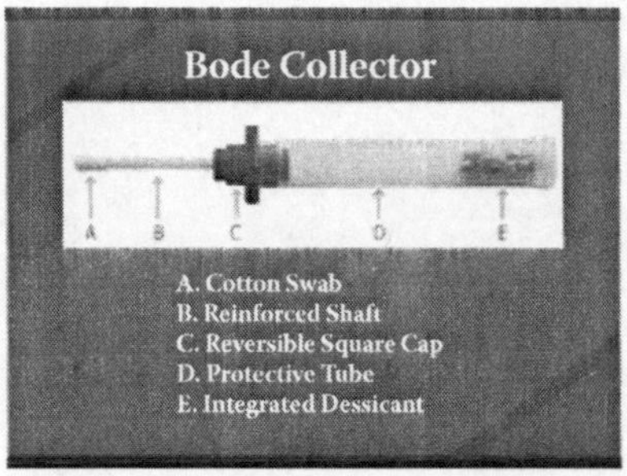

Figure 1.4 The Bode Technology Group's SecurSwab™ Collector consists of a cotton-tipped swab, a reinforced shaft, reversible square cap, a protective tube, and an integrated desiccant. (Photograph from Bode Technology Group Web site, http://www.bodetech.com/, assessed June 6, 2012.)

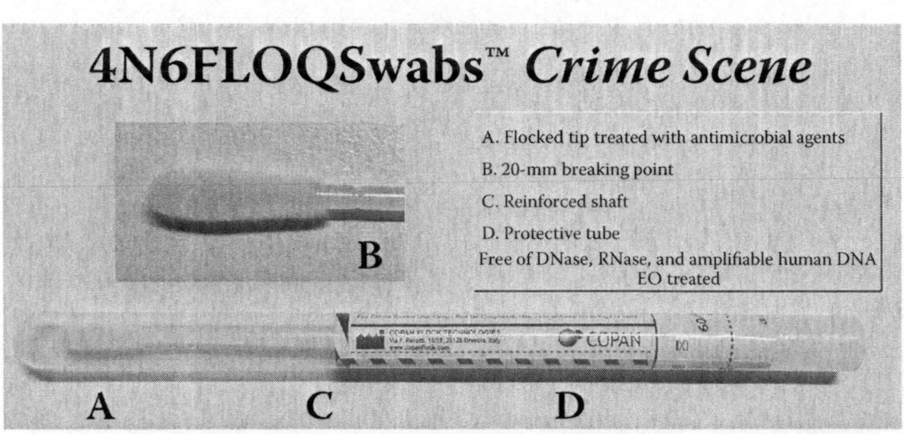

Figure 1.5 4N6FLOQSwabs™ *Crime Scene* swabs utilize patented flock technology to maximize DNA collection and elution efficiency. The swabs are certified as free of DNase, RNase, and amplifiable human DNA and are treated with antimicrobial agents to prevent microbial contamination. (Photograph from Copan Flock Technologies, Brescia, Italy. With permission.)

1.4.8 Swabbing Techniques for Touch DNA

In processing a crime scene for touch DNA, proper swabbing techniques maximize the ability to recover as much DNA as possible from particular substrates. The following procedures are a guide for proper swabbing:

i. Don personal protective equipment.
ii. Withdraw distilled water from the vial with a sterile plastic pipette.
iii. Remove a sterile cotton-tipped swab from a sealed container. Use each swab only once.
iv. Place a drop of distilled water on the swab's side. Avoid saturating the swab. Avoid dipping the swab in distilled water.
v. Swab the area. Rotate the swab so the entire swab surface is used. Avoid reusing areas of swab if possible (may redeposit samples onto substrate). Use one swab for approximately every 15 sq cm of surface area. After using a hydrated swab, it is permissible to use a dry swab when swabbing the same area. Confer with your laboratory concerning the usage of wet and dry swabs.
vi. Use additional swabs as necessary and use a unique identifier to label the swabs accordingly. When swabbing an irregular grainy surface, swab with the grain using a back-and-forth motion while rotating the swab surface to ensure that the same area of the swab is not reused. Reswab the area with a dry sterile swab. After the swabs are air-dried, it is permissible to place both the swabs into one paper enclosure.

1.4.9 Recovery of Biological DNA Evidence from Various Substrates

The following guidelines pertain to the recovery of wet and dried biological samples. A biological sample such as a dried droplet of blood can be recovered by using the moistened portion of a swab as described in Section V. Be cognizant that, for proper swab saturation, only a small amount of water is required for the swab to air-dry quickly and ensure that the sample will be of a proper concentration (Figures 1.6 and 1.7). Improper saturation occurs

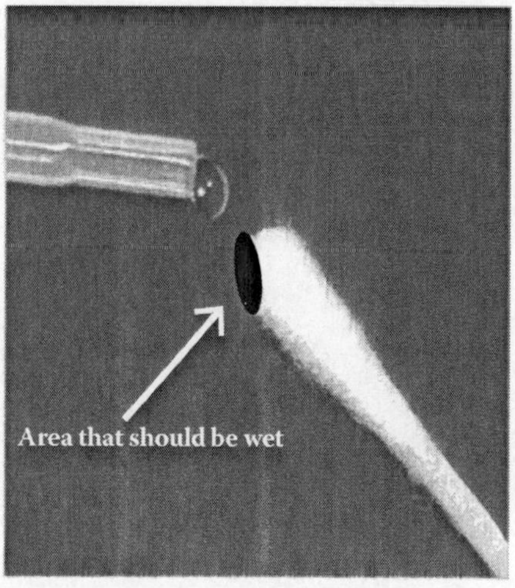

Figure 1.6 (See color insert.) For proper hydration and swabbing, a single drop of distilled water should be applied to the side of a cotton-tipped swab. (Photograph from the New York City Office of the Chief Medical Examiner.)

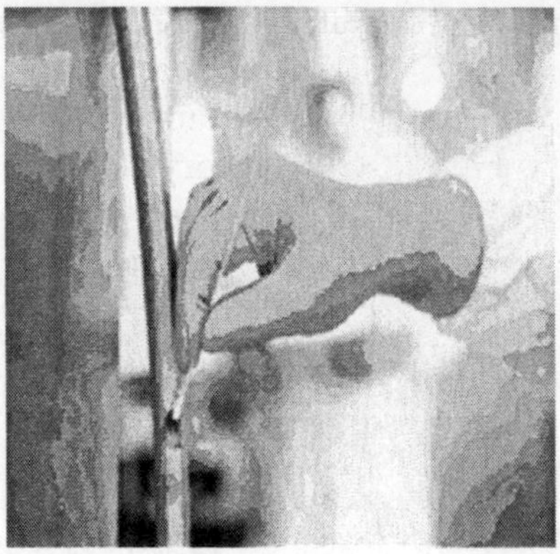

Figure 1.7 (See color insert.) The proper swabbing technique for the recovery of a dried blood sample from the side handle of a refrigerator door. Note the use of personal protective equipment. (Photograph by the author.)

when too much water is applied to the swab; this causes a diluted sample concentration. Use multiple swabs when necessary to collect the entire stain. Reswab the area with a dry sterile swab and subsequently air-dry all swabs. It is permissible to package all swabs used for the same biological sample into one paper enclosure.

Biological evidence such as skin tissue, bone fragments, teeth, and nails have a high likelihood of yielding a full DNA profile. These items must be packaged in paper or cardboard and forwarded directly to the laboratory for DNA profiling.

Prior to recovering a blood sample from glass or other substrates, use a magnifying glass to examine the blood for the presence of ridge detail. Ridge detail in blood indicates a patent fingerprint, which must be processed accordingly. If at all possible, collect the entire substrate that the bloody patent print is on, package it, and forward the entire item to the laboratory. If the substrate cannot be packaged and sent to the laboratory, photograph the print using a scale, then proceed with the swabbing process.

Swabbing is not the only way to collect evidence for DNA analysis. Additional DNA collection methods include scraping dried biological evidence, cutting a swatch from the substrate that contains the unknown stain, and submitting the entire substrate to the laboratory.

Dried biological samples on substrates other than clothing can be scraped with a sterile scalpel and collected. The scrapings are then placed in filter paper, which is folded and inserted into a paper envelope. Scalpel blades must be sterilized or replaced after each DNA sample is recovered. In addition, it is much easier and safer to swab a sample than to scrape a sample and possibly cut yourself. Confer with the laboratory personnel and establish a protocol to determine which collection method is preferred. Submitting the entire substrate to the laboratory often can be advantageous. In the event that the DNA results are negative or the quantity is insufficient for a full DNA profile, the evidence is easily accessible for reprocessing. Processing a substrate at the crime scene eliminates the opportunity for a second analysis. When it is not practical to submit the entire substrate, a swatch can be cut and submitted instead. Clothing should not be processed at the crime scene. Instead, document, package, and submit the clothing to the laboratory, which will process it for stains, hair, skin cells, and so forth. Additional DNA evidence recovery procedures include:

i. Blood samples found in snow or water should be collected immediately to avoid further dilution. The largest possible quantity of these samples should be collected in a clean, suitable container, avoiding contamination as much as possible. Label the samples and submit them to the laboratory directly.

ii. The crime scene should be carefully examined for hair, since hair is difficult to detect and can be overlooked easily. Hair is a potential source of DNA evidence and should be documented, recovered, packaged, and forwarded to the laboratory. Hair should be placed in filter paper, which is then folded and inserted into a coin envelope or similar-type envelope. At the laboratory, a microscopist will examine the hair and determine whether it is suitable for DNA analysis. Tweezers often are used by evidence collectors to recover hair from a crime scene. Tweezers are not recommended in most cases because they can cause damage to the hair structure. Tweezers also must be sterilized between uses. When tweezers are a necessity, an alternative method to sterilizing the tweezers is to use disposable tweezers. However, collecting hair using a gloved hand is the fastest, easiest, and safest method. Another safe and easy collection method is the use of a gel lifter. Confer with your laboratory for proper submission standards.

iii. Burglary tools, such as pry bars and hammers, can be good sources of touch DNA. When these tools are used as weapons, they can carry not only touch DNA, but also blood, hair, and skin. Tools should be sent to the laboratory for analysis rather than processed at the scene so potential trace evidence is not missed. A laboratory has the proper lighting and equipment to process the tool for DNA, wound comparison, and the like. Communicate with your laboratory and establish protocols as to what types of evidence are to be processed in the field and what types are to be submitted directly to the laboratory.

iv. DNA can be typed from epithelial cells found in saliva. Saliva may be deposited on drink containers, cigarette butts, bottles, telephones, cell phones, envelope flaps, stamps, and bitemarks. If a drinking glass is involved, swab the rim, air-dry the swabs, and package them. Following swabbing, the glass can be processed for fingerprints. Alternatively, the entire glass can be sent to the laboratory for processing.

v. Ligatures are any items used by a suspect to tie or bind victims during the commission of a criminal act. Examples of ligatures include duct tape, rope, electric cords, belts, scarves, bandanas, and wire ties. Because these items were touched and perhaps handled roughly by the perpetrator, there is a strong possibility that skin cells were deposited in or on the ligature. Since the same likelihood exists that the ligature will contain the victim's DNA, elimination samples must be obtained from the victim and submitted to the laboratory. A ligature should be packaged in paper and submitted to the laboratory. Although a ligature is considered touch DNA evidence, it should not be swabbed at the crime scene.

vi. Paper items such as letters and envelopes should be submitted to the laboratory for analysis, although studies have shown that processing paper items for DNA has a low success rate. Processing glossy magazine covers for DNA has a slightly higher success rate, but the chances of recovering DNA still are relatively low. If saliva has been applied to the paper, such as when an envelope flap has been licked, then the likelihood of obtaining a full DNA profile increases dramatically. When a paper item is swabbed for DNA, it is possible that the swab will inadvertently destroy a latent fingerprint. Paper items chemically processed with ninhydrin often yield fingerprints of value. Therefore, it is highly recommended that paper items be chemically processed for fingerprints.

vii. Clothing found at the crime scene must be sent to the laboratory to be processed for stains, hair, skin cells, and so forth; this includes hats and masks. A ski mask, for example, likely will contain saliva and skin cells that cannot be adequately recovered at the scene. Bloody sheets, towels, clothing, and other fabrics containing biological evidence also should be collected whole and sent to the laboratory, not processed at the scene.

viii. Cigarette butts, cigar ends, and other smoked items can contain both skin cells and saliva. They have a high success rate for DNA recovery. These items should be packaged individually in separate paper containers such as coin envelopes or similar-type envelopes and sent to the laboratory for processing. Be sure to document the location where they were recovered.

ix. Firearms can be an excellent source of DNA evidence. Most of a firearm's surface area is not conducive to recovering fingerprints due to uneven surfaces. However, the uneven surfaces make parts of the firearm very conducive to DNA recovery because such irregular surfaces tend to collect skin cells from their handlers. Prior to processing a firearm for DNA, the firearm must be rendered safe. *Safety is paramount.* Before handling the firearm, make sure you are familiar with that type of firearm and know how to unload it. Then, while wearing gloves, remove all of the cartridges from the firearm and visually inspect it to confirm no cartridges are present. Once the firearm has been rendered safe, then it should be handled minimally. DNA processing now can be performed. Firearms can be swabbed for touch DNA at the scene using a sterile hydrated swab, followed by a sterile dry swab. Prime areas to swab include grips, trigger, front sight, and slide.

In addition to swabbing the firearm for touch DNA, be sure to swab the magazine, if the firearm has one. Be sure to swab the lips and the floor plate of the magazine. (A magazine is a storage device for ammunition. It is removed from a firearm when adding cartridges. The lip area of a magazine (Figure 1.8) is the top area where cartridges are inserted; the floor plate is the bottom of the magazine.) Both discharged shells and cartridges are common firearms-related evidence that may hold the suspect's skin cells. Discharged shells generally do not yield fingerprints. Although the surface area of a cartridge is small, there is a possibility that ridge detail could be present. The head stamps, or ends, of discharged shells and cartridges can be a source of touch DNA and should be swabbed.

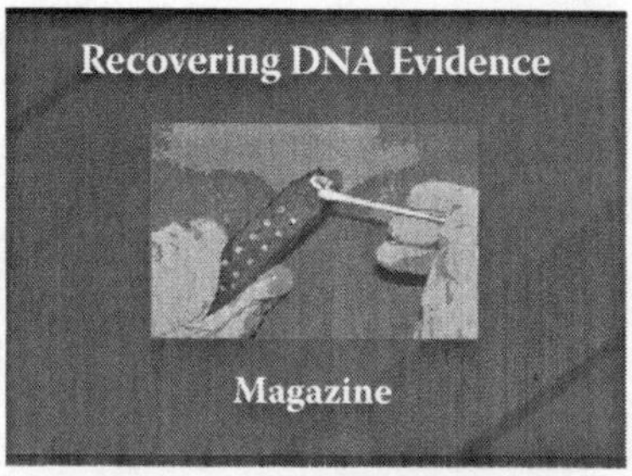

Figure 1.8 **(See color insert.)** The lip area of a firearm's magazine is a potential source of DNA. (Photograph by the author.)

1.4.10 Types of DNA Samples

There are three basic categories of DNA samples. Crime scene samples are unknown forensic DNA samples recovered from crime scenes. An evidence collector recovers the sample, but (as with fingerprints) does not know who deposited the sample. Crime scene samples are searched in the DNA database against other crime scene samples and convicted offenders.

Elimination samples are samples from individuals such as victims, victims' family members, or any other persons who had prior legitimate access to the crime scene. Elimination samples also should include evidence collectors and laboratory personnel. Elimination samples are also referred to as "known samples," "reference samples," or "buccal swabs." Buccal swabs are elimination samples obtained by swabbing the interior of an individual's mouth. They are used for comparison purposes during DNA analysis to eliminate known DNA profiles from the unknown DNA profiles recovered from the crime scene.

Abandoned samples are DNA samples abandoned by an individual known to law enforcement. This can be as simple as an individual discarding a cigarette butt in a public domain.

1.4.11 Packaging DNA Evidence

DNA and other biological evidence must be allowed to air-dry before packaging. Package items separately in paper bags or cardboard boxes, making sure that the packaging is adequate to hold the item. Packaging materials that will be needed on the scene include:

i. Coin envelopes for packaging dried swabs used to collect DNA samples.
ii. Paper bags for packaging clothing or other lightweight items containing DNA evidence.
iii. Cardboard boxes for guns, knives, or any item of DNA evidence too cumbersome for a paper bag.
iv. Evidence tape for sealing the bags and boxes. Paper packaging is breathable and allows the item to dry completely. Plastic and airtight containers create conditions favorable for the growth of bacteria and mold, which are detrimental to the sample. Always package DNA in some form of paper container. The packaging process can be summarized in the following steps:
    a. Select the proper packaging (envelope, bag, or box) based on the size and weight of the item.
    b. Place the item into the package.
    c. Seal the package with evidence tape.
    d. Initial and date the taped seal.
    e. Add a unique identifier and add other pertinent case information.
    f. Prepare the chain-of-custody reports.
v. Air-drying a DNA evidence swab before packaging: swabs must be allowed to dry before packaging. The proper method is to allow the swabs to air-dry. Depending on the concentration of the sample and the temperature, drying can take anywhere from seconds to several minutes. It is recommended that the investigator not blow air on the swab to decrease drying time, as this can cause fragile DNA evidence to become detached and lost from the swab and because it can introduce contaminants from the air into the evidence. There are two suggested methods used to dry swabs collected at the crime scene. The first method is to use a block of styrofoam prenumbered with the corresponding numbers on the evidence log.

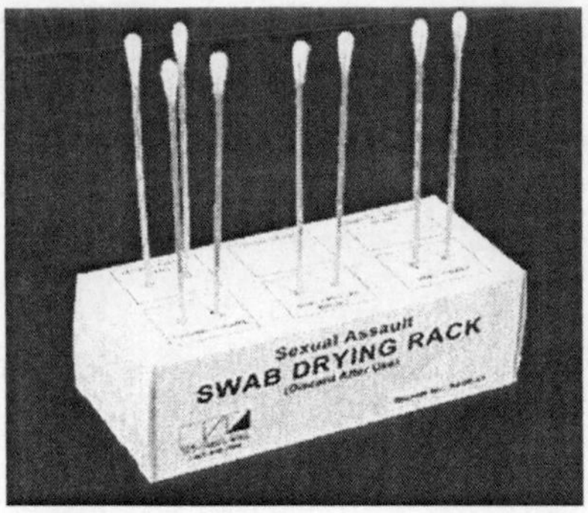

Figure 1.9 An example of a drying rack for air-drying swabs. (Photograph from Tri-Tech Forensics Web site, http://tritechforensics.com, accessed June 6, 2012.)

> The wood-stick end of the wet swab is inserted into the styrofoam block at the proper number as it is collected. As the investigator finishes another task, he or she returns to the styrofoam block and evaluates the swab for dryness. If it is dry enough, it is packaged. If not, it should remain in the block for a longer period of time (Figure 1.9).

The second method is to tape the wood-stick end of the swab to the edge of a tabletop and let the suspended swab air-dry while other tasks are attended to. When using this method, *extreme caution must be used not to contaminate the swabs.* There are also forensic packaging containers that permit swabs to dry while packaged. The swabs are inserted directly into a tube after processing. The tubes contain desiccants to enhance the drying time. However, these tubes are expensive and some cost-conscious agencies prefer to use the air-drying methods described above. After swabs are air-dried, they should be packaged in a paper envelope. If two swabs were used on the same surface area, they may be packaged together and labeled as such. Otherwise, swabs should be packaged in separate envelopes.

1.4.12 Transporting and Storing DNA Evidence

DNA evidence should be transported to the laboratory expeditiously. Once crime scene processing has concluded, place all sealed individual packages of DNA evidence into a carrying box or bag. Make sure that all evidence is accounted for. Place the box into a vehicle and transport the evidence to the laboratory without delay. Be sure to keep the evidence cool and dry. Transport liquid samples in refrigerated or insulated containers. If someone other than the investigator is transporting the evidence, chain of custody must be transferred; it is transferred again when the evidence reaches the laboratory. Degradation is the breaking down of DNA into smaller fragments by chemical or physical processes. Degradation of DNA may limit its use as evidence. Factors that promote DNA degradation include ultraviolet rays (prolonged exposure); heat, humidity, and moisture; bacteria and fungi (often found in foliage and soil); and acids or chemical cleaning solutions (such as bleach).

Actions such as storing evidence in vehicle trunks, vans, office desks, direct sunlight, frost-free refrigerators, and nontemperature/humidity-controlled facilities subject the biological evidence to increased heat, humidity fluctuations, and ultraviolet rays—all factors that can accelerate degradation. Extended exposure to heat or humidity causes degradation of biological evidence. To reduce this threat, move packaged items from the crime scene to a suitable storage facility as soon as possible. Storage of DNA evidence is the last link in the chain of custody. Whether DNA evidence is stored in the laboratory or in a storage facility, certain conditions must be adhered to in order to prevent degradation. DNA evidence should be stored in a spacious, cool, and dry environment. Although it is preferred, most DNA evidence does not require refrigeration and may be stored at ambient room temperature or cooler.

1.5 Conclusions

Evidence collectors assigned to law enforcement agencies around the world perform a vital investigative role. The power of the evidence they detect and collect is used to identify offenders or, more importantly, to exonerate the innocent. Within a forensic investigation, evidence collectors are part of a team whose goal it is to solve a case. Other members of the team are the forensic scientists who meticulously analyze the evidence received to reach a scientific conclusion. Agencies for both evidence collectors and forensic scientists establish protocols and training procedures to ensure that the citizens they serve remain safe and that justice prevails for all.

Acknowledgments

Within the forensic community a host of renowned scientists have provided me with knowledge, training, support, and friendship. I would like to thank the following agencies and corporations that I have been privileged to serve and/or be affiliated with throughout my career: the New York City Police Department; Forensic Investigations Division Assistant Commissioner Peter A. Pizzola, Ph.D.; New York City Office of the Chief Medical Examiner; Director for the Department of Forensic Biology Mechthild Prinz, Ph.D.; Applied Biosystems, Genetic Systems Human Identification Division President Leonard Klevan, Ph.D.; University of North Texas Center for Human Identification Co-Director Arthur J. Eisenberg, Ph.D.; The Bode Technology Group, Vice President for Sales and Marketing Randolph J. Nagy; the Federal Bureau of Investigation National Academy Associates; the National Institute of Justice President's DNA Initiative; and the U.S. Department of State's Anti-Terrorism Assistance Program.

Optimizing Storage and Handling of DNA Extracts*

2

STEVEN B. LEE
CECELIA A. CROUSE
MARGARET C. KLINE

Contents

2.1 Introduction 20
2.1.1 Forensic DNA Storage Issues 20
2.1.2 Factors Influencing DNA Stability 20
2.1.3 Typing Strategies of Nonoptimal Samples 21
2.2 Importance of Sample Storage 22
2.2.1 Forensic DNA Databanks and Casework Samples 22
2.2.2 Nonforensic DNA Databanks and Biobanks 22
2.2.3 Mechanisms of DNA Loss 22
2.3 DNA Storage and Handling Strategies 23
2.3.1 Tube Characteristics 23
2.3.2 Cold Storage 24
2.3.3 Dry Storage Comparisons 25
2.3.4 Additional DNA Storage Modalities 32
Acknowledgments 33
References 34

Abstract: Nucleic acid sample storage is of paramount importance in forensic science as well as in epidemiological, clinical, and genetic laboratories. Millions of biological samples, including cells, viruses, and DNA/RNA, are stored every year for diagnostics, research, and forensic science. PCR has permitted the analysis of minute sample quantities. Samples such as bone, teeth, touch samples, and some sexual assault evidence may yield only low-quality and low-quantity DNA/RNA. Efficient storage of the extracted DNA/RNA is needed to ensure the stability of the sample over time for retesting of the CODIS STRs, mtDNA, YSTRs, mRNA, and other future marker-typing systems.

Amplification of some or all of these markers may fail because the biological material has been highly degraded, contains inhibitors, is too low in quantity, or is contaminated with contemporary DNA. Reduction in recovery has been observed with refrigerated liquid

* Contribution of the U.S. National Institute of Standards and Technology. Not subject to copyright. Certain commercial equipment, instruments, and materials are identified in order to specify experimental procedures as completely as possible. In no case does such identification imply a recommendation or endorsement by the National Institute of Standards and Technology nor does it imply that any of the materials, instruments, or equipment identified are necessarily the best available for the purpose. Points of view in this document are those of the authors and do not necessarily represent the official position or policies of the U.S. Department of Justice.

DNA extracts and also those exposed to multiple freeze-thaw cycles. Therefore, the development of optimal storage and amplification methods is critical for successful recovery of profiles from these types of samples since, in many cases, retesting is necessary.

This chapter is divided into three sections. The Introduction (Section 2.1) reviews forensic DNA storage, factors that influence DNA stability, and a brief review of molecular strategies to type nonoptimal DNA. Section 2.2 discusses the importance of DNA extract storage in forensic and nonforensic DNA databanks and the mechanisms responsible for loss during storage. Finally, Section 2.3 describes strategies and technologies being utilized to store DNA.

2.1 Introduction

2.1.1 Forensic DNA Storage Issues

Millions of biological samples, including cells, viruses, and DNA/RNA, are stored every year for diagnostics, research, and forensic science. DNA extracts from forensic evidence samples such as hair, bones, teeth, and sexual assault evidence may contain less than 100 pg of DNA (Gill 2001; Gill et al. 2000; Phipps and Petricevic 2007; Smith and Ballantyne 2007). Low DNA yields may be due to damage (Budowle et al. 2005; Coble and Butler 2005; Irwin et al. 2007) or degradation (Eichmann and Parson 2008; Hill et al. 2008; Irwin et al. 2007); small cell numbers found in low copy number (LCN) or "touch" samples (Balogh et al. 2003; Budowle et al. 2009; Gill 2001; Gill et al. 2000; Hanson and Ballantyne 2005; Irwin et al. 2007; Kita et al. 2008; Kloosterman and Kersbergen 2003); oligospermic (Sibille et al. 2002) or aspermic perpetrators (Shewale et al. 2003); or low male DNA from extended interval postcoital samples in sexual assault cases (Hall and Ballantyne 2003). Trace biological evidence (e.g., fingerprints and touch evidence) may provide low yields (Balogh et al. 2003; Kita et al. 2008; Lagoa et al. 2008; Schulz and Reichert 2002; van Oorschot and Jones 1997; Wickenheiser 2002). Biological evidence may be consumed with the result that the DNA extracts may be the only remaining genomic resource to retest and test with new technologies for retrospective and prospective testing. Optimal storage of DNA is therefore critical to retrospective (retesting) or prospective (downstream analysis with additional or new genetic markers) testing (Clabaugh et al. 2007; Larsen and Lee 2005; Lee et al. 2012).

2.1.2 Factors Influencing DNA Stability

Degradation is a major factor in the ability to analyze low-quantity samples such as those derived from ancient or degraded bones and teeth and those from mass disasters (Budowle et al. 2005; Budowle et al. 2009; Irwin et al. 2007; O'Rourke et al. 2000; Paabo et al. 2004). Degradation results in the reduction or loss of the structural integrity of cells and the quantity and quality of genomic DNA. Many laboratories store DNA extracts frozen in Tris-Ethylenediaminetetraacetic acid (TE) buffer. However, reduction in DNA recovery may occur with refrigerated liquid DNA extracts and those repeatedly frozen and thawed (Davis et al. 2000; Shikama 1965) or stored in certain microcentrifuge tubes (Gaillard and Strauss 1998, 2001; Larsen and Lee 2005).

Low yields or loss of DNA due to these factors may preclude or diminish the ability to test LCN crime scene samples using current STR methods; therefore, other methods such as mini-amplicon STRs (Asamura et al. 2008; Coble and Butler 2005; Hill et al. 2008; Mulero et al. 2008; Opel et al. 2006, 2007) or less discriminating mtDNA testing

(Eichman and Parson 2008; Lee et al. 2008) are typically dictated for low-quantity samples in advanced states of degradation. The quantity and quality of template DNA from many low-copy forensic samples falls below recommended thresholds (0.5 to 1.25 ng; Collins et al. 2004) and ineffective storage only exacerbates further sample loss. Poor sample quality and the presence of inhibitors may lead to incomplete genetic profiles or no profile, reducing the probative value of the results.

In addition to sample quantity and intrinsic differences in sample types resulting in differences in quality and quantity, extrinsic differences resulting from (a) the effectiveness of the extraction method utilized, (b) the type and effectiveness of preservatives and storage buffers (e.g., presence of antimicrobial agents and nuclease inhibitors in the storage matrices and buffers), (c) purity, especially regarding the amount of nuclease contamination, (d) ionic strength, (e) tube material and quality, (f) exposure to UV, (g) temperature and humidity range and duration in short- or long-term storage, and (h) exposure to multiple freeze-thaw cycles (as occurs with repeated sampling or unexpected power loss), may all lead to differences in the ability to recover and retest the samples.

2.1.3 Typing Strategies of Nonoptimal Samples

Modifications to existing amplification and typing protocols such as mini-amplicons, whole genome amplification (Ballantyne et al. 2007; Hanson and Ballantyne 2005), and LCN protocols (Budowle et al. 2009; Gill 2001; Gill et al. 2000; Phipps and Petricevic 2007; Smith and Ballantyne 2007) to increase the DNA signal and consequently, the analytical success rate of challenged samples, are currently being investigated (Prinz et al. 2006; Roeder et al. 2009). Other approaches have been adopted that include addition of more *Taq* polymerase and Bovine Serum Albumin (J. Wallin of California Department of Justice, personal communication) and increasing cycle number or injection time (Forster et al. 2008). Amplifying DNA with over 28 cycles is widely used (Forster et al. 2008). Nested primer amplification and increased time and voltage for electrokinetic injection of samples have also improved profiling success (Lagoa et al. 2008). Post-PCR purification to remove any ionic components that compete with PCR products during electrokinetic injection has also been used to enhance results (Lederer et al. 2002).

New PCR enhancement reagents have also recently been reported (Le et al. 2008). PCRboost™ has been reported to enhance amplification of low-quality and low-quantity samples and those containing inhibitors such as hematin and humic acid (Le et al. 2008) as well as indigo dye (Wang et al. 2010). Although these approaches have resulted in some success, they have not been universally adopted by forensic DNA laboratories due to inconsistent results (especially on highly degraded, inhibited, or low-quantity samples), high cost, and/or additional validation requirements. The mini-amplicon multiplex AmpℓSTR® Minifiler™ (Mulero et al. 2008) has greatly improved the ability to amplify degraded samples; however, it does not contain all of the CODIS core loci. In addition, new STR multiplexes are continually being developed and optimized with the goal of enhancing amplification and improving results for highly degraded, inhibited, low-quantity samples.

These new developments in typing strategies of nonoptimal samples underscore the importance of DNA storage. New methods that push the lower limit of detection expand applications to extremely low-quantity and low-quality samples. Stable DNA storage and handling over time are therefore especially important when the amount of sample is limited.

2.2 Importance of Sample Storage

2.2.1 Forensic DNA Databanks and Casework Samples

The importance of DNA storage is obvious in the global growth and expansion of forensic DNA databases and repositories. The Combined DNA Index System (CODIS) currently has 9,875,100 offender profiles and 447,300 forensic profiles (Federal Bureau of Investigation 2012 http://www.fbi.gov/about-us/lab/codis/ndis-statistics). The European Network of Forensic Science Institutes, which includes 36 countries, reports offender profiles and 9,770,475 "stains" as of December 2011 (European Network of Forensic Science Institutes 2012, http://www.enfsi.eu/sites/default/files/documents/enfsi_survey_on_dna-databases_in_europe_december_2011_0.pdf). The Armed Forces DNA Identification Laboratory (AFDIL) provides worldwide scientific consultation, research, and education services in the field of forensic DNA analysis to the Department of Defense and other agencies. AFDIL provides DNA reference specimen collection, accession, and storage of United States military and other authorized personnel and processes thousands of samples in casework each year. Forensic DNA laboratories around the world also process thousands of samples each year. All of these samples need to be properly stored and maintained. In addition to these samples, many laboratories conducting forensic DNA casework and data banking also store casework extracts as well as dilutions of the extracted DNA samples. Finally, new international forensic DNA databases, expansion of database laws to include arrestees, missing persons databases, and additional DNA samples from new property-crime casework programs collectively increase the rate of growth and expansion of the number of stored DNA extracts.

2.2.2 Nonforensic DNA Databanks and Biobanks

In addition to the growth and expansion of forensic DNA databanks, several other types of DNA biobanks have been established. These include clinical biobanks to assist in the development of new medicines and drugs (Roden et al. 2008). For example, the United Kingdom Biobank has set a goal to collect, store, and eventually distribute half a million samples with related medical information from 30 to 35 clinics in Great Britain (Blow 2009). Additional efforts are underway at the Vanderbilt University School of Medicine in Nashville, Tennessee, where they are planning a 250,000-person DNA study, and the Oakland, California, Kaiser Permanente DNA Biobank of 500,000 samples (Blow 2009).

Several DNA banks have been established for studying human evolution. These include the worldwide Genographic project run by Dr. Spencer Wells and the National Geographic Society (Zalloua et al. 2008). The goal of this study is to analyze historical patterns in DNA from participants around the world to better understand our human genetic roots (Zalloua et al. 2008). Biodiversity DNA databanks have also been established supporting research on global diversity in response to extinctions. One such group hosts the DNA Bank Network (Zetzsche et al. 2009) with 10,448 taxa containing 32,532 DNA samples.

2.2.3 Mechanisms of DNA Loss

Understanding the different mechanisms of DNA loss provides a foundation for developing the most optimal methods for efficient storage of DNA. There is a body of literature that describes how DNA may be damaged with exposure to temperature fluctuations such as freeze-thaw cycles (Davis et al. 2000; Shikama 1965). In addition, it is well known that both water and oxygen may damage DNA through hydrolysis and oxidative damage (Bonnet et al. 2010). Many laboratories have therefore explored other options for storage, including dry state storage.

The assumption is often made that if nucleic acids are dried they are then stable for long periods of time. However, it is becoming increasingly evident that degradation can occur during storage that can irreversibly damage the samples. For example, Lindahl (1993) reviewed evidence that DNA can undergo chemical changes such as depurination, hydrolysis, and oxidation even at low moisture content. Hofreiter et al. (2001) suggest that such chemical degradation might be responsible for the difficult recovery of DNA from aged samples. Although dried DNA is stable in the short term, it is nevertheless imperative to prevent detrimental chemical changes for optimal recovery.

More recently, dry-storage DNA damage has been studied and it was found that solid-state DNA degradation is greatly affected by atmospheric water and oxygen at room temperature (Bonnet et al. 2010). DNA may be lost by aggregation. As pointed out by Bonnet et al. (2010), loss by aggregation is highly significant since laboratory plastic tubes and plate seals generally are not airtight and therefore both water and oxygen may adversely react with DNA. In this study the authors also tested the stabilizing effects of the additive trehalose. In the presence of trehalose, solid-state natural DNA, heated to 120°C, does not denature (Zhu et al. 2007). This stabilization effect of trehalose may be explained by its ability to block the negative charges on the phosphates (water replacement) or by hydrogen bonding between trehalose and DNA, which may reduce the DNA structural fluctuations (vitrification hypothesis) (Alkhamis 2008; Zhu et al. 2007).

Mechanisms for DNA damage during storage have recently been reviewed (Bonnet et al. 2010). In addition to chemical damage, loss may also occur by the co-extraction and then subsequent action of nucleases that may not have been removed in the purification procedures. This is an important consideration in crude DNA extraction procedures such as Chelex (Walsh et al. 1991) that are then stored over time. In addition, loss may also occur during any additional manipulations of the DNA via purification through additional phase separations and column purifications. Finally, dilutions of DNA and subsequent storage in distilled water may also result in loss through damage by water (Bonnet et al. 2010).

2.3 DNA Storage and Handling Strategies

2.3.1 Tube Characteristics

It has been well documented that loss of DNA may occur due to the material and quality of the tubes used to store the samples (Kline et al. 2005; Larsen and Lee 2005). Polypropylene plastic microcentrifuge tubes that are routinely utilized in forensic DNA laboratories may retain DNA (Gaillard and Strauss 1998, 2001; Larsen and Lee 2005) with the amount of adsorbed DNA as high as 5 ng/mm^2 of tube wall (Gaillard and Strauss 1998). In addition, different tube lots from the same manufacturer have been reported to retain variable amounts (5–95%) of DNA (Gaillard and Strauss 2001). The authors suggest the use of polyallomar tubes or introducing 0.1% detergent, Triton-X 100, to prevent the retention of DNA on polypropylene tubes (Gaillard and Strauss 2001).

Polytetrafluoroethylene (PTFE, known commercially as Teflon®) tubes have also been compared to polypropylene for DNA storage (Kline et al. 2005). The researchers conducted an interlaboratory blind quantification study and reported that recovery of the low-target DNA concentration samples (50 pg), stored in PTFE tubes was 73% versus only 56% from samples stored in polypropylene. This suggests that at this low DNA concentration, a significant proportion of the sample DNA binds to the polypropylene walls and greater DNA recovery can be achieved with storage in PTFE-coated tubes (Kline et al. 2005).

2.3.2 Cold Storage

Among the most common strategies for DNA preservation are cold and dry storage strategies that include: (1) 4°C refrigeration, (2) –20°C, (3) –80°C, (4) –196°C (liquid nitrogen), and (5) dry storage on a solid matrix. Protection in the "dry state" and cryopreservation at –196°C both maintain the DNA in the glassy or vitreous state. In the glassy state, DNA and other molecules lose the ability to diffuse. This results in very little movement at the molecular level. In fact, "movement of a proton (the hydrogen ion) has been estimated to be approximately one atomic diameter in 200 years"; this in turn makes any chemical reactions highly unlikely over hundreds of years (Baust 2008, p. 251). If, however, moisture is reintroduced to the "dry state" or an increase in temperature occurs above the glass transition temperature of water (nominally –135°C), chemical reactions may start again resulting in DNA instability (Baust 2008).

Storage at –20°C to –80°C may provide adequate conditions depending on the quality and quantity of DNA needed for further testing and the duration of storage. Most forensic DNA laboratories utilize –20°C to –80°C freezers for storage. Forensic DNA research efforts have focused on developing new methods of amplification and typing with low-quality and low-quantity samples due in part to the observation that current storage methods are not optimal. Neither –20°C nor –80°C conditions have been shown to provide long-term storage quality equivalent to maintenance at liquid nitrogen temperatures (Baust 2008). Unfortunately, the storage of all forensic DNA extracts at liquid nitrogen temperature is not practical with over 15 million samples in U.S. and European forensic databanks alone.

As stated by Baust, "There are few studies that provide definitive answers to the question of optimal storage conditions for DNA" (Baust 2008, p. 251). The National Institute of Standards and Technology (NIST) and the National Cancer Institute have published data that suggest that "colder is better" and NIST has shown humidity control to be an important factor in stable storage. This is consistent with the fact that cryopreservation and dry-state storage both reduce DNA chemical reactivity.

Forensic DNA scientists face additional variables in optimizing DNA storage protocols. These variables include the initial contaminants that might be co-extracted with the DNA from crime scene samples, different DNA purities and final dilution buffers utilized in DNA extraction methodologies, the integrity of storage conditions including exposure to different temperatures, humidity and light, the tube material and efficiency of the seal, and downstream sample requirements. According to Baust, "Dry matrix storage should be dry and devoid of changes in moisture content … and cold conditions should rely on stable, noncycling temperatures" (Baust 2008, p. 251). That is, when storing samples, there should be no temperature fluctuations such as those found in frost-free cycles of most modern refrigerators and there is a need to conduct comparative tests on DNA storage methodologies on forensic DNA samples over time using different storage approaches.

2.3.3 Dry Storage Comparisons

Trehalose: Smith and Morin (2005) conducted a comparison of different storage conditions with the addition of potential preserving agents. Dilutions of known concentrations of human placental DNA and gorilla fecal DNA were stored under four conditions (4°C, -20°C, -80°C, dry at room temperature), with three additives (TE buffer, Hind III digested Lambda DNA, and trehalose). The effectiveness of the different methods was tested periodically using qPCR and PCR assay of a 757 bp fragment. The highest quantity of DNA remained in samples stored at −80°C, regardless of storage additives, and those dried at room temperature in the presence of trehalose (Smith and Morin 2005). DNA quality was best preserved in the presence of trehalose, either dried or at −80°C; significant quality loss occurred with -20°C and 4°C storage (Smith and Morin 2005). These results indicate that dry storage with an additive such as trehalose may improve recovery of low-quantity and low-quality DNA versus traditional liquid extract freezer storage.

DNA storage tests under different conditions and a literature review has been conducted by the DNA Bank Network of Germany (Zetzsche et al. 2009). This organization was established in spring 2007 and is currently funded by the German Science Foundation (DFG) and was initiated by GBIF Germany (Global Biodiversity Information Facility). DNA bank databases of all their partners are linked and are accessible via a central Web portal (DNA Bank Network 2012 http://www.dnabank-network.org/Index.php) providing DNA samples of complementary collections (microorganisms, protists, plants, algae, fungi, and animals) to support biodiversity applications. In their reviews and the results of their tests they determined that long-term storage of DNA samples in buffer should be carried out at −80°C or below. Furthermore, as expected, dried, lyophilized DNA must be stored at low relative humidity to avoid DNA aggregation (DNA Bank Network 2012). They also determined that energy and environmental costs were the main reasons to support dry storage at ambient temperature (DNA Bank Network 2012).

FTA® Technology: FTA cards contain chemicals that lyse cells, denature proteins, and protect nucleic acids from nucleases, oxidative, and UV damage. Others have evaluated treated filter paper for collection and storage of buccal cells. The treated filter paper technology FTA (Sigurdson et al. 2006) is used in a room-temperature storage product, offered by GenVault (Carlsbad, California). Following 7 years of storage, the researchers found only modest DNA yields and reduced recovery that was insufficient for WGA (Sigurdson et al. 2006). However, others have shown good recovery from FTA paper for forensic DNA analysis resulting in full DNA profiles following years of dry storage (Fujita and Kubo 2006; Park et al. 2008; Smith and Burgoyne 2004; Tack et al. 2007).

SampleMatrix®/QiaSafe®: Biomatrica Inc. has developed a proprietary technology for the dry storage of biological materials at ambient temperatures. The key component of this technology is SampleMatrix (SM, also known as QiaSafe), a synthetic chemistry storage medium that was developed based on anhydrobiosis ("life without water"), a natural protective mechanism that enables survival of some multicellular organisms in extremely dry environments (Crowe et al. 1998). Such organisms can produce high concentrations of disaccharides, particularly trehalose, a nonreducing disaccharide of glucose, to protect their cellular structures during prolonged droughts and can be revived by simple rehydration (Crowe et al. 1998). Recent evidence suggests that trehalose can preserve intact cells *in vitro* in the dry state (Wolkers et al. 2002). Trehalose disaccharides are predicted to interact with DNA molecules through minor groove interactions based on hydrogen bonding (Figure 2.1A). Biomatrica has developed proprietary synthetic compounds that mimic the protective properties of anhydrobiotic molecules with additional improvements that are especially pertinent to protecting DNA during dry storage. SM, a much improved synthetic formulation, is predicted to form similar interactive patterns with DNA as naturally occurring anhydrobiotic molecules (Figure 2.1B).

The protective properties of SM are based on its ability to form a stabilizing structure via glass formation at a higher temperature than natural disaccharides and therefore to provide improved protective properties as compared to trehalose (Clabaugh et al. 2007).

Storage of samples at different amounts demonstrated the protective properties of SM on DNA when PCR amplicons were detected in essentially all SM-protected samples at 70°C whereas unprotected samples showed more variable results. This is especially apparent in samples containing limited amounts of DNA (≤10 ng). It is noteworthy that the 4 ng samples stored at −20°C for 24 h resulted in markedly less amplicon than an identical sample stored dry in SM at 70°C (Figure 2.1C).

Stabilization of low-concentration DNA samples in SM has also been observed over 1 year (Ahmad et al. 2009; Lee et al. 2012). For this study, purified male and female DNA was extracted from buccal swabs using DNA IQ™ (Promega, Madison, Wisconsin) followed by quantification using the Quantifiler® Human DNA Quantification Kit (Applied Biosystems, Foster City, California). DNA samples from the male and female donors were serially diluted and added in replicates into SM multiwell plates and tubes for final DNA concentrations ranging from 4 ng to 0.0625 ng in a total of 20 μL of water. Replicate DNA samples ($n = 4$) at each concentration were applied into SM multiwell plates and then dried overnight in a laminar flow hood at room temperature. Samples were maintained inside a storage cabinet with desiccant included to create a humidity-controlled environment (SM+D samples). A separate set of samples was stored inside an identical storage cabinet without desiccant to assess the effects of uncontrolled humidity on sample stability when stored in SM (SM-D samples). Identical samples were also aliquoted into empty polypropylene microcentrifuge tubes and stored frozen at −20°C as standard in-house controls (Control) for comparison. The samples were either processed immediately (0 d), or 1 d, 1 week, 2 weeks, 1.5 months, 2 months, 3 months, 6 months, and 1 year prior to recovery and analysis by quantitative PCR followed by PowerPlex®16 (Promega, Madison, Wisconsin) STR analysis. Samples stored dry in SM were rehydrated with 20 μL of water and used directly in downstream applications without further purification to remove matrix components (Ahmad et al. 2009; Lee et al. 2012).

Recovered samples were quantified to determine the yield of DNA following dry storage in SM versus frozen control reference samples and also compared to initial quantification values obtained from the original DNA stock solutions at the time of sample preparation (0 d). Based on these quantification values, the average yield of DNA recovered following dry storage in SM under controlled humidity (SM +D-Red) conditions was dramatically improved as compared to samples stored without humidity control (SM-D-Beige) and in-house control (Control-Blue) samples stored frozen for 1 year (Figure 2.2A) (Ahmad et al. 2009; Lee et al. 2012). Similar results were detected after 4 months of storage where recovery of SM-stored DNA was significantly improved versus conventional polypropylene microfuge tubes (Figure 2.2B) (Clabaugh et al. 2007).

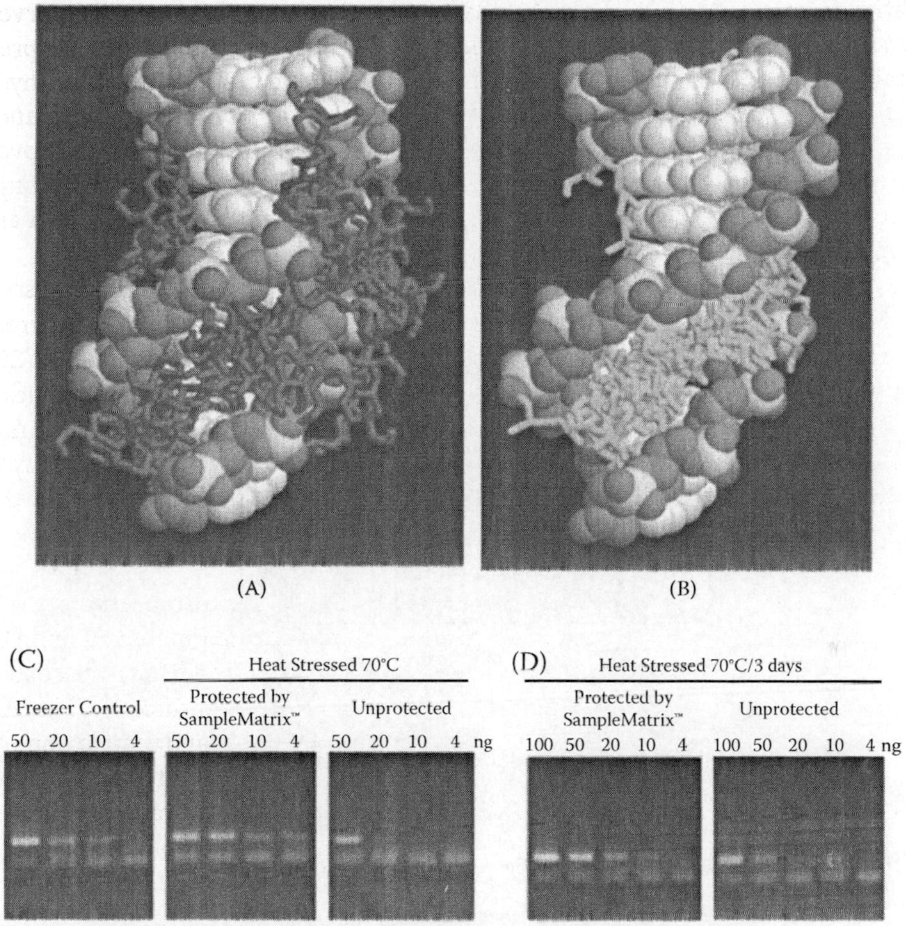

Figure 2.1 (See A and B in color insert.) Protective properties of SM with nucleic acid molecules: (A) Three-dimensional depiction of trehalose disaccharides used in nature predicted to interact with nucleic acid molecules through minor groove interactions based on hydrogen bonding. (B) Three-dimensional depiction of SM as it is predicted to form similar interaction patterns as trehalose. (C) PCR results on human gDNA at 50 ng, 20 ng, 10 ng, and 4 ng with and without SM frozen, after 24 h at 70°C or 3 d at 70°C are shown in Figure 2.1D. Various amounts of human genomic DNA (100 ng, 50 ng, 20 ng, 10 ng, and 4 ng) were applied in microcentrifuge tubes containing SM or into empty tubes (control samples). The samples were allowed to air-dry overnight in a laminar flow hood. Samples were then placed on a heat block maintained at 70°C. Samples were removed from the heat block at 24 h or 72 h and the DNA was hydrated in 10 µL water for 15 min on the benchtop prior to use in downstream applications without further purification. Aliquots of rehydrated samples were used to amplify the fibroblast growth factor 13 (FGF13) gene by PCR using 2.5 U Taq DNA polymerase (NEB), 3 µL 10x thermopol reaction buffer (NEB), 0.5 µL dNTPs (10 mmol/L each nucleotide), FGF13 forward (5′gaatgttaacaacatgctggc3′) and FGF13 reverse (5′agaagctttaccaatgttttcca3′) in a final volume of 30 µL. Cycling parameters were: 94°C for 5 min followed by 40 cycles of 94°C for 15 sec, 55°C for 30 sec, and 72°C for 30 sec. A 10 µL aliquot of each PCR reaction was run on a 0.8% agarose gel stained with ethidium bromide. (From Clabaugh K. et al., 2007, Storage of DNA samples at ambient temperature using DNA-SampleMatrix, *18th Annu Meet Int Symp Human Identification*, Hollywood, CA, October 2007.)

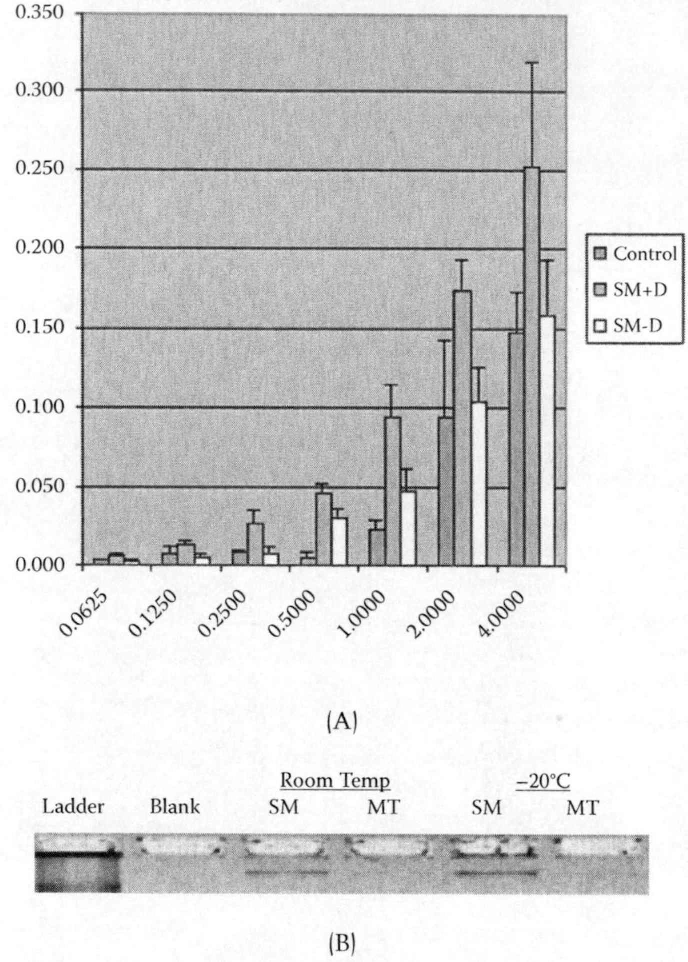

Figure 2.2 (See A in color insert.) DNA recovery from samples stored in SampleMatrix versus –20°C: (A) Average DNA recovery of SampleMatrix-stored DNA versus –20°C frozen controls. Replicate DNA samples at seven different concentrations were stored at ambient room temperature in SampleMatrix with dessicant (SM+D), without dessicant (SM–D), or at –20°C as frozen liquid controls in polypropylene microfuge tubes. Quantification was performed utilizing the ABI Human Quantifiler kit as per manufacturer's recommendations. Recovery from SM+D-stored samples at room temperature was higher than that of frozen controls for every concentration. (From Ahmad T. et al. 2009, Biomatrica DNA SampleMatrix®, A new prospect for forensic DNA sample storage, *Proc 2009 AAFS Annu Meet* 15:108–9; Lee S. B. et al., 2012, Assessing a novel room temperature DNA storage medium for forensic biological samples, *Forensic Sci Int Genet* 6(1):31.40.) (B) 10 ng of Control DNA (K562) was stored in SM and MT for 4 months at room temp and –20°C; 1% agarose gel electrophoresis in 1X TBE followed by EtBr staining and laser flatbed scanning with an FMBIO III Plus was used to detect the recovered DNA. (From Clabaugh K. et al., 2007, Storage of DNA samples at ambient temperature using DNA-SampleMatrix, *18th Annu Meet Int Symp Human Identification*, Hollywood, CA, October 2007.)

A consortium of DNA biodiversity laboratories also conducted research on DNA storage that focused on new and suitable protective substances, storage at higher temperatures, rehydration of lyophilized DNA, and the usage of special cryotubes (DNA Bank Network 2010, http://www.dnabank-network.org/publications/Workshop_Long-term_DNA_storage-Summary_and_Abstracts.pdf). A workshop for long-term storage of DNA samples was held August 12, 2009, at the Systematics 2009 conference in Leiden, Netherlands (DNA Bank Network 2010). At this workshop, it was reported that the characteristics of four commercial dry-storage systems at ambient temperature were tested: GenPlate, QIAsafe (also known as SampleMatrix), GenTegra, and DNAshell. It is worth noting that GenPlate storage is based on FTA paper and GenTegra is designed to protect RNA samples (see http://www.genvault.com/html/products/gentegra-RNA.html) with both being products of the same manufacturer (GenVault). DNAshell is a product from Imagene. Preliminary results of DNA storage experiments based on q-RT PCR data were presented by the DNA Bank Network. QIAsafe/SampleMatrix showed good storage performance at ambient temperature and better DNA recovery when stored in SampleMatrix at –20°C and –80°C. Unbuffered DNA in water (RT, 4°C and –20°C) was subject to fast degradation as would be expected from hydrolysis. The samples stored in QIAsafe/SampleMatrix provided comparable results to the theoretical best practice of sample storage in liquid nitrogen (see DNA Bank Network 2010, http://www.dnabank-network.org/publications/Workshop_Long-term_DNA_storage-Summary_and_Abstracts.pdf).

Shipping Study: The ability of SM to protect samples during shipment and storage conditions was evaluated in studies performed at NIST (Clabaugh et al. 2007). Aliquots of 20 μL of human genomic DNA samples at 1.0 ng/μL, 0.25 ng/μL, and 0.05 ng/μL in Tris-EDTA (TE) buffer were placed into multiple individual wells in four separate 96-well plates containing SM. The plates were dried overnight in a laminar flow hood and then sealed with aluminum foil seals provided by the manufacturer (Clabaugh et al. 2007).

Two plates were shipped continuously across the country for 208 d in a cardboard shipping container (with no insulation or cold source). The remaining two plates were maintained at ambient temperature laboratory conditions. A temperature and humidity monitor was included in the shipping container to continuously monitor the environmental conditions experienced during shipping and storage. Control DNA samples of identical concentration in TE were stored at 4°C in PTFE containers.

The shipped samples were exposed continuously to environmental conditions (some extreme) during the summer and early fall of 2007. In total, the package traveled six times between Gaithersburg, Maryland and San Diego, California, via the U.S. Postal Service. This series of shipping events was followed by an additional 14 d in a car trunk in Maryland. The package was then mailed across the country two more times and then maintained at ambient environmental conditions in a Maryland home attic for an additional 56 d including the month of August. The plates were then removed from the shipping container and stored at laboratory ambient conditions for the remainder of the study (208 d total).

The other two plates were maintained at ambient temperature within the laboratory for the entire 208 d study without humidity control. During the summer, energy conservation efforts used to control the building HVAC system were detected and can be observed in the resulting temperature and humidity plots (Figure 2.3) (Clabaugh et al. 2007). Positive control DNA samples of identical concentration in TE were stored at 4°C in PTFE containers for the duration of the study.

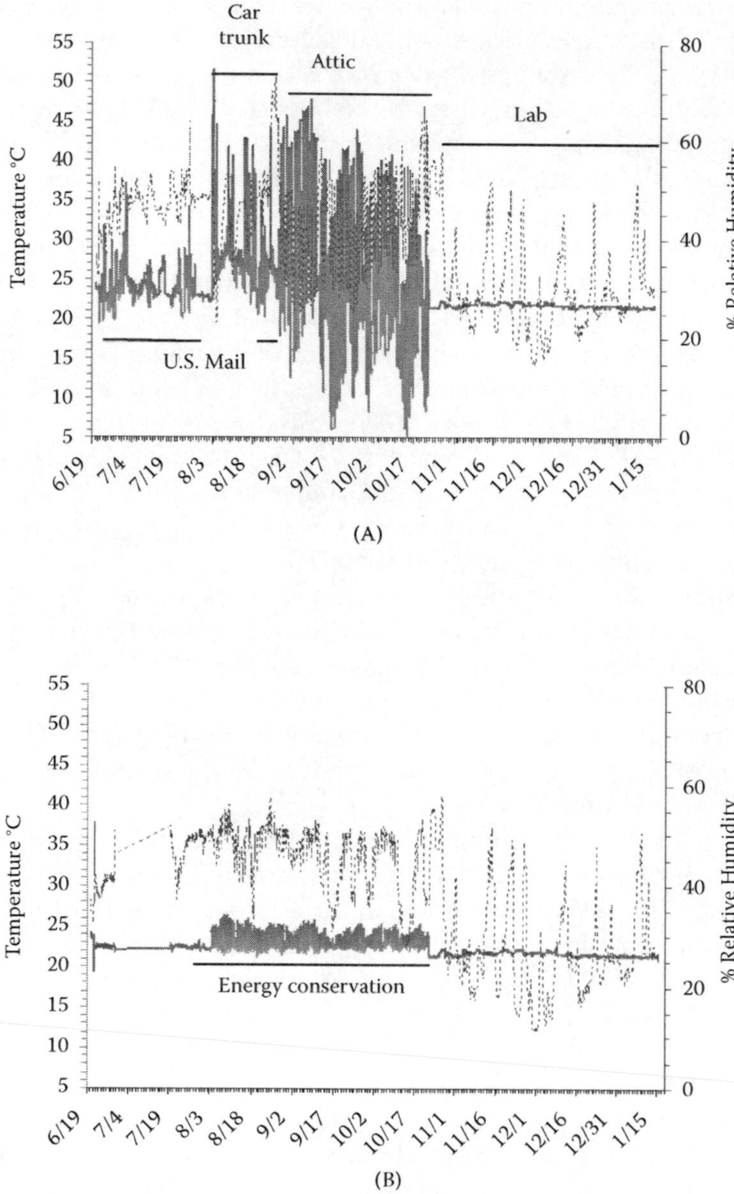

Figure 2.3 (See color insert.) Temperature and humidity plots over 208 d of DNA storage and shipping at ambient temperature in the laboratory and in shipping containers: (A) temperature and humidity plots from shipped sample plate monitor and (B) temperature and humidity plots for room ambient temperature plate monitor. (From Clabaugh K. et al., 2007, Storage of DNA samples at ambient temperature using DNA-SampleMatrix, *18th Annu Meet Int Symp Human Identification*, Hollywood, CA, October 2007.)

Data obtained from the temperature and humidity monitor indicate that the mailed samples were exposed to a maximum of 51.6°C, 73% RH, and a minimum of 5.3°C, 15% RH, with median and average conditions of 22.1°C, 40% RH, and 23.6°C, 39% RH, respectively (Figure 2.3). Laboratory ambient conditions reached a maximum of 26.4°C, 58% RH, and a minimum of 19.4°C, 11% RH; median and average conditions were 22.2°C, 41% RH, and 22.4°C, 38% RH, respectively. Energy conservation measures were implemented during the months of August through November 2007 (that caused the fluctuations seen in the ambient temperature and humidity plots in Figure 2.3 [Clabaugh et al. 2007]).

In order to assess any effects on DNA stability during long-term storage, DNA samples from four time points (0 d, 8 d, 23 d, and 56 d) were rehydrated and analyzed using quantitative PCR (Quantifiler, Applied Biosystems, Foster City, California). Identical control DNA samples that were stored dry in SM, but maintained at room temperature in the laboratory, were also quantified in the same manner. Results (comparison of Ct values) indicate that overall there was little to no difference in the amount of DNA recovered following storage and shipment in SM as compared with samples maintained dry in SM at room temperature resulting in 0 to 10% difference in recovery (Figure 2.4). These results indicate successful protection of dried DNA samples in SM during shipment in a standard cardboard shipping box without extra insulation or cold source.

To assess the yield of recovery following storage in SM for 208 d, quadruplicate DNA samples subjected to shipping stress were analyzed by qPCR as described in the paragraph above and compared with identical control samples stored in SM in the laboratory. Also included for analysis were the original DNA samples maintained in buffer and maintained at 4°C in PTFE containers for the same time period. Results of recovery of DNA following shipment and storage for 208 d indicate that the amount of DNA recovered from samples protected dry in SM following exposure to shipping stress and storage is comparable with

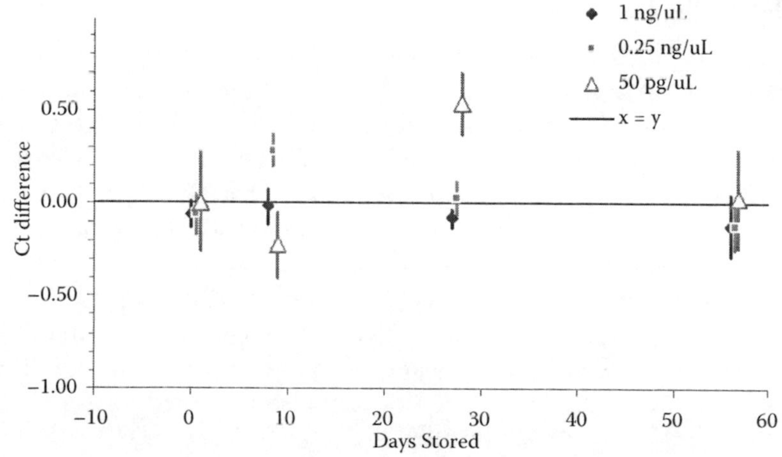

Figure 2.4 Delta Ct values from qPCR of shipped versus ambient-stored samples over 56 d (the X-axis displays time in days and the Y-axis displays the delta Ct value). Well-characterized DNA of a known concentration was diluted to target concentrations of 1.0, 0.25, and 0.05 ng/μL (50 pg/μL). TE–4 buffer was used as the fourth sample set. Data show that in general there is slight to no difference (displayed when Y = 0) in the relative value of DNA quantity and quality between the samples that were held at ambient temperature and humidity and the samples that were shipped and held in extreme environmental conditions.

Table 2.1 Quantification Results ($n = 4$) for Shipped Samples versus Samples Stored at Lab Ambient Temperatures and Those Stored at 4°C in Polytetrafluoroethylene (PTFE) after 208 Days

	Condition		
DNA (ng/μL)	Shipped[a]	Lab Ambient[b]	4°C PTFE[c]
1.00	0.65 ± 0.06	0.69 ± 0.03	1.01 ± 0.02
0.25	0.18 ±0.03	0.20 ± 0.01	0.30 ± 0.01
0.05	0.04 ±0.00	0.04 ± 0.06	0.05 ± 0.00

[a] Shipped in SampleMatrix.
[b] Stored at ambient temperature and humidity in SampleMatrix.
[c] Stored at 4°C in polytetrafluoroethylene container.

identical DNA samples that were maintained at ambient temperatures in the laboratory for the same time period (Table 2.1). There was a slight decrease in the amount of DNA as compared to samples maintained in solution in PTFE containers and kept at 4°C (Table 2.1). The use of PTFE containers is unfortunately not cost effective for the number of samples that need to be stored (Clabaugh et al. 2007).

Samples subjected to shipping stress and those stored for the identical time period at ambient laboratory conditions were also analyzed by multiplex STR analysis using the AmpFℓSTR Identifiler system (Applied Biosystems). Rehydrated samples were used directly in amplification reactions without further purification. Profiles obtained from DNA stored in SM (1 ng used for amplification) were compared with samples maintained in solution at 4°C in PTFE-coated tubes. Slightly lower RFU were obtained from samples subjected to shipping stress while stored dry in SM. However, all samples gave complete Identifiler profiles above 200 RFU, even samples amplified from 250 pg input DNA (Clabaugh et al. 2007).

2.3.4 Additional DNA Storage Modalities

Genomic DNA is also stored in Southern blots from restriction fragment length polymorphism (RFLP) with a variable number of tandem repeat membranes, slot blots, and dot blots (both DQ alpha and PM), as well as in the liquid PCR products. Extraction and typing was recently reported from DNA extracted from VNTR RFLP Membranes (Steadman et al. 2008). In addition, genomic DNA can be isolated from PCR products such as DQ Alpha products (Hochmeister et al. 1995), PM, D1S80, and CTT STR products (Patchett et al. 2002), and the isolated genomic DNA can be typed utilizing new genetic markers. It is expected that the new STR multiplex products can also be utilized in the same manner—that is, as a source of genomic DNA as the template is not consumed in the PCR reaction. This suggests that in cases where all the evidence and extracted DNA and dilutions have been consumed, the remaining PCR products may also be stored for future typing and retesting. Finally, it has been well documented that a portion of the DNA is lost in columns such as Centricon-100s and Microcon 100s (Meinenger et al. 2008). If the entire sample is consumed in the extraction, then any remaining DNA extracted from the columns and eluates could also be stored under the most optimal conditions for future testing.

Finally, there are recently developed forensic methods that are based on RNA, indicating the need to optimize RNA storage as well as DNA. Different genetic expression patterns exist in different tissue types and detection of differential expression has been utilized to determine the origin of biological evidence based on determining relative abundance of messenger RNA. Body fluid identification has been reported based on their mRNA profiles (Bauer 2007; Juusola and Ballantyne 2003, 2005, 2007; Nussbaumer et al. 2006; Zubakov et al. 2008). In addition, estimating the age of a bloodstain was reported using analysis of mRNA: rRNA ratios (Anderson et al. 2005). This type of information may be a useful tool in establishing when the evidence was deposited and may help in determining when a crime was committed. Advantages of the mRNA-based approach versus the conventional biochemical tests include greater specificity, simultaneous and semiautomatic analysis, rapid detection, decreased sample consumption, and compatibility with DNA-extraction methodologies. The quantification of the amounts of the mRNA species relative to housekeeping genes is a critical aspect of the assays (Bauer 2007; Juusola and Ballantyne 2007). These relatively new RNA-based methods require sensitivity and a high degree of quantitative accuracy and highlight the need to stabilize the storage of both RNA and DNA in extracts. Stable dry storage of RNA for gene expression analysis out to 11 d (Wan et al. 2009) and for microarray expression analysis out to 4 weeks (Hernandez et al. 2009) has recently been reported.

The future of biological evidence analysis will continue to require the development, validation, and adoption of new methods and genetic markers (Budowle and van Daal 2009). The optimization of storage of nucleic acid extracts is a pivotal step toward maximizing the potential of the new methods and markers in future testing.

Acknowledgments

The authors would like to thank the following individuals who provided data and/or assisted in some way in the preparation of this document: Russ Miller, Taha Ahmad, and Karen Crenshaw of the Palm Beach Sheriff's Office for assistance in performance and analysis of data shown in Figure 2.2A; Kimberly Clabaugh, Keri Larsen, Brie Silva, and Sal Murillo of San Jose State University for assistance in preparing data shown in Figure 2.2B; David L. Duewer of NIST for assistance with data presentation in Figures 2.3 and 2.4 and Table 2.1, and Dr. Rolf Muller of Biomatrica for data in Figures 2.1 and support for research on SampleMatrix. The authors would also like to acknowledge *Forensic Science Review* editors Dr. Ray Liu and Dr. Jaiprakash Shewale for their support, collaboration, and leadership in the field of forensic DNA, and Ines Iglesias-Lee and Beverly Lee-Carroll for their many contributions in scientific discussions on this manuscript.

Research conducted at SJSU was supported in part by a California State University Program for Education and Research in Biotechnology Joint Venture grant to S. Lee and R. Muller, the National Science Foundation NSF-REU grant #DBI-0647160 to J. Soto, C. Ouverney, and S. Lee of SJSU, and by the California Association of Criminalists A. Reed and V. McGlaughlin Endowment Fund for scholarships to K. Clabaugh and B. Silva at SJSU and support provided by the Chair, Dr. Mark Correia (Justice Studies Department, SJSU), Associate Dean Dr. Barbara Conry, and Dean Dr. Charles Bullock (College of Applied Sciences and Arts, SJSU).

The shipping study described here was supported by the National Institute of Justice through interagency agreement 2003-IJ-R-029 to the Office of Law Enforcement Standards, National Institute of Standards and Technology.

References

Ahmad T., R. W. Miller, A. B. McGuckian, J. Conover-Sikorsky, and C. A. Crouse. 2009. Biomatrica DNA SampleMatrix®. A new prospect for forensic DNA sample storage. *Proc 2009 AAFS Annu Meet* 15:108–9.

Alkhamis K. A. 2008. Influence of solid-state acidity on the decomposition of sucrose in amorphous systems (I). *Int J Pharm* 362:74–80.

Allison S. D., B. Chang, T. W. Randolph, and J. F. Carpenter. 1999. Hydrogen bonding between sugar and protein is responsible for inhibition of dehydration-induced protein unfolding. *Arch Biochem Biophys* 365:289–98.

Anderson S., B. Howard, G. R. Hobbs, and C. P. Bishop. 2005. A method for determining the age of a bloodstain. *Forensic Sci Int* 148:37–45.

Asamura H., S. Fujimori, M. Ota, T. Oki, and H. Fukushima. 2008. Evaluation of miniY-STR multiplex PCR systems for extended 16 Y-STR loci. *Int J Legal Med* 122:43–9.

Ballantyne K. N., R. A. van Oorschot, and R. J. Mitchell. 2007. Comparison of two whole genome amplification methods for STR genotyping of LCN and degraded DNA samples. *Forensic Sci Int* 166:3–41.

Balogh M. K., J. Burger, K. Bender, P. M. Schneider, and K. W. Alt. 2003. STR genotyping and mtDNA sequencing of latent fingerprint on paper. *Forensic Sci Int* 137:188–95.

Bauer M. 2007. RNA in forensic science. *Forensic Sci Int Genet* 1:69–74.

Baust J. G. 2008. Strategies for the storage of DNA. *Biopreserv Biobank* 6:251–2.

Belotserkovskii B. P. and B. H. Johnston. 1997. Denaturation and association of DNA sequences by certain polypropylene surfaces. *Anal Biochem* 251:251–62.

Blow N. 2009. Biobanking: Freezer burn. *Nat Methods* 6:173–8.

Bonnet J., M. Colotte, D. Coudy et al. 2010. Chain and conformation stability of solid-state DNA: Implications for room temperature storage. *Nucleic Acids Res* 38:1531–46.

Budowle B. and A. van Daal. 2009. Extracting evidence from forensic DNA analyses: Future molecular biology directions. *BioTechniques* 46:339–50.

Budowle B., F. R. Bieber, and A. J. Eisenberg. 2005. Forensic aspects of mass disasters: Strategic considerations for DNA-based human identification. *Leg Med* (Tokyo) 7:230–43.

Budowle B., A. J. Eisenberg, and A. van Daal. 2009. Validity of low copy number typing and applications to forensic science. *Croat Med J* 50:207–17.

Clabaugh K., S. Silva, K. Odigie et al. 2007. Storage of DNA samples at ambient temperature using DNA-SampleMatrix. *18th Annu Meet Int Symp Human Identification*, Hollywood, CA; October 2007.

Coble M. D. and J. M. Butler. 2005. Characterization of new miniSTR loci to aid analysis of degraded DNA. *J Forensic Sci* 50:43–53.

Collins P. J., L. K. Hennessy, C. S. Leibelt, R. K. Roby, D. J. Reeder, and P. A. Foxall. 2004. Developmental validation of a single-tube amplification of the 13 CODIS STR Loci, D2S1338, D19S433 and amelogenin: The AmpFlSTR Identifiler PCR amplification kit. *J Forensic Sci* 49:1265–77.

Crowe J. H., J. F. Carpenter, and L. M. Crowe. 1998. The role of vitrification in anhydrobiosis. *Annu Rev Physiol* 60:73–103.

Davis D. L., E. P. O'Brien, and C. M. Bentzley. 2000. Analysis of the degradation of oligonucleotide strands during the freezing/thawing processes using MALDI-MS. *Anal Chem* 72:5092–6.

DNA Bank Network. 2010. Long-Term DNA Storage Workshop Proceedings. http://www.dnabank-network.org/publications/Workshop_Long-term_DNA_storage-Summary_and_Abstracts.pdf (accessed October 20, 2012).

DNA Bank Network. 2012. http://www.dnabank-network.org/Index.php (accessed October 15, 2012).

Eichmann C. and W. Parson. 2008. "Mitominis": Multiplex PCR analysis of reduced size amplicons for compound sequence analysis of the entire mtDNA control region in highly degraded samples. *Int J Legal Med* 122:385–8.

European Network of Forensic Science Institutes. 2012. Survey on DNA-databases in Europe December 2011. http://www.enfsi.eu/sites/default/files/documents/enfsi_survey_on_dna-databases_in_europe_december_2011_0.pdf (accessed October 21, 2012).

Federal Bureau of Investigation. 2012 Combined DNA Index System Statistics. http://www.fbi.gov/about-us/lab/codis/ndis-statistics (accessed October 15, 2012).

Forster L., J. Thomson, and S. Kutranov. 2008. Direct comparison of post-28-cycle PCR purification and modified capillary electrophoresis methods with the 34-cycle "low-copy number" (LCN) method for analysis of trace forensic DNA samples. *Forensic Sci Int Genet* 2:318–28.

Fujita Y. and S. Kubo. 2006. Application of FTA technology to extraction of sperm DNA from mixed body fluids containing semen. *Leg Med* (Tokyo) 8:43–7.

Gaillard C. and F. Strauss. 1998. Avoiding adsorption of DNA to polypropylene tubes and denaturation of short DNA fragments; Elsevier Technical Tips online 1998; http://frstrauss.free.fr/reprints/gaillard_TTO98.pdf (accessed July 3, 2004).

Gaillard C. and F. Strauss. 2001. Eliminating DNA loss and denaturation during storage in plastic microtubes. *Am Clin Lab* 20:52–4.

Gill P. 2001. Application of low-copy number DNA profiling. *Croat Med J* 42:229–32.

Gill P., J. Whitaker, C. Flaxman, N. Brown, and J. Buckleton. 2000. An investigation of the rigor of interpretation rules for STRs derived from less than 100 pg of DNA. *Forensic Sci Int* 112:17–40.

Hall A. and J. Ballantyne. 2003. Novel Y-STR typing strategies reveal the genetic profile of the semen donor in extended interval post-coital cervicovaginal samples. *Forensic Sci Int* 136:58–72.

Hanson E. K. and J. Ballantyne. 2005. Whole genome amplification strategy for forensic genetic analysis using single or few cell equivalents of genomic DNA. *Anal Biochem* 346:246–57.

Hernandez G. E., T. S. Mondala, and S. R. Head. 2009. Assessing a novel room-temperature RNA storage medium for compatibility in microarray gene expression analysis. *BioTechniques* 47:667–8.

Hill C. R., M. C. Kline, M. D. Coble, and J. M. Butler. 2008. Characterization of 26 miniSTR loci for improved analysis of degraded DNA samples. *J Forensic Sci* 53:73–80.

Hochmeister M. N., B. Budowle, U. V. Borer, and R. Dirnhofer. 1995. A method for the purification and recovery of genomic DNA from an HLA DQA1 amplification product and its subsequent amplification and typing with the AmpliType PM PCR Amplification and Typing Kit. *J Forensic Sci* 40:649–53.

Hofreiter M., D. Serre, H. N. Poinar, M. Kuch, and S. Paabo. 2001. Ancient DNA. *Nat Rev Genet* 2:353–9.

Irwin J. A., M. D. Leney, O. Loreille et al. 2007. Application of low-copy number STR typing to the identification of aged, degraded skeletal remains. *J Forensic Sci* 52:1322–7.

Juusola J. and J. Ballantyne. 2003. Messenger RNA profiling: A prototype method to supplant conventional methods for body fluid identification. *Forensic Sci Int* 135:85–96.

Juusola J. and J. Ballantyne. 2005. Multiplex mRNA profiling for the identification of body fluids. *Forensic Sci Int* 152:1–12.

Juusola J. and J. Ballantyne. 2007. mRNA profiling for body fluid identification by multiplex quantitative rt-PCR. *J Forensic Sci* 52:1252–62.

Kita T., H. Yamaguchi, M. Yokoyama, T. Tanaka, and N. Tanaka. 2008. Morphological study of fragmented DNA on touched objects. *Forensic Sci Int Genet* 3:32–6.

Kline M. C., D. L. Duewer, J. W. Redman, and J. M. Butler. 2005. Results from the NIST 2004 DNA Quantitation Study. *J Forensic Sci* 50:570–8.

Kloosterman A. and D. P. Kersbergen. 2003. Efficacy and limits of genotyping low copy number (LCN) DNA samples by multiplex PCR of STR loci. *J Soc Biol* 197:351–9.

Lagoa A. M., T. Magalhães, and M. F. Pinheiro. 2008. Genetic analysis of fingerprints—Could WGA or nested-PCR be alternatives to the increase of PCR cycles number? *Forensic Sci Int Genet* 1:48–9.

Larsen K. and S. Lee. 2005. Optimization strategies for DNA storage. *16th Int Symp Human Identification*, Grapevine, TX.

Le L., K. C. Clabaugh, A. Chang et al. 2008. Recovering DNA profiles from low quantity and low quality forensic samples. *19th Int Symp Human Identification*, Hollywood, CA.

Lederer T., G. Braunschweiger, P. Betz, and S. Seidl. 2002. Purification of STR-multiplex amplified microsamples can enhance signal intensity in capillary electrophoresis. *Int J Legal Med* 116:165–9.

Lee H. Y., N. Y. Kim, M. J. Park, W. I. Yang, and K. J. Shin. 2008. A modified mini-primer set for analyzing mitochondrial DNA control region sequences from highly degraded forensic samples. *BioTechniques* 44:555–6.

Lee S. B., K. C. Clabaugh, S. Silva et al. 2012. Assessing a novel room temperature DNA storage medium for forensic biological samples. *Forensic Sci Int Genet* 6(1):31–40.

Lindahl T. 1993. Instability and decay of the primary structure of DNA. *Nature* 362:709–15.

Meinenger M. R. D., B. Barloewen, J. Jones, and S. B. Lee. 2008. Evaluation of purification columns for forensic DNA extraction. *19th Int Symp Human Identification*, Hollywood, CA.

Mulero J. J., C. W. Chang, R. E. Lagacé et al. 2008. Development and validation of the AmpFℓSTR MiniFiler PCR Amplification Kit: A MiniSTR multiplex for the analysis of degraded and/or PCR inhibited DNA. *J Forensic Sci* 53:838–52.

Nussbaumer C., E. Gharehbaghi-Schnell, and I. Korschineck. 2006. Messenger RNA profiling: A novel method for body fluid identification by real-time PCR. *Forensic Sci Int* 157:181–6.

Opel K. L., D. T. Chung, J. Drabek, J. M. Butler, and B. R. McCord. 2007. Developmental validation of reduced-size STR Miniplex primer sets. *J Forensic Sci* 52:1263–71.

Opel K. L., D. T. Chung, J. Drabek, N. E. Tatarek, L. M. Jantz, and B. R. McCord. 2006. The application of miniplex primer sets in the analysis of degraded DNA from human skeletal remains. *J Forensic Sci* 51:351–6.

O'Rourke D. H., M. G. Hayes, and S. W. Carlyle. 2000. Ancient DNA studies in physical anthropology. *Annu Rev Anthropol* 29:217–42.

Paabo S. P., H. Poinar, D. Serre et al. 2004. Genetic analyses from ancient DNA. *Annu Rev Genet* 38:645–79.

Park S. J., J. Y. Kim, Y. G. Yang, and S. H. Lee. 2008. Direct STR amplification from whole blood and blood- or saliva-spotted FTA without DNA purification. *J Forensic Sci* 53:335–41.

Patchett K. L., K. J. Cox, and D. M. Burns. 2002. Recovery of genomic DNA from archived PCR product mixes for subsequent multiplex amplification and typing of additional loci: Forensic significance for older unsolved criminal cases. *J Forensic Sci* 47:786–96.

Phipps M. and S. Petricevic. 2007. The tendency of individuals to transfer DNA to handled items. *Forensic Sci Int* 168:162–8.

Prinz M., L. Schiffner, and J. Sebestyen et al. 2006. Maximization of STR DNA typing success for touched objects. *Progress Forensic Genet* 11: 651–3.

Roden D. M., J. M. Pulley, M. A. Basford et al. 2008. Development of a large-scale de-identified DNA biobank to enable personalized medicine. *Clin Pharmacol Ther* 84:362–9.

Roeder, A., P. Elsmore, M. Greenhalgh, and A. McDonald. 2009. Maximizing DNA profiling success from sub-optimal quantities of DNA: A staged approach. *Forensic Sci Int Genet* 3:128–37.

Schulz, M. M. and W. Reichert. 2002. Archived or directly swabbed latent fingerprints as a DNA source for STR typing. *Forensic Sci Int* 127:128–30.

Shewale, J. G., S. C. Sikka, E. Schneida, and S. K. Sinha. 2003. DNA profiling of azoospermic semen samples from vasectomized males by using YPLEX 6 amplification kit. *J Forensic Sci* 48:127–9.

Shikama, K. 1965. Effect of freezing and thawing on the stability of double helix of DNA. *Nature* 207:529–30.

Sibille, I., C. Duverneuil, G. Lorin de la Grandmaison et al. 2002. Y-STR DNA amplification as biological evidence in sexually assaulted female victims with no cytological detection of spermatozoa. *Forensic Sci Int* 125:212–6.

Sigurdson, A. J., M. Ha, M. Cosentino et al. 2006. Long-term storage and recovery of buccal cell DNA from treated cards. *Cancer Epidemiol Biomarkers Prev* 15:385–8.

Smith, L. M. and L. A. Burgoyne. 2004. Collecting, archiving and processing DNA from wildlife samples using FTA databasing paper. *BMC Ecol* 4:4.

Smith, P. J. and J. Ballantyne. 2007. Simplified low-copy-number DNA analysis by post-PCR purification. *J Forensic Sci* 52:820–9.

Smith, S. and P. A. Morin. 2005. Optimal storage conditions for highly dilute DNA samples: A role for trehalose as a preserving agent. *J Forensic Sci* 50:1101–8.

Steadman, S. A., J. D. McDonald, J. S. Andrews, and N. D. Watson. 2008. Recovery and STR amplification of DNA from RFLP membranes. *J Forensic Sci* 53:349–58.

Tack, L. C., M. Thomas, and K. Reich. 2007. Automated forensic DNA purification optimized for FTA card punches and identifiler STR-based PCR analysis. *Clin Lab Med* 27:183–91.

van Oorschot, R. A. and M. K. Jones. 1997. DNA fingerprints from fingerprints. *Nature* 387:767.

Walsh, P. S., D. A. Metzger, and R. Higuchi. 1991. Chelex 100 as a medium for simple extraction of DNA for PCR-based typing from forensic material. *BioTechniques* 10:506–13.

Wan, E., M. Akana, J. Pons et al. 2009. Green technologies for room temperature. *Curr Issues Mol Biol* 12:135–42.

Wang, C., C. Trogdon, L. Le, M. Meinenger, and S. B. Lee. 2010. Comparison of different amplification reagents for alleviating inhibitory effects of indigo dye in PCR. *Proc Vol 16 Annu Meet Am Acad Forensic Sci*, Seattle, WA.

Wickenheiser, R. A. 2002. Trace DNA: A review, discussion of theory, and application of the transfer of trace quantities of DNA through skin contact. *J Forensic Sci* 47:442–50.

Wolkers, W. F., F. Tablin, and J. H. Crowe. 2002. From anhydrobiosis to freeze-drying of eukaryotic cells. *Comp Biochem Physiol A Mol Integr Physiol* 131:535–43.

Zalloua, P. A., D. E. Platt, M. El Sibai et al. 2008. Genographic Consortium. Identifying genetic traces of historical expansions: Phoenician footprints in the Mediterranean. *Am J Hum Genet* 83:633–42.

Zetzsche, H., G. Dröge, and B. Gemeinholzer. 2009. DNA Bank Network—Webkatalog und Referenzdatenbank für organismische DNA. *GfBS Newsletter* 22:4, available at http://www.dnabank-network.org/Index.php (accessed February 4, 2010).

Zhu, B., T. Furuki, T. Okuda, and M. Sakurai. 2007. Natural DNA mixed with trehalose persists in B-form double-stranding even in the dry state. *J Phys Chem B* 111:5542–4.

Zubakov, D., E. Hanekamp, M. Kokshoorn, W. van Ijcken, and M. Kayser. 2008. Stable RNA markers for blood and saliva identification revealed from whole genome expression analysis of timewise degraded stains. *Int J Legal Med* 122:135–42.

Extraction of DNA from Forensic Biological Samples for Genotyping

3

JAMES E. STRAY
JASON Y. LIU
MAXIM G. BREVNOV
JAIPRAKASH G. SHEWALE

Contents

3.1 Introduction: Importance of DNA Extraction 40
3.2 DNA Extraction Methods 42
3.2.1 One-Tube DNA Extraction Protocols 43
3.2.1.1 Chelex® Method 43
3.2.1.2 FTA Paper 43
3.2.1.3 Thermostable Proteinases 44
3.2.2 Two-Step DNA Extraction Protocols 44
3.2.2.1 Lysis 44
3.2.2.2 DNA Separation and Isolation 46
3.2.3 Off-the-Shelf Kits for Isolation of DNA from Forensic Samples 48
3.2.4 PCR Compatible Reagents 48
3.3 Automated Extraction of DNA 49
3.3.1 Benchtop Systems 50
3.3.2 High-Throughput Systems 51
3.4 Comparison Studies 51
3.5 Differential Extraction of DNA 55
3.5.1 Sample Collection 56
3.5.2 Cell Elution 56
3.5.3 Utility of DNase in Removal of Contaminating DNA 57
3.5.4 Laser Microdissection 57
3.6 Conclusions 58
Acknowledgments 59
References 59

Abstract: Biological forensic samples constitute evidence with probative organic matter. Evidence believed to contain DNA is typically processed for extraction and purification of its nucleic acid content. Forensic DNA samples are composed of two things, a tissue and the substrate it resides on. Compositionally, a sample may contain almost anything and for each, the type, integrity, and content of both tissue and substrate will vary, as will the contaminant levels. This fact makes the success of extraction one of the most unpredictable steps in genotypic analysis. The development of robust genotyping systems and analysis platforms for short tandem repeat (STR) and mitochondrial DNA sequencing and the acceptance of results generated by these methods in the court system resulted in a high demand for DNA testing. The increasing variety of sample submissions created a need to isolate DNA from forensic samples that may be compromised or contain low levels of biological material. In the past decade, several robust chemistries and isolation methods have been developed to safely and reliably recover DNA from a wide array of sample types in high yield and free of polymerase chain reaction (PCR) inhibitors. In addition, high-throughput automated workflows have been developed to meet the demand for processing increasing numbers of samples. This review summarizes a number of the most widely adopted methods and the best practices for DNA isolation from forensic biological samples, including manual, semi-automated, and fully automated platforms.

3.1 Introduction: Importance of DNA Extraction

Genotyping of biological samples is now routinely performed in human identification (HID) laboratories for applications like paternity testing, forensic casework, DNA databasing, missing persons testing, lineage studies, identification of remains, and so forth. The genotyping protocol comprises sample collection, extraction of DNA from the biological sample, quantification of the DNA, amplification of target loci, identification of amplified products, and the analysis of results. Extraction of DNA is the first and most critical step of sample processing simply because a better quality and higher quantity of DNA leads to a higher quality, and likely more complete, genotype. Isolation of DNA from forensic samples is a challenging process that creates bottlenecks in the sample-processing workflow. The challenges are primarily due to the nature of biological samples. Unlike most clinical samples, the biological samples processed in forensic laboratories vary widely and include biological fluids (saliva, blood, and semen), tissues, hair, bone, tooth, nail, and so forth. Moreover, body fluids can be deposited on a wide range of substrates (rather unlimited), mixed with inhibitors of the polymerase chain reaction (PCR), exposed to varying environmental conditions, or subjected to uncontrolled degradation, and they are often present in limited quantities. Even the single-source samples processed for DNA databasing exhibit variation in the type of body fluid (typically blood or saliva) and substrates. The spectrum of substrates on which biological samples get deposited expands dramatically for forensic evidence samples. Examples of the most common PCR inhibitors found in forensic samples include hematin from blood, humic acids from soil, textile dyes such as indigo from denim, calcium from bones, melanin from soil, and tannins from leather. The amount of DNA isolated from forensic samples cannot be predicted ahead of extraction because of the nature of these samples.

DNA in the cells is associated with a number of physiological components and other macromolecules that synthesize, replicate, maintain, and catabolize the DNA molecule *in vivo*. DNA in forensic samples may also bear irreversible covalent modifications prior to collection or due to improper storage conditions (e.g., base adducts, nicks, base removal [abasic sites], or DNA-protein cross-links). Further, the chemical moieties and physical particles from the substrate where cells are embedded can be released during the extraction procedure. If these substances are not removed during the DNA extraction procedure, they can interfere in the downstream processes of DNA analysis such as PCR. Yet another dimension that adds to the challenge is the shifts and advancements in genotyping technologies. In the past two decades, forensic laboratories have transitioned from restriction fragment length polymorphism (RFLP) analysis to highly sensitive genotyping methods based on autosomal short tandem repeats (STRs), Y-chromosome STRs (Y-STRs), miniSTRs, mitochondrial DNA (mtDNA) sequencing, and single nucleotide polymorphisms (SNPs) (Bauer and Patzelt 2003; Butler et al. 2003; Chong et al. 2005; Collins et al. 2004; Holland and Parson 1999; Kidd et al. 2006; Krenke et al. 2002, 2005; Mulero et al. 2006, 2008; Phillips et al. 2007; Shewale et al. 2004). With the advancements in multiplex capabilities, PCR technology, detection technology, and software for data analysis, it is possible to obtain complete genotypes from as little as 100–250 pg of DNA, equivalent to ~15–35 diploid genomes. There is a demonstrable benefit for the collection and processing of evidence samples collected for crimes such as burglary (Bing et al. 2000). In order to obtain a maximally interpretable genotype in the first attempt and to handle high volumes of evidence samples, efficient and high-throughput DNA extraction methodology has become a prerequisite. It is desirable to have an extraction methodology that:

1. Achieves extraction of DNA from a variety of biological samples.
2. Enables isolation of DNA from samples that contain small quantities of biological material.
3. Isolates DNA at a high concentration so that the volume of extract used for PCR is minimal.
4. Removes substances that interfere with PCR.
5. Does not introduce any contamination.
6. Does not introduce inhibitors of PCR.
7. Maintains sample integrity.
8. Is rapid.
9. Is amenable to automation.

The emphasis in the evaluation of extraction chemistries gets distributed almost equally between the quantity and quality of DNA; higher quantity so that minimal sample is consumed or the ability to obtain all DNA present in a small quantity of biological material, and the quality of the DNA extract, inasmuch as the ultimate goal of DNA analysis is to obtain a full STR profile or genotype devoid of any PCR artifacts.

The workflow for extraction of DNA is dependent on the type of sample being processed. Challenges in extraction of DNA for paternity and DNA database applications are comparatively less due to multiple reasons: the samples are present in abundant quantities, are stored in a controlled environment, and are relatively less degraded; only a fraction of the extract is used for genotyping, resulting in a lower load of compounds that, if present, inhibit PCR; less stringent criteria for data quality since mixture interpretation is not applicable; and current genotyping systems can tolerate single-stranded or partially degraded samples. The extraction methodologies for processing single-source samples are, therefore, driven by high throughput, first-pass success rate, and lower cost. On the other hand, typical forensic casework samples are composed of reference samples from the victim and the suspect and evidence sample(s) from the crime scene. The quantity and state (or quality) of the evidence samples is not within the control of the collection agency or the laboratory analysts. Samples from mass disaster recovery, missing persons, ancient samples, and so forth, are similar in nature to evidence samples. However, utmost care for handling and storage is taken after the samples are collected to protect their integrity and minimize degradation. Robotic platforms are now adopted for sample processing. In general, evidence samples and the samples containing low quantities of biological materials are processed for DNA extraction in separate rooms or at separate times. The improvements made in the past decade in the extraction chemistries enable the genotype to be obtained from small quantities of biological samples. Nagy et al. (2005) documented a complete STR profile using as low as 3, 15–18, and 24 cells per PCR reaction when using cells in suspension, cells on denim and leather, or cells on cotton, respectively.

3.2 DNA Extraction Methods

A wide variety of methods based on different principles are available for extraction of DNA. The reader may refer to the review articles and methods published earlier for detailed discussions (Bing et al. 2000; Brettell et al. 2005; Budowle et al. 2000; Butler et al. 2003; Gill 2001; Lincoln and Thomson 1998, Rechsteiner 2006). It is rather difficult to describe all the DNA extraction methodologies published and used in all forensic laboratories for different types of samples. In this chapter, we have attempted to cover the most commonly used DNA extraction methods in forensic laboratories; exclusion of any specific method or commercial kit is unintentional. Extraction of DNA is achieved in multiple ways. DNA extraction methods are means to obtain DNA preparation in a form that is compatible for downstream applications. Such methods can be classified into two major groups: methods that extract (or release) DNA from cells but do not involve purification steps, and methods that process the cell lysate further for purification of DNA. Some basic steps in DNA extraction and purification are:

1. Cell disruption or cell lysis: The objectives of lysis protocols are to make the biological material accessible to lysis reagents, disrupt cell membranes, release the DNA from proteins like the histones, and bring DNA into a homogeneous liquid phase. This is commonly achieved by a combination of enzymatic, physical, and chemical treatments such as alkaline treatment, proteolysis, detergent treatment, and incubation at elevated temperature.
2. Separating the DNA-containing fraction from cell debris by physicochemical methods: Precipitation, organic extraction, solid-phase binding, electrophoresis.
3. Isolating the DNA: Bringing DNA into an aqueous phase compatible with downstream applications.

The efficiency of each step contributes to the overall recovery of DNA. For simplicity, we have grouped the DNA extraction methods into two categories: one-tube, wherein lysate is processed for downstream applications without purification; and two-step, wherein lysis and purification of DNA are differentiated. Yet another approach that has been developed in recent years is either direct amplification of substrate or lysate obtained using PCR-compatible reagent with robust STR kits such as Identifiler® Direct (Applied Biosystems, part of Life Technologies, Foster City, California) and PowerPlex 18D (Promega Corporation, Madison, Wisconsin). Thus, the DNA extraction or cell lysis step is integrated with amplification (Applied Biosystems 2012; Oostdik et al. 2011).

3.2.1 One-Tube DNA Extraction Protocols

During single-tube extraction procedures, lysis of cells is achieved by chemical or physical agents and the cell free extract is used for genotyping without any purification step. The advantages of such methods are simple, quick, high throughput, and minimal chances of contamination. The most simple method is alkaline lysis of cells with dilute NaOH (up to 0.2 M) followed by neutralization (Klintschar and Neuhuber 2000; Rechsteiner 2006).

3.2.1.1 Chelex® Method

Extraction procedures using Chelex 100 resin (Bio-Rad Laboratories, Hercules, California) are popular in the forensic science community because they save time, reduce costs, simplify extractions, reduce safety risks, and minimize potential for contamination. The method developed by Walsh et al. (1991) involves rupture of cells by osmotic shock in boiling water for about 8 min in the presence of the chelating resin Chelex 100 resin. Paired iminodiacetate ions on the resin bind polyvalent metal ions like magnesium, thereby preventing the degradation of DNA by DNase. The DNA is sheared; however, it is suitable for genotyping methods like STR anslysis. The Chelex method is used to successfully extract DNA from many forensic samples, including bloodstains, tissue, hair, and bone (Bing et al. 2000; Suenaga and Nakamura 2005; Sweet et al. 1996; Tsuchimochi et al. 2002; Vandenberg et al. 1997; Walsh et al. 1991). The Chelex procedure does not include a purification step. Therefore, if the samples contain inhibitors or contaminants, by increasing the sample size one also increases the concentration of these substances in the DNA extract, which can inhibit PCR.

3.2.1.2 FTA Paper

FTA cards (Whatmann, part of GE Healthcare, Buckinghamshire, United Kingdom) are used for easy collection, shipment, and long-term storage at room temperature of blood and saliva samples (Harty et al. 2000; Milne et al. 2006). FTA paper is a cellulose paper treated with proprietary chemicals that lyse the cells, protect DNA from nuclease degradation, and prohibit microbial growth (Burgoyne 1996). DNA on FTA paper is stable for several years. A portion of the card after repeated washings can be directly used for amplification or DNA can be eluted using standard procedures. FTA Elute cards are differentiated from classical FTA cards in that DNA bound to the matrix is released into water. Zhou et al. (2006) described a two-step protocol for extraction of DNA from small (1.2 mm) punches of bloodstains on FTA and 3-mm filter paper (Whatman, Piscataway, New Jersey). A small punch size contains a sufficient number of cells for genotyping and limits the quantity of the inhibitors in the sample used for extraction. The discs in the PCR plate are washed with 20 mM NaOH followed by Tris buffer, pH 8.0, air-dried, and used directly for amplification. Recently, Wolfgramm et al. (2009) demonstrated a two-step procedure for elution of DNA wherein a 3-mm punch of sample is washed with 500 μL of ultrapure water and transferred to another tube containing 30 μL of ultrapure water. The tube is then heated at 94°C for 1–10 min. The yield of DNA was more dependent on the drying time in the preparation of the sample than the incubation time at 94°C. The DNA from FTA paper can also be purified by using conventional DNA extraction methods (Kline et al. 2002; Lorente et al. 1998a).

3.2.1.3 *Thermostable Proteinases*

Purification of DNA from cell lysate following proteinase K digestion is necessary since the lysis is performed typically in the presence of detergents like sodium dodecyl sulfate (SDS) and reducing agents like dithiothreitol (DTT). The presence of additional disruptive chemicals is required because the temperature of maximum activity of proteinase K is relatively low (~56°C). Thermostable proteinases that are active at high temperature when the cell membrane breaks, without incorporation of detergents or reducing agents, offer an alternate single-step method for extraction of DNA from some, if not all, forensic-type samples. Biological samples are incubated with EA 1 proteinase from a thermophilic Bacillus strain EA 1 (ZyGEM, Hamilton, New Zealand) at pH 7.0 and 75°C for 15 min followed by heating at 96°C for 15 min for inactivation of the EA 1 (Coolbear et al. 1992; Moss et al. 2003). At such an elevated temperature for lysis, nucleases released from the cells are inactive and are hydrolyzed by the EA 1. The method is effective in the extraction of samples like blood, bloodstains on different substrates, saliva, swabs from beer bottles, touch trace samples, and so forth (Moss et al. 2003). However, the method has not been successful for extraction of DNA from samples like cigarette butts and bloodstains on black denim. The failure of obtaining interpretable STR profiles is attributed to the presence of inhibitors. EA 1 enzyme is now available and has been formulated by ZyGEM (Hamilton, New Zealand) as the *forensic*GEM™ kits family. The forensicGEM products can be used in a 96-well format or for any number of samples using PCR tubes and a thermal cycler.

One-tube extraction kits with proprietary lysis solutions are also available commercially. QuickExtract™ DNA Extraction Solution and MasterAmp™ Buccal Swab Kit from Epicentre (Madison, Wisconsin) and Extract-N-Amp™ Blood PCR Kit from Sigma Aldrich (St. Louis, Missouri) are examples.

3.2.2 Two-Step DNA Extraction Protocols

3.2.2.1 *Lysis*

Commonly used lysis methods include proteinase K digestion in the presence of detergents like SDS and N-lauryl sarcosine (NLS), reducing agents like DTT, and chelating agents like ethylenediamine tetraacetic acid (EDTA); chemical lysis using reagents such as guanidinium thiocyanate (GuSCN), guanidinium hydrochloride (GuHCl), detergent cocktails, and sodium hydroxide; osmotic shock; and heat treatment (Budowle et al. 2000; Castella et al. 2006; Kishore et al. 2006; Lincoln and Thomson 1998; Nagy et al. 2005). Lysis solutions based on GuSCN and GuHCl are widely used for extraction of DNA because of their combined ability to disrupt cells and inactivate nucleases (Ciulla et al. 1988; Cox 1968). DTT is often incorporated for lysis of samples like sperm and hair. EDTA is incorporated for lysis of calcified tissues like bone, tooth, nail, and so forth. Sewell et al. (2008) postulated that cellulases present in the lysis buffer of the DNeasy® Kit (Qiagen, Valencia, California) enable the release of DNA from paper. Sutlovic et al. (2007) recommended addition of polyvinyl-polypyrrolidone (PVPP) to the samples prior to lysis for removal of humic acid and other phenolic compounds from soiled samples such as human remains found in mass graves and postmortem tissues.

Hair is a more complex tissue; hair shaft and hair with follicles have distinct histology. Hair cells, analogous to cancer cells, experience rapid growth after differentiation from germ cells in the hair follicle requiring additional mitochondria. Thus, hairs are natural candidates for mtDNA typing rather than STR typing. However, the keritinization process may lead to degradation of DNA. Different aspects of STR typing from hair are reviewed by McNevin et al. (2005). Telogen hairs, although found at crime scenes, do not provide very high yields of DNA; from a sample set of 510 telogen hairs from 60 donors (6–9 hairs per donor), 30% of the samples yielded <100 pg, 41.6% yielded 100–550 pg, and 28.6% yielded >550 pg of nuclear DNA (Opel et al. 2008). In general, a pretreatment with detergents and mild lysis buffers is recommended to clean the hair and remove extraneous DNA (McNevin et al. 2005; Opel et al. 2008). In contrast, McNevin et al. (2005) demonstrated that genomic DNA is localized within the outer cuticle, either within cuticle cell nuclei or in cells trapped between cuticle scales. Extensive pretreatment procedures adopted for cleaning of hair removes the genomic DNA present on the exterior of the shaft and results in poor yield. The investigators recommend rinsing the hair with water to obtain a greater yield of genomic DNA.

Bone is a good source of DNA for aged samples. This is because the DNA is well protected within the matrix compared to other soft tissues (Hochmeister et al. 1991; Lassen et al. 1994). Although this is an advantage, it poses a challenge in the lysis of biological material. Typically bone, after cleaning of exterior tissue, is pulverized in a cryomill or grinder. A portion of the pulverized bone powder is subjected to demineralization and lysis in the presence of chelating agents, detergents, and proteinase K for a longer period (24–48 h) (Loreille et al. 2007). PrepFiler® BTA lysis buffer (Applied Biosystems, part of Life Technologies, Foster City, California) is developed for effective lysis of bone and tooth samples; lysis of pulverized bone powder is accomplished in 2 h (Liu et al. 2012; Stray et al. 2009). Stray and Shewale (2010) have discussed the different aspects of DNA extraction from ancient samples.

Deposition of saliva may occur on envelopes and stamps. Various approaches such as steaming to separate the envelope flap or stamp, removal of cuttings, and swabbing are used for preparation of envelope or stamp samples for lysis. Interference of glue in the extraction procedure is dependent on the type of chemistry and reagent composition. The PrepFiler BTA lysis buffer developed by Stray et al. (2009) for lysis of bones is also effective in the lysis of samples containing adhesives like envelopes, stamps, chewing gum, and cigarette butts. The use of cuttings of envelopes sampled directly without steam-opening or swabbing, which are cut into small pieces, achieves high extraction efficiency (Ng et al. 2007; Sinclair and McKechnie 2000).

Paraffin-embedded tissues offer unique challenges for the extraction of DNA. Formalin generates protein-protein and protein-nucleic acid cross-links and methylene bridges between purine and pyrimidine bases (Duval et al. 2009). The yield of DNA is typically low and the DNA is degraded. Deparaffinization is achieved by a series of washes with xylene, ethanol, and phosphate-buffered saline (PBS). Dewaxed samples are then processed for lysis using proteinase K or other methods. Shi et al. (2002) proposed a heat-induced antigen retrieval method for simultaneous deparaffinization and lysis. Tissue sections are heated in a universal buffer solution at alkaline pH and elevated temperatures; optimal results were achieved at pH 9.0 and 120°C. It is postulated that alkaline conditions break the cross-links caused by formalin fixation, denaturate proteins, and facilitate cell lysis. Duval et al. (2009) developed a pretreatment with Hemo-DE solvent solution (Scientific Safety Solvents, Keller, Texas) followed by a three-step protocol comprising proteinase K digestion, heat treatment for reversal of cross-links, and another proteinase K digestion for effective lysis of tissues preserved in buffered formalin and alcohol-based fixatives.

3.2.2.2 DNA Separation and Isolation

Denatured nuclear proteins that package DNA, cellular constituents and metabolites, and substances released from the substrate during lysis make up the total lysate. Nucleic acids are purified from this mixture using methods based on numerous principles.

Organic extraction: The organic or phenol-chloroform method, often regarded as the "gold standard," was one of the earliest developed methods for DNA isolation, and is still a predominant method used for forensic casework samples, particularly those containing scarce amounts of biological materials like hair shafts, bones, and compromised samples (Drobnic 2003; Lorente et al. 1998b; Marmur 1961; Primorac et al. 1996; Schmerer et al. 1999; Zehner et al. 2004). The procedures used by different laboratories are not identical. In general, cells are lysed using a buffered proteinase K and detergent (SDS or NLS). Addition of phenol-chloroform-isoamyl alcohol (PCI) to a sample lysate promotes separation of nonpolar (organic) and polar (aqueous) phases. The phenol is not miscible with water, denatures protein, and sequesters the denatured hydrolyzed protein in the organic phase. During this process, the DNA remains in the aqueous phase in its double-stranded state. The DNA that partitions into the aqueous phase is concentrated by ethanol precipitation or by membrane filtration (Microcon® [Merck KGAA] or similar). Major advantages of organic extraction methods are high yields of unfragmented nuclear and mitochondrial DNA, and the ability to process many different sample types. The procedure is, however, tedious and requires use of toxic chemicals. Although phenol-chloroform yields very pure DNA, it has a lower extraction efficiency than solid-phase techniques due to the multiple phases and buffers required for extraction and ethanol precipitation.

DNA separation and isolation using a solid phase: Generally, all practical solid-phase methods of DNA separation are based on the principle that DNA has a different affinity to specific chemical groups located on the surface of a solid support under different conditions. The process of DNA separation can be divided into three major steps: bind, wash to remove contaminants, and release. During the binding step, under specific conditions, DNA is attracted from crude lysate to the solid phase and stays bound during wash steps while remaining non-DNA contaminants (such as proteins, cellular material, and compounds from substrates that inhibit PCR) that have lower affinity for the solid surface are removed. The DNA release step is performed under different conditions when DNA attains a low affinity to the solid phase and is eluted into the surrounding liquid. There are several chemistries for solid-phase extraction of DNA, but all of them use unique properties of the DNA molecule such as a negative charge and the tendency to form hydrogen bonds involving sugar-phosphate backbone as well as hydrophobic properties of heterocyclic bases.

The silica-based extraction method is commonly used and based on the principle of high affinity of DNA to silica in the presence of chaotropic agents such as NaI, $NaClO_4$, GuSCN, or GuHCl. The combination of GuSCN-based lysis and silica-based extraction was described in 1990 (Boom et al. 1990). The silica-based method allows elution of purified DNA in a low-salt solution such as water or Tris-EDTA (TE) buffer in a form ready for downstream PCR applications. The most commonly used forms of silica are membranes (e.g., spin columns in QIAamp® Blood kits from Qiagen, Valencia, California), silica beads within a spin column (e.g., UltraClean® Forensic DNA Isolation Kit from MoBio Laboratories, Carlsbad, California), and silica magnetic particles (e.g., DNA IQ Systems from Promega, Madison,

Wisconsin, and EZ1® DNA Investigator Kit from Qiagen (Valencia, California). The capacity of DNA bonding to silica is proportional to surface area and becomes saturated with certain quantities of DNA. This property is explored as a means of obtaining a constant concentration of DNA in the extracts from samples that contain large quantities of DNA (Promega 2009a). However, low recovery of DNA due to the presence of silica-binding impurities or compounds in the lysate is a possibility; for example, Poone et al. (2009) demonstrated that one or more chemicals, 3,3',5,5'-tetramethylbenzidine (TMB) in particular, in the Hemastix® reagent strip (Bayer, Elkhart, Indiana) binds strongly to the silica DNA IQ beads limiting the binding of DNA. A similar observation is reported by Petersen and Kaplan (2011) for isolation of DNA from samples tested with Hemastix processed on BioRobot® EZ1 (Qiagen, Valencia, California) using the EZ1 DNA Investigator Kit. The yield of DNA from Hemastix tested samples were improved by addition of MTL buffer to the sample prior to extraction. Apparently, MTL buffer prohibits the binding of components of Hemastix to the silica beads, thereby increasing the DNA yield.

DNA separation and isolation by charge is used in the ChargeSwitch® Technology (CST®, Invitrogen, Carlsbad, California). The solid phase used in this technology contains functional groups providing a switchable surface charge dependent on the pH of the surrounding buffer to facilitate nucleic acid purification. In low pH conditions, the CST beads have a positive charge that binds the negatively charged nucleic acid backbone. Proteins and other contaminants that do not possess net positive charge under the binding conditions are not bound and are washed away in an aqueous wash buffer. To elute nucleic acids, the charge on the surface of the bead is neutralized by raising the pH to 8.5 using a low-salt elution buffer. Purified nucleic acids are eluted instantly into the elution buffer, and are ready for use in downstream applications.

DNA extraction using a solid phase that contains embedded multifunctional polymer is commercialized in the PrepFiler Forensic DNA Extraction kits (Applied Biosystems, part of Life Technologies, Foster City, California). In the presence of a certain concentration of isopropanol, DNA binds selectively to the surface of magnetic particles with a highly developed surface area. After a simple wash step purified DNA is eluted in TE buffer. The PrepFiler Kit is suitable for a broad range of forensic samples (Brevnov et al. 2009b,c; Liu et al. 2012).

Kopka et al. (2011) developed the Fingerprint DNA Finder Kit based on a reversed approach compared to traditional adsorption and desorption of nucleic acids to a matrix. In this approach, inhibitors like proteins and low-molecular-weight components in the cell lysate are retained on the substrate while nucleic acids are not bound. Silica particles with optimized porous structure are coated with a thin film of polymer that selectively binds different substances. Thus, small molecules present in the lysate enter the pore and proteins bind to the polymer leaving DNA molecules unabsorbed (Lauk and Schaaf 2007). The purified DNA can be used for downstream genotyping applications. The kit is useful for STR analysis of a variety of forensic samples, particularly fingerprints on forensic evidentiary materials (Kopka et al. 2011). Dieltjes et al. (2011) adopted a modified protocol for the QIAamp DNA Mini Kit for samples collected from cartridges, bullets, and casings. During 7 years after implementation, reliable and reproducible DNA profiles were obtained from 26.5% of criminal cases (163 of 616) and 6.9% of cartridges, bullets, and casing items (283 of 4,085).

Broemeling et al. (2008) described a multidimensional electrophoresis method termed synchronous coefficient of drag alteration (SCODA) for purification and concentration of

DNA from lysate without removal of substrate. The SCODA is selective for long-charged polymers like DNA, thereby rejecting smaller charged molecules. The method uses rotating electric fields to concentrate nucleic acids in the center of an electrophoresis gel, without the need for electrodes in the concentration region. The method is low-throughput and can be useful for certain sample types.

Centrifugal filter units separate molecules by size through a series of washing and centrifugation steps. Millipore (Bedford, Massachusetts) produces two centrifugal filter units under the brand names Centricon® and Microcon filters. Removal of detergents and other low-molecular-weight substances by microfiltration is suggested by Schiffner et al. (2005) for obtaining DNA from low copy number samples like fingerprints and handled objects. The samples are lysed by proteinase K and 0.01% SDS at 56°C for 30 min followed by heating at 100°C for 10 min. Poly A RNA is added to the lysate and the DNA in the lysate is purified by Microcon 100 filtration. The quantity of SDS used in this procedure is very low (0.2%) compared to traditional proteinase K digestion methods.

3.2.3 Off-the-Shelf Kits for Isolation of DNA from Forensic Samples

Several commercial kits are now available for the extraction of DNA from forensic samples from companies such as Applied Biosystems (part of Life Technologies, Foster City, California), Qiagen (Valencia, California), Promega (Madison, Wisconsin), Invitrogen (Carlsbad, California), MoBio Laboratories (Carlsbad, California), ZyGEM (Hamilton, New Zealand), BioSystems S.A. (Buenos Aires, Argentina), Epicentre (Madison, Wisconsin), and so forth. Significant effort and evaluation work is invested in the development of current off-the-shelf kits; the development of magnetic particles that enable ease of separation is an example. During the 1980s and 1990s, the majority of commercial DNA extraction kits were developed for isolation of DNA from blood or samples containing relatively large quantities of biological samples. Examples include commercial kits like Dynabeads® DNA Direct (Invitrogen), QIAamp Blood Kit (Qiagen), Ready Amp™ Genomic DNA Purification System and Wizard® Genomic DNA Purification Kit (Promega). The developments in highly sensitive genotyping systems and capillary electrophoresis instruments created a need for extraction chemistries suitable for forensic-type samples. In the past decade, second-generation DNA extraction kits designed for forensic genotyping applications have been developed. Examples of some commonly used commercial DNA extraction kits in the forensic laboratories are: PrepFiler kits (Applied Biosystems 2008; Brevnov et al. 2009b; Brevnov et al. 2009c; Liu et al. 2012; Stray et al. 2009; Zimmermann et al. 2009), QIAamp Micro DNA kits (Qiagen 2003), Qiagen EZ1 Investigator kits (Anslinger et al. 2005; Kishore et al. 2006; Montpetit et al. 2005; Qiagen 2008), DNA IQ (Crouse et al. 2005; Fregeau et al. 2006; Greenspoon et al. 1998; Promega Corporation 2009a,b), Fingerprint DNA Finder (Kopka et al. 2011), ChargeSwitch (Invitrogen 2005; Zimmermann et al. 2009), *forensic*GEM kits (Moss et al. 2003), and so forth.

3.2.4 PCR Compatible Reagents

In the past few years, the workflow for genotyping, particularly STR profiling, of reference samples in forensic laboratories has undergone a transition. Reference samples are different from the evidence samples in the sense that the samples contain ample quantities of biological material, stored in controlled conditions, not extensively degraded, and are devoid of excessive PCR inhibitors. The new workflow is based on the substrate on which the sample is collected (Applied Biosystems 2012; Oostdik et al. 2011). Samples on treated substrates like FTA cards and NUCLEIC-CARD™ are processed directly for amplification. PCR master mix (or reaction mix) from robust amplification kits like Identifiler Direct, NGM SElect™ Express or PowerPlex 18D is added to a small portion of the card (~1.2-mm punch) and PCR is performed. Successful amplification without DNA isolation is achieved because (a) proprietary reagents on such substrates lyse the deposited cells while DNA remains bound to the substrate, (b) ample quantities of DNA are present, and (c) the master mix of the amplification kits is formulated to tolerate the PCR inhibitors carried from small quantities of substrate.

Direct amplification of biological samples deposited on swabs or untreated paper does not yield desired results since the cells do not get lysed. PCR-compatible reagents like Prep-N-Go™, Bode PunchPrep™, and Lyse and Go PCR Reagent have been developed. These reagents contain proprietary detergents that do not inhibit the PCR amplification when limited quantity is incorporated. The protocol for swabs involves lysis of cells by the reagent either at room temperature or by heating and transferring a small quantity of lysate (2–4 μL) to the PCR master mix (or reaction mix) of the robust STR kit. For samples deposited on untreated paper, a small punch (~1.2 mm) of the sample is transferred to ~2 μL of the PCR-compatible reagent followed by addition of the PCR master mix (or reaction mix) of the robust STR kit. Linacre et al. (2010) demonstrated direct amplification of biological samples deposited on different fabrics such as polyester, cotton, nylon, denim, and cotton swab. The sample resembled touch evidence samples since the volunteers rubbed the fabrics for about 5 sec between their thumbs and first fingers. A higher number of alleles with higher RFU values were obtained for all tested substrates with direct PCR amplification compared to extraction of DNA by standard methods followed by amplification. Use of direct amplification or PCR-compatible reagents reduces the losses incurred during the DNA extraction protocol and minimizes the chances of contamination due to minimal sample handling.

3.3 Automated Extraction of DNA

In the 1990s there was a trend to move from manual processing to automated processing of forensic samples. Almost all steps of the genotyping process—including DNA extraction, quantification setup, normalization, PCR setup, and preparation of plates for genetic analyzers—are now automated (Brevnov et al. 2009b; Crouse et al. 2005; Fang et al. 2010; Greenspoon et al. 2004, 2006; Nagy et al. 2005; Parson and Steinlechner 2001). Automation offers several advantages such as maintenance of sample integrity, high throughput, consistent performance, workflow integration, audit trail, compliance with laboratory information management system (LIMS), minimal sample-transfer and data-entry errors, and saving the time of trained forensic scientists.

It is important to achieve similar efficiency in terms of yield of DNA and quality of DNA for downstream applications using automated and manual methods. It is not advisable to compromise on these parameters at the cost of high throughput by automation. However, it should be remembered that no automated system can inspect, critically judge, and adjust process parameters to accommodate the peculiarities of any given sample during the extraction process. Hence, a major challenge for automation is to deliver high-quality DNA in comparable yield to the same chemistry used by manual means without such adjustment. Extraction chemistries suitable for automation must work well for nearly all samples. One way this is achieved even for manual methods is to provide variable lysis buffers and protocols upstream of a standard or common purification protocol. Lysis modules that require minimal sample transfer, provide high throughput, and are easy to operate are developed. PrepFiler LySep™ columns (Brevnov et al. 2009a; Stray et al. 2009), PrepFiler Filter Plate (Brevnov et al. 2009b), and Slicprep™ 96 Device (Promega, Madison, Wisconsin) (Tereba et al. 2005) are some examples. The common phase of DNA extraction typically marks the start of the fully automated method, and extraction performance and consistency becomes a function of the robustness of the automated platform and the downstream isolation chemistry. Automated systems for extraction of DNA, in general, run the isolation of DNA operations without any manual steps between the instrument loading and collection of the extracted DNA. Development of magnetic beads or particles technology is probably the most important innovation in the development of automated protocols for solid-phase–based extraction of DNA. A major advantage of magnetic separation technology is the ease of separation of the particles from the liquid phase using the simple application of a magnetic field. Placing the tubes or 96-well plate on a magnet attracts the magnetic particles toward the magnet so that the liquid can be pipetted easily with minimal loss of the particles. Capture of the magnetic particles within the pipette tip is also achieved by placing the magnet to the side of the tip and controlling the flow of the liquid; benchtop instruments like Automate Express™ (Applied Biosystems, part of Life Technologies, Foster City, California), iPrep™ (Invitrogen, Carlsbad, California), BioRobot EZ1 (Qiagen, Valencia, California), and so forth, adopt this principle. The size of particles may vary across a broad range (0.4–10 μm) as will magnetic content, delivering a broad range of physical properties that are critical for development and optimization of automated applications. The surface of the particles can be modified with a variety of functional groups such as silica, ion exchangers, molecules with certain ionic charges, polymers, surfaces with hydrophobic moieties, and so forth (Deggerdal and Larsen 1997).

3.3.1 Benchtop Systems

Low-throughput systems like the Automate Express (Applied Biosystems, part of Life Technologies, Foster City, California), BioRobot EZ1 and BioRobot EZ1 XL (Qiagen, Velencia, California), Maxwell® 16 (Promega, Madison, Wisconsin), and iPrep (Invitrogen, Carlsbad, California) are becoming popular in forensic casework analysis due to their advantages over manual extraction protocols. In the absence of reported estimates of the installation of such benchtop systems, and given that more laboratories are coming on board to automate, it may be safe to predict that approximately 30% of forensic laboratories utilize one or more of these systems for DNA extraction from casework samples. In general, samples are lysed and substrates are removed manually. Liu et al. (2012) developed a LySep column for lysis and removal of substrate without tube transfer. The LySep column is inserted inside a sample tube to form a LySep column assembly. The substrate and lysis buffer are retained in the LySep column due to a membrane placed at the bottom of the LySep column. At the end of lysis, the LySep column assembly is centrifuged, during which the membrane deforms, resulting in collection of the lysate in the sample tube while retaining the substrate in the LySep column.

Isolation and purification of DNA from clear lysate is now a hands-off operation using the benchtop instrument. The reagent cartridges are filled with proprietary reagents. Based on the configuration of the instrument, up to 24 samples are processed in 20–35 min. With minimal hands-on time required after sample lysis, it is possible for one analyst to extract ~100 samples in a single shift from units having a small footprint, low capital cost, and consistently high performance. A one-cartridge-per-sample configuration minimizes the loss of reagents and ensures uniformity of the working reagents from extraction to extraction. Kishore et al. (2006) observed that incorporation of carrier RNA to the cell lysate increased the yield of DNA by 4- to 20-fold from low-level and degraded samples using the BioRobot EZ1 and M48 systems. Further, being closed systems, the benchtop systems can achieve cleaner operation compared to high-throughput open systems.

3.3.2 High-Throughput Systems

High-throughput systems based on adaptable liquid-handling robots by manufacturers such as Tecan (Männedorf, Switzerland), Beckman (Brea, California), Hamilton (Reno, Nevada), PerkinElmer (Waltham, Massachusetts), Qiagen (Valencia, California), and Becton Dickinson (Franklin Lakes, New Jersey) have been employed for forensic DNA extraction. As mentioned, extraction methods developed on open liquid-handling platforms are more prone to external contamination risk. Early systems examined for use in forensic DNA extraction proved that contamination-free operation could be achieved. Several factors contribute to contamination-free operation including precise control over the movement of liquid-handling tips, tight control over liquid aspiration and dispensing parameters, and, for some, systems that establish dedicated pathways movement of liquid-handling tips and avoidance of risky overcrossings of reagent reservoirs, processing plastics, and sample tubes. Unlike benchtop units, open systems do not have a fixed configuration. This flexibility allows adaptation of worktable layouts to accommodate different workflows, extraction chemistries, or plastics.

All major commercial chemistries have been successfully integrated into high-throughput platforms. The most commonly used methods include using the QIAamp, DNA IQ, PrepFiler, and ChargeSwitch systems. Several laboratories have implemented automated protocols for extraction of DNA using high-throughput systems for reference as well as casework samples; see individual papers for detailed discussions (Brevnov et al. 2009b; Chapman et al. 2008; Crouse et al. 2003, 2005; Fregeau et al. 2010; Gehrig et al. 2009; Gleizes et al. 2006; Greenspoon et al. 2004; Lima et al. 2004; Mandrekar et al. 2002; Nagy et al. 2005, 2009; Pizzamiglio et al. 2006; Stangegaard et al. 2009; Stray et al. 2009; Witt et al. 2012).

3.4 Comparison Studies

Vandenberg et al. (1997) compared phenol/chloroform, Chelex resin, *Definitive* DNA preparation kit, Dynabeads DNA Direct, QIAamp Blood Kit, ReadyAmp™ Genomic DNA Purification System, and Wizard Genomic DNA Purification Kit for extraction of DNA from blood and bloodstains. The Chelex kit chemistry provided better results with respect to economical recovery of amplifiable DNA in a quick, simple, and safe manner. The sample size was very high; 25 μL liquid blood, and saturated stains on cotton cloth. In another study, Greenspoon et al. (1998) observed that Chelex kit–extracted DNA preparations stored at −20°C exhibited loss of amplification of a locus in the CCTV quadruplex system compared to the QIAamp kit protocol. Hoff-Olsen et al. (1999) compared five methods for extraction of DNA from decomposed human tissue. The success rate for STR profiling was highest when using a silica-based method followed by phenol/chloroform, glass fiber, InstaGene Kit (Bio Test), and Chelex kit. Castella et al. (2006) compared three extraction methods; the Chelex kit, phenol/chloroform, and the QIAamp DNA Mini Kit with inclusion of the QIAshredder method. A total of 128 samples comprising blood and saliva on swabs, muscle tissue, cigarette butts, saliva on foods, and epidermal cells on clothes were tested. The success rate of STR profiling using the QIAshredder/QIAamp method was 82% compared to 61% by Chelex kit or phenol/chloroform methods.

Davis et al. (2012) compared the performance of AutoMate Express, EZ1 Advanced XL, and Maxwell 16 benchtop DNA extraction systems using 14 forensic type samples. The yield and success of STR profile was dependent on the sample types; none of the systems provided a greater yield of DNA for all tested sample types. The AutoMate Express system offered seamless preprocessing using the LySep column for lysis and separation of substrate, and consumed the least number of consumables.

The HID EVOlution™ Extraction System (Applied Biosystems, part of Life Technologies, Foster City, California) was developed and validated to automate DNA extraction from biological samples using the PrepFiler Automated Forensic DNA Extraction Kit on the Tecan Freedom EVO® 150 robot (Brevnov et al. 2009b). Scripts for automation of the PrepFiler Kit on other platforms such as Beckman, Hamilton, Qiagen Symphony, and Perkin-Elmer Janus are in development. The PrepFiler Kit is a recently developed DNA-extraction technology for extraction of DNA from forensic-type samples that enables consistent recovery of DNA in high yields and quality, providing good quality of STR profiles from most forensic-type samples (Figures 3.1 and 3.2) while effectively removing the most commonly encountered potent inhibitors of PCR (Figures 3.3 and 3.4). The total time required for extraction of DNA from lysate using the HID EVOlution Extraction System for 96 samples was 2.5 h. In one 8-h working shift, 288 samples can be processed without manual intervention. Manual operations take 1 h for a batch of 12 samples and 96 samples can be processed in a day. Thus, the automated workflow provides a threefold increase in sample output and the operator is free to perform other tasks in the laboratory. The time savings are similar to other automated extraction systems.

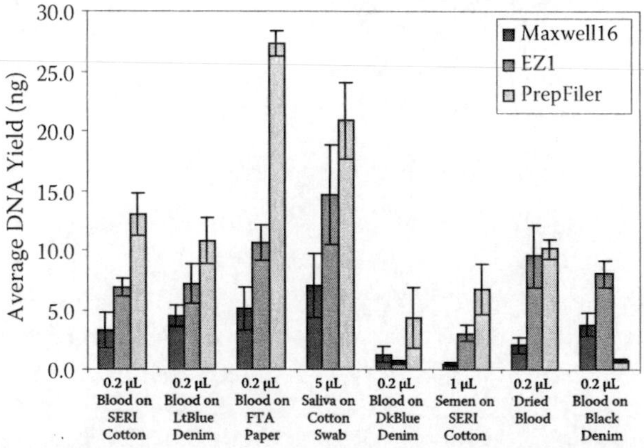

Figure 3.1 Comparison of DNA yield obtained from mock forensic-type samples extracted with DNA IQ® Trace DNA Kit (Maxwell® 16), EZ1® DNA Investigator Kit (EZ1 Advanced XL) and PrepFiler® Automated Forensic DNA Extraction Kit (Tecan EVO® 150). The samples were extracted in replicates of four.

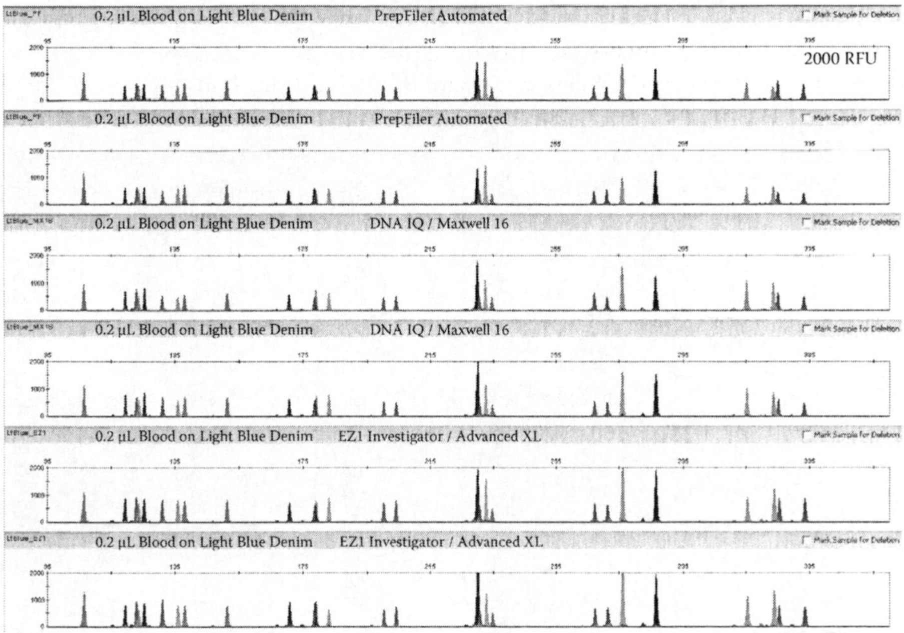

Figure 3.2 (See color insert.) Representative STR profiles for 1 ng of DNA extracted from 0.2 µL blood spotted on light blue denim (4-mm punch) obtained using the Identifiler® PCR Amplification Kit. The DNA was extracted using either the PrepFiler® Automated Kit (top two panels), DNA IQ® Trace DNA Kit (middle two panels), or EZ1® Investigator kit (bottom two panels). Yield of DNA from respective kits is summarized in Figure 3.1.

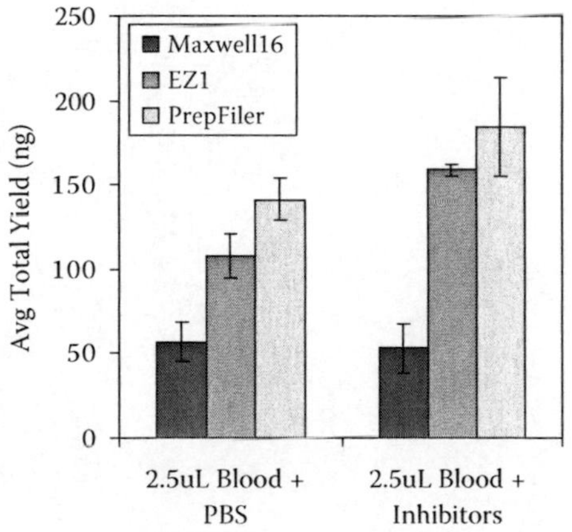

Figure 3.3 Comparison of DNA yield obtained from 2.5 µL blood spiked with an inhibitor mix (indigo, hematin, humic acid, and urban dust extract) or PBS and spotted on a 4-mm punch of white cotton cloth (Serological Research Institute, Richmond, California) extracted with DNA IQ® Trace DNA Kit (Maxwell® 16), EZ1® DNA Investigator Kit (EZ1 Advanced XL) and PrepFiler® Automated Forensic DNA Extraction Kit (Tecan EVO® 150). The samples were extracted in replicates of four.

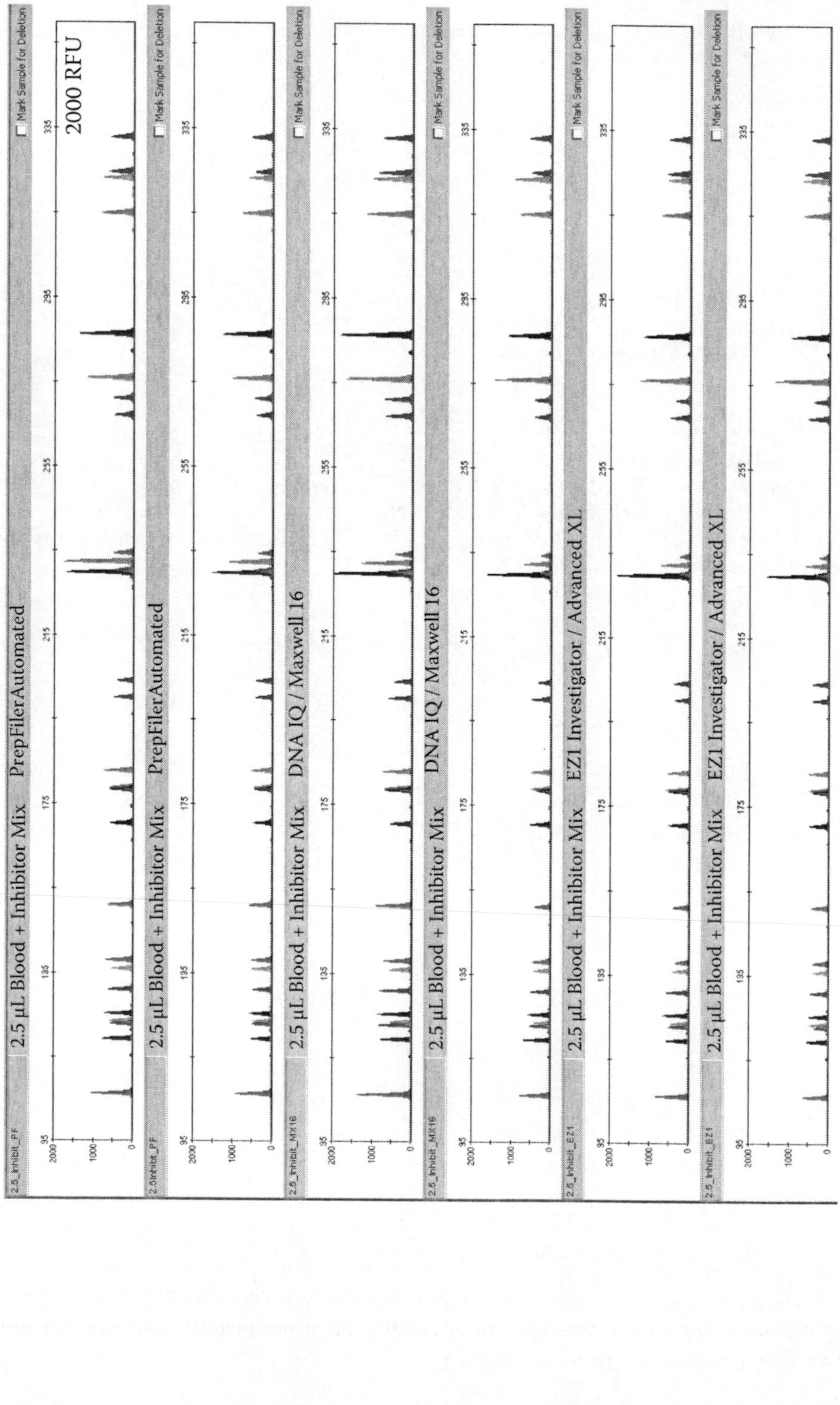

Figure 3.4 **(See color insert.)** Representative STR profiles for 1 ng of DNA obtained from 2.5 µL blood spiked with an inhibitor mix (indigo, hematin, humic acid, and urban dust extract) using the Identifiler® PCR Amplification Kit. The DNA was extracted using either the PrepFiler® Automated Kit (top two panels), DNA IQ® Trace DNA Kit (middle two panels), or EZ1® Investigator kit (bottom two panels). Yield of DNA from respective kits is summarized in Figure 3.3.

3.5 Differential Extraction of DNA

Forensic DNA analysis of sexual assault evidence often involves analysis of DNA from spermatozoa (sperm) and from other cells such as epithelial cells. Sperm are normally obtained from a rape victim by rubbing a swab against a mucous membrane. Because epithelial cells may out number sperm cells in the sample by at least an order of magnitude, the former can cause contamination of sperm cell DNA when the sperm DNA is purified. Consequently, it is often desirable to separate, as cleanly as possible, the sperm cells and epithelial cells, or the sperm DNA and the epithelial DNA, prior to analysis. The current widely used differential extraction method for obtaining sperm DNA from a sexual assault sample is based on preferential lysis by cell type. This method explores the differential stability between sperm and the cell membranes of other cell types that may be present in forensic samples (Gill et al. 1985). In this method, cells on a vaginal swab are first eluted off the swab in phosphate-buffered saline (PBS) or water. Epithelial cells in the sample are preferentially lysed first by using proteinase K and SDS. Intact sperm (sperm heads) are then separated from epithelial cell lysate and lysed in digestion buffer containing proteinase K, SDS, and DTT. The repetitive centrifugation and washing to separate intact sperm from epithelial cell lysate is not only time consuming and tedious but also contributes to sperm loss and undesired female DNA carryover. The success of differential extraction depends on the condition of the sample and the skills of the operator. The DNA from the epithelial cell and sperm lysates (commonly referred to as fractions) is isolated using standard DNA isolation procedures. A protocol comparison study of nine Swiss laboratories for extraction of DNA from simulated sexual-assault samples revealed that the yield of male and female DNA was highly variable, which is expected (Vuichard et al. 2011). However, surprisingly enough it was observed that >90% of the male DNA present in the samples was lost after differential DNA extraction. The protocols used by these laboratories are segregated in five groups.

Wiegand et al. (1992) attempted to improve the method of Gill et al. (1985) for samples having low sperm counts by using mild lysis conditions and by avoiding the washing steps. The Differex™ Kit (Promega, Madison, Wisconsin) utilizes nonaqueous fluid to separate the sperm pellet from epithelial cell DNA during the first centrifugation step to reduce the number of sperm pellet washes (Promega Corporation 2004). Chen et al. (1998) and Garvin (2003) separated the sperm from the epithelial cells before differential lysis by gravitational or mild vacuum filtration or by use of a filter material that can withstand strong vacuum or centrifugal forces without having the pores increase in size. An automated protocol for differential extraction was recently developed on QIAcube (Qiagen, Valencia, California) (Scherer and Pasquale 2012). Schoell et al. (1999) demonstrated the separation of sperm and vaginal cells with flow cytometry after immunostaining based on their differences in ploidy, major histocompatibility class I, CD45, and cytokeratin expression.

Continued efforts have been made over the years to either improve the existing differential extraction workflow or develop physical cell-separation techniques to isolate sperm from sexual assault samples before cell lysis and sperm DNA genotyping. Some of the most recent progress made toward improving current differential extraction workflow and efforts on achieving physical isolation of sperm cells before cell lysis and male DNA genotyping are described here. Bienvenue and Landers (2010) have reviewed different microfluidic approaches developed for extraction of DNA.

3.5.1 Sample Collection

Postcoital vaginal sampling is typically carried out using cotton swabs and it is well recognized by the forensic community that a significant portion of the sperm cells are not recovered from the cotton vaginal swab for differential extraction. Use of a nylon flocked swab for better collection and subsequent elution of sperm and epithelial cells has been investigated by Benschop et al. (2010). Nylon flocked swabs have an open fiber structure, which allows rapid absorption, better sample release, and no sample entrapment such as occurs for traditional cotton swabs. In the study, standard cotton and nylon flocked swabs for postcoital vaginal sampling were compared. The vaginal swabs were sorted into categories on the basis of the time since intercourse (TSI) and subjected to microscopic examination for the presence of cells of male origin, presumptive tests for the detection of seminal fluid, and DNA typing. Cellular elution was sixfold more efficient from the nylon flocked swabs. Similarly, microscopic analysis of nylon swab samples was less time consuming because the higher cell yield and better cell morphology simplify detection of cells of male origin. Both swab types reveal similar results regarding presumptive tests and male DNA typing. The DNA contents from each fraction of a swab between cotton or nylon flocked swab were also compared. Four different DNA-containing fractions were obtained from each swab: the portion of the eluted cells saved for microscopic analysis (microscopic fraction, MF), the nonsperm fraction (NF), the sperm fraction (SF), and the remnants of the swab possibly retaining cell material (retained fraction, RF). RF exhibited approximately 30% of the total DNA yield for the cotton swabs but only 9% for nylon flocked swabs. When comparing the two swab types for MF, NF, and SF, nylon flocked swabs show higher yields for MF and SF and a lower yield for NF.

3.5.2 Cell Elution

A typical workflow for processing a sexual-assault vaginal swab sample starts with elution of sperm and epithelial cells off the swab. This is usually accomplished using one of two basic methods. The first method involves soaking the cotton swab or a portion of the swab in PBS, followed by mechanical shaking to aid in the release of sperm into solution. However, the sperm recovery is typically less than 10% of what was originally present on the cotton swab (Voorhees et al. 2006). Higher recovery of about 30–40% of the sperm is achieved by processing the swab directly. However, sperm surface antigens are destroyed by proteinase K digestion and hence any subsequent physical cell-separation techniques that require intact sperm antigens are not possible. Voorhees et al. (2006) investigated the use of enzymatic digestion of a cotton swab to enhance the elution of sperm. Based on SEM imaging of sperm affixed to a cotton fiber surface, it is evident that the average size of a cotton fiber is much larger than the dimensions of sperm and that the interaction of sperm with cotton fibers is better represented as an adsorption process rather than entanglement. The authors reported that samples digested with *Aspergillus niger* cellulase yielded a twofold enhancement in sperm elution over buffer alone, and preliminary testing of higher-activity cellulases from *Trichoderma reesei* showed even greater enhancement. These results indicate that cellulose-digesting enzymes enhance the release of sperm and epithelial cells from cotton swab over buffer alone, providing a higher recovery of DNA. Norris et al. (2007) developed a method for chemically induced enhancement of cell elution and recovery from cotton swabs. The method exploits the exclusive use of detergents for intact cell removal, and can be utilized in conjunction with, or to circumvent, conventional differential extraction buffer (DE buffer). Samples treated with Sarkosyl and SDS yielded 54.4% and 78.5% recovery of sperm, respectively, compared to 39.4% by conventional DE buffer. It was observed that anionic detergents have the greatest effect. A higher concentration of detergent was necessary for samples stored for longer times.

3.5.3 Utility of DNase in Removal of Contaminating DNA

Sexual assault samples contain extraneous DNA mainly due to degradation of epithelial cells and to some extent DNA released from compromised sperm. The extraneous DNA gets into the epithelial cell lysate (fraction) during the traditional differential extraction protocol. Garvin et al. (2009) described the use of nuclease to selectively remove the soluble female DNA. DNase I reduces the amount of soluble DNA by over 1,000-fold, while having virtually no effect on the DNA remaining in the sperm head and therefore inaccessible to the enzyme. Nuclease inactivation and sperm lysis then yield a soluble, pure male DNA fraction. An aliquot of soluble DNA is removed prior to nuclease addition to provide the female DNA fraction. Vaginal swabs taken at defined time points following consensual sex and taken from rape victims were processed using the nuclease method or the standard method. The nuclease method gave clear male profile at more loci (Garvin et al. 2009).

Hudlow and Buoncristiani (2012) developed a high-throughput protocol for processing sexual-assault samples based on differential lysis of epithelial cells and sperm under alkaline conditions and removal of excessive nonsperm DNA using DNaseI. The samples are heated under mild alkaline conditions (0.1 M NaOH) to generate nonsperm lysate. The nonsperm lysate is separated from the substrate and neutralized with 2 M Tris, pH 7.5. Residual nonsperm DNA on the substrate is removed by treatment with DNaseI. Sperm DNA is then released from the substrate by heating in presence of 1.0 N NaOH followed by neutralization with glacial acetic acid, cleanup using NucleoSpin XS column and purification. It is possible to obtain DNA extracts from ~2–6 swabs in 2 h and up to 96 swabs in about 4 h without the use of a robotic system and with a minimum of sample handling. The method is tested only with the swabs stored under frozen conditions. Adoption of this method to other sample types or samples stored under different conditions may require further optimization.

3.5.4 Laser Microdissection

Laser microdissection (LM) or laser capture microdissection (LCM or LMD) physically separates sperm from epithelial cells in order to obtain a cleaner sperm genotype. There are several different LM methods reported but all involve the use of a laser beam to remove selected cells from a glass slide (Anoruo et al. 2007; Budimlija et al. 2005; Di Martino et al. 2004). The cell-capture method, system configuration, applications, and type of laser involved vary between companies. However, all systems offer an inspection option enabling the user to examine both the excised material and the area from which it was cut. Elliott et al. (2003) reported that LM outperformed preferential lysis in 15 out of 16 samples. One important finding was a significant association between the quality of male profiles recovered and the time since intercourse, which was independent of the number of sperm analyzed, and suggested that DNA was degraded even though sperm heads were intact. Sanders et al. (2006) evaluated various histological stains for use with LMD and

DNA analysis and found that hematoxylin/eosin staining performed best in its ability to differentiate sperm and epithelial cells while exhibiting the least negative effect on further downstream analysis. They also evaluated different DNA isolation methods on LMD-collected cells. The results indicated that both QIAamp and Lyse-N-Go™ (ThermoFisher [Pierce], Rockford, Illinois) kit methods were useful for recovery of DNA from LMD-collected sperms. The STR analysis performed on LMD-separated sperm from mixtures of semen and female buccal epithelial cells provided clear STR profiles of the male donor with the absence of any additional alleles from the female donor. Meredith et al. (2012) developed a one-tube extraction and amplification method for profiling of sperm and epithelial cells recovered by LM. Lysis of epithelial cells was achieved by proteinase K digestion in the presence of Tween 20 (Uniqema Americas, LLC) as detergent; DTT was incorporated for lysis of sperm. Proteinase K was inactivated by heating at 95° C for 10 min. It was possible to obtain full STR profile from as few as 30 sperm and 15 epithelial cells (Meredith et al. 2012).

LMD is effective for recovering sperm present in the sexual assault samples. However, in a small but significant proportion of the total sexual assault cases, the assailants are azoospermic. This means sperm would not be recovered. However, nonsperm cells found in semen, such as leukocytes and epithelial cells from the ejaculatory duct and urethra, may be present on the swab along with cells sloughed from the penis during intercourse (Pollak 1943). Murray et al. (2007) describe a method of visually identifying diploid male cells in such samples using fluorescence *in situ* hybridization (FISH), and selectively harvesting them by means of LM. FISH was carried out using the CEP1 X SpectrumOrange™ Y SpectrumGreen™ DNA Probe Kit (Vysis, Downers Grove, Illinois). The probes are specific to alpha satellite DNA sequences in the centromeric region of chromosome X and satellite III DNA at the Yq12 region of chromosome Y. This technique was tested on 26 postcoital vaginal swabs taken at a range of times after intercourse and useful DNA profiles were generated from samples taken up to 24 h after intercourse.

3.6 Conclusions

There have been multiple advances in the preparation of DNA for forensic analysis over the past decade. Improvements in the solid-phase, paramagnetic magnetic bead, and particle-based methods for capture, cleanup, and release of DNA ready for downstream analysis have led a transition away from traditional organic extraction, silica membrane-based, and resin-based (Chelex kit) protocols. Magnetic particles are easily pulled down and resuspended without the need for centrifugal or vacuum filtration, both of which increase the potential for aerosol formation. Because of this, magnetic separation techniques are more easily adapted to an automated workflow. The result is the expansion of semi- and fully automated platforms with medium (benchtop) to high-throughput (liquid handlers) capabilities for DNA isolation. These developments overcome the bottlenecks in the extraction of DNA and it is possible to process evidence samples that contain small quantities of biological materials such as tape lifts, touch evidence, and stains of minute quantities of body fluid for genotyping. Improved DNA purification methods are integrated with the differential extraction protocols to achieve increased efficiency of DNA isolation from sexual assault samples. In addition to methods developed explicitly for forensic DNA isolation, there is an expansion of RNA-based methods that similarly need chemistries and methods for the unique demands of evidence sample processing. Indeed, messenger RNA (mRNA) analysis for applications like body fluid identification, anatomical origin of a tissue, determining the age of a sample, and so forth, are being investigated (Bauer and Patzelt 2003; Juusola and Ballantyne 2007). Combined methods that achieve total nucleic acid (DNA and RNA) recovery have the benefit that they reduce the amount of sample consumed for analysis (Bauer and Patzelt 2003; Bowden et al. 2011).

Acknowledgments

We thank Heidi Kijenski, Lisa Calandro, Lori Hennessy, and Manohar Furtado from Life Technologies for useful discussions.

References

Anoruo, B., R. van Oorschot, J. Mitchell, and D. Howells. 2007. Isolating cells from non-sperm cellular mixtures using the PALMA® microlaser micro dissection system. *Forensic Sci Int* 173: 93–6.

Anslinger, K., B. Bayer, B. Rolf, W. Keil, and W. Eisenmenger. 2005. Application of the BioRobot EZ1 in a forensic laboratory. *Leg Med* 7: 164–8.

Applied Biosystems. 2008. PrepFiler™ Forensic DNA Extraction Kit User, guide PPN 4390932 Rev. B.

Applied Biosystems. 2012. AmpF/STR® Identifiler® Direct PCR Amplification Kit User Guide. PPN 4415125 Rev G.

Bauer, M. and D. Patzelt. 2003. A method for simultaneous RNA and DNA isolation from dried blood and semen stains. *Forensic Sci Int* 136: 76–78.

Benschop, C. C. G., D. C. Wiebosch, A. D. Kloosterman, and T. Sijen. 2010. Post-coital vaginal sampling with nylon flocked swabs improves DNA typing. *Forensic Sci Int Genet* 4: 115–21.

Bienvenue, J. M. and J. P. Landers. 2010. DNA extraction on microfluidic devices. *Forensic Sci Rev* 22: 187–97.

Bing, D. H., F. R. Bieber, M. M. Holland, and E. F. Huffine. 2000. Isolation of DNA from forensic evidence. *Curr Protoc Hum Genet* Suppl 26: 14.3.1.

Boom, R., C. J. Sol, M. M. Salimans, C. L. Jansen, P. M. Wertheim-van Dillen, and J. van der Noordaa. 1990. Rapid and simple method for purification of nucleic acids. *J Clin Microbiol* 28: 495–503.

Bowden, A., R. Fleming, and S. Harbison. 2011. A method for DNA and RNA co-extraction for use on forensic samples using the Promega DNA IQ system. *Forensic Sci Int Genet* 5: 64–8.

Brettell, T. A., J. M. Butler, and R. Saferstein. 2005. Forensic science. *Anal Chem* 77: 3839–60.

Brevnov, M. G., H. S. Pawar, J. Mundt, L. M. Calandro, M. R. Furtado, and J. G. Shewale. 2009c. Developmental validation of the PrepFiler Forensic DNA Extraction Kit for extraction of genomic DNA from biological samples. *J Forensic Sci* 54: 599–607.

Brevnov, M. G., J. Y. Liu, D. Donovan et al. 2009a. In *20th Int Symp Human Identification,* Las Vegas, NV.

Brevnov, M. G., J. Mundt, J. Benfield et al. 2009b. Automated extraction of DNA from forensic sample types using the PrepFiler Automated Forensic DNA Extraction Kit. *J Assoc Lab Automation* 14: 294–302.

Broemeling, D. J., J. Pel, D. C. Gunn et al. 2008. An instrument for automated purification of nucleic acids from contaminated forensic samples. *JALA Charlottesville Va* 13: 40–8.

Budimlija, Z. M., M. Lechpammer, D. Popiolek, F. Fogt, M. Prinz, and F. R. Bieber. 2005. Forensic applications of laser capture microdissection: Use in DNA-based parentage testing and platform validation. *Croat Med J* 46: 549–55.

Budowle, B., J. Smith, T. R. Moretti, and J. DiZinno. 2000. *DNA typing protocols: Molecular biology and forensic analysis.* Eaton Publishing: Natick, MA.

Burgoyne, B. 1996. US Patent 5 496 562.

Butler, J. M., Y. Shen, and B. R. McCord. 2003. The development of reduced size STR amplicons as tools for analysis of degraded DNA. *J Forensic Sci* 48: 1054–64.

Castella, V., N. Dimo-Simonin, C. Brandt-Casadevall, and P. Mangin. 2006. Forensic evaluation of the QIAshredder/QIAamp DNA extraction procedure. *Forensic Sci Int* 156: 70–3.

Chapman, B. R., L. Tack, and D. Pitcher. 2008. An excursion down an outback automation highway: Progression of an Australian forensic biology laboratory toward an automated high-throughput facility. In *19th Int Symp Human Identification*, Hollywood, CA.

Chen, J., L. Kobilinsky, D. Wolosin et al. 1998. A physical method for separating spermatozoa from epithelial cells in sexual assault evidence. *J Forensic Sci* 43: 114–8.

Chong, M. D., C. D. Calloway, S. B. Klein, C. Orrego, and M. R. Buoncristiani. 2005. Optimization of a duplex amplification and sequencing strategy for the HVI/HVII regions of human mitochondrial DNA for forensic casework. *Forensic Sci Int* 154: 137–48.

Ciulla, T. A., R. M. Sklar, and S. L. Hauser. 1988. A simple method for DNA purification from peripheral blood. *Anal Biochem* 174: 485–8.

Collins, P. J., L. K. Hennessy, C. S. Leibelt, R. K. Roby, D. J. Reeder, and P. A. Foxall. 2004. Developmental validation of a single-tube amplification of the 13 CODIS STR loci, D2S1338, D19S433, and amelogenin: The AmpFℓSTR Identifiler PCR Amplification Kit. *J Forensic Sci* 49: 1265–77.

Coolbear, T., J. M. Whittaker, and R. M. Daniel. 1992. The effect of metal ions on the activity and thermostability of the extracellular proteinase from a thermophilic Bacillus, strain EA.1. *Biochem J* 287 (Pt 2): 367–74.

Cox, R. A. 1968. In *Nucleic Acids Part B* (Ed, Lawrence Grossman, K. M.) Academic Press, pp. 120–129.

Crouse, C. A. and J. Conover. 2003. Successful validation and implementation of (semi)-automation in a small caseworking laboratory. In *14th Int Symp Human Identification*, Phoenix, AZ.

Crouse, C. A., S. Yeung, S. Greenspoon et al. 2005. Improving efficiency of a small forensic DNA laboratory: Validation of robotic assays and evaluation of microcapillary array device. *Croat Med J* 46: 563–77.

Davis, C. P., J. L. King, B. Budowle, A. J. Eisenberg, and M. A. Turnbough. 2012. Extraction platform evaluations: A comparison of AutoMate Express, EZ1(R) Advanced XL, and Maxwell(R) 16 Bench-top DNA extraction systems. *Leg Med (Tokyo)* 14: 36–9.

Deggerdal, A. and F. Larsen. 1997. Rapid isolation of PCR-ready DNA from blood, bone marrow and cultured cells, based on paramagnetic beads. *Biotechniques* 22: 554–7.

Di Martino, D., G. Giuffre, N. Staiti, A. Simone, M. Le Donne, and L. Saravo. 2004. Single sperm cell isolation by laser microdissection. *Forensic Sci Int* 146 Suppl: S151–3.

Dieltjes, P., R. Mieremet, S. Zuniga, T. Kraaijenbrink, J. Pijpe, and P. de Knijff. 2011. A sensitive method to extract DNA from biological traces present on ammunition for the purpose of genetic profiling. *Int J Legal Med* 125: 597–602.

Drobnic, K. 2003. Analysis of DNA evidence recovered from epithelial cells in penile swabs. *Croat Med J* 44: 350–4.

Duval, K., R. A. Aubin, J. Elliott et al. 2009. Optimized manual and automated recovery of amplifiable DNA from tissues preserved in buffered formalin and alcohol-based fixative. *Forensic Sci Int Genet* 4: 80–8.

Elliott, K., D. S. Hill, C. Lambert, T. R. Burroughes, and P. Gill. 2003. Use of laser microdissection greatly improves the recovery of DNA from sperm on microscope slides. *Forensic Sci Int* 137: 28–36.

Fang, R., J. Y. Liu, H. L. Kijenski et al. 2010. The HID EVOlution System for Automation of DNA Quantification and Short Tandem Repeat Analysis. *J Assoc Lab Automation* 15: 65–73.

Fregeau, C. J., C. M. Lett, and R. M. Fourney. 2010. Validation of a DNA IQ-based extraction method for TECAN robotic liquid handling workstations for processing casework. *Forensic Sci Int Genet* 4: 292–304.

Fregeau, C. J., M. Lett, J. Elliott, K. L. Bowen, T. White, and R. M. Fourney. 2006. Adoption of automated DNA processing for high volume DNA casework: A combined approach using magnetic beads and real-time PCR. *Int Cong Ser* 1288: 688–690.

Garvin, A. M. 2003. Filtration based DNA preparation for sexual assault cases. *J Forensic Sci* 48: 1084–7.

Garvin, A. M., M. Bottinelli, M. Gola, A. Conti, and G. Soldati. 2009. DNA preparation from sexual assault cases by selective degradation of contaminating DNA from the victim. *J Forensic Sci* 54: 1297–303.

Gehrig, C., D. Kummer, and V. Vastella. 2009. Automated DNA extraction of forensic samples using the QIAsymphony platform: Estimations of DNA recovery and PCR inhibitor removal. *Forensic Sci Int Genet* Suppl 2:85–6.

Gill, P. 2001. Application of low copy number DNA profiling. *Croat Med J* 42: 229–32.

Gill, P., A. J. Jeffreys, and D. J. Werrett. 1985. Forensic application of DNA "fingerprints." *Nature* 318: 577–9.

Gleizes, A., L. Pene, S. Fanget, and A. Paleologue. 2006. Automation of Invitrogen ChargeSwitch forensic DNA purification kit using Hamilton STARlet workstation to achieve optimal DNA recovery from forensic casework samples. *22nd Congr Int Soc Forensic Genet,* Copenhagen, Denmark.

Greenspoon, S. A., J. D. Ban, K. Sykes et al. 2004. Application of the BioMek 2000 Laboratory Automation Workstation and the DNA IQ System to the extraction of forensic casework samples. *J Forensic Sci* 49: 29–39.

Greenspoon, S. A., K. L. Sykes, J. D. Ban et al. 2006. Automated PCR setup for forensic casework samples using the Normalization Wizard and PCR Setup robotic methods. *Forensic Sci Int* 164: 240–8.

Greenspoon, S. A., M. A. Scarpetta, M. L. Drayton, and S. A. Turek. 1998. QIAamp spin columns as a method of DNA isolation for forensic casework. *J Forensic Sci* 43: 1024–30.

Harty, L. C., M. Garcia-Closas, N. Rothman, Y. A. Reid, M. A. Tucker, and P. Hartge. 2000. Collection of buccal cell DNA using treated cards. *Cancer Epidem Biomar Prev* 9: 501–6.

Hochmeister, M. N., B. Budowle, U. V. Borer, U. Eggmann, C. T. Comey, and R. Dirnhofer. 1991. Typing of deoxyribonucleic acid (DNA) extracted from compact bone from human remains. *J Forensic Sci* 36: 1649–61.

Hoff-Olsen, P., B. Mevag, E. Staalstrom, B. Hovde, T. Egeland, and B. Olaisen. 1999. Extraction of DNA from decomposed human tissue. An evaluation of five extraction methods for short tandem repeat typing. *Forensic Sci Int* 105: 171–83.

Holland, M. M. and T. J. Parson. 1999. Mitochondrial DNA sequence analysis—Validation and use for forensic casework. *Forensic Sci Rev* 11: 21–50.

Hudlow, W. R. and M. R. Buoncristiani. 2012. Development of a rapid, 96-well alkaline based differential DNA extraction method for sexual assault evidence. *Forensic Sci Int Genet* 6: 1–16.

Invitrogen. 2005. ChargeSwitch® Forensic DNA Purification Kit Instruction Manual version A; #25-0825.

Juusola, J., and J. Ballantyne. 2007. mRNA profiling for body fluid identification by multiplex quantitative RT-PCR. *J Forensic Sci* 52: 1252–62.

Kidd, K. K., A. J. Pakstis, W. C. Speed et al. 2006. Developing a SNP panel for forensic identification of individuals. *Forensic Sci Int* 164: 20–32.

Kishore, R., W. Reef Hardy, V. J. Anderson, N. A. Sanchez, and M. R. Buoncristiani. 2006. Optimization of DNA extraction from low-yield and degraded samples using the BioRobot EZ1 and BioRobot M48. *J Forensic Sci* 51: 1055–61.

Kline, M. C., D. L. Duewer, J. W. Redman, J. M. Butler, and D. A. Boyer. 2002. Polymerase chain reaction amplification of DNA from aged blood stains: Quantitative evaluation of the "suitability for purpose" of four filter papers as archival media. *Anal Chem* 74: 1863–9.

Klintschar, M. and F. Neuhuber. 2000. Evaluation of an alkaline lysis method for the extraction of DNA from whole blood and forensic stains for STR analysis. *J Forensic Sci* 45: 669–73.

Kopka, J., M. Leder, S. M. Jaureguiberry, G. Brem, and G. O. Boselli. 2011. New optimized DNA extraction protocol for fingerprints deposited on a special self-adhesive security seal and other latent samples used for human identification. *J Forensic Sci* 56: 1235–40.

Krenke, B. E., A. Tereba, S. J. Anderson et al. 2002. Validation of a 16-locus fluorescent multiplex system. *J Forensic Sci* 47: 773–85.

Krenke, B. E., L. Viculis, M. L. Richard et al. 2005. Validation of male-specific, 12-locus fluorescent short tandem repeat (STR) multiplex. *Forensic Sci Int* 151: 111–24.

Lassen, C., S. Hummel, and B. Herrmann. 1994. Comparison of DNA extraction and amplification from ancient human bone and mummified soft tissue. *Int J Legal Med* 107: 152–5.

Lauk, C. and J. Schaaf. 2007. A new novel approach for the extraction of DNA from postage stamps. *Forensic Sci Commun* 9: 1–7.

Lima, S., A. Marignani, L. Hulse-Smith et al. 2004. Validation and optimization of a robotic system used to fully automate forensic DNA extraction, quantification, DNA normalization and STR amplification analysis. *Proc 15th Int Symp Human Identification,* Phoenix, AZ.

Linacre, A., V. Pekarek, Y. C. Swaran, and S. S. Tobe. 2010. Generation of DNA profiles from fabrics without DNA extraction. *Forensic Sci Int Genet* 4: 137–41.

Lincoln, P. J. and J. Thomson. 1998. *Forensic DNA Profiling Protocols, Methods in Molecular Biology Vol 198.* Humana Press: Totowa, NJ.

Liu, J. Y., C. Zhong, A. Holt et al. 2012. AutoMate Express forensic DNA extraction system for the extraction of genomic DNA from biological samples. *J Forensic Sci* 57: 1022–30.

Loreille, O. M., T. M. Diegoli, J. A. Irwin, M. D. Coble, and T. J. Parsons. 2007. High efficiency DNA extraction from bone by total demineralization. *Forensic Sci Int Genet* 1: 191–5.

Lorente, J. A., M. Lorente, M. J. Lorente et al. 1998a. Newborn genetic identification: Expanding the fields of forensic haemogenetics. *Prog Forensic Genet* 7: 114–6.

Lorente, M., C. Entrala, J. A. Lorente, J. C. Alvarez, E. Villanueva, and B. Budowle. 1998b. Dandruff as a potential source of DNA in forensic casework. *J Forensic Sci* 43: 901–2.

Mandrekar, P., L. Flanagan, R. McLaren, and A. Tereba. 2002. Automation of DNA isolation, quantitation, and PCR setup. *Proc 13th Int Symp Human Identification,* Phoenix, AZ.

Marmur, J. 1961. A procedure for the isolation of decarboxyribonucleic acid from microorganisms. *J Mol Biol* 3: 208–18.

McNevin, D., L. Wilson-Wilde, J. Robertson, J. Kyd, and C. Lennard. 2005. Short tandem repeat (STR) genotyping of keratinised hair. Part 1. Review of current status and knowledge gaps. *Forensic Sci Int* 153: 237–46.

Meredith, M., J. A. Bright, S. Cockerton, and S. Vintiner. 2012. Development of a one-tube extraction and amplification method for DNA analysis of sperm and epithelial cells recovered from forensic samples by laser microdissection. *Forensic Sci Int Genet* 6: 91–6.

Milne, E., F. M. van Bockxmeer, L. Robertson et al. 2006. Buccal DNA collection: Comparison of buccal swabs with FTA cards. *Cancer Epidem Biomark Prev* 15: 816–9.

Montpetit, S. A., I. T. Fitch, and P. T. O'Donnell. 2005. A simple automated instrument for DNA extraction in forensic casework. *J Forensic Sci* 50: 555–63.

Moss, D., S. A. Harbison, and D. J. Saul. 2003. An easily automated, closed-tube forensic DNA extraction procedure using a thermostable proteinase. *Int J Legal Med* 117: 340–9.

Mulero, J. J., C. W. Chang, L. M. Calandro et al. 2006. Development and validation of the AmpFℓSTR Yfiler PCR amplification kit: A male specific, single amplification 17 Y-STR multiplex system. *J Forensic Sci* 51: 64–75.

Mulero, J. J., C. W. Chang, R. E. Lagace et al. 2008. Development and validation of the AmpFℓSTR MiniFiler PCR Amplification Kit: A MiniSTR multiplex for the analysis of degraded and/or PCR inhibited DNA. *J Forensic Sci* 53: 838–52.

Murray, C., C. McAlister, and K. Elliott. 2007. Identification and isolation of male cells using fluorescence *in situ* hybridisation and laser microdissection, for use in the investigation of sexual assault. *Forensic Sci Int Genet* 1: 247–52.

Nagy, M., C. Hahne, B. Henske et al. 2009. The fully automated DNA extraction with the QIAsymphony SP-Validation and first experiences in forensic casework. *Proc 23rd World Congr Int Soc Forensic Genet*, Buenos Aires, Argentina.

Nagy, M., P. Otremba, C. Kruger et al. 2005. Optimization and validation of a fully automated silica-coated magnetic beads purification technology in forensics. *Forensic Sci Int* 152: 13–22.

Ng, L. K., A. Ng, F. Cholette, and C. Davis. 2007. Optimization of recovery of human DNA from envelope flaps using DNA IQ System for STR genotyping. *Forensic Sci Int Genet* 1: 283–6.

Norris, J. V., K. Manning, S. J. Linke, J. P. Ferrance, and J. P. Landers. 2007. Expedited, chemically enhanced sperm cell recovery from cotton swabs for rape kit analysis. *J Forensic Sci* 52: 800–5.

Oostdik K., M. Ensenberger, B. Krenke, C. Sprecher, and D. Storts. 2011. The PowerPlex™ 18D system: A direct-amplification STR system with reduced thermal cycling time. *Profiles in DNA*. Available at http://www.promega.com/resources/articles/profiles-in-dna/2011/the-powerplex-18d-system-a-direct-amplification-str-system-with-reduced-thermal-cycling-time/ (Accessed on 10/04/2012).

Opel, K. L., E. L. Fleishaker, J. A. Nicklas, E. Buel, and B. R. McCord. 2008. Evaluation and quantification of nuclear DNA from human telogen hairs. *J Forensic Sci* 53: 853–7.

Parson, W. and M. Steinlechner. 2001. Efficient DNA database laboratory strategy for high throughput STR typing of reference samples. *Forensic Sci Int* 122: 1–6.

Petersen, D. and M. Kaplan. 2011. The use of Hemastix® severely reduces DNA recovery using the BioRobot® EZ1. *J Forensic Sci* 56: 733–5.

Phillips, C., R. Fang, D. Ballard et al. 2007. Evaluation of the Genplex SNP typing system and a 49plex forensic marker panel. *Forensic Sci Int Genet* 1: 180–5.

Pizzamiglio, M., A. Marino, D. My et al. 2006. A new way to mean low copy number processing DNA from sweat evidences. *Proc 4th Eur Acad Forensic Sci*, Helsinki, Finland.

Pollak, O. J. 1943. Semen and seminal stains. *Arch Pathol* 35: 140–84.

Poone, H., J. Elliott, J. Modler, and C. Fregeau. 2009. The use of Hemastix® and the subsequent lack of DNA recovery using the Promega DNA IQ™ System. *J Forensic Sci* 54: 1278–86.

Primorac, D., S. Andelinovic, M. Definis-Gojanovic et al. 1996. Identification of war victims from mass graves in Croatia, Bosnia, and Herzegovina by use of standard forensic methods and DNA typing. *J Forensic Sci* 41: 891–4.

Promega Corporation. 2004. Differex™ System; Part #TBD020 1–7.

Promega Corporation. 2009a. DNA IQ™ System - Database Protocol; Part #TB297.

Promega Corporation. 2009b. DNA IQ™ System - Small Sample Casework Protocol; Part #TB296.

Qiagen. 2003. QIAamp® DNA Micro Handbook.

Qiagen. 2008, QIAamp® Investigator BioRobot Kit Handbook.

Rechsteiner, M. 2006. Applying revolutionary technologies to DNA extraction for forensic studies. *Forensic Magazine*, April/May.

Sanders, C. T., N. Sanchez, J. Ballantyne, and D. A. Peterson. 2006. Laser microdissection separation of pure spermatozoa from epithelial cells for short tandem repeat analysis. *J Forensic Sci* 51: 748–57.

Scherer, M. and F. D. Pasquale. 2012. How to improve analysis of sexual assault samples using novel tools: Differential wash and human:male quantification. *Green Mountain DNA Conf*, Burlington, VT.

Schiffner, L. A., E. J. Bajda, M. Prinz, J. Sebestyen, R. Shaler, and T. A. Caragine. 2005. Optimization of a simple, automatable extraction method to recover sufficient DNA from low copy number DNA samples for generation of short tandem repeat profiles. *Croat Med J* 46: 578–86.

Schmerer, W., M. S. Hummel, and B. Herrmann. 1999. Optimized DNA extraction to improve reproducibility of short tandem repeat genotyping with highly degraded DNA as target. *Electrophoresis* 20: 1712–6.

Schoell, W. M., M. Klintschar, R. Mirhashemi, and B. Pertl. 1999. Separation of sperm and vaginal cells with flow cytometry for DNA typing after sexual assault. *Obstet Gynecol* 94: 623–7.

Sewell, J., I. Quinones, C. Ames et al. 2008. Recovery of DNA and fingerprints from touched documents. *Forensic Sci Int Genet* 2: 281–5.

Shewale, J. G., H. Nasir, E. Schneida, A. M. Gross, B. Budowle, and S. K. Sinha. 2004. Y-chromosome STR system, Y-PLEX 12, for forensic casework: Development and validation. *J Forensic Sci* 49: 1278–90.

Shi, S. R., R. J. Cote, L. Wu et al. 2002. DNA extraction from archival formalin-fixed, paraffin-embedded tissue sections based on the antigen retrieval principle: Heating under the influence of pH. *J Histochem Cytochem* 50: 1005–11.

Sinclair, K. and V. M. McKechnie. 2000. DNA extraction from stamps and envelope flaps using QIAamp and QIAshredder. *J Forensic Sci* 45: 229–30.

Stangegaard, M., T. G. Froslev, R. Frank-Hansen et al. 2009. Automated extraction of DNA and PCR setup using a Tecan Freedom EVO® liquid handler. *Forensic Sci Int Genet* Suppl 2: 74–6.

Stray, J. E., A. Holt, M. Brevnov, L. M. Calandro, M. R. Furtado, and J. G. Shewale. 2009. Extraction of high quality DNA from biological materials and calcified tissues. *Forensic Sci Int Genet* Suppl 2: 159–60.

Stray, J. E. and J. G. Shewale. 2010. Extraction of DNA from human remains. *Forensic Sci Rev* 22: 177–85.

Suenaga, E. and H. Nakamura. 2005. Evaluation of three methods for effective extraction of DNA from human hair. *J Chromatog Biol Anal Technol Biomed Life Sci* 820: 137–41.

Sutlovic, D., M. D. Gojanovic, and S. Andelinovic. 2007. Rapid extraction of human DNA containing humic acid. *Croat Chem Acta* 80: 117–20.

Sweet, D., M. Lorente, A. Valenzuela, J. A. Lorente, and J. C. Alvarez. 1996. Increasing DNA extraction yield from saliva stains with a modified Chelex method. *Forensic Sci Int* 83: 167–77.

Tereba, A., J. Krueger, R. Olson, P. Mandrekar, and B. McLaren. 2005. High-throughput sample processing on solid supports using the Slicprep™ 96 device. *Profiles in DNA* 8: 3–5.

Tsuchimochi, T., M. Iwasa, Y. Maeno et al. 2002. Chelating resin-based extraction of DNA from dental pulp and sex determination from incinerated teeth with Y-chromosomal alphoid repeat and short tandem repeats. *Am J Forensic Med Pathol* 23: 268–71.

Vandenberg, N., R. A. van Oorschot, and R. J. Mitchell. 1997. An evaluation of selected DNA extraction strategies for short tandem repeat typing. *Electrophoresis* 18: 1624–6.

Voorhees, J. C., J. P. Ferrance, and J. P. Landers. 2006. Enhanced elution of sperm from cotton swabs via enzymatic digestion for rape kit analysis. *J Forensic Sci* 51: 574–9.

Vuichard, S., U. Borer, M. Bottinelli et al. 2011. Differential DNA extraction of challenging simulated sexual-assault samples: A Swiss collaborative study. *Invest Genet* 2: 11–7.

Walsh, P. S., D. A. Metzger, and R. Higuchi. 1991. Chelex 100 as a medium for simple extraction of DNA for PCR-based typing from forensic material. *Biotechniques* 10: 506–13.

Wiegand, P., M. Schurenkamp, and U. Schutte. 1992. DNA extraction from mixtures of body fluid using mild preferential lysis. *Int J Legal Med* 104: 359–60.

Witt, S., J. Neumann, H. Zierdt, G. Gebel, and C. Roscheisen. 2012. Establishing a novel automated magnetic bead-based method for the extraction of DNA from a variety of forensic samples. *Forensic Sci Int Genet* 6: 539–47.

Wolfgramm Ede, V., F. M. de Carvalho, V. R. Aguiar et al. 2009. Simplified buccal DNA extraction with FTA Elute Cards. *Forensic Sci Int Genet* 3: 125–7.

Zehner, R., J. Amendt, and R. Krettek. 2004. STR typing of human DNA from fly larvae fed on decomposing bodies. *J Forensic Sci* 49: 337–40.

Zhou, H., J. G. Hickford, and Q. Fang. 2006. A two-step procedure for extracting genomic DNA from dried blood spots on filter paper for polymerase chain reaction amplification. *Anal Biochem* 354: 159–61.

Zimmermann, P., K. Vollack, B. Haak, et al. 2009. Adaptation and evaluation of PrepFiler™ extraction technology in an automated forensic DNA analysis process with emphasis on DNA yield, inhibitor removal and contamination security. *Forensic Sci Int Genet* Suppl 2: 62–4.

Extraction of DNA from Human Remains

4

JAMES E. STRAY
JAIPRAKASH G. SHEWALE

Contents

4.1	Introduction	65
	4.1.1 Bone and Tooth as Sources of DNA	65
	4.1.2 Extraction of DNA	67
4.2	Precautions	67
4.3	Preparation of Samples	68
4.4	Lysis	68
4.5	Isolation of DNA	70
4.6	Kits and Protocols	71
4.7	Success Rate	73
4.8	Conclusions	73
Acknowledgments		75
References		75

Abstract: Improvements to analytical methods have made it possible for highly discriminative genotypic information to be gleaned from smaller and smaller amounts of sample material. This fact makes it practical to genotype samples or remains consisting of bone and tooth samples that likely would not have yielded interpretable genotypic results a short time ago. In parallel, there have been improvements to protocols specifically designed to recover DNA from very old calcified tissues (i.e., of an ancient or compromised nature). This review discusses the current best practices for isolating and purifying DNA from bones and teeth with a focus on the processes of lysis and DNA purification linked together to yield DNA from these challenging samples. The mitochondrial and genomic DNA recovered from more recently developed techniques for isolation from skeletal remains and teeth, even from very old samples, is surprisingly amenable to genotypic analysis.

4.1 Introduction

4.1.1 Bone and Tooth as Sources of DNA

Cellular material in bone and tooth samples is well preserved due to the high complexity of the matrix components such as hydroxyapatite, collagen, osteocalcin, and minerals. About 70% of bone is comprised of hydroxyapatite, a complex structure containing several salts of calcium: calcium phosphate, calcium carbonate, calcium fluoride, calcium hydroxide, and citrate. Because of the protective quality of this tough matrix or shell, bone and tooth samples are better and, many times, constitute the only source for human identity in aged or compromised evidence samples, or in cases of severe degradation (e.g., mass disaster, archived samples, historical cases, etc.). Teeth are the hardest tissue in the human body, due to dental enamel, and can remain intact under extreme environmental conditions. The International Commission on Missing Persons (ICMP) was established in 1996 with the objective of resolving the fate of the estimated 40,000 individuals missing from the conflicts in the former Yugoslavia from 1991 to 1995, and 1999 in Kosovo (Parsons et al. 2007). These types of cases illustrate the important role that DNA analysis methods specific to bone and teeth play in the identification of human remains. In addition to mass atrocities, it is estimated that in the United States alone there are more than 40,000 skeletal remains stored at medical examiners, coroners, and law enforcement agencies, which cannot be identified by conventional methods (Gonzalez et al. 2009).

The commonly used genotyping systems applied in the DNA profiling of human remains include mitochondrial DNA (mtDNA) sequencing, autosomal short tandem repeat (STR) profiling, Y-STR profiling, and single nucleotide polymorphism (SNP) typing (Alonso et al. 2001; Coble et al. 2009; Edson et al. 2004, 2009; Parsons et al. 2007; Ricaut et al. 2005).

Many important studies are reported that utilize the mtDNA sequencing and/or STR profiling of human remains in resolving missing persons, mass disasters, family incidences, plane crashes, archeological samples, natural disasters, terrorist attacks, victims of war or genocide, as well as forensic cases. The following are examples of some studies:

- Based on historical reports, the Romanov family of Tsar Nicholas II, which ruled Russia for over 300 years, was killed on July 17, 1918. Nine members of the family were reportedly buried in one mass grave while two of the children were buried in a separate grave. The identities of remains buried in a mass grave recovered in 1991 were confirmed to be those of Romanov ancestry by Gill et al. (1994) by employing DNA analysis methods. Allaying controversy, Ivanov et al. (1996), using mtDNA sequencing, confirmed that bone presumed to be that of Tsar Nicholas II, but which contained a single T/C mismatch, was indeed authentic by virtue of an identical heteroplasmy revealed in bone exhumed from the grave of his brother the Grand Duke of Russia, Georgij Romanov, at Sts. Peter and Paul Cathedral in St. Petersburg. In 2007, archeologists discovered a collection of remains from the second grave approximately 70 m from the mass grave. These remains were identified by Coble et al. (2009) to be those of the two missing Romanov children; the identification was also conducted by way of mitochondrial DNA sequencing and profiling of autosomal and Y-STRs.
- Irwin et al. (2007) identified the remains of James B. McGovern Jr., a World War II fighter pilot who was killed on May 6, 1954, in Laos while airlifting supplies to French troops during the First Indochina War.
- Schilz et al. (2006) genotyped the remains of three individuals comprising a prehistoric cave-dwelling family. The bone and tooth samples were from the Bronze Age, approximately 3,000 years old, and originally recovered from the Lichtenstein cave in Germany. It is postulated that the cave temperature, which remains steady at 6–8°C, favored the preservation of these skeletal remains.
- Mundorff et al. (2009) examined 19,970 sets of human remains from the World Trade Center (WTC) disaster that were exposed to a variety of taphonomic factors

including UV radiation, humidity, moisture, heat, fire, and mold. The samples were analyzed for STR profiling, mtDNA sequencing, and SNP typing.
- Milos et al. (2007) profiled 25,361 skeletal elements recovered from mass graves originating from conflicts that occurred in the former Yugoslavia (Bosnia and Herzegovina and Kosovo) during the 1990s.

4.1.2 Extraction of DNA

Processing of bone and tooth samples, particularly from ancient samples, has been a challenging task. Bone samples bring complexity to the DNA extraction process due to the restricted access of reagents used to lyse cells and denature protein-DNA complexes in the embedded cells. In addition, there is nonuniformity of sampling even from pulverized bone powder, compounded by the generally low quantity of cellular material and age-related degradation of DNA in the bone and the presence of polymerase chain reaction (PCR) inhibitors. Besides DNA hydrolysis by cellular nucleases and microbial degradation, the DNA molecules are prone to chemical degradation such as hydrolytic and oxidative damage (Lindahl 1993). Hydrolytic damage results in deamination of bases as well as depurination and depyrimidination: the formation of abasic sites more prone to strand cleavage at low pH. Oxidative damage results in modified bases. The major site of oxidative damage is cleavage of the 5-6 C = C double bonds of both pyrimidines and imidazole ring of purines leading to ring fragmentation of both moieties. Yet another challenge in isolation of DNA from bone is the tendency of DNA to complex with the hydroxyapatite present in bone. In fact, the dentine cement from powdered tooth is the sample of choice for isolating either genomic or mitochondrial DNA, since it remains well structured and has minimal contamination with nonhuman DNA (Presecki et al. 2000).

The quantity of endogenous DNA available in human remains often is limited and hence recovery of as much DNA as possible by extraction techniques is of immense importance. The emphasis is to maximize DNA yields and minimize the co-extraction of PCR inhibitors. Optimization of the protocol for extraction of DNA is a compromise between DNA release, degradation of DNA during extraction, and minimal co-extraction of PCR inhibitors. The physical and chemical structure of macromolecules that protect the DNA in bone from environmental and microbial degradation also hinders the accessibility of biological material to the lysis reagents. The focus of extraction studies has been the release of DNA from biological material embedded in the calcified tissues. Anderung et al. (2008) have grouped the extraction protocols into three categories: release of DNA by degrading the hydroxyapatite; release of DNA from bone apatite by incorporating competing ions; and those that do not consider the bone apatite.

4.2 Precautions

Cross-contamination is a major concern in the processing of bone samples. This mainly is because the samples contain small quantities of degraded biological material and the grinding/pulverization process creates circulating fine particles that can settle on and contaminate equipment and work surfaces at a level not seen, but detectable by highly sensitive genotyping methods. The importance of a clean environment is emphasized by several investigators (Capelli and Tschentscher 2005; Davoren et al. 2007; Edson et al. 2004; Keyser-Tracqui and

Ludes 2005; Ricaut et al. 2005). Sterilization of all tools used in cleaning and pulverization of bones and performing the processing under a laminar hood is general practice. The ICMP laboratory designated six separate areas for bone STR profiling: physical cleaning, chemical cleaning, grinding, extraction of DNA, PCR setup, and PCR amplification and fragment analysis (Davoren et al. 2007). Poinar (2003) emphasized 10 criteria to ensure meticulous authentication of the genotyping results from ancient bone samples: (i) physically isolated work areas, (ii) amplification of controls, (iii) test the molecular behavior, (iv) quantitation, (v) reproducibility, (vi) cloning, (vii) independent replication, (viii) biochemical preservation, (ix) associated remains, and (x) phylogenetic sense. In addition to these, other precautions such as multiple DNA extraction from different skeletal elements from the same individual when possible and PCR amplification separated by time, comparison of results against the database of staff who handled the samples, and contamination tracking were suggested (Edson et al. 2004; Keyser-Tracqui and Ludes 2005).

4.3 Preparation of Samples

The surface of bone and tooth samples very often is coated with unwanted material, residual tissue, and can become contaminated during collection. The samples are cleaned by one or more of the physical or chemical methods: shaving by sterile scalpel blade, buffing by mechanical means (e.g., a rotary sanding tool) to remove several millimeters of surface, washing with 5–10% commercial bleach (0.5% sodium hypochlorite) followed by rinsing with 96% ethanol, and washing with detergents followed by double-distilled water. Kemp and Smith (2005) demonstrated that endogenous embedded bone and tooth DNA was quite stable to extreme bleaching (i.e., soaking in 6% bleach for 21 h). The cleaned bone or tooth is then pulverized using freezer mills, grinders, or a drill press (Capelli and Tschentscher 2005; Davoren et al. 2007; Keyser-Tracqui and Ludes 2005). Loreille et al. (2007) sonicated the cleaned bone samples in 20% bleach for 5 min followed by bleach wash, rinsing with UV-irradiated water, sonication for additional 5 min in UV-irradiated water, wash with 100% ethanol, and overnight air-drying under a sterilized fume hood prior to pulverization.

Selection of the best bone for DNA analysis can be debatable since each bone or portion of a bone is itself a unique specimen, due to differential degradation and varying biologic content. Keyser-Tracqui and Ludes (2005) recommended using heavy bones rather than more brittle bone that has lost lipid and collagen; long bones (e.g., femur, tibia, and humerus) are preferred over rib or other thin bones, and compact bone is preferred to spongy bone. Tooth can be processed by preparing powder or removal of pulp after sectioning (Keyser-Tracqui and Ludes 2005). Commonly used samples from tooth for extraction of DNA can range from crushing of the entire tooth, or removing the pulp material by conventional endodontic access, vertical split, and horizontal section through the cervical root (Presecki et al. 2000).

4.4 Lysis

Most commonly used methods for lysis are based on the use of ethylenediamine tetra acetic acid (EDTA) as a chelating agent, sodium dodecyl sulfate (SDS) as a detergent, and proteinase K (Anderung et al. 2008; Capelli and Tschentscher 2005; Edson et al. 2004; Keyser-Tracqui and Ludes 2005; Presecki et al. 2000; Ricaut et al. 2005). Lysis is performed by incubation at 56°C for a period of 16 to 48 h. Some in-house protocols call for a second aliquot of proteinase K to be added to boost the lysis. Proteinase K–based lysis protocols provided by commercial kits like QIAamp DNA Blood Maxi Kit (Qiagen, Valencia, California) have been modified for lysis of bone and tooth samples by some investigators (Davoren et al. 2007; Parsons et al. 2007).

Hoss and Paabo (1993) described chemical lysis using guanidinium thiocynate, EDTA, and Triton-X-100 for lysis of ground Pleistocene bones. Rohland and Hofreiter (2007) investigated bear bones and tooth samples of Pleistocene age from nine different caves using different lysis conditions; lysis using a buffer containing EDTA, dithiothreitol, polyvinylpyrrolidone, N-phenacyl thiazolium bromide, N-lauroyl sarcosine, and proteinase K was effective in obtaining a maximum amount of amplifiable DNA.

Loreille et al. (2007) developed a total demineralization protocol for complete dissolution of the bone samples for achieving high yields of DNA. The extraction buffer was comprised of 0.5 M EDTA and 1% lauryl-sarcosinate. Lysis of bone samples was performed by incubating 0.6 to 1.21 g of bone powder in 9–18 mL of extraction buffer and 200 μL of proteinase K (20 mg/mL) in a rotary shaker at 56°C for ~18 h (overnight). The yields of DNA were on an average 4.6 times higher than a standard protocol wherein lysis was achieved using the EDTA, SDS, and proteinase K. Further, using the total demineralization protocol, the yields of DNA using the bone powder prepared using a freezer mill and blender cup were similar; the freezer mill tends to grind bone more finely than the blender cup. The protocol is useful for isolating nuclear and mitochondrial DNA (Coble et al. 2009; Irwin et al. 2007; Loreille et al. 2007). Amory et al. (2012) improved the demineralization protocol to process <0.5 g of bone powder, decreasing the lysis time to 5–6 h, incorporating an additional purification step using the QIAquick® column (postextraction cleanup) (Qiagen, Valencia, California), and using automation on the QIAcube® robotic platform (Qiagen, Valencia, California). The total demineralization protocol offers several advantages:

- Bone powder is completely dissolved, which minimizes the loss of DNA in residual bone powder.
- The effect of one grinding method over the other is minimal.
- Needs as little as 0.2 g of bone powder compared to 1–2 g for traditional protocols.
- Reduced sample input minimizes the quantity of inhibitors co-extracted.
- Complete dissolution enables the release of DNA embedded in dense crystalline aggregates of the bone matrix.

Gonzales et al. (2009) described the utility of pressure-cycling technology (PCT) for enhancing the lysis of bone samples. The samples after proteinase K lysis were processed for additional physical disruption using cycles of hydrostatic pressure between atmospheric and ultra-high levels (up to 35,000 psi) in a pressure-generating instrument, the Barocycler NEP3229 (Pressure BioSciences Inc, South Easton, Massachusetts). An increase in the DNA yield of ~2.3-fold was achieved for the bone samples tested.

4.5 Isolation of DNA

DNA from lysates is isolated by commonly used methods such as phenol-chloroform (Alonso et al. 2001; Edson et al. 2004; Keyser-Tracqui and Ludes 2005; Kochl et al. 2005; Loreille et al. 2007; Presecki et al. 2000), binding to silica-based materials (Hoss and Paabo 1993; Rohland and Hofreiter 2007; Yang et al. 1998), Chelex method (Faerman et al. 1995), Centricon filters (Millipore Corp., Bedford, Massachusetts) (Anzai et al. 1999), or dextran blue (Kalmar et al. 2000). Commercial DNA isolation systems such as QIAamp DNA Blood Maxi Kit (Davoren et al. 2007; Parsons et al. 2007), CleanMix Kit (Talent, France) (Keyser-Tracqui and Ludes 2005; Ricaut et al. 2005), DNA IQ (Promega Corporation, Madison, Wisconsin) (Mandrekar et al. 2002), BioRobot EZ1 (Qiagen, Valencia, California) (Schilz et al. 2006), and PrepFiler Forensic DNA Extraction Kit (Applied Biosystems, part of Life Technologies, Foster City, California) (Barbaro et al. 2009; Stray et al. 2009) are also used for the isolation of DNA from bone and tooth lysates.

Rohland and Hofreiter (2007) compared silica, Centricon (Millipore Corp., Bedford, Massachusetts), phenol/chloroform, DNeasy Tissue Kit (Qiagen, Valencia, California), All-tissue Kit, and DNA IQ system for extraction of DNA using Pleistocene-age bear bone samples; silica (binding in presence of high concentration of guanidinium thiocynate) and the DNA IQ system provided a higher yield of DNA among the methods tested. Davoren et al. (2007) compared silica-based and phenol-chloroform methods for extraction of DNA using 24 femur samples from people killed in the Balkans during the armed conflicts between 1992 and 1995 and buried in mass graves. The yield of DNA using the silica-based method was higher, at varying degrees, compared to the phenol-chloroform method for all tested samples.

Rucinski et al. (2012) compared complete demineralization coupled with the organic extraction method with the Qiagen Blood Maxi Kit for extraction of 39 exhumed bone samples. The yield of DNA and success rate for obtaining STR profile were higher using the organic extraction method; complete profiles were obtained from 37 of the 39 tested samples with the organic extraction method whereas only 9 samples exhibited complete profile using the Qiagen Blood Maxi Kit.

Lee et al. (2010) developed a protocol comprising a demineralization procedure, QIAamp Blood Maxi spin columns, and buffers from QIAquick PCR purification kit. Feasibility of the protocol was demonstrated using 55-year-old skeletal remains.

Anderung et al. (2008) described a fishing method to isolate DNA from samples containing small quantities of biological material. The salts and low molecular impurities were removed by diafiltration of clear lysate using 30,000 Da cut-off Amicon® Ultra-4 filter (Millipore Corp., Bedford, Massachusetts). The DNA was then hybridized to biotinylated probes and isolated by binding to streptavidin-coated Dynabeads M-280 beads (Invitrogen, Carlsbad, California). The method offers certain advantages such as the eluate contains targeted fragments and minimal co-extraction of inhibitors.

The quantity of DNA in the extract obtained from human remains is generally low. The extracts are processed for concentration of DNA by using Microcon or Centricon (both from Millipore Corp., Bedford, Massachusetts) filtration devices (Keyser-Tracqui and Ludes 2005; Ricaut et al. 2005).

4.6 Kits and Protocols

Manufacturers of DNA extraction kits such as Applied Biosystems (part of Life Technologies, Foster City, California), Qiagen (Valencia, California) and Promega Corporation (Madison, Wisconsin) have developed proprietary buffers and/or protocols for extraction of DNA from bone and tooth samples.

The protocol described by Promega involves lysis of bone using proprietary Bone Incubation Buffer and proteinase K at 56°C for appropriate times followed by isolation of DNA using the DNA IQ system either by manual operation or by using the Maxwell 16 system (Promega, Madison, Wisconsin) (Gonzalez et al. 2009; Mandrekar et al. 2002; Promega Corporation 2008).

Qiagen (Valencia, California) provides user-developed protocols for extraction of DNA from bone and tooth samples using the QIAamp DNA Micro Kit, EZ1 DNA Tissue Kit, and MagAttract® DNA Mini M48 Kit (Qiagen 2005a,b,c).

Stray et al. (2009) developed PrepFiler BTA Lysis Buffer (Applied Biosystems, part of Life Technologies, Foster City, California) for the lysis of biological material in calcified tissues like bone and tooth. The protocol utilizes a minimal quantity of bone powder, as low as 50 mg, which is much lower than other described methods in the literature. Further, lysis was achieved in a 2-h time frame compared to 16–48 h for traditional protocols. Thus, it was possible to obtain the STR profile from a bone sample within 2 days. This protocol conserved valuable evidence samples, increased the throughput of the laboratory, and saved valuable time of trained forensic analysts. Lysis was performed by adding 230 μL of freshly prepared lysis reagent containing 220 μL of proprietary PrepFiler BTA Lysis Buffer, 7 μL of proteinase K, and 3 μL of 1.0 M dithiothreitol (DTT) to 50–200 mg of bone and incubating the sample at 56°C in an Eppendorf Thermomixer R (VWR, Batavia, Illinois) at 1,100 rpm for 2 h. At the end of incubation, the tube was centrifuged at 10,000 x g for 5 min and clear lysate (maximal of 200 μL) was processed for isolation of DNA using the procedure described for the PrepFiler Forensic DNA Extraction kit (Applied Biosystems 2008; Stray et al. 2009). The yield of DNA from bone and tooth samples is summarized in Table 4.1.

Table 4.1 Yield of Human DNA from Bone and Tooth Samples[a]

Sample[b]	Description	Total Yield (ng)
Bone-1	100 mg long bone	1.9
Bone-2	100 mg long bone	0.65
Bone-3	100 mg long bone	0.45
Bone-4	100 mg long bone	1.6
Bone-5	100 mg long bone	0.8
Bone-6	100 mg long bone	0.25
Bone-7	100 mg long bone	2.15
Bone-CA	50 mg clavicle bone	8
Tooth	10 mg	450

[a] Extractions were performed using the PrepFiler® BTA Lysis Buffer and PrepFiler® kit reagents.

[b] Bone samples were of unknown age. Tooth sample was 1 year old.

Table 4.2 Effect of Quantity of Bone Powder on DNA Yield[a]

Bone-CA (mg)	Total Yield (ng)	Yield (ng/g bone)
25	4.95 ± 1	0.2
50	9 ± 3	0.18
100	12 ± 3	0.12
200	22 ± 6	0.11

[a] Extractions were run in triplicate using the PrepFiler® BTA Lysis Buffer and PrepFiler® kit reagents.

The yield of DNA from 100 mg of long bone samples ranged from 0.25–2.15 ng, which is typical of such sample types, since each bone sample is unique in the content and state of biological material. The quantity of bone sample was varied from 50–200 mg. The lysate volume collected was adjusted, whenever necessary, to 200 μL using PrepFiler BTA lysis buffer prior to isolation of DNA. The total yield of DNA was almost doubled when the quantity of bone powder was increased from 25 mg to 50 mg powder (Table 4.2). Further increase in the quantity of bone powder increased the yield of DNA; however, the incremental increase in the DNA yield was not proportional to the quantity of bone powder. An STR profile generated using the MiniFiler Kit for the Bone-7 sample is presented in Figure 4.1. A complete and conclusive profile indicates high purity of the extracted DNA. Liu et al. (2010, 2012) further developed the PrepFiler Express BTA Forensic DNA Extraction Kit for isolation of bone, tooth, and samples containing adhesives using the AutoMate Express system (Applied Biosystems, part of Life Technologies, Foster City, California), a benchtop automated system for isolation of DNA from as many as 13 samples in about 30 min.

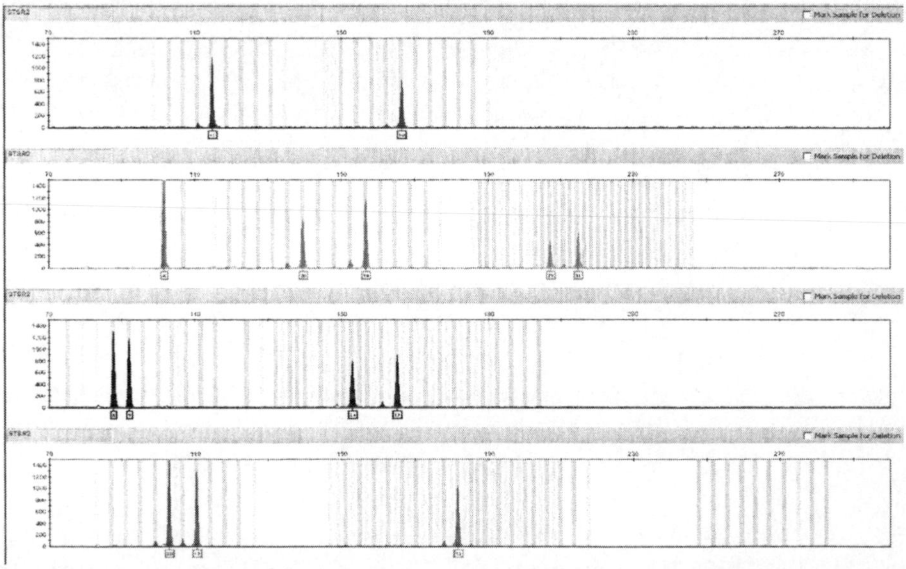

Figure 4.1 STR profile for DNA extracted from Bone-7. DNA was extracted using the PrepFiler® BTA Lysis Buffer and PrepFiler kit reagents. 300 pg of human DNA was amplified using the AmpFℓSTR® MiniFiler™ PCR Amplification Kit. The dye channels (from top to bottom) are 6-FAM™, VIC®, NED™, and PET®, respectively.

4.7 Success Rate

Several factors play a role in the success of acquiring a genotype from human remains. Each bone is a unique sample in terms of content of biological material, extent of degradation, nature of contaminants present, and extent of environmental exposure. In addition, the methods for extraction of DNA, buffer composition of PCR, length of amplicons, and the like play determinative roles. mtDNA sequencing has been a method of choice for such degraded samples. However, the development of shorter amplicons for STR loci (miniSTRs) and PCR buffers that tolerate a relatively high concentration of PCR inhibitors have made autosomal STR profiling, which achieves a higher power of discrimination, a method of choice for bone samples (Mulero et al. 2008, 2009; Parsons et al. 2007).

Edson et al. (2009) processed 558 cranial fragments for mtDNA sequencing. The success rate for temporal (n = 149), unspecified skull (n = 31), frontal (n = 40), occipital (n = 117), and parietal (n = 221) was 90%, 52%, 68%, 65%, and 52%, respectively. Edson et al. (2004) observed >90% success rate for rib (n = 26) and femur (n = 192) samples for mtDNA sequencing.

The studies by Milos et al. (2007) and Mundorff et al. (2009) are examples of a large number of human remains processed for DNA typing. The success rates for different skeletal elements from these studies are summarized in Table 4.3. The results of Milos et al. (2007) demonstrate that DNA is best preserved in femur and teeth; highest success rate was observed from dense cortical bones of weight-bearing leg bones, while long arm bones showed a significantly lower success rate. The foot phalanx and patella exhibited the highest success rate in the WTC disaster samples, followed by metatarsal, femur, and tibia, respectively (Mundorff et al. 2009). These findings follow previous findings of higher success rates with denser and weight-bearing bones.

4.8 Conclusions

Whether due to severe catastrophic circumstances or the end result of natural decomposition, skeletal elements often survive when all other tissues have vanished. When only bone and teeth remain, the ability to genotype depends entirely on whether DNA can be recovered from these calcified tissues. It is often the case that only small fragments of bone or teeth can be recovered and thus methods of extraction must be efficient and non-damaging to the endogenous DNA available. This review summarized the most successful methods for rescuing the mitochondrial and genomic DNA from tooth and bone. The importance of these methods cannot be overstated, for these resilient genetic time capsules hold answers that can confirm the identity and whereabouts of persons lost in war, mass disasters, or simply with time. Indeed, anthropology and archeology alike have benefited from the genetic analyses of DNA contained in fragments of well-preserved bone or tooth. Such DNA can provide invaluable links to the past and help to determine indigenous foundations and reveal ancestral flow.

Bones and teeth are not an abundant reservoir for nucleic acids. When other tissues are absent or compromised, however, they prove to be a reliable source for genomic and mitochondrial DNA. Bone and teeth DNA are obtained from several sources. For bone, this includes the mineralized osteocytes, the most abundant cells in compact bone, osteoblasts and osteoclasts, which drive mineralization and demineralization of bone at its surface,

Table 4.3 Success Rate of DNA Typing from Different Skeletal Elements

	WTC Victims[a]		Bosnia and Herzegovina 1992[b]		Sreberenica 1995[b]		Kosovo 1999[b]	
Skeletal Element	n	Success Rate (%)	n	Success Rate (%)	n	Success Rate (%)	n	Success Rate (%)
Foot phalanx	25	80	NA[c]	NA	NA	NA	NA	NA
Patella	83	80	NA	NA	NA	NA	NA	NA
Metatarsal	257	72	91	30.77	0	0	29	37.93
Femur	143	71	5729	82.77	3459	90.49	2168	92.11
Tibia	125	70	821	77.22	398	73.62	110	74.55
Mandible	46	65	NA	NA	NA	NA	NA	NA
Rib	1301	64	NA	NA	NA	NA	NA	NA
Innominate	62	63	NA	NA	NA	NA	NA	NA
Vertebra	72	61	51	43.14	0	0	95	72.63
Humerus	110	61	1329	43.79	887	46.67	199	60.30
Ulna	87	61	218	16.97	144	24.31	82	35.37
Fibula	159	60	90	54.44	38	81.58	32	62.50
Radius	120	60	232	17.24	152	29.61	85	35.29
Sacrum	27	59	NA	NA	NA	NA	NA	NA
Hand phalanx	83	57	NA	NA	NA	NA	NA	NA
Scapula	92	54	0	0	0	0	35	57.14
Clavicle	97	54	81	16.05	0	0	47	42.55
Tarsal	37	51	NA	NA	NA	NA	NA	NA
Skull	494	47	NA	NA	NA	NA	NA	NA
Metacarpal	211	44	0	0	0	0	18	61.11
Teeth	NA	NA	4197	78.91	1659	89.51	1107	86.99
Mandibular body	NA	NA	66	50.00	41	73.17	24	45.83
Illum	NA	NA	107	38.32	0	0	78	73.08
Cranium	NA	NA	544	36.58	77	45.45	136	53.68

[a] Success rate for identification of human remains from WTC disaster. The success rate presented is "Combined Elements Dataset (CED)." Data adapted from Mundorff, A. Z. et al., 2009, DNA preservation in skeletal elements from the World Trade Center disaster: Recommendations for mass fatality management, *J Forensic Sci* 54:739–45.

[b] Success rate for skeletal remains from different years of conflict in Yugoslavia during 1990s. Data adapted from Milos, A. et al., 2007, Success rates of nuclear short tandem repeat typing from different skeletal elements, *Croat Med J* 48:486–93.

[c] Not available.

and arterial, venous, and capillary networks that nourish the hard bone and hematopoietic progenitor and adipose cells of the marrow within. The DNA source in teeth is a similar mix of cells beneath the enamel and beneath the fibrous matrix (dentin) in the pulp. The pulp is lined by odontoblasts, which replenish the collagenous dentin, composed primarily of fibroblasts, nerve, and vascular tissues. Because of their density, architecture, and composition, bone and teeth are natural insulators, and as such, protect these various cell types from being consumed by animals, insects, microbes, or fire, as well as from chemical, radiation, or erosive damage.

At present, there is no single best practice for isolating DNA from bones or teeth. The first step in any recovery protocol involves a lysis step. As highlighted above, the most successful lysis techniques utilize a combination of an effective chelator of divalent metals, ionic detergents, and proteolytic or chemical dissociation to break down the matrix of the preserved cells. More variability is expected from the lysis phase compared to recovery of the DNA from the lysate. Since the upstream lysis and downstream purification/recovery processes for DNA are separable, it is possible to test different lysis conditions to find one that works best for a given sample. The advantage is that products of most lysis protocols discussed can be fed into any number of the well-established and reliable methods for DNA purification, including homebrew- and commercial-kit–based chemistries. Ultimately, the best combination of techniques will depend on the nature of the bone or tooth sample itself. Improvements in the lysis conditions for bone and teeth have led to increased yield from reduced sample input. Thus, it has become more practical to test multiple extraction methods on one sample. When combined with the increased power and sensitivity of currently available genotyping assays, there is an improved success rate for cells preserved in nondecomposed bone and teeth.

Acknowledgments

We thank Heidi Kijenski, Lisa Calandro, and Lori Hennessey from Life Technologies for useful discussions.

References

Alonso, A., S. Andelinovic, P. Martin, D. Sutlovic, et al. 2001. DNA typing from skeletal remains: Evaluation of multiplex and megaplex STR systems on DNA isolated from bone and teeth samples. *Croat Med J* 42:260-6.

Amory, S., R. Huel, A. Bilic, O. Loreille, and T. J. Parson. 2012. Automatable full demineralization DNA extraction procedure from degraded skeletal remains. *Forensic Sci Int Genet* 6:398–406.

Anderung, C., P. Persson, A. Bouwman, R. Elburg, and A. Gotherstrom. 2008. Fishing for ancient DNA. *Forensic Sci Int Genet* 2:104–7.

Anzai, T., T. K. Naruse, K. Tokunaga, T. Homma, H. Baba, T. Akazawa, and H. Inoka. 1999. HLA genotyping of 5,000- and 6,000-year-old ancient bones in Japan. *Tissue Antigens* 54:53–8.

Applied Biosystems. 2008. PrepFiler™ Forensic DNA Extraction Kit User Guide. PPN 4390932 Rev. B. Available at http://www3.appliedbiosystems.com/cms/groups/applied_markets_support/documents/general-documents/cms_053966.pdf.

Barbaro, A., P. Cormaci, and A. Agostino. 2009. Validation of PrepFiler™ forensic DNA extraction kit (Applied Biosystems). *Forensic Sci Int Genet Suppl Ser* 2:176–7.

Capelli, C. and F. Tschentscher. 2005. Protocols for ancient DNA typing. *Methods Mol Biol* 297:265–78.

Coble, M. D., O. M. Loreille, M. J. Wadhams, et al. 2009. Mystery solved: The identification of the two missing Romanov children using DNA analysis. *PLoS One* 4:e4838.

Davoren, J., D. Vanek, R. Konjhodzic, J. Crews, E. Huffine, and T. J. Parson. 2007. Highly effective DNA extraction method for nuclear short tandem repeat testing of skeletal remains from mass graves. *Croat Med J* 48:478–85.

Edson, S. M., A. F. Christensen, S. M. Barritt, A. Meehan, M. D. Leney, and L. N. Finelli. 2009. Sampling of the cranium for mitochondrial DNA analysis of human skeletal remains. *Forensic Sci Int Genet Suppl* 2:269–70.

Edson, S. M., J. P. Ross, M. D. Coble, T. J. Parson, and S. M. Barritt. 2004. Naming the dead—Confronting the realities of rapid identification of skeletal remains. *Forensic Sci Rev* 16:63–90.

Faerman, M., D. Filon, G. Kahila, C. L. Greenbaltt, P. Smith, and A. Oppenheim. 1995. Sex identification of archaeological human remains based on amplification of X and Y amelogenin alleles. *Gene* 167:327–32.

Gill, P., P. L. Ivanov, C. Kimpton, et al.1994. Identification of the remains of the Romanov family by DNA analysis. *Nat Genet* 6:130–5.

Gonzalez, S., E. Feller, D. Peters, B. Budowle, and A. Eisenberg. 2009. Pressure cycling technology (PCT) applications for DNA extractions from challenging forensic samples. *Presentation—20th Int Symp Human Identification,* Las Vegas, NV.

Hoss, M. and S. Paabo. 1993. DNA extraction from Pleistocene bones by a silica-based purification method. *Nucleic Acid Res* 21:3913–4.

Irwin, J. A., S. M. Edson, O. Loreille, et al. 2007. DNA identification of "Earthquake McGoon" 50 years postmortem. *J Forensic Sci* 52:1115–9.

Ivanov, P., M. Wadhams, R. Roby, M. Holland, V. Weedn, and T. Parsons. 1996. Mitochondrial sequence heteroplasmy in the Grand Duke of Russia Georgij Romanov establishes the authenticity of the remains of Tsar Nicholas II. *Nat Genet* 12:417–20.

Kalmar, T., C. Z. Bachrati, A. Marcsik, and I. Rasko. 2000. A simple and efficient method for PCR amplifiable DNA extraction from ancient bones. *Nucleic Acid Res* 28(12):E67.

Kemp, B. and D. Smith. 2005. Use of bleach to eliminate contaminating DNA from the surface of bones and teeth. *Forensic Sci Int* 154:53–61.

Keyser-Tracqui, C. and B. Ludes. 2005. Methods for studying ancient DNA. *Methods Mol Biol* 297:253–64.

Kochl, S., H. Niederstatter, and W. Parson. 2005. DNA extraction and quantitation of forensic samples using phenol-chloroform method and real-time PCR. *Methods Mol Biol* 297:13–30.

Lee, H. Y., M. J. Park, N. Y. Kim, J. E. Sim, W. I. Yang, and K.-J. Sim. 2010. Simple and highly effective DNA extraction method from old skeletal remains using silica columns. *Forensic Sci Int Genet* 4:275–80.

Lindahl, T. 1993. Instability and decay of the primary structure of DNA. *Nature* 362:709–15.

Liu, J., M. Brevnov, A. Holt, et al. 2010. Automated extraction of high quality genomic DNA from forensic evidence samples using a cartridge-based system. *Proc Annu Meet Amer Acad Forensic Sci,* Seattle, WA.

Liu, J. Y., C. Zhong, A. Holt, et al. 2012. Automate Express forensic DNA extraction system for the extraction of genomic DNA from biological samples. *J Forensic Sci* 57:1022–30.

Loreille, O. M., T. M. Diegoli, J. A. Irwin, M. D. Coble, and T. J. Parsons. 2007. High efficiency DNA extraction from bone by total demineralization. *Forensic Sci Int Genet* 1:191–5.

Mandrekar, P. V., L. Flanagan, and A. Tereba. 2002. Forensic extraction and isolation of DNA from hair, tissue and bone. *Profiles in DNA* 5:11–13.

Milos, A., A. Selmanovic, L. Smajlovic, R. L. M., et al. 2007. Success rates of nuclear short tandem repeat typing from different skeletal elements. *Croat Med J* 48:486–93.

Mulero, J., R. Green, N. Oldroyd, and L. Hennessy. 2009. Development of a new forensic STR multiplex with enhanced performance for degraded and inhibited samples. *Presentation—23rd World Congr Int Soc Forensic Genet*, Buenos Aires, Argentina.

Mulero, J. J., C. W. Chang, R. E. Lagacé, et al. 2008. Development and validation of the AmpFℓSTR® MiniFiler™ PCR amplification kit: A MiniSTR multiplex for the analysis of degraded and/or PCR inhibited DNA. *J Forensic Sci* 53:838–52.

Mundorff, A. Z., E. J. Bartelink, and E. Mar-Cash. 2009. DNA preservation in skeletal elements from the World Trade Center disaster: Recommendations for mass fatality management. *J Forensic Sci* 54:739–45.

Parsons, T.J., R. Huel, J. Davoren, et al. 2007. Application of novel "mini-amplicon" STR multiplexes to high volume casework on degraded skeletal remains. *Forensic Sci Int Genet* 1:175–9.

Poinar, H. N. 2003. The top 10 list: Criteria of authenticity for DNA from ancient and forensic samples. *Int Congr Ser* 1239:575–9.

Presecki, Z., H. Brkic, D. Primorac, and I. Drmic. 2000. Methods of preparing the tooth for DNA isolation. *Acta Stomatol Croat* 34:21–4.

Promega Corporation. 2008. Bone extraction protocol to be used with the DNA IQ™ system. Application note. Available at
http://www.promega.com/applications/hmnid/referenceinformation/boneprotocol.pdf.

Qiagen. 2005a. Purification of genomic DNA from bones using the QIAamp® DNA Micro Kit. User developed protocol. Available at http://www1.qiagen.com/literature/render.aspx?id=510.

Qiagen. 2005b. Purification of genomic DNA from bones or teeth using the EZ1 DNA Tissue Kit. User developed protocol. Available at http://www1.qiagen.com/literature/render.aspx?id=647.

Qiagen. 2005c. Purification of DNA from bones or teeth using the MagAttract® DNA Mini M48 Kit. User developed protocol. Available at http://www1.qiagen.com/literature/render.aspx?id=683.

Ricaut, F.-X., C. Keyser-Tracqui, E. Crubezy, and B. Ludes. 2005. STR-genotyping from human medieval tooth and bone samples. *Forensic Sci Int* 151:31–5.

Rohland, N. and M. Hofreiter. 2007. Comparison and optimization of ancient DNA extraction. *BioTechniques* 42:343–52.

Rucinski, C., A. L. Malaver, E. J. Yunis, and J. J. Yunis. 2012. Comparison of two methods for isolating DNA from human skeletal remains for STR analysis. *J Forensic Sci* 57:706–12.

Schilz, F., D. Schmidt, and S. Hummel. 2006. Automated purification of DNA from bones of a bronze age family using the BioRobot® EZ1 workstation. Application Note # 1031410. Available at http://www1.qiagen.com/literature/render.aspx?id=2283.

Stray, J., A. Holt, M. Brevnov, L. M. Calandro, M. R. Furtado, and J. G. Shewale. 2009. Extraction of high quality DNA from biological materials and calcified tissues. *Forensic Sci Int Genet Suppl Ser* 2:159–60.

Yang, D. Y., B. Eng, J. S. Waye, J. C. Duder, and S. R. Saunders. 1998. Improved DNA extraction from ancient bones using silica-based spin columns. *Am J Phys Anthropol* 105:539–43.

Sample Assessment

II

RNA Profiling for the Identification of the Tissue Origin of Dried Stains in Forensic Biology

5

ERIN K. HANSON
JACK BALLANTYNE

Contents

5.1	Introduction	82
5.2	Messenger RNA	82
	5.2.1 Biology	82
	5.2.2 Messenger RNA Profiling	83
	5.2.2.1 Tissue mRNA Markers	83
	5.2.2.2 Analytical Methodologies	86
	5.2.2.3 Validation for Casework	87
5.3	MicroRNA	93
	5.3.1 Biology	93
	5.3.2 MicroRNA Profiling	94
5.4	Conclusions	96
References		97

Abstract: Examination of crime scene items for biological evidence typically begins with a preliminary screening for the presence of biological fluids in order to identify possible sources of DNA. Conventional biochemical and immunological assays employed for this screening require multiple tests to be performed in a serial manner, can consume a significant amount of valuable evidentiary material, and can require a significant amount of time and labor for completion. Moreover, the presence of several biological fluids and tissues—such as saliva, vaginal secretions, menstrual blood and skin—cannot be conclusively identified using current methods. Due to the disadvantages of conventional body fluid testing, some operational crime laboratories have chosen to bypass the body fluid identification process and proceed directly to DNA analysis. However, while reducing the time spent on each case, this "shortcut" could result in a failure to provide important probative information regarding the nature of the crime as well as result in increased cost to crime laboratories if unnecessary DNA testing is performed. In the past several years, a number of forensic researchers have attempted to develop molecular-based approaches to body fluid identification that would provide operational crime laboratories with significantly improved specificity. This has resulted in an increased interest in the use of RNA profiling strategies for the identification of forensically relevant biological fluids and tissues. This review provides an overview of studies carried out on the use of both messenger RNA and small (micro) RNA profiling. The results of these studies are encouraging and presage the routine identification of the tissue source(s) of forensic evidence using molecular-based approaches.

5.1 Introduction

In the past, the standard practice in forensic biology casework analysis typically included a preliminary screening of evidentiary items recovered during the investigation of criminal offenses in order to identify the presence, and possible tissue origin, of biological material. Typically, conventional methods for body fluid stain analysis are carried out in a serial manner, with a portion of the stain being tested for only one body fluid at a time. Frequently, multiple tests are required to first presumptively identify the presence of biological fluids, followed by additional testing in order to confirm the presence of the fluid or identify the species of origin. Therefore, these methods are costly not only in terms of the time and labor required for their completion, but also in the amount of sample consumed. Although these conventional methods can confirm the presence of human blood and semen, none of the routinely used serological and immunological tests can definitively identify the presence of human saliva, vaginal secretions, or menstrual blood. With the large volume of cases that operational crime laboratories are faced with processing every year, significant time and resources are devoted solely to the screening of evidentiary items for the presence of biological materials. The inability to positively confirm the presence of certain biological fluids, the consumption of portions of valuable, limited samples, and the time and labor required, have resulted in a trend to bypass conventional body fluid identification methods and proceed straight to the analysis of any DNA present in forensic samples. The disuse or infrequent use of body fluid identification methods could prevent the recovery of probative information crucial to the investigation and prosecution of the case. For example, consider a sexual assault on a female victim with an object (recovered from the suspect) where the victim's DNA is recovered from the object. He could claim that the victim handled the item during previous casual encounters and this would be why her DNA was present. However, the significance of this evidence would increase if the source of the DNA could be shown to originate from vaginal epithelial cells, a circumstance that would be consistent with a sexual encounter but not with casual handling. Currently there is no test available to positively identify the presence of vaginal secretions. Therefore, the routine use of highly specific body fluid identification methods prior to DNA analysis awaits the development of suitable molecular-genetics–based methods that are fully compatible with the current DNA analysis pipeline.

The use of a molecular-genetics–based approach using messenger RNA (mRNA) and microRNA (miRNA) profiling has been proposed to supplant conventional methods for body fluid identification. This review provides an overview of current progress in the development of RNA-based profiling strategies for the identification of forensically relevant biological fluids and tissues.

5.2 Messenger RNA

5.2.1 Biology

Terminal differentiation of cells involves a series of developmentally regulated processes during which certain genes are turned off (i.e., are transcriptionally silent) and others are turned on (i.e., are actively transcribed and translated into protein) (Alberts et al. 1994). Each cell type will possess a unique pattern of gene expression with differences in mRNA composition and abundance (Alberts et al. 1994). Biological fluids are a complex mixture of cells and secretions from numerous tissues and glands. Therefore, each tissue or biological fluid will also have a unique pattern of gene expression or a "multicellular transcriptome" with contributions from the constellation of differentiated cells that comprise an individual fluid. A biological fluid should therefore be able to be identified if the type and abundance of the mRNAs present within the sample could be determined. This is the basis for mRNA profiling for body fluid identification.

5.2.2 Messenger RNA Profiling

5.2.2.1 Tissue mRNA Markers

Numerous putative tissue-specific genes have been identified for forensically relevant biological fluids and are listed in Table 5.1. From the summary provided below, the current list of tissue-specific markers reflects the known anatomy and physiological processes of the respective biological fluids.

Blood: Reported blood-specific markers include transcripts from the hemoglobin protein subunits (hemoglobin-beta chain, HBB; hemoglobin-alpha, HBA), enzymes of the heme biosynthesis pathway (aminolevulinate δ synthease 2, ALAS2; hydroxymethylbilane synthase, HMBS [also known as PBGD]), cell-surface and membrane proteins (glycophorin A, GYPA; ankyrin 1, ANK1; beta-spectrin, SPTB; CD93 molecule, C1QR1; adhesion molecule AMICA1), others reported to be expressed in peripheral leukocytes (aquaporin 9, AQP9; neutrophil cystolic factor 2, NCF2; caspase 2, CASP2; complement component 5a receptor 1, C5R1; C1QR1; arachindonate 5-lipoxygenase-activating protein, ALOX5AP), and myelocytes and hematopoietic cells (myeloid cell nuclear differentiation antigen, MNDA; rho GTPase activating protein 26, ARGHAP26; adhesion molecule AMICA1) (Fang et al. 2006; Fleming and Harbison 2010; Haas et al. 2009; Juusola and Ballantyne 2005, 2007; Noreault-Conti and Buel 2007; Nussbaumer et al. 2006; Zubakov et al. 2008, 2009).

Semen: Semen-specific markers include transcripts encoding DNA-binding proteins during spermatogenesis (protamine 1, PRM1; protamine 2, PRM2), gel matrix formation proteins (semenogelin 1 and 2, SEMG1 and SEGM2), prostate enzymes (transglutaminase 4, TGM4), prostate specific antigen (PSA or KLK3), and sperm mitochondria-associated cysteine-rich protein (MCSP) (Bauer and Patzelt 2003; Fang et al. 2006; Fleming and Harbison 2010; Haas et al. 2009; Juusola and Ballantyne 2005, 2007; Noreault-Conti and Buel 2007; Nussbaumer et al. 2006; Sakurada et al. 2009).

Saliva: The reported saliva-specific mRNA markers include genes involved in host defense (histatin 3, HTN3; statherin, STATH), proline-rich salivary proteins (proline-rich protein BstNI subfamily 4, PBR4; small proline-rich protein 3, SPRR3; small proline-rich protein 1A, SPPR1A), and oral epithelial and mucosal proteins (keratin 4, KRT4; keratin 6A, KRT6A; keratin 13, KRT13) (Fang et al. 2006; Fleming and Harbison 2010; Haas et al. 2009; Juusola and Ballantyne 2003, 2005, 2007; Sakurada et al. 2009; Zubakov et al. 2008, 2009).

Vaginal Secretions: The reported vaginal-specific mRNA markers include genes involved in antimicrobial defense (mucin 4, MUC4; human beta defensin 1, HBD1) and genes involved in hormone regulation (estrogen receptor 1, ESR1) (Cossu et al. 2009; Fang et al. 2006; Haas et al. 2009; Juusola and Ballantyne 2005; Nussbaumer et al. 2006). However, some studies indicate that these candidates may not be entirely specific to vaginal

Table 5.1 List of mRNA Gene Markers Reported in the Forensic Literature

Biological Fluid	Gene	Reference
BLOOD	ALAS2	(Juusola and Ballantyne 2007)
	SPTB	(Haas et al. 2009; Juusola and Ballantyne 2005, 2007; Noreault-Conti and Buel 2007; Richard et al. 2012)
	HMBS (PBGD)	(Haas et al. 2009; Juusola and Ballantyne 2005)
	CD3G	(Noreault-Conti and Buel 2007)
	HBB	(Haas et al. 2009; Lindenbergh et al. 2012; Noreault-Conti and Buel 2007)
	CASP2	(Zubakov et al. 2008, 2009)
	AM1CA1	(Lindenbergh et al. 2012; Zubakov et al. 2008, 2009)
	C1QR1	(Zubakov et al. 2008, 2009)
	ALOX5AP	(Zubakov et al. 2008, 2009)
	AQP9	(Zubakov et al. 2008, 2009)
	C5R1	(Zubakov et al. 2008, 2009)
	NCF2	(Zubakov et al. 2008, 2009)
	MNDA	(Zubakov et al. 2008, 2009)
	ARHGAP26	(Zubakov et al. 2008, 2009)
	GYPA	(Fleming and Harbison 2009)
	ANK1	(Fang et al. 2006)
	HBA	(Nussbaumer et al. 2006)
SEMEN	PRM1	(Bauer and Patzelt 2003; Haas et al. 2009; Juusola and Ballantyne 2005, 2007; Lindenbergh et al. 2012)
	PRM2	(Bauer and Patzelt 2003; Fleming and Harbison 2009; Haas et al. 2009; Juusola and Ballantyne 2005, 2007; Noreault-Conti and Buel 2007; Richard et al. 2012; Sakurada et al. 2009)
	MSP	(Fang et al. 2006)
	TGM4	(Fang et al. 2006; Fleming and Harbison 2009; Richard et al. 2012)
	PSA (KLK3)	(Fang et al. 2006; Noreault-Conti and Buel 2007; Nussbaumer et al. 2006)
	SEMG1	(Fang et al. 2006; Lindenbergh et al. 2012; Noreault-Conti and Buel 2007; Sakurada et al. 2009)
	SEMG2	(Fang et al. 2006)
SALIVA	HTN3	(Fleming and Harbison 2009; Haas et al. 2009; Juusola and Ballantyne 2003, 2005, 2007; Lindenbergh et al. 2012; Richard et al. 2012; Sakurada et al. 2009)
	STATH	(Fleming and Harbison 2009; Haas et al. 2009; Juusola and Ballantyne 2003, 2005, 2007; Lindenbergh et al. 2012; Richard et al. 2012; Sakurada et al. 2009)
	PRB4	(Fang et al. 2006)
	SPRR3	(Zubakov et al. 2008, 2009)
	SPRR1A	(Zubakov et al. 2008, 2009)
	KRT4	(Lindenbergh et al. 2012; Zubakov et al. 2008, 2009)
	KRT6A	(Zubakov et al. 2008, 2009)
	KRT13	(Lindenbergh et al. 2012; Zubakov et al. 2008, 2009)
VAGINAL SECRETIONS	MUC4	(Cossu et al. 2009; Haas et al. 2009; Juusola and Ballantyne 2005; Lindenbergh et al. 2012; Nussbaumer et al. 2006; Richard et al. 2012)

(*continued*)

Table 5.1 List of mRNA Gene Markers Reported in the Forensic Literature (Continued)

Biological Fluid	Gene	Reference
	HBD1	(Cossu et al. 2009; Haas et al. 2009; Juusola and Ballantyne 2005; Lindenbergh et al. 2012)
	ESR1	(Fang et al. 2006)
	SFTA2	(Hanson and Ballantyne 2012)
	FUT6	(Hanson and Ballantyne 2012)
	DKK4	(Hanson and Ballantyne 2012)
	IL19	(Hanson and Ballantyne 2012)
	MYOZ1	(Hanson and Ballantyne 2012)
	CYP2B7P1	(Hanson and Ballantyne 2012)
MENSTRUAL BLOOD	MMP7	(Bauer and Patzelt 2008; Haas et al. 2009; Juusola and Ballantyne 2005, 2007; Lindenbergh et al. 2012; Richard et al. 2012)
	MMP10	(Bauer and Patzelt 2002; Juusola and Ballantyne 2007)
	MMP11	(Bauer and Patzelt 2002, 2008; Fleming and Harbison 2009, Haas et al. 2009; Lindenbergh et al. 2012)
	CK19	(Bauer et al. 1999)
	PR	(Bauer et al. 1999)
SKIN	LCE1C	(Hanson et al. 2011, 2012)
	LCE1D	(Hanson et al. 2011, 2012)
	LCE2D	(Hanson et al. 2011, 2012)
	CCL27	(Hanson et al. 2011, 2012)
	IL1F7	(Hanson et al. 2011, 2012)
	CDSN	(Lindenbergh et al. 2012; Visser et al. 2011)
	LOR	(Lindenbergh et al. 2012; Visser et al. 2011)
	KRT9	(Visser et al. 2011)

secretions and demonstrate some cross-reactivity with other biological fluids (Cossu et al. 2009). Therefore, additional candidates are needed for the definitive identification of vaginal secretions. Recently, six additional vaginal-specific mRNA markers have been identified through the use of whole transciptome profiling (RNA-Seq) of vaginal swabs (surfactant-associated protein 2, SFTA2; fucosyltransferase 6, FUT6; dickkopf homolog 4, DKK4; interleukin 19, IL19; myozenin 1, MYOZ1; and cytochrome P450, family 2, subfamily B, polypeptide 7 pseudogene 1, CYP2B7P1) (Hanson and Ballantyne 2012). Two of these genes, CYP2B7P21 and MYOZ1, consistently demonstrated high specificity and sensitivity for vaginal secretions (Hanson and Ballantyne 2012). The function of these two gene transcripts in vaginal secretions is unknown.

Menstrual Blood: Menstrual blood markers include genes involved in the breakdown of the extracellular matrix (matrix metallopeptidase 7, MMP7; matrix metallopeptidase 10, MMP10; matrix metallopeptidase 11, MMP11), epithelial cell markers (keratin 19, CK19), and hormone regulation (progesterone receptor, PR) (Bauer et al. 1999; Bauer and Patzelt 2002, 2008; Fleming and Harbison 2010; Haas et al. 2009; Juusola and Ballantyne 2005, 2007).

Skin: Skin markers include members of the late cornified envelope (late cornified envelope proteins 1C, LCE1C; 1D, LCE1D; and 1E, LCE1E) (Hanson et al. 2011, 2012), genes involved in suppression of inflammatory responses (interleukin 1 family member 7, IL1F7, also known as IL37) (Hanson et al. 2011, 2012), in recruitment of T cells to normal or inflamed skin (chemokine [c-c motif] ligand 27, CCL27) (Hanson et al. 2011, 2012), in the assembly of the epidermal cornified cell envelope (loricrin, LOR; corneodesmosin, CDSN) (Visser et al. 2011), and in the development of epithelial cell cytoskeleton (keratin 9, KRT9) (Visser et al. 2011).

Housekeeping Genes: The success and accuracy of any biological assay involving the use of quantitative expression analysis depends on proper normalization of data. The purpose of normalization is to minimize potential variation that can mask or exaggerate biologically meaningful changes. Quantitative assessments of total RNA in a sample can be affected by various factors including extraction efficiencies of RNA from different body fluids and the substrates on which they were deposited, as well as potential RNA degradation. The currently available RNA quantitation methods are not human-specific and therefore RNA quantity estimations can also be affected by the presence of contaminating nonhuman species. Normalization of mRNA quantitative expression assays involves the use of ubiquitously expressed housekeeping genes responsible for cell maintenance functions. The housekeeping genes used in the mRNA assays include S15 (Juusola and Ballantyne 2003; Setzer et al. 2008), β-actin (Juusola and Ballantyne 2003; Setzer et al. 2008), glyceraldehyde-3-phosphate dehydrogenase (GAPDH) (Juusola and Ballantyne 2003, 2007; Setzer et al. 2008; Zubakov et al. 2008), transcription elongation factor 1α (TEF) (Fleming and Harbison 2010), ubiquitin conjugating enzyme (UCE) (Fleming and Harbison 2010), and glucose-6-phosphate dehydrogenase (G6PDH) (Fleming and Harbison 2010).

5.2.2.2 Analytical Methodologies

Total RNA of sufficient quality and quantity for analysis can be isolated from dried biological stains (Juusola and Ballantyne 2003) using a standard guanidinium–isothiocyanate-based lysis solution followed by phenol chloroform and isoproponal precipitation or using commercial silica-based spin columns. Using these extraction methods, typically hundreds of nanograms of total RNA can be recovered from 50-μL body fluid stains or whole swabs. Smaller input amounts, from 1 to 50 ng of total RNA, are sufficient for subsequent reverse-transcription reactions. However, individual assays may require less input depending on the analytical platform used and the assay efficiency. Attempts are also being made to develop RNA and DNA co-extraction methods in order to simultaneously recover DNA and RNA from forensic samples (Alvarez et al. 2004; Bauer and Patzelt 2003; Haas et al. 2009). The use of DNA-RNA co-extraction approaches would permit RNA analysis to be integrated into the DNA analysis pipeline and eliminate the need for separate samplings for tissue identification and DNA profiling.

Total RNA extracts are then typically treated with DNase in order to eliminate any residual contaminating DNA. Complementary DNA (cDNA) can then be synthesized using a reverse transcriptase–polymerase chain reaction (RT-PCR). This can be accomplished using an oligo-dT primer that will anneal to the poly(A) tail (3' end) or random primers (i.e., decamers, hexamers) that will anneal to random locations along the RNA transcript, creating the DNA complement. Target genes from what is effectively a cDNA library can then be amplified using primers designed to be specific for that gene (now cDNA). Since mRNA sequences will also be found in the exons of genes in genomic DNA, additional primer-design strategies are often employed in order to ensure that amplification of RNA, and not trace levels of contaminating DNA (present as a result of incomplete DNase digestion), is achieved. In order to eliminate amplification of genomic DNA, primers can be designed to span exon-exon boundaries and therefore anneal to a chimeric sequence that does not exist in genomic DNA since parts of it would be separated by an intron. Alternatively, primers can be designed in exons spanning at least one intron in order to produce differently sized products between the mRNA and corresponding genomic DNA. This approach is largely only beneficial when the analytical method distinguishes between differently sized amplification products.

Two main methodological approaches are used for the analysis of gene-expression patterns in different biological fluids: (a) multiplex PCR followed by capillary electrophoresis–laser-induced fluorescence (CE–LIF) analysis, and (b) quantitative (real-time) PCR (qPCR). Akin to standard DNA profiling, multiplex RT-PCR PCR using CE–LIF detection allows for the inclusion of numerous (>3) biomarkers per reaction. Therefore, individual samples can be assayed for all relevant body fluids simultaneously. This could be of particular benefit when admixed biological fluid samples are encountered. However, multiplex PCR CE–LIF-based approaches are usually based on a digital interpretation of the data into either presence or absence. Therefore, if the specificity of a candidate gene is based on more subtle differential expression levels among tissue types, it would not be a suitable marker for inclusion in a multiplex PCR system. Real-time PCR-based methods, on the other hand, do allow for a quantitative assessment of expression data and can be particularly useful if the uniqueness of a gene expression pattern is based on the relative abundance of a particular mRNA that is detectable in multiple body fluids but in significantly different abundances. Real-time PCR assays allow for a determination of the actual amount of PCR product during each amplification cycle as indicated by the intensity of the fluorescence signal, provided the detection threshold is within the log-linear phase of amplification. Detection of products using end-point analysis (i.e., CE–LIF-based systems) does not permit accurate quantitation assessments because measurements are taken after reactions have largely reached the plateau phase of amplification, where no significant product formation is occurring (due to PCR reagent depletion and template renaturation, which prevent primer hybridization). Despite the presence of differing amounts of starting material, it may be possible to achieve similar product levels during the plateau phases, making accurate quantitative assessments of end-point PCR assays challenging. However, real-time PCR assays are limited in the number of fluorescent dyes that can be used and only a single marker can be labeled with each of these dyes. Unlike CE–LIF systems that can distinguish products labeled with the same fluorescent dye based on size, multiple signals from the same fluorescent dye cannot be distinguished in real-time PCR assays and therefore the number of markers that can be included in a single reaction is limited to three or four. The need for multiple assays for the analysis of a single sample can lead to consumption of already limited amounts of sample as well as increased sample variation between the individual reactions. Even though both approaches have their advantages, they are each compatible with the current capabilities of forensic laboratories.

5.2.2.3 Validation for Casework

Stability in Environmentally Compromised Samples: There might be some reluctance to utilize RNA profiling assays in the forensic community due to concerns over the perceived instability of RNA in biological samples. Several studies have been conducted in order to assess the stability of RNA in dried forensic stains (Ferri et al. 2004; Haas et al. 2009; Setzer et al. 2008; Zubakov et al. 2008, 2009). Many of these studies have involved the examination of samples stored at room temperature with and without exposure to normal light for varying lengths of time (3 months to 16 years) (Haas et al. 2011a,b; Zubakov et al. 2008, 2009).

Successful detection of RNA markers in these samples demonstrates the chemical stability of RNA in dried and aged samples. Another study conducted by Setzer et al. (2008) examined biological fluid samples (blood, semen, saliva, vaginal secretions) that were exposed to a range of environmental conditions including heat, humidity, and rain for up to 547 days. Detection of tissue-specific mRNAs was observed in samples stored outside but protected from rain for 7 days (saliva and semen), 30 days (blood), and 180 days (vaginal secretions) (Setzer et al. 2008). As expected, rain had a detrimental effect on mRNA detection for all biological fluids tested (blood—3 days, saliva—1 day, semen—7 days, vaginal—3 days) (Setzer et al. 2008). It is unclear at this time whether there is bona fide differential stability of body-fluid–specific mRNA biomarkers or whether the varying detection limits were wholly or partially due to differential adsorption/desorption properties of the body fluids on the particular substrates. Additionally, the apparent differential stability of body fluid mRNA biomarkers is reminiscent of that observed with conventional protein markers. The results of these studies indicate that RNA of sufficient quantity and quality can be recovered from biological samples exposed to a range of environmental insults but that, as with DNA, heat, humidity, and rain are detrimental to RNA stability. Additional studies should focus on determining the most stable mRNA candidates for use in mRNA profiling assays as well as improving the analytical sensitivity of the RT-PCR methodology, especially for the challenging samples frequently encountered in forensic casework.

Multiplex development: In order for RNA profiling strategies to be useful in forensic casework, it is important to develop multiplex systems for the simultaneous analysis of numerous body-fluid–specific genes. Five reports describe the design and evaluation of multiplex reverse-transcription–end-point PCR systems for the identification of forensically relevant fluids (blood, semen, saliva, vaginal secretions, and menstrual blood) (Fleming and Harbison 2010; Haas et al. 2009; Juusola and Ballantyne 2005). Juusola and Ballantyne (2005) reported the development of the first multiplex system for body fluid identification. The octaplex system was designed to include two mRNA markers per fluid (blood, semen, saliva, vaginal secretions) with a redundancy of markers to account for possible variation in expression (Juusola and Ballantyne 2005) (Figure 5.1). A 9-plex system was also described, including a marker allowing for the identification of menstrual blood (Juusola and Ballantyne 2005). The sensitivity of the octaplex system ranged from <200 pg (semen markers) to 12 ng (vaginal secretions) (Juusola and Ballantyne 2005). The study also demonstrated an ability to detect multiple fluids in admixed samples even with one body fluid present in excess (Juusola and Ballantyne 2005). Haas et al. (2009) also reported the development of a multiplex system for body fluid identification. Detection of the mRNA markers required as little as 0.001 μL of blood (obtained by serial dilution), and 1 μL of semen and saliva (Haas et al. 2009). The menstrual blood markers MMP7 and MMP11 were evaluated over a whole menstrual cycle (days 1–28) and demonstrated a higher expression during the first 4 days of the cycle (Haas et al. 2009). A comparison between standard serological testing (blood—tetramethylbenzidine, Hexagon OBTI test; saliva—α amylase; sperm—*Baecchi* staining and microscopy; semen—acid phosphatase) and the RNA multiplex system on casework samples demonstrated concordant results (Haas et al. 2009). Fleming and Harbison (2010) also recently reported the development of a multiplex system for the detection of blood, menstrual blood, semen, and saliva using a single marker per fluid as well as an incorporation of three housekeeping genes. The latter suggest that a single-marker approach is sufficient if coupled with existing screening tests (Fleming and Harbison 2010). The specificity of the multiplex system was demonstrated

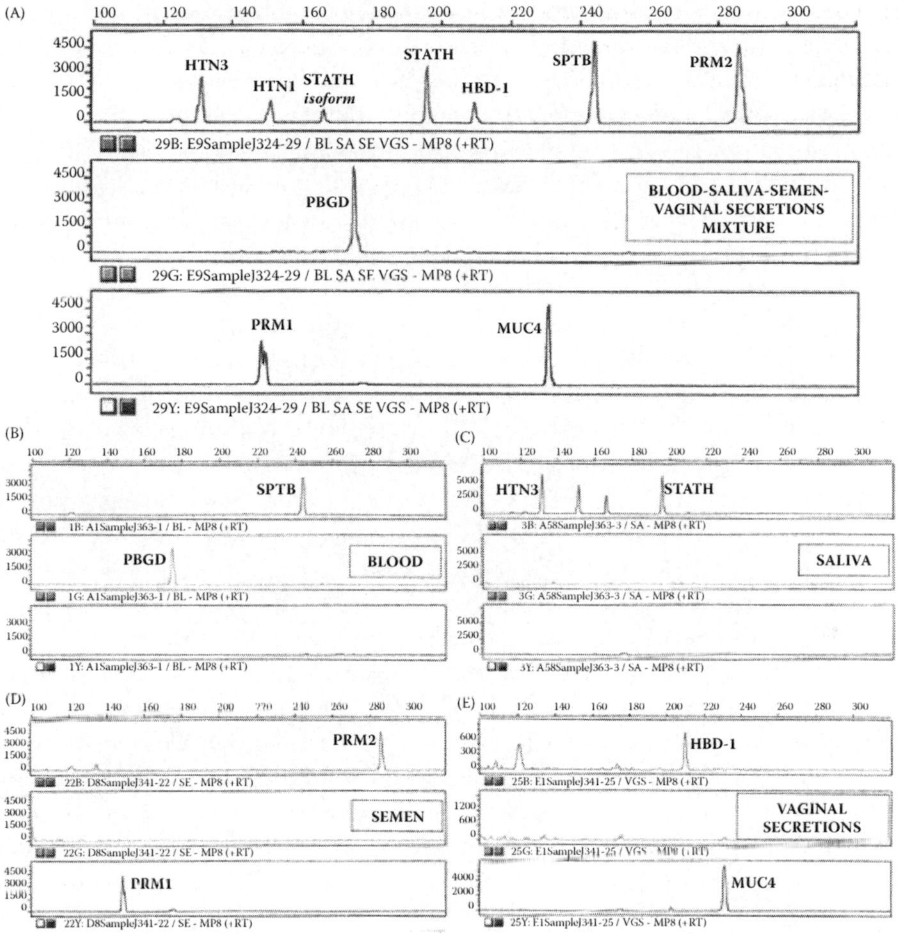

Figure 5.1 Octaplex assay for the identification of blood, saliva, semen, and vaginal secretions. RT-PCT products for the octaplex using RNA extracted from a bloodstain, a saliva stain, a semen stain, and a vaginal swab (A) simultaneously and (B–E) individually are shown. Controls without RT were run in parallel with the RT reaction (not shown). The octaplex with SPTB, PBGD, STATH, HTN3, PRM1, PRM2, HBD-1, and MUC4 was amplified subsequent to the RT reaction. The peaks at approximately 152 bp and 165 bp in the blue (top) channel (in [A] and [C]) are products from HTN1 and STATH isoform, respectively. (From Juusola, J. and J. Ballantyne, 2005, *Forensic Sci Int* 152:1. Reproduced with permission.)

with no cross-reactivity between body fluids (Fleming and Harbison 2010). As little as 0.1–0.3 ng of input total RNA was sufficient to detect the body-fluid–specific markers (Fleming and Harbison 2010). Richard et al. (2012) also reported the development of a multiple system for the analysis of five human body fluids: saliva (two markers), vaginal secretions (one marker), menstrual blood (one marker), blood (one marker), and semen (two markers). While all mRNA markers were present in their target fluids, a high level of cross-reactivity was observed between MUC4 (vaginal secretions) and saliva samples as well as cross-reactivity with other markers in semen samples (Richard et al. 2012). Lindenbergh et al. (2012) also recently reported the development of a multiplex system that simultaneously amplifies 19 mRNA markers: three housekeeping, three blood, two saliva, two semen, two menstrual secretion, two vaginal mucosa, three general mucosa, and two skin markers. Successful results (both DNA and RNA profiling) were obtained from swabs of human skin (e.g., hands, feet, back) as well as samples stored for many years (4–28 years) (Lindenbergh et al. 2012).

The results of these studies demonstrate the feasibility of developing multiplex systems for the simultaneous identification of forensically relevant biological fluids. The sensitivities, specificities, and performance of the mRNA profiling assays with admixed and casework samples augur well for their possible future use in forensic casework. Widespread adoption of the technology, however, may depend on the commercial sector providing appropriately validated test kits.

In addition to end-point PCR multiplex systems, a number of real-time PCR assays for the detection of body-fluid–specific mRNA markers have been reported (Bauer and Patzelt 2008; Fang et al. 2006; Haas et al. 2009; Juusola and Ballantyne 2007; Noreault-Conti and Buel 2007; Nussbaumer et al. 2006; Sakurada et al. 2009). A study conducted by Juusola and Ballantyne (2007) described the development of four triplex real-time PCR systems for the identification of blood, semen, saliva, and menstrual blood using, in each assay, two body-fluid–specific markers and a housekeeping gene as an internal control (Juusola and Ballantyne 2007). A delta Ct (dCt) metric was employed to interpret the expression data from the triplex assays whereby the Ct of the body fluid gene is subtracted from the Ct of the housekeeping gene (Juusola and Ballantyne 2007). Positive dCt results indicate that the body-fluid–specific gene is present in higher abundance than the housekeeping and negative dCt results indicate that the body fluid gene is present in lower levels than the housekeeping gene or not present at all. Such an approach with a two-gene assay and the dCts plotted in a two-dimensional (2D) scatter plot results in the individual sample data being clustered into one of four quadrants (+/+ (quadrant I) – upper right, –/+ (quadrant II) – upper left, +/– (quadrant III) – lower left, –/– (quadrant IV) – lower right) (Figure 5.2). A positive result would be indicated by a sample's data point appearing in quadrant I,

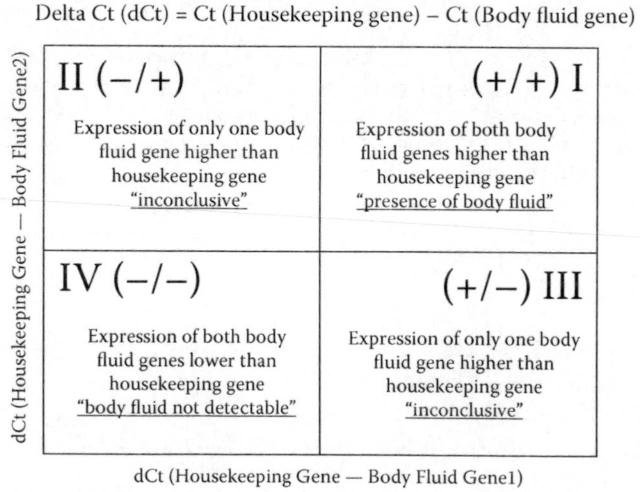

Figure 5.2 Schematic diagram of dCt quadrant plot data interpretation. A delta Ct (dCt) is obtained by subtracting the Ct value of the body fluid gene from the Ct value of the included housekeeping gene. Positive values indicate a higher expression of the body fluid gene compared to the housekeeping gene. Possible results include (+/+) indicating the presence of a particular body fluid, (+/–) or (–/+) indicating an inclusive result, and (–/–) indicating that the presence of a particular body fluid is not detected.

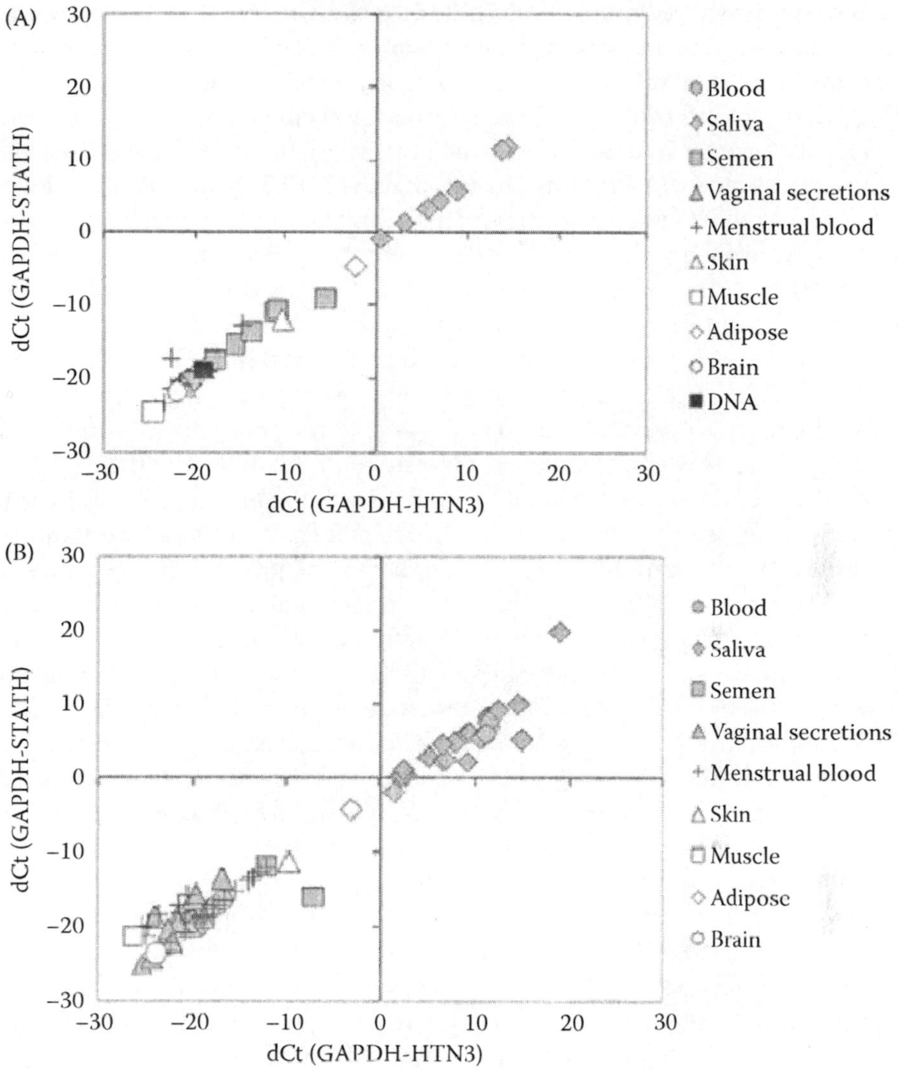

Figure 5.3 Specificity of real-time PCR triplex for saliva identification. (A) To test the specificity of the qRT-PCR assay for the identification of saliva, body fluid samples from multiple individuals (blood, $n = 17$; saliva, $n = 7$; semen, $n = 6$; vaginal secretions, $n = 5$; and menstrual blood, $n = 6$) were analyzed with the saliva triplex assay. Other samples tested included RNA samples from skin ($n = 1$), muscle ($n = 1$), adipose tissue ($n = 1$), brain ($n = 1$), and DNA ($n = 1$). The results are displayed as scatter plots in which each sample's dCt (GAPDH-HTN3) and dCt (GAPDH-STATH) are plotted. (B) To determine whether specificity was maintained over a range of RNA input concentrations, the saliva triplex assay was tested using different input quantities of body fluid RNA (20 fg–500 ng) and tissue RNA (25 and 50 ng). The results are displayed as scatter plots in which each sample's dCt (GAPDH-HTN3) and dCt (GAPDH-STATH) are plotted. (From Juusola, J. and J. Ballantyne, 2007, *J Forensic Sci* 52:1252. Reproduced with permission.)

whereas inconclusive results would be indicated by data in quadrants II and III, and a negative result indicated by data points in quadrant IV. The triplex assays used by Juusola and Ballantyne (2007) demonstrated a high degree of specificity for the body fluid of interest (data points located in upper right quadrant, +/+) (Figure 5.3). There was a reported 5- to 40-fold increase in sensitivity compared to a CE-based system with as little as 1 pg of total RNA input required for analysis (Juusola and Ballantyne 2007). The results from the triplex assays indicate that it may also be possible to develop multiplex real-time PCR-based assays for routine identification of forensic biological samples.

EDNAP exercise: Recently, a collaborative exercise on mRNA profiling for the identification of blood was organized by the Institute of Legal Medicine, University of Zurich, on behalf of the European DNA Profiling Group (EDNAP) (Haas et al. 2011a,b). This was the first study conducted by members of the forensic community to collectively evaluate the reproducibility of mRNA profiling for blood identification for potential future use in forensic casework. The primary aim of the study was to evaluate the ability of laboratories to successfully implement RNA extraction and profiling methodologies and to compare the relative abundance and/or sensitivity of the tested blood markers (HBB, SPTB, and PBGD) (Haas et al. 2011a,b). Participating laboratories were asked to evaluate seven blood samples (two 50-μL bloodstains and five 10-μL bloodstains, four human blood samples, and one cow blood sample) as well as a blood-dilution series (5, 1, 0.1, 0.01, and 0.001 μL, prepared by dilution of EDTA-blood in 0.9% NaCl to a final volume of 5 μL per swab). Fifteen of the 16 participating laboratories were able to implement the methodologies and generate profiling results despite many of the laboratories having no prior RNA experience (Haas et al. 2011a,b). The versatility of the experimental approaches to mRNA profiling was demonstrated through the use of various different extraction kits and methods, reverse-transcription kits, PCR mixes, and instrument platforms (Haas et al. 2011a,b). Hemoglobin beta (HBB) was detected by all 15 laboratories in the two 50-μL and four human 10-μL stains. It was found to be the most sensitive blood marker with detection in as little as 0.001 μL of blood (nine laboratories). Beta-spectrin was detected in the two 50-μL and four 10-μL human bloodstains by a majority of laboratories, but was found to be less sensitive than HBB, with detection down to 0.1-μL stains by only two laboratories and 1-μL stains by 11 laboratories. Similar sensitivity was observed for PBGD with variable successful detection in the 10- and 50-μL bloodstains and detection only in ≥1 μL of blood for the majority of participating laboratories (Haas et al. 2011a,b). A second EDNAP collaborative exercise on RNA profiling for body fluid identification was conducted again for the analysis of blood (Haas et al. 2012). In this second study, two multiplex systems were utilized for the identification of blood: a highly sensitive duplex (HBA, HBB) and a moderately sensitive pentaplex (ALAS2, CD3G, ANK1, SPTB, and PBGD) (Haas et al. 2012). Additionally, the use of DNA/RNA co-extraction methods were evaluated (Haas et al. 2012). As with the original study, all participating laboratories were able to successfully isolate and detect mRNA in dried bloodstains (Haas et al. 2012). Thirteen laboratories utilized co-extraction methodologies and demonstrated the ability to successfully confirm the presence of blood and to obtain autosomal STR profiles from the bloodstain donor (Haas et al. 2012). A third EDNAP collaborative exercise has been completed and included an evaluation of semen (PRM1, PRM2, TGM4, PSA, and SEMG1) and saliva (HTN3, STATH, and MUC7) mRNA markers (Haas et al. 2011a,b). The results of this study have been submitted for publication.

The EDNAP studies demonstrate the ease with which mRNA profiling may be integrated into operational laboratories using existing equipment and facilities and, more importantly, the success to be expected using an mRNA profiling approach for body fluid identification. These studies provided laboratories, most with no prior RNA experiments, with an opportunity to evaluate the potential suitability of RNA profiling assays in their current workflows. Additionally, the results of this study, as well as subsequent EDNAP studies that will be performed to evaluate additional body fluids, will provide these laboratories with supporting validation material including peer-reviewed publications. Thus far the EDNAP studies have evaluated markers for blood, semen, and saliva (Haas et al. 2011a,b, 2012). Recently, an additional study evaluating markers of menstrual blood as well as various housekeeping genes has been completed and a subsequent study evaluating markers for the identification of vaginal secretions is also being conducted.

5.3 MicroRNA

5.3.1 Biology

Although the mRNA profiling strategies described above (see Section 5.2.2) appear to be suitable for use in forensic casework and can be performed with existing hardware, the size of the amplification products used in these mRNA assays (~200–300 nt) may not be ideal for use with the degraded or compromised samples frequently encountered in forensic casework. The results of some of the mRNA stability studies indicate that it may be necessary to develop additional strategies for severely degraded or compromised samples. In an attempt to improve the analysis of degraded DNA samples, reduced-size amplicons for both autosomal STR testing and mitochondrial DNA analysis have been developed (Asamura et al. 2007a,b; Butler et al. 2003; Coble and Butler 2005; Eichmann and Parson 2008; Opel et al. 2007; Park et al. 2007; Wang et al. 2006). Therefore, it would seem appealing to adopt the same principle to mRNA profiling assays in order to improve the success of analysis with highly degraded samples. However, as a result of the technical requirements needed to ensure the specificity of mRNA assays, it will be challenging to develop reduced-size mRNA amplicons. Another way to reduce the amplicon size would be to employ small RNA biomarkers instead of mRNA.

Recently, there has been an explosion of interest in a class of small (~20–30 nt in length) noncoding RNAs, microRNAs (miRNAs) involved in the regulation of various developmental and biological processes (Babar et al. 2008; Baehrecke 2003; Bagnyukova et al. 2006; Carrington and Ambros 2003; Carthew 2006; Carthew and Sontheimer 2009; Di et al. 2006; Garzon et al. 2006; Kloosterman and Plasterk 2006; Lee et al. 2006; Zhao and Srivastava 2007). Mature miRNAs have complementary sequences to binding sites typically located in the 3' UTR of mammalian genes on which they will bind to as part of the miRNA-induced silencing complex (miRISC) (Carthew and Sontheimer 2009). There are various proposed mechanisms for how miRNAs exert their regulatory control, including (i) degradation of mRNA through deadenylation, decapping, and exonucleolytic digestion; (ii) prevention of preinitiation complex formation (recruitment of initiation factors and ribosomal subunits); (iii) competition between miRISC and eIF4E for binding to mRNA 5' cap; (iv) promotion of premature ribosome dissociation from mRNAs; (v) deadenylation of mRNA tail to prevent mRNA circularization (Carthew and Sontheimer 2009). Additionally, numerous studies have reported tissue-specific miRNA expression (Beuvink et al. 2007; Lagos-Quintana et al. 2003; Landgraf et al. 2007; Liang et al. 2007). Therefore, it may be possible to identify body-fluid–specific miRNA biomarkers. Recent studies demonstrate that miRNA profiling may be useful for identifying forensically relevant biological fluids (Hanson et al. 2009; Zubakov et al. 2010).

5.3.2 MicroRNA Profiling

Isolation of miRNA can be performed using many of the same methodologies described for mRNA with no requirement for further enrichment of the small RNA fractions. Several commercially available extraction kits are designed specifically for the recovery of small RNAs such as the *mir*Vana™ miRNA isolation kit (Ambion), the miRNeasy Mini Kit (Qiagen), miRACLE™ miRNA isolation kit (Strategene), and the PureLink™ miRNA isolation kit (Ivitrogen). While some applications may require a small RNA-enriched fraction, it has not yet been demonstrated that this would be required for forensic assays. Additionally, enrichment of separate small RNA fractions may not be advantageous in some cases if the use of housekeeping genes is required for normalization or it is desirable to perform subsequent mRNA profiling in order to confirm the presence of an individual biological fluid.

Two main approaches have been used for the reverse transcription of small RNAs. The first approach involves the use of miRNA-specific primers for reverse transcription. An example of such an approach is the TaqMan® microRNA reverse-transcription kit (Applied Biosystems, part of Life Technologies, Foster City, California). A miRNA-specific step-loop primer is utilized in the reverse-transcription reaction, which provides specificity for the mature miRNA and results in a product of sufficient length for downstream real-time PCR analysis (Chen et al. 2005). However, a disadvantage to such an approach is that only a single miRNA is reverse-transcribed in each reaction. This may not be ideal for use in forensic casework when there are often limited quantities of sample. Primer assays can potentially be multiplexed within a single assay but would require additional optimization to ensure optimal reaction conditions and sensitivity. An alternative approach involves the use of a universal primer for reverse transcription of all RNA species (mRNA, miRNA, and other small RNAs) and a miRNA-specific primer during subsequent real-time PCR. An example of this type of approach is the miScript PCR system (Qiagen). Reverse transcription is performed using an oligo-dT and random primers. Since mature miRNAs do not possess a poly(A) tail, polyadenylation and reverse transcription are performed in the same reaction. The oligo-dT primer has a universal sequence at the 5' end that is then utilized as the binding site for a universal primer in subsequent real-time PCR detection assays (miRNA or mRNA). This approach allows for multiple miRNAs to be examined from a single sample.

Two recent studies have demonstrated the potential of miRNA profiling for body fluid identification (Hanson et al. 2009; Zubakov et al. 2010). Hanson et al. (2009) provided the first comprehensive evaluation of miRNA expression (67% of the miRNA*ome*) in forensically relevant biological fluids (blood, semen, saliva, vaginal secretions, and menstrual blood). No truly body-fluid–specific miRNAs were identified (i.e., expression detected in a single body fluid) (Hanson et al. 2009). However, numerous miRNAs were identified that were differentially expressed among the various biological fluids. Normalization of expression data was performed using RNU6B (U6b) in order to overcome any potential sources of variation (e.g., presence of contaminating nonhuman species, extraction inefficiencies among the various biological fluids). Normalized expression data for two miRNAs per biological fluid was used to develop 2-D scatter plots (Hanson et al. 2009). Distinct clusters of the body fluid of interest were observed for each of the 2-D scatter plots (Figure 5.4). It was demonstrated that a panel of miRNAs (blood—miR451, miR16; semen—miR135b, miR10b; saliva—miR205, miR658; vaginal secretions—miR124a, miR372; menstrual

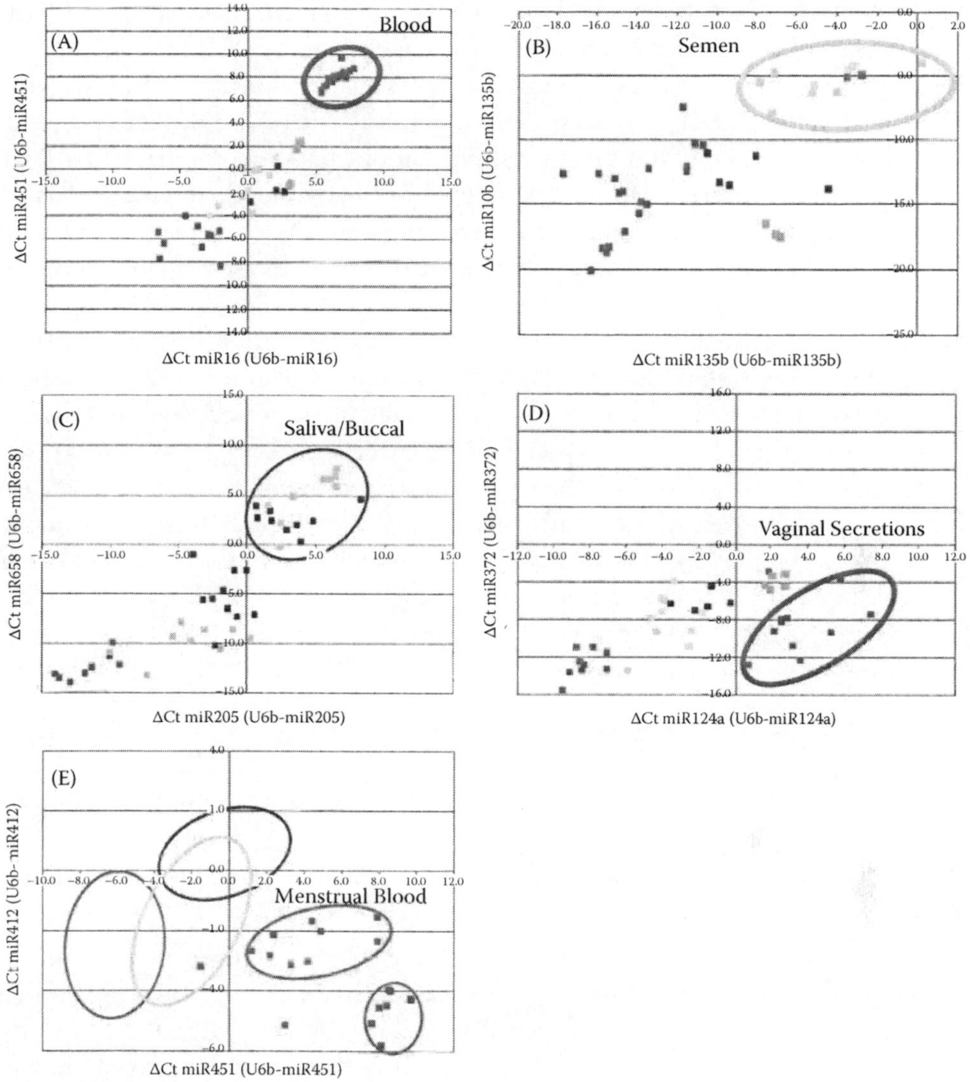

Figure 5.4 (See color insert.) Evaluation of additional samples with miRNA body-fluid identification assays. To further demonstrate the ability of the miRNA assay to identify individual body fluids, additional samples were analyzed (blood, n = 19; semen, n = 11, including two vasectomized males; saliva, n = 18, including buccal and saliva samples; vaginal secretions, n = 11; menstrual blood, n = 11): (A) blood assay; (B) semen assay; (C) saliva assay; (D) vaginal secretions assay; (E) menstrual blood assay. Individual body-fluid data points are presented by colored squares: red, blood; yellow, semen; orange, semen from vasectomized males; dark blue, saliva; light blue, buccal swabs; green, vaginal secretions; pink, menstrual blood. (From Hanson, E. K., H. Lubenow, and J. Ballantyne, 2009, *Anal Biochem* 387:303. Reproduced with permission.)

blood—miR451, miR412) was differentially expressed to such a degree to permit the identification of all forensically relevant biological fluids (Hanson et al. 2009). The developed miRNA assays appear to be highly specific for the individual body fluids since no cross-reactivity was observed with 21 human tissues.

A second study by Zubakov et al. (2010) similarly performed a screening of 718 miRNAs using a microarray platform in the attempt to identify miRNA candidates for body fluid identification. Fourteen potential candidates with differential expression patterns were identified from the microarray data (menstrual blood—miR185*, miR144; venous blood—miR20a, miR106a, miR185; semen—miR943, miR135a, miR10a, miR507; saliva—miR583, miR518c*, miR208b; vaginal secretions—miR617, miR891a) (Zubakov et al. 2010). The specificity of the blood and semen miRNA candidates were confirmed by RT-PCR testing (Zubakov et al. 2010). However, discrepancies between the microarray data and subsequent confirmation testing with RT-PCR testing were observed for the semen, saliva, and menstrual blood candidates, thus requiring further work to identify successful candidates (Zubakov et al. 2010).

The results of the two above-mentioned studies provide an initial indication that small RNA profiling may be suitable for body fluid identification (Hanson et al. 2009; Zubakov et al. 2010). Further work will be needed in order to evaluate the performance of the miRNA assays with highly degraded and compromised biological fluid samples as well as bonafide casework samples. It will also be necessary to investigate possible method-specific effects resulting in inconsistent expression data between analysis methods (Zubakov et al. 2010). It appears that the complexities of miRNA expression will warrant further and extensive validation regarding miRNA candidate selection, analytical methodologies employed, and normalization strategies in order to develop robust and reliable miRNA assays for use in forensic casework.

5.4 Conclusions

Advancements in forensic biology have been focused largely on DNA analysis and the improved recovery of STR profiles from biological evidence. This chapter has summarized efforts being made to improve other areas of forensic biology, specifically that of body fluid identification. Identification of the fluid or tissue source of origin of forensic biological material with a high degree of specificity not only identifies potential samples for DNA analysis needed to identify the donor of the material, but it can also provide valuable investigative information regarding the circumstances or the nature of the criminal offense being investigated. Current methods for the identification of biological fluids are costly in terms of time and sample, and are lacking the specificity to allow for the identification of all forensically relevant biological fluids. The improved specificity of the described m/miRNA profiling strategies and their compatibility with the DNA analysis pipeline make them ideally suited for use in routine forensic casework. Although the initial results are promising, additional studies will be necessary prior to widespread routine use of RNA profiling methods. Continual efforts will be needed in order to identify novel, highly specific RNA candidates and to evaluate the sensitivity and stability of all identified markers in order to determine which are most suitable for use with challenging forensic casework samples. Additionally, the development of commercially available typing kits would facilitate the implementation of RNA profiling strategies into routine use in operational crime laboratories. A few laboratories are currently utilizing RNA profiling assays for body fluid identification (e.g., NFI, Netherlands; ESR, New Zealand) and several U.S. laboratories are evaluating their use in forensic casework.

References

Alberts, B., D. Bray, J. Lewis et al. 1994. *Molecular Biology of the Cell.* New York: Garland Publishing.

Alvarez, M., J. Juusola, and J. Ballantyne. 2004. An mRNA and DNA co-isolation method for forensic casework samples. *Anal Biochem* 335: 289–98.

Asamura, H., S. Fujimori, M. Ota et al. 2007a. MiniSTR multiplex systems based on non-CODIS loci for analysis of degraded DNA samples. *Forensic Sci Int* 173: 7–15.

Asamura, H., H. Sakai, M. Ota et al. 2007b. MiniY-STR quadruplex systems with short amplicon lengths for analysis of degraded DNA samples. *Forensic Sci Int Genet* 1: 56–61.

Babar, I. A., F. J. Slack, and J. B. Weidhaas. 2008. miRNA modulation of the cellular stress response. *Future Oncol* 4: 289–98.

Baehrecke, E. H. 2003. miRNAs: Micro managers of programmed cell death. *Curr Biol* 13: R473–R475.

Bagnyukova, T. V., I. P. Pogribny, and V. F. Chekhun. 2006. MicroRNAs in normal and cancer cells: A new class of gene expression regulators. *Exp Oncol* 28: 263–9.

Bauer, M., A. Kraus, and D. Patzelt. 1999. Detection of epithelial cells in dried blood stains by reverse transcriptase-polymerase chain reaction. *J Forensic Sci* 44: 1232–6.

Bauer, M. and D. Patzelt. 2002. Evaluation of mRNA markers for the identification of menstrual blood. *J Forensic Sci* 47: 1278–82.

Bauer, M. and D. Patzelt. 2003. A method for simultaneous RNA and DNA isolation from dried blood and semen stains. *Forensic Sci Int* 136: 76–8.

Bauer, M. and D. Patzelt. 2003. Protamine mRNA as molecular marker for spermatozoa in semen stains. *Int J Legal Med* 117: 175–9.

Bauer, M. and D. Patzelt. 2008. Identification of menstrual blood by real time RT-PCR: Technical improvements and the practical value of negative test results. *Forensic Sci Int* 174: 55–9.

Beuvink, I., F. A. Kolb, W. Budach et al. 2007. A novel microarray approach reveals new tissue-specific signatures of known and predicted mammalian microRNAs. *Nucleic Acids Res* 35: e52.

Butler, J. M., Y. Shen, and B. R. McCord. 2003. The development of reduced size STR amplicons as tools for analysis of degraded DNA. *J Forensic Sci* 48: 1054–64.

Carrington, J. C. and V. Ambros. 2003. Role of microRNAs in plant and animal development. *Science* 301: 336–8.

Carthew, R. W. 2006. Gene regulation by microRNAs. *Curr Opin Genet Dev* 16: 203–8.

Carthew, R. W. and E. J. Sontheimer. 2009. Origins and mechanisms of miRNAs and siRNAs. *Cell* 136: 642–55.

Chen, C., D. A. Ridzon, A. J. Broomer et al. 2005. Real-time quantification of microRNAs by stem-loop RT-PCR. *Nucleic Acids Res* 33: e179.

Coble, M. D. and J. M. Butler. 2005. Characterization of new miniSTR loci to aid analysis of degraded DNA. *J Forensic Sci* 50: 43–53.

Cossu, C., U. Germann, A. Kratzer et al. 2009. How specific are the vaginal secretion mRNA-markers HBD1 and MUC4? *Forensic Sci Int Genet* Suppl Ser 2: 536–7.

Di, L. G., G. A. Calin, and C. M. Croce. 2006. MicroRNAs: Fundamental facts and involvement in human diseases. *Birth Defects Res C Embryo Today* 78: 180–9.

Eichmann, C. and W. Parson. 2008. "Mitominis": Multiplex PCR analysis of reduced size amplicons for compound sequence analysis of the entire mtDNA control region in highly degraded samples. *Int J Legal Med* 122: 385–8.

Fang, R., C. F. Manohar, C. Shulse et al. 2006. Real-time PCR assays for the detection of tissues and body fluid specific mRNAs. *Prog Forensic Genet* 11: 685–7.

Ferri, G., C. Bini, S. Ceccardi et al. 2004. Successful identification of two years old menstrual bloodstain by using MMP-11 horter amplicons. *J Forensic Sci* 49: 89–93.

Fleming, R. and S. Harbison. 2010. The development of a mRNA multiplex RT-PCR assay for the definitive identification of body fluids. *Forensic Sci Int Genet* 4:244–56.

Garzon, R., F. Pichiorri, T. Palumbo et al. 2006. MicroRNA fingerprints during human megakaryocytopoiesis. *Proc Natl Acad Sci USA* 103: 5078–83.

Haas, C., E. Hanson, M. J. Anjos et al. 2012. RNA/DNA co-analysis from blood stains—Results of a second collaborative EDNAP exercise. *Forensic Sci Int Genet* 6: 70–80.

Haas, C., E. Hanson, W. Bar et al. 2011a. mRNA profiling for the identification of blood—Results of a collaborative EDNAP exercise. *Forensic Sci Int Genet* 5: 21–6.

Haas, C., E. Hanson, N. Morling et al. 2011b. Collaborative EDNAP exercises on messenger RNA/DNA co-analysis for body fluid identification (blood, saliva, semen) and STR profiling. *Forensic Sci Int Genet* Suppl Ser 3: e5–e6.

Haas, C., B. Klesser, C. Maake et al. 2009. mRNA profiling for body fluid identification by reverse transcription end-point PCR and real-time PCR. *Forensic Sci Int Genet* 3: 80–8.

Hanson, E. and J. Ballantyne. 2012. Highly specific mRNA biomarkers for the identification of vaginal secretions in sexual assault investigations. *Sci Justice;* In press: doi:10.1016/j.scijus.2012.03.007:

Hanson, E., C. Haas, R. Jucker et al. 2011. Identification of skin in touch/contact forensic samples by messenger RNA profiling. *Forensic Sci Int Genet* Suppl Ser 3: e306.

Hanson, E., C. Haas, R. Jucker et al. 2012. Specific and sensitive mRNA biomarkers for the identification of skin in "touch DNA" evidence. *Forensic Sci Int Genet* 6: 548–58.

Hanson, E. K., H. Lubenow, and J. Ballantyne. 2009. Identification of forensically relevant body fluids using a panel of differentially expressed microRNAs. *Anal Biochem* 387: 303–14.

Juusola, J. and J. Ballantyne. 2003. Messenger RNA profiling: A prototype method to supplant conventional methods for body fluid identification. *Forensic Sci Int* 135: 85–96.

Juusola, J. and J. Ballantyne. 2005. Multiplex mRNA profiling for the identification of body fluids. *Forensic Sci Int* 152: 1–12.

Juusola, J. and J. Ballantyne. 2007. mRNA profiling for body fluid identification by multiplex quantitative RT-PCR. *J Forensic Sci* 52: 1252–62.

Kloosterman, W. P. and R. H. Plasterk. 2006. The diverse functions of microRNAs in animal development and disease. *Dev Cell* 11: 441–50.

Lagos-Quintana, M., R. Rauhut, J. Meyer et al. 2003. New microRNAs from mouse and human. *RNA* 9: 175–9.

Landgraf, P., M. Rusu, R. Sheridan et al. 2007. A mammalian microRNA expression atlas based on small RNA library sequencing. *Cell* 129: 1401–14.

Lee, C. T., T. Risom, and W. M. Strauss. 2006. MicroRNAs in mammalian development. *Birth Defects Res C Embryo Today* 78: 129–39.

Liang, Y., D. Ridzon, L. Wong et al. 2007. Characterization of microRNA expression profiles in normal human tissues. *BMC Genomics* 8: 166.

Lindenbergh, A., M. de Pagter, G. Ramdayal et al. 2012. A multiplex (m)RNA-profiling system for the forensic identification of body fluids and contact traces. *Forensic Sci Int Genet* 6: 565–77.

Noreault-Conti, T. L. and E. Buel. 2007. The use of real-time PCR for forensic stain identification. *Profiles in DNA* 10: 3–5.

Nussbaumer, C., E. Gharehbaghi-Schnell, and I. Korschineck. 2006. Messenger RNA profiling: A novel method for body fluid identification by real-time PCR. *Forensic Sci Int* 157: 181–6.

Opel, K. L., D. T. Chung, J. Drabek et al. 2007. Developmental validation of reduced-size STR Miniplex primer sets. *J Forensic Sci* 52: 1263–71.

Park, M. J., H. Y. Lee, U. Chung et al. 2007. Y-STR analysis of degraded DNA using reduced-size amplicons. *Int J Legal Med* 121: 152–7.

Richard, M. L., K. A. Harper, R. L. Craig et al. 2012. Evaluation of mRNA marker specificity for the identification of five human body fluids by capillary electrophoresis. *Forensic Sci Int Genet* 6: 452–60.

Sakurada, K., H. Ikegaya, H. Fukushima et al. 2009. Evaluation of mRNA-based approach for identification of saliva and semen. *Legal Med (Tokyo)* 11: 125–8.

Setzer, M., J. Juusola, and J. Ballantyne. 2008. Recovery and stability of RNA in vaginal swabs and blood, semen, and saliva stains. *J Forensic Sci* 53: 296–305.

Visser, M., D. Zubakov, K. N. Ballantyne et al. 2011. mRNA-based skin identification for forensic applications. *Int J Legal Med* 125: 253–63.

Wang, H. P., C. Liu, and H. Y. Sun. 2006. Recent advancement in miniSTR research. *Fa Yi Xue Za Zhi* 22: 159–2.

Zhao, Y. and D. Srivastava. 2007. A developmental view of microRNA function. *Trends Biochem Sci* 32: 189–97.

Zubakov, D., A. W. Boersma, Y. Choi et al. 2010. MicroRNA markers for forensic body fluid identification obtained from microarray screening and quantitative RT-PCR confirmation. *Int J Legal Med* 124: 217–26.

Zubakov, D., E. Hanekamp, M. Kokshoorn et al. 2008. Stable RNA markers for identification of blood and saliva stains revealed from whole genome expression analysis of time-wise degraded samples. *Int. J Legal Med* 122: 135–42.

Zubakov, D., M. Kokshoorn, A. Kloosterman et al. 2009. New markers for old stains: Stable mRNA markers for blood and saliva identification from up to 16-year-old stains. *Int J Legal Med* 123: 71–4.

Assessment of DNA Extracted from Forensic Samples Prior to Genotyping

6

MAURA BARBISIN
JAIPRAKASH G. SHEWALE

Contents

6.1 Introduction 102
6.2 Need for Assessment 102
6.3 Methods for DNA Assessment 104
6.3.1 Spectrophotometry and Fluorescence Spectroscopy 104
6.3.2 Hybridization 105
6.3.3 End-Point PCR 105
6.3.4 Real-Time PCR 107
6.3.4.1 Real-Time PCR Chemistries 111
6.3.4.2 Real-Time PCR Assays 112
6.3.4.3 Important Concepts in Real-Time PCR Assay Design 116
6.4 Standard Reference Material 2372 123
6.5 Conclusions 123
Acknowledgments 124
References 124

Abstract: Quantification of human DNA has been an integral part of forensic DNA analysis. Hybridization-based methods such as QuantiBlot® kits were used extensively in the 1990s. These methods fulfilled the need at the time, since their sensitivity range was similar to the genotyping methods in use, such as restriction fragment length polymorphism (RFLP). Later, the development of robust and more sensitive megaplex genotyping systems such as short tandem repeats (STRs), mitochondrial DNA typing, and single nucleotide polymorphisms (SNPs), created the need not only for quantification and/or detection of DNA at the picogram level, but also for assessment of the quality of the DNA extract to make informed decisions and ensure the success of downstream analysis. Real-time PCR-based quantification methods fulfilled this need. The different real-time polymerase chain reaction (PCR) methods developed so far range from singleplex reactions for quantification of human or mitochondrial DNA to multiplex systems that enable analysis of up to four targets for quantification of human DNA, human male DNA, mitochondrial DNA, detection of PCR inhibitors, or determination of the extent of DNA degradation. Incorporation of these systems into the workflow enables selection of appropriate genotyping systems and increases the first-pass success rate for obtaining a genotype using a minimal amount of evidence sample. The real-time PCR methods described here would also be useful as DNA assessment tools prior to other genotyping methods like copy number variation, insertion/deletion detection, sequencing, *Alu* dimorphism, and so forth, that are currently being investigated as additional informative tools for human identification purposes.

6.1 Introduction

DNA profiling of biological samples has become a standard technology in forensic analysis. The trend toward DNA analysis methods began soon after the demonstration of the ability to detect many highly variable genetic loci simultaneously to generate an individual-specific DNA "fingerprint" for use in human genetic analysis (Jeffreys et al. 1985). During the past two decades, the methodology for DNA genotyping witnessed a number of enhancements starting from restriction fragment length polymorphism (RFLP) and continuing to evolve to the currently used genotyping methods, including short tandem repeat (STR) profiling (Butler et al. 2003; Collins et al. 2004; Krenke et al. 2002, 2005; Mulero et al. 2006, 2008; Shewale et al. 2004), mitochondrial DNA (mtDNA) sequencing (Chong et al. 2005; Holland and Parson 1999), and single nucleotide polymorphism (SNP) genotyping (Kidd et al. 2006; Phillips et al. 2007). Even though the individual methods have advanced, the major steps in the workflow for DNA typing, such as DNA extraction, DNA quantification, genotyping, and data analysis, remained consistent over time.

The utility of the second step, assessment of the quantity and quality of DNA, was realized in the early 1990s; Walsh et al. (1992) developed a chemiluminescent method for quantification and evaluation of the quality of human DNA. The quantification of DNA was adopted early on for determining the efficiency of DNA extraction and the optimal amount of extract to be used for downstream analysis, which ranged from simple electrophoresis to multiplex genotyping applications. Quantification methods like UV absorbance spectroscopy that measure total DNA with no ability to distinguish human DNA from other species (animal, microbial, etc.) are not suitable for forensic DNA analysis since this often results in insufficient human DNA being added to the amplification reaction. This limitation is evident from the recommendation for evaluation of the quantity of human DNA in the forensic evidence sample released by the U.S. Federal Bureau of Investigation's (FBI) standards (DNA Advisory Board 2000) at the time when relatively simple multiplex systems (3- to 10-plex) were available. With the development of megaplex STR genotyping systems (20- to 25-plex) and the introduction of Y-STRs, miniSTRs, SNPs, and mtDNA genotyping assays, the expectations for the accuracy and consistency of the human DNA quantification systems have increased. At first, quantification was used to monitor the initial amount of DNA extract used as template for the genotyping amplification reaction. Later, the need to generate more information about the sample, in addition to the quantity of human DNA, to enable the selection of the appropriate genotyping system became evident. Thus, the methods for quantification and assessment of the DNA extract have evolved in the past decade. The topic of DNA quantification has been the subject of various reviews in recent years (Alonso et al. 2003; Nicklas and Buel 2003c). Readers may refer to these publications for more details.

6.2 Need for Assessment

There are several unknowns when a forensic sample is processed and the DNA extract is obtained. Since the quantity and quality of the DNA in the extract can impact the genotyping results, the assessment of the DNA extract is an important step in the workflow for the following reasons:

- DNA quantity in forensic samples varies considerably due to different types and quality of evidence material.
- Samples may be contaminated with nonhuman DNA.
- Samples may be available in limited quantities.
- Samples may be degraded.
- Samples may contain inhibitors of PCR that are co-extracted with the DNA. This may lead to complete inhibition of PCR providing false negative results, partial inhibition of PCR resulting in an imbalanced or partial profile, need for repurification, or need for dilution of the extract.
- Selection of an appropriate genotyping method to obtain a conclusive profile in the first attempt. The methods may include autosomal STRs (most commonly used genotyping method in the forensic laboratory), miniSTRs (STR genotyping systems with shorter amplicons, useful for degraded samples), Y-STRs (designed for profiling of male DNA in mixture samples containing small amounts of male DNA and large excess of female DNA), SNPs (useful in mass disaster, complex paternity, missing persons, and degraded samples), and mtDNA (useful for hair, ancient bone, and degraded samples containing small quantities of nuclear DNA).
- Achievement of high-quality interpretable genotyping profiles. Low input quantity of DNA template may result in profiles with low peak heights, peak imbalance at heterozygous loci, and allele dropout due to stochastic effects. High input quantity of DNA template may result in off-scale peak heights, stutter peaks, pull-up peaks, and incomplete adenylation (-A peaks).

Thus, the DNA genotyping systems require a defined range of DNA template quantities to obtain interpretable and conclusive profiles as well as to generate maximal probative information. Use of too little or too much template DNA may yield inconclusive results, thereby wasting critical evidence samples, expensive reagents, and analysts' time, thereby decreasing the overall efficiency of the laboratory. In general, the information desired by a forensic analyst includes:

- *Quantity of human nuclear DNA*: Concentration of human DNA determines the volume of DNA extract used for amplification. At times, a DNA extract containing a low concentration of DNA can be concentrated by using spin column methods (e.g., Centricon). If the overall quantity of nuclear DNA is very low, the probability of success of STR analysis is also low and mtDNA sequencing may be an option due to the much greater number of copies of the mitochondrial genome present in a single cell compared to the chromosomal DNA.
- *Quantity of human male nuclear DNA*: Concentration of human male DNA determines the volume of DNA extract used for amplification of Y-STRs.
- *Mixture ratio for male and female nuclear DNA*: Evidence samples are often a mixture of male and female contributors. Knowing the ratio of total human and human male DNA is useful for choosing between autosomal and Y-STR profiling. Samples containing high quantities of female DNA compared to male DNA provide undetectable male profiles when amplified for autosomal STRs and, therefore, Y-STR analysis is a better option to ensure recovery of information from such samples.
- *Presence of PCR inhibitors*: Samples containing PCR inhibitors can be repurified to obtain interpretable profiles. Alternatively, robust genotyping systems like the

AmpFℓSTR MiniFiler, AmpFℓSTR Identifiler Plus, or Powerplex 16 HS systems can be used when the sample contains relatively low quantities of PCR inhibitors.
- *Quantity of mtDNA*: Evaluation of mtDNA copies determines the volume of DNA extract used for amplification in mtDNA typing.
- *Extent of DNA degradation*: The extent of degradation is useful information for the selection of genotyping systems like SNPs or miniSTRs over traditional STR multiplexes due to their higher success rate with compromised samples.

For the past decade, highly sensitive methods like real-time PCR have provided a platform for development of assays that determine the quantity of human and human male nuclear DNA, mitochondrial DNA, the presence of PCR inhibitors, and the extent of DNA degradation. These tools allow forensic scientists to select appropriate genotyping methods or modify their analysis procedures to achieve desired results in the first attempt, leading to higher success rates, increased productivity, cost savings, and reduced sample consumption. For example, Cupples et al. (2009) demonstrated the utility of Quantifiler kits in predicting the success of STR profiles: 73% of the samples exhibiting "undetected" quantification results did not provide STR profiles.

6.3 Methods for DNA Assessment

6.3.1 Spectrophotometry and Fluorescence Spectroscopy

UV absorbance at 260 nm is the most commonly used traditional method for quantification of nucleic acids. Typically, homogenous aqueous solutions of purified double-stranded DNA with an optical density (OD) of 1.0 at 260 nm in a 1-cm-length cuvette contain 50 ng/μL of total DNA (Sambrook and Russell 2001). This method is not suitable for forensic casework for multiple reasons: lack of specificity for human DNA, inability to distinguish between DNA and RNA and between intact and fragmented DNA, interference due to protein and other biomolecules released from the substrate, increase in absorbance due to single nucleotides and single stranded DNA, and low sensitivity (≥ 2 ng/μL) (Georgiou and Papapostolou 2006). More sensitive methods utilizing fluorescent dyes, such as Hoechst-33258, PicoGreen®, and OliGreen® dyes have been developed for quantification of DNA. These dyes enhance the fluorescence after binding to DNA molecules. Hoechst-33258 and PicoGreen dyes bind to dsDNA whereas OliGreen dye binds to ssDNA. Hoechst-33258 dye binds to AT-rich regions, but does not enhance the fluorescence significantly, thereby providing sensitivity only in the nanogram range. PicoGreen and OliGreen dye-based assays provide higher sensitivity down to the picogram level. See Nicklas and Buel (2003c) for a detailed discussion of these methods. These fluorometric methods, although highly sensitive, do not differentiate between human and nonhuman DNA and therefore are not useful for assessment of forensic casework samples. However, some high-throughput laboratories use PicoGreen dye quantification methods for database samples. In addition, Hoechst-33258 and PicoGreen dye-based measurements are significantly affected by the degree of fragmentation in the DNA sample: up to a 70% difference in fluorescence is observed when a highly fragmented sample is analyzed compared to its intact form. Using this characteristic of the two dyes, standard curves for intact and fragmented DNA can be generated to obtain accurate measurements of dsDNA and infer the degree of fragmentation of an unknown sample (Georgiou and Papapostolou 2006). Hoechst dye binds to the AT-rich regions whereas DNA base preference of PicoGreen dye is not reported to our knowledge. The mechanism of decrease in the enhancement of fluorescence of fragmented DNA compared to intact DNA cannot be speculated since the binding mechanism of these dyes with DNA is not yet fully understood (Georgiou and Papapostolou 2006).

6.3.2 Hybridization

A commonly used method by forensic laboratories for quantification of human DNA in the 1990s was the QuantiBlot® kit, which was originally developed by Walsh et al. (1992). The method is based on hybridization of a primate-specific, highly repetitive alpha satellite DNA sequence on chromosome 17 (D17Z1). DNA is bound to a membrane hosted in a slot-blot apparatus. The bound DNA is hybridized to the biotinylated D17Z1 probe. Subsequent binding of streptavidin-horseradish peroxidase to the bound probe enables chemiluminescent detection using a luminol-based reagent; the oxidation of luminol by horseradish peroxidase results in the emission of photons that is detected on autoradiography film. The intensity of the signal is proportional to the quantity of DNA. A series of standards is run along with the samples, and the quantity of the DNA is estimated by visual comparison or computer image analysis. The results of the assays require a degree of interpretational skills. In addition, they are subjective, time-consuming, labor-intensive, and not amenable for automation. Further, the hybridization-based quantification methods like QuantiBlot kits are comparatively less sensitive (150 pg to 10 ng) compared to the sensitivity of recently developed genotyping kits.

Mandrekar et al. (2001) described the AluQuant® Human DNA Quantification System based on hybridization coupled with an enzymatic reaction for the detection. A nucleotide probe specific for repeated sequences in the human genome is mixed with the DNA sample, which initiates a series of coupled enzymatic reactions. READase™ polymerase recognizes the hybrid as a substrate for phosphorylation at the 3' terminus of the double-stranded DNA reaction wherein one dNTP is generated. Subsequently, READase kinase transfers the terminal phosphate from the dNTP to adenosine diphosphate (ADP) to form adenosine triphosphate (ATP). ATP released from this reaction is detected colorimetrically by using a third enzyme, luciferase. The light produced by luciferase is proportional to the amount of ATP generated and is directly correlated to the quantity of human DNA present in the sample. The assay determined as low as 50 pg of DNA.

Recently, Tak et al. (2012) developed a highly sensitive quantum dot (Qdot)-based hybridization assay for quantification of human DNA. The assay can detect as low as 2.5 fg of human DNA due to strong emission and photostability of Qdot particles. Nevertheless, the assay is relatively laborious compared to currently used real-time PCR-based assays.

Although some compounds may interfere with the hybridization methods, these methods are not capable of detecting the presence of PCR inhibitors and therefore cannot predict the success of the PCR-based genotyping systems.

6.3.3 End-Point PCR

Yield gels—agarose gels of amplified products followed by ethidium bromide (EtBr) staining—are a classical example of end-point PCR quantification. This is a traditional and still used method in laboratories for monitoring the quality and quantity of DNA. The method enables approximate estimation of the quantity of DNA. End-point PCR quantification methods for human DNA measure the amplified product formed after a fixed number of cycles. They are more sensitive than slot-blot methods, but have poor precision compared to real-time PCR assays. In end-point PCR quantification methods, results are variable from sample to sample or across replicates because the measurements are taken in the plateau phase of the amplification. At this point the reaction is not in the exponential phase anymore, has slowed down, and has probably reached the saturation point due to the depletion of some reaction components. This depletion occurs at different rates in different wells and may result in different final quantities even though the initial template amount was the same in each well. Further, after the PCR, these methods require the additional step of loading and staining the agarose gel.

Matsuda et al. (2005) described a qualitative method for identification of as low as 1 pg of human DNA in a sample by amplification of a 157-bp fragment from the human mitochondrial cytochrome *b* gene and detection of the amplified product by electrophoresis on an agarose gel. Hiroshige et al. (2009) developed a screening assay for detection of as low as 0.01 ng of human DNA in a biological sample by amplification of the *FOXP1* gene. The amplified product was detected by gel electrophoresis. Nicklas and Buel (2003a) developed a fluorescent end-point assay for quantification of human DNA. The primers for amplification of the Ya5 subfamily of the *Alu* marker were labeled with QSY 7 quencher and the PCR mix contained SYBR® green dye. Prior to amplification, the plate is denatured and read on a fluorescent plate reader to detect the background signal. Fluorescence is measured again at the end of the PCR. The difference in fluorescence is correlated to the quantity of DNA in a sample and standard DNA of known concentrations is also run to generate a calibration curve. The quantification values ranged from 10 pg to 10 ng. The use of QSY 7-labeled primers reduces the fluorescence resulting from intercalation of SYBR Green I dye with primers and primer-dimers, which are shorter than the targeted amplified product, due to its close proximity. Von Wurmb-Schwark et al. (2004) developed a duplex end-point PCR comprising amplification of a 164-bp amplicon from the betaglobin gene for nuclear DNA and a 260-bp amplicon from the NADH dehydrogenase subunit 1 (ND1) for mtDNA. The amplified fragments, after completion of PCR, are separated by capillary electrophoresis (CE) and the quantity of target DNA (nuclear and mitochondrial) is determined from the intensity of the peak. Quantitative template amplification technology (Q-TAT)-based duplex and triplex assays are developed for quantification of DNA (Allen and Fuller 2006; Wilson et al. 2010). The duplex assay amplifies the amelogenin target gene to generate X-chromosome–specific and Y-chromosome–specific amplicons and the triplex assay includes amplification of additional nonhuman template to detect presence of PCR inhibitors. The amplified products are separated by CE. The quantity of human DNA in the sample is estimated by comparing the fluorescence of the X and Y amplicons produced from unknown samples with the fluorescence obtained from the amplification of known quantities of reference DNA in a standard curve.

Another duplex assay involving simultaneous amplification of cytochrome b and hypervariable D-loop mitochondrial DNA generating amplicons of 309 and 259 bp, respectively, is described by Bataille et al. (1999) for detection of nonhuman DNA and human DNA in a sample. The PCR products are separated by gel electrophoresis. The presence of only the 309-bp band indicates that the sample is of nonhuman origin, while the presence of both bands indicates human origin. The nonhuman species tested are chicken, horse, beef, and pig. The assay was not tested with mixed samples containing human and nonhuman DNA.

6.3.4 Real-Time PCR

The real-time PCR methodology monitors the progress of amplification continuously as the reaction progresses (Applied Biosystems 2003, 2008b; Nicklas and Buel 2003c; Promega Corporation 2007). The technology has found application in many disciplines such as food, pharmaceutical, environmental, and clinical testing; bioterrorism monitoring; and human identification. Applications involve detection as well as relative and absolute quantification of a target gene in a sample. The term qPCR is also used for the real-time PCR quantification assays. Quantitative real-time PCR, unlike other chemical and physical methods, determines the quantity of amplifiable DNA segments in a sample; using a standard curve for a well-characterized target of interest, copy number and then quantification values for any unknown sample can be determined. The real-time PCR quantification assays described in the literature employ different approaches for measuring the accumulated amplified product. The underlying principle is to monitor the quantity of accumulated amplified product at every cycle by detecting the signal emitted by a fluorescent reporter dye. The cycle threshold (C_T) value is the cycle at which the fluorescence signal crosses the threshold at the beginning of the exponential phase of the amplification curve. The threshold is an arbitrary value based on the variability of the baseline data at the initial cycles of amplification (Applied Biosystems 2008a, 2003). Thus, the lower the C_T value, the higher the quantity of DNA present in the original sample. A linear relationship between the C_T values and the quantity of DNA template is observed. Varying quantities (in general eight) of purified DNA preparation of known concentration, referred to as standard DNA or a quantification standard, are run on each qPCR plate. A standard curve is generated by plotting the C_T values against log concentration of DNA. The amount of DNA present in a sample is computed by measuring the C_T value and comparing it with the standard curve C_T values. The well-to-well variation in the signal strength is minimized by normalizing the fluorescence of the reporter dye to a passive reference dye present in each well.

Availability of different fluorescent dyes, quenchers, probe design approaches, robust master mixes, and real-time instruments with multicolor detection capabilities have enabled the development of more and more complex real-time PCR assays, from the original singleplexes to the more recent multiplexes. Critical factors in designing multiplex real-time PCR quantification assays include species specificity, primer and probe specificity, primer and probe cross-reactivity, amplicon size, dynamic range, sensitivity, compatibility with different extraction methods, and similar amplification efficiencies of the targets. The different real-time PCR assays developed for assessment of DNA samples are summarized in Table 6.1.

Advantages of real-time PCR: Real-time PCR assays for quantification of DNA offer advantages over the traditional UV, fluorometric, and hybridization-based assays such as: specificity for the target being amplified, higher sensitivity, quantitative relationship between the amounts of target template and the PCR product generated at any given cycle prior to reaching saturation, greater dynamic range of quantification, multiplexing capabilities, ease of adoption, less hands-on time, homogenous assay format, minimal use of sample, measurement of amplifiable DNA, quantification of specific genomes of interest,

Table 6.1 Real-Time Assays for Assessment of DNA Extracts

Real-Time Assay	Application	Name of the Assay	Target Gene and Amplicon Size	Standard Curve Dynamic Range	Reference
Singleplex	Human DNA quantification	—	Yb8 *Alu*: 226 bp.	1 pg to 10 ng	(Walker et al. 2003)
		—	Yd6 *Alu*: 200 bp.	0.1 to 100 ng	(Walker et al. 2003)
		—	Ya5 subfamily *Alu*: 124 bp.	1 pg to 16 ng	(Nicklas and Buel 2003b)
		—	Ya5 subfamily *Alu*: 113 bp.	1 pg to 256 ng	(Nicklas and Buel 2005)
		H-Quant	*Alu* Yb8: 216 bp.	10 pg to 50 ng	(Shewale et al. 2007)
	mtDNA quantification	—	ND1: 262 bp.	10 to 1 million copies of mtDNA	(von Wurmb-Schwark et al. 2002)
		—	Target sequence 13,288 to 12,392 of revised CRS	10 to 100 million copies of mtDNA	(Kavlick et al. 2011)
	mtDNA quantification and extent of degradation	—	HV I: 113 and 287 bp.	60 to $6x10^6$ copies of mtDNA	(Alonso et al. 2004)
Duplex	Human DNA quantification and detection of PCR inhibitors	Quantifiler® Human DNA Quantification Kit	hTERT for human: 62 bp. Synthetic nucleotide for IPC: 79 bp.	46 pg to 100 ng	(Applied Biosystems 2003; Green et al. 2005)
		Investigator® Quantiplex	4NS1C for human: 146 bp. Internal Control: 200 bp.	4.8 pg to 20 ng	(Qiagen 2011; Scherer and Di Pasquale 2012)
	Human male DNA quantification and detection of PCR inhibitors	Quantifiler® Y Human Male DNA Quantification Kit	SRY for human male: 64 bp. Synthetic nucleotide for IPC: 79 bp.	46 pg to 100 ng	(Applied Biosystems 2003; Green et al. 2005)
	Nuclear and mtDNA quantification	nuTH01-mtND1 Duplex qPCR	TH01 for human: ~170–190 bp. ND1 region for mtDNA: 69 bp.	40 pg to 100 ng for nuclear DNA and 16 to 8 million copies for mtDNA	(Timken et al. 2005)

(continued)

Table 6.1 Real-Time Assays for Assessment of DNA Extracts (Continued)

Real-Time Assay	Application	Name of the Assay	Target Gene and Amplicon Size	Standard Curve Dynamic Range	Reference
		—	*Alu* Yb8 for nuclear: 71 bp. mtDNA for mito: 79 bp.	1 pg to 100 ng	(Walker et al. 2005)
		—	RB1 for nuclear: 79 bp. Coding region of mitochondria between nucleotides 8294 to 8436: 143 bp.	0.1 to 10^4 DNA copies	(Andréasson et al. 2002, 2006)
		Modular Assays	RB1 for nuclear: 79, 156, or 246 bp. mtDNA: 102, 143, 283, or 404 bp. IPC for nuclear DNA: 156 bp. IPC for mtDNA:143 bp.	10 pg to 100 ng for nuclear DNA. 10 to 6,400,000 mtGE for mtDNA	(Niederstätter et al. 2007)
	Human and human male DNA quantification	—	Amelogenin: 106 bp (AMGX) and 112 bp (AMGY)	8 pg to 1 ng (AMGX) and 40 pg to 1.25 ng (AMGY)	(Alonso et al. 2004)
		—	TPOX for human: 63 bp. SRY for male: 70 bp.	46 pg to 100 ng	(Horsman et al. 2006)
		—	Ya5 subfamily *Alu* for human: 127 bp. DYZ5 for male: 137 bp.	7.8 pg to 128 ng	(Nicklas and Buel 2006)
	DNA degradation assay	—	Two amplicons from *Alu*. Long: 246 bp. Short: 63 bp.	7.8 pg to 32 ng	(Nicklas et al. 2012)
Triplex	Human and human male DNA quantification and detection of PCR inhibitors	Quantifiler® Duo DNA Quantification Kit	RPPH1 for human: 140 bp. SRY for human male: 130 bp. Synthetic nucleotide embedded in a plasmid for IPC: 130 bp.	46 pg to 100 ng	(Applied Biosystems 2008b; Barbisin et al. 2008, 2009)

(continued)

Table 6.1 Real-Time Assays for Assessment of DNA Extracts (Continued)

Real-Time Assay	Application	Name of the Assay	Target Gene and Amplicon Size	Standard Curve Dynamic Range	Reference
		Plexor® HY System	RNU2 for human: 99 bp. TSPY for human male: 133 bp. Novel IPC sequence: 150 bp.	6.4 pg to 100 ng	(Krenke et al. 2008; Promega Corporation 2007)
		Investigator Quantiplex® HYres	4NS1C for human: 146 bp. Y-chromosome specific: 129 bp. Internal Control: 200 bp.	4.9 pg to 20 ng	(Qiagen 2012; Scherer and Di Pasquale 2012)
	DNA degradation and detection of PCR inhibitors	nuTH01-nuCSF-IPC triplex qPCR assay	TH01 for human: ~170–190 bp. CSF1PO for determining extent of degradation: 67 bp. Synthetic nucleotide for IPC: 77 bp.	32 pg to 32 ng; 7.2 pg to 100 ng	(Swango et al. 2006, 2007)
	Human nuclear, mitochondrial, and male Y-chromosome DNA quantification	—	*Alu* Yb8 for nuclear: 71 bp. mtDNA for mito: 79 bp. Homologous region of sex chromosomes for human male: 69 bp.	0.1 to 100 ng	(Walker et al. 2005)
Quadruplex	Human and human male DNA quantification, DNA degradation and detection of PCR inhibitors	nuTH01-nuSRY-nuCSF-IPC quadruplex qPCR assay	TH01 for human: ~170-190 bp. SRY for male: 137 bp. CSF1PO for determining extent of degradation: 67 bp. Synthetic nucleotide for IPC: 77 bp.	7.2 pg to 100 ng	(Hudlow et al. 2008)

and amenability for automation. Further, real-time PCR-based assays provide better correlation with the current genotyping systems since both methodologies are PCR-based.

6.3.4.1 Real-Time PCR Chemistries

Intercalating dyes: Intercalating dyes like SYBR Green I dye emit low fluorescence when free in solution, but form a highly fluorescent complex when bound to double-stranded DNA. Thus, the increase in fluorescence is proportional to the quantity of amplified product generated. However, in addition to the target amplicons, the intercalating dyes bind to primer-dimers, nonhuman DNA, and products of nonspecific amplification. Because of this lack of specificity, SYBR-based assays often require a secondary test (e.g., melting curve analysis) to confirm the accuracy of the results. The intensity of fluorescence also varies with the amplicon length. Due to their nature, intercalating dyes like SYBR Green I dyes are suitable for singleplex reactions.

TaqMan chemistry: TaqMan chemistry-based technology is widely regarded as the gold standard for DNA and RNA quantitation (Holland et al. 1991; Schmittgen and Livak 2008). TaqMan assays take advantage of the inherent 5' nuclease activity of Taq DNA polymerases (Holland et al. 1991). Typically, the assay comprises a set of forward and reverse primers specific for the target DNA sequence and a nonextendable hybridization probe homologous to the region between the two PCR primers. The TaqMan probe is labeled with a 5' fluorescent reporter dye and a 3' nonfluorescent quencher. The TaqMan probes in some assays like the Quantifiler assay possess an additional minor groove binder (MGB) chemical moiety at the 3' end for increasing the melting temperature (T_m) and, therefore, signal specificity (Afonina et al. 1997; Barbisin et al. 2008; Green et al. 2005). The incorporation of the MGB moiety achieves higher efficiency of amplification using shorter-length probes. At the start of PCR cycling, the fluorescence is minimal since most of the probes are intact and the quencher is in close proximity to the fluorescent reporter dye. During the course of amplification, in the extension phase of PCR, Taq DNA polymerase cleaves the fluorescent dye from the TaqMan probe annealed to the template using its 5' to 3' nuclease activity. Reporter dye and quencher are now decoupled and fluorescence is emitted. The rate of increase in the fluorescence is proportional to the quantity of cleaved probe and therefore to the quantity of DNA template. Thus, the fluorescence crosses the threshold value at a different cycle number (C_T) depending on the amount of DNA template present in the amplification reaction. TaqMan assays ensure high specificity at both the primer and the probe level, high sensitivity, and multiplex capabilities. Multiplex systems as high as quadruplexes are designed to obtain maximal information about the DNA sample using minimal sample consumption (Table 6.1). The multiplex capabilities are restricted by the optical capacity of the real-time PCR instruments. For example, a 7500 Real-Time PCR System can deconvolute the fluorescent signals of five dyes. In general, one dye channel is used for passive reference and therefore a maximum of four targets can be detected in the same well on this instrument.

Molecular beacons: Like TaqMan probes, molecular beacons are designed to exploit the concept of quenching the fluorescent dye with a quencher moiety. However, the unique structure of molecular beacons comprises the nucleotide sequence complimentary to the target sequence flanked by two self-complimentary sequences that can form a hairpin structure. A fluorescent dye is linked at one end of the probe and the quencher moiety is linked at the other end (Tyagi et al. 1998). Thus, in the absence of the target sequence, the complimentary sequences within the beacon form a hairpin structure bringing the fluorescent dye in close proximity to the quencher, resulting in absence of signal. In the presence of the target sequence (amplified product in the real-time assays), the probe hybridizes to it to form a linear molecule, thereby separating the dye from the quencher and emitting fluorescence. The fluorescence is a result of the hybridization of the molecular beacon to the target sequence, but, unlike TaqMan probes, in this case the probe is not cleaved. However, as in TaqMan assays, fluorescence increases with the progress in the amplification and is proportional to the DNA template quantity.

Scorpions chemistry: Scorpions primers are bifunctional molecules wherein the primer and probe are covalently linked. Scorpions primer carries a 5′ extension comprising a blocker, a quencher, a fluorophore, a pair of self-complementary sequences, and a probe element specific for target amplicon (Whitcombe et al. 1999). After a round of PCR extension from the primer, a newly synthesized target region gets attached to the same strand as the probe. After a second round of denaturation and annealing, the probe and target hybridize, thereby separating fluorophore and quencher to generate fluorescence (Whitcombe et al. 1999). Thus, the fluorescence increases with the progress in the amplification and is a measure of the DNA template quantity.

Plexor® chemistry: Unlike TaqMan, molecular beacon, or Scorpions assays, Plexor chemistry measures reduction in the fluorescence of the reporter dye (Johnson et al. 2004; Moser and Prudent 2003; Sherrill et al. 2004). The fluorescence of the reporter dye is turned off by the incorporation of a quencher during the progress of amplification, thereby reducing the fluorescent signal as the quantity of the amplified product increases. A modified cytosine moiety, iso-dC, linked to a fluorescent dye is incorporated at the 5' end of one of the PCR primers. The second PCR primer is unlabeled. The PCR reagent mix contains dabcyl-iso-dGTP along with other deoxynucleotides. During the PCR, dabcyl-iso-dGTP is incorporated opposite to the iso-dC residue (complementary pairing). Thus, the incorporated dabcyl moiety is in close proximity to the fluorescent dye present in the primer, thereby resulting in reduction of its fluorescence. The extent of the reduction in fluorescence is proportional to the extent of PCR and hence to the quantity of the template DNA. The C_T value is the cycle at which the fluorescence drops below the threshold (Krenke et al. 2008). In addition to the measurement of fluorescence, melting or dissociation curve analysis may be performed to characterize the PCR product. The specificity of the reaction can be ascertained by determining the melting temperature (T_m) of the amplicons.

6.3.4.2 *Real-Time PCR Assays*

6.3.4.2.1 Singleplex Real-Time PCR Assays

Human DNA quantification: The *Alu* family of interspersed repeats is the most successful mobile genetic element in terms of integration and copy number within primate genomes. The high copy number of *Alu* repeats in the human genome makes these assays sensitive and ideal for human DNA detection and quantification. The *Alu* markers used for quantification of human DNA include Yb8, Yd6, and Ya5 and an inter-*Alu* sequence (Table 6.1) (Nicklas and Buel 2003b, 2005; Shewale et al. 2007; Sifis et al. 2002; Walker et al. 2003). The extent of amplification is monitored either by the SYBR Green I dye (Nicklas and Buel 2003b; Shewale et al. 2007; Walker et al. 2003) or molecular beacon MGB Eclipse® Reagent (Nicklas and Buel 2005).

mtDNA quantification: Von Wurmb-Schwark et al. (2002) developed a singleplex assay for quantification of mtDNA using the NADH dehydrogenase subunit 1 (ND1) target gene and SYBR Green I dye for measurement of the amplification. Alonso et al. (2004) determined the extent of mtDNA degradation by performing two singleplex real-time PCRs targeting HV I region and generating fragment sizes of 113 and 287 bp in separate wells of the same plate to assess mtDNA preservation (copy number and degradation state). Duplex design for this assay was not optimized since a decrease in the PCR efficiency of the 287-bp target was observed in duplex reactions. Kavlick et al. (2011) proposed the use of a synthesized DNA standard for quantification of mtDNA. The assay enables detection of as low as 10 copies of mtDNA. Synthetic DNA standards lack topological constraints as well as provide better quality and contamination control, which leads to greater accuracy of quantification.

6.3.4.2.2 Duplex Real-Time PCR

Human DNA quantification and detection of PCR inhibitors: The Quantifiler Human DNA Quantification Kit enables quantification of human DNA using the human telomerase reverse transcriptase (hTERT) target gene in conjunction with a 79-base synthetic oligonucleotide included in the assay that serves as an internal PCR control (IPC) (Applied Biosystems 2003; Green et al. 2005). The TaqMan® probes are labeled with FAM™ and VIC® dyes, respectively. ROX™ dye is included in the reaction mix as reference dye. The dynamic range of quantification is from 46 pg (23 pg/μL) to 100 ng (50 ng/μL) (Green et al. 2005). A reduction in the amplification of the IPC template with resulting increase in the IPC C_T value indicates the presence of PCR inhibitors. The number of copies of IPC template is identical in all wells. However, the resulting IPC C_T value is not always uniform. We observe slightly higher C_T values in wells containing higher amounts of human DNA template due to competition for amplification reagents. Nonetheless, the IPC C_T value is a reliable piece of information to guide the operator in choosing the downstream genotyping systems (discussed later in Section 6.3.4.3).

Qiagen developed an Investigator® Quantiplex system based on Scorpions chemistry for quantification of human DNA and detection of PCR inhibitors (Qiagen 2011; Scherer and Di Pasquale 2012). The duplex assay amplifies a 146-bp amplicon from 4NS1C target gene for human DNA and a 200-bp internal control. The multicopy nature of the human target (~40 copies per cell) enables quantification below 4.9 pg/μL and detection to <1 pg/μL of human DNA.

Human male DNA quantification and detection of PCR inhibitors: The Quantifiler Y Human Male DNA Quantification Kit is analogous to the Quantifiler Human DNA Quantification Kit except for the target gene that in this case is the human male sex-determining region Y (SRY) (Green et al. 2005).

Human and male DNA quantification: Nicklas and Buel (2006) developed a duplex TaqMan assay for quantification of total human and male DNA. The amplification targets for human and male DNA are the Ya5 subfamily of the *Alu* marker and DYZ5, respectively. The TaqMan probes are labeled with VIC and FAM dyes, respectively.

Alonso et al. (2004) developed an assay that amplifies a segment of the X-Y homologous amelogenin gene enabling not only human and human male gene quantification, but also sex determination. The MGB probe that specifically detects the AMGX-fragment is labeled with FAM; the one for the AMGY-fragment is labeled with VIC.

Horsman et al. (2006) described a duplex assay using TPOX and SRY as target genes for the detection of total human and human male DNA, respectively. The probes were labeled with VIC (human target) and FAM (male target) dyes using TAMRA™ as quencher.

Nuclear and mtDNA quantification: Andréasson et al. (2002, 2006) have described a duplex assay for quantification of nuclear and mtDNA and demonstrated its utility in the analysis of challenging samples such as body hairs and epithelial cells recovered from different articles like rings, watches, glasses, earrings, and so forth. The assay targets the human retinoblastoma susceptibility gene (RB1) on chromosome 13 for nuclear DNA and the mitochondrial target spans over the genes for tRNA lysine and ATP synthetase 8 in the coding region of the mitochondrial genome.

The assay described by Timken et al. (2005) comprises a TH01 target for nuclear DNA (nuTH01, ~170–190 bp) and the NADH dehydrogenase subunit 1 (ND1, 69 bp) for mtDNA. Separate standard curves are generated for nuclear DNA and mtDNA. Nuclear DNA copy numbers are estimated using the ratio of one haploid nuclear copy per 3.3 pg genomic DNA and mitochondrial copy numbers are estimated using a ratio of 400 or 450 mitochondrial copies per 3.3 pg of respective standard DNA used. The assay exhibited sensitivity of ~15 haploid nuclear copies and ~12 mitochondrial copies.

Walker et al. (2005) developed another duplex assay for quantification of human nuclear and mtDNA. The targets for this assay are a region on Yb8 *Alu* with a diagnostic base within the *Alu* Y family and a conserved region (positions 8250–8550) of human mtDNA, respectively.

Niederstätter et al. (2007) described a modular concept for quantification of nuclear DNA, mtDNA, detection of PCR inhibitors, and extent of DNA degradation in a modular duplex real-time PCR. The scientists have designed primers to amplify three amplicons of 79, 156, and 246 bp from the RB1 gene, exon 25 for nuclear DNA; four amplicons of 102, 143, 283, and 404 bp for mtDNA; one amplicon of 156 bp for IPC for nuclear DNA (plasmid containing 246 fragment from RB1), and one amplicon of 143 bp for IPC for mtDNA (plasmid containing 404 bp mtDNA fragment). Duplex real-time PCR reactions are performed in various combinations to obtain desired information; for example, nuclear DNA and mtDNA duplex for quantification of nuclear and mtDNA, nuclear DNA and nuclear IPC for quantification of nuclear DNA and detection of PCR inhibitors, and duplex PCR with different amplicon sizes to help determine the extent of DNA degradation. The $RB1_{156\,bp}/pMt_{IPC}$ duplex assay was used in 12,000 casework samples over 2 years' time and the sample rerun rate was decreased from 18% to 7%, demonstrating its utility in assessing the DNA extracts.

DNA degradation assay: Nicklas et al. (2012) developed a duplex assay that simultaneously amplifies two amplicons, 63 bp and 246 bp, using a common forward primer and two reverse primers for a single *Alu* target. The assay enables determining the extent of DNA degradation in a biological sample.

6.3.4.2.3 Triplex Real-Time PCR

Human and human male DNA quantification and detection of PCR inhibitors: Barbisin et al. (2008, 2009) developed the Quantifiler Duo DNA Quantification Kit that enables quantification of total human DNA and human male DNA and detection of the presence of PCR inhibitors in a biological sample in a single amplification reaction. The assay utilizes TaqMan chemistry. The target genes are ribonuclease P RNA component H1 (RPPH1), sex-determining region Y (SRY), and a synthetic oligonucleotide sequence for human, human male, and IPC measurement, respectively. The TaqMan probes are labeled with VIC, FAM, and NED™ dyes, respectively. ROX dye is included in the reaction mix as reference dye. The size of the amplicons for the three targets is similar (140 bp for RPPH1 and 130 bp for SRY and IPC targets) to minimize any preferential amplification (Table 6.1). In addition to the quantification of DNA, the assay provides a qualitative assessment of forensic evidence samples in terms of determining male-to-female DNA mixture ratios and estimating the extent of PCR inhibition. When a single-source male DNA sample is tested, the observed C_T values for the male DNA target are in general about 1 C_T higher than those for the human target, which is attributed to the haploid nature of the male target on the Y-chromosome compared to the diploid nature of the human target on an autosomal chromosome (Barbisin et al. 2008, 2009). Nevertheless, the C_T values are also determined by the amplification efficiencies and therefore sometimes the delta C_T between the two targets may be slightly affected, as observed in other multiplex assays (Horsman et al. 2006).

Krenke et al. (2008) developed the Plexor HY System for simultaneous quantification of human DNA and human male DNA and detection of the presence of PCR inhibitors. As the name indicates, the assay is based on the Plexor chemistry previously described in Section 6.3.4.1. The human target is the RNU2 locus that encodes a small nuclear RNA involved in pre-mRNA processing (multicopy gene, 99-bp amplicon); the human male target is TSPY locus on the Y-chromosome that encodes a testis-specific protein (multicopy gene, 133-bp amplicon), and a novel nucleotide sequence is the target for the IPC assay (150-bp amplicon) (Table 6.1). The primers are labeled with fluorescein, CAL Fluor® Orange 560, and CAL Fluor® Red 610, respectively. IC5 is included in the reaction mix as passive reference dye. As with Quantifiler Duo, this assay can provide information on the male-to-female mixture ratio and extent of PCR inhibition.

Qiagen developed an Investigator Quantiplex HYres system for quantification of human DNA and human male DNA, as well as detection of PCR inhibitors (Qiagen 2012; Scherer and Di Pasquale 2012). The triplex assay is based on Scorpions chemistry and amplifies a 146-bp amplicon from multicopy 4NS1C target gene (~40 copies per cell) for human DNA, a 129-bp amplicon from a multicopy target gene (~18 copies per cell) located on human Y-chromosome, and a 200-bp internal control. The multicopy nature of the human and human male targets enables quantification below 4.9 pg/μL and detection to <1 pg/μL of human DNA.

DNA degradation and detection of PCR inhibitors: Swango et al. (2006, 2007) described a triplex assay for determining the extent of DNA degradation and presence of PCR inhibitors. The assay comprises two targets for human DNA, TH01 (nuTH01, ~170–190 bp) and CSF1PO (nuCSF, 67 bp), and contains a synthetic oligonucleotide as the IPC target sequence (77 bp). The extent of DNA degradation is determined by monitoring the extent of amplification of the long and short amplicons generated from human DNA. The degradation ratio is computed by dividing the quantity of short nuCSF amplicon by the quantity of long nuTH01 amplicon and is used for evaluating the level of degradation in the DNA of a given sample.

Human nuclear, mitochondrial, and male DNA quantification: Walker et al. (2005) developed a triplex assay for quantification of human nuclear, mitochondrial, and male DNA. The targets for these assays were a region on Yb8 *Alu* with a diagnostic base within the *Alu* Y family, a conserved region (positions 8250–8550) of human mtDNA, and a homologous region of human sex chromosomes containing 90-bp deletion on X-chromosome. The assay provides unique sample assessment features for selection of autosomal STR, Y-STR, or mtDNA systems for genotyping.

6.3.4.2.4 Quadruplex Real-Time PCR Hudlow et al. (2008) developed a quadruplex qPCR assay for assessment of total human DNA, human male DNA, DNA degradation, and detection of the presence of PCR inhibitors. The assay amplifies a ~170–190-bp target sequence that spans the TH01 STR locus (nuTH01), a 137-bp region within SRY locus (nuSRY), a 67-bp target sequence flanking the CSF1PO STR locus (nuCSF), and a 77-bp synthetic DNA template as IPC target. The amplification of nuTH01 and nuSRY is monitored for determining the quantity of human and human male DNA, respectively; the C_T value for the IPC enables detection of PCR inhibitors; and the nuCSF:nuTH01 ratio provides indication on the DNA degradation in the sample. Although both nuCSF and nuTH01 amplify human DNA, the size of the amplicons is different. The difference in signal between the shorter nuCSF amplicon (67 bp) and the longer nuTH01 amplicon (~170–190 bp) is utilized in determining the extent of degradation. Degraded samples are expected to provide higher signal for the short amplicon than for the long amplicon, whereas intact samples should exhibit similar results for both amplicons.

6.3.4.3 Important Concepts in Real-Time PCR Assay Design

Size of amplicons: To obtain better estimates of the quantity of amplifiable DNA, the size of the amplicons used for quantification should be in the range of those used in the targeted genotyping systems. The length of amplicons in commonly used STR systems, mtDNA sequencing, and SNP typing generally ranges from 80 to 400, 100 to 400, and 20 to 150 bp, respectively. The size of the amplicons for the different real-time PCR assays summarized in Table 6.1 ranges from 62 to 287 bp, which falls within the genotyping ranges that were mentioned above. Quantification assays with smaller amplicons tend to overestimate the quantity of DNA for STR typing, for example, when working with degraded samples.

Species specificity: As mentioned earlier, the quantification of human DNA in forensic samples is critical since the DNA extracts are contaminated with DNA from other nonhuman sources like microorganisms, domestic pets, farm animals, and so forth. Thus, the primers and probes used in the assays need to exhibit desired specificity. The qPCR assays described in Table 6.1 exhibit high specificity for human DNA with some cross-reactivity with higher primates and very low cross-reactivity, if any, with other nonhuman species tested. (See the single original papers for the details on the species-specificity studies conducted by the authors.)

NTC signal: No template controls (NTC) are included in the run plates to monitor the potential assay contamination due to reagents, processes, plasticware, or equipment. A C_T value other than undetermined for the specific target in the NTC wells indicates possible contamination of the reagents or the equipment used to set up the plate and, in general, repeating the experiment is highly recommended. However, at times, a C_T value is detected in the NTC wells and it is not the result of a true contamination. Possible explanations are the presence of a very low level of nonspecific amplification or dye spectral cross-talk between channels in the real-time instrument. These factors can be controlled with good assay design and appropriate instrument maintenance (Applied Biosystems 2008a).

Limit of quantification and detection: It is useful to differentiate between the two following concepts: limit of quantification and limit of detection. The limit of quantification of a certain assay may be described as the lowest quantifiable amount that typically is encompassed by the lowest quantity of DNA present in the quantification standard curve. At this level the measurements have a fairly good confidence interval determined by a low standard deviation. The limit of detection, on the other hand, may be considered as the quantity of DNA that yields a detectable signal above the threshold (Horsman et al. 2006).

Limits of detection as low as 1-15 pg/μL are reported for different assays (Barbisin et al. 2009; Horsman et al. 2006; Hudlow et al. 2008; Krenke et al. 2008; Nicklas and Buel 2003b, 2005; Shewale et al. 2007; Swango et al. 2006). Typically, at this low DNA template concentration, the assay is in the stochastic effect range and in some cases the reproducibility of the signal is low. In general, several replicates are required and the amplification efficiencies can vary greatly from well to well. In this situation the quality of the STR profile based on the quantity of human DNA can vary. The ultimate goal of DNA quantification is to determine the volume of extract to be used in the genotyping amplification reaction. The amount of DNA recommended for STR typing ranges from 0.5 to 2.0 ng for different commercial kits. Thus, for samples containing DNA at concentrations lower than 0.05 ng/μL (0.1 ng/μL for some STR kits), in most cases it is highly recommended to add the maximum recommended volume of extract to the amplification reaction for the respective genotyping kit. Measurements obtained at very low levels of template DNA, therefore, can be informative and may not affect the downstream STR reaction setup. Alternatively, an analyst can decide to either concentrate the sample or perform mtDNA sequencing. The limit of quantification for the Quantifiler Duo DNA Quantification kit is 0.023 ng/μL and the limit of detection is 0.006 ng/μL (Barbisin et al. 2009).

Physicochemical properties of the plastics used in performing the assay play a critical role in the accuracy of DNA quantification, particularly for samples containing low quantities of DNA. Some plastics materials bind DNA due to hydrophobic interactions. Ellison et al. (2006) recommend using low-retention plastic tubes for real-time PCR assays for quantification of samples below 100 genome equivalents (660 pg).

Detection of PCR inhibitors: A synthetic oligonucleotide template is incorporated in the multiplex assays for detection of PCR inhibitors. The concept is demonstrated using known PCR inhibitors like hematin, humic acid, calcium chloride, indigo dye, carmine dye, soil extract, and so forth (Barbisin et al. 2009; Green et al. 2005; Hudlow et al. 2008; Krenke et al. 2008; Nicklas and Buel 2006). The IPC C_T increases with increasing concentrations of inhibitor and the failure of the amplification is observed at higher concentrations. An increase in the IPC C_T value (ΔC_T or $\Delta IPC = \Delta C_{T\,inhibitor} - \Delta C_{T\,no\text{-}inhibitor}$) of more than 0.75 (Hudlow et al. 2008; Swango et al. 2006, 2007) or more than 2 (Promega Corporation 2007) is attributed to the presence of a PCR inhibitor. Hudlow et al. (2008) proposed to calculate a normalized inhibitor factor (NIF) to estimate the degree of inhibition per nanogram of sample using the formula

$$NIF = \frac{\Delta IPC\ (\text{cycles})}{\text{Quantity of DNA (ng)}}$$

with the implication that the larger the NIF, the more likely the sample is to exhibit PCR inhibition effects in STR amplification. It is noted that an increased IPC C_T value is merely an indication of the presence of PCR inhibitors and may not perfectly correlate to the performance of the STR amplification for the following reasons: the concentration of the inhibitor in the qPCR assay and in the STR amplification reaction is not identical (e.g., 2 and 10 μL of extract are used for Quantifiler and Identifiler systems, respectively); different multiplex PCR reactions are affected in different ways by the same inhibitor; the effect of the inhibitory compound is unknown (e.g., certain compounds may not cause any shift in IPC C_T value but may inhibit the genotyping system); and inhibition may be due to factors that are not detectable by the IPC (e.g., DNA cross-links). Moreover, some next-generation

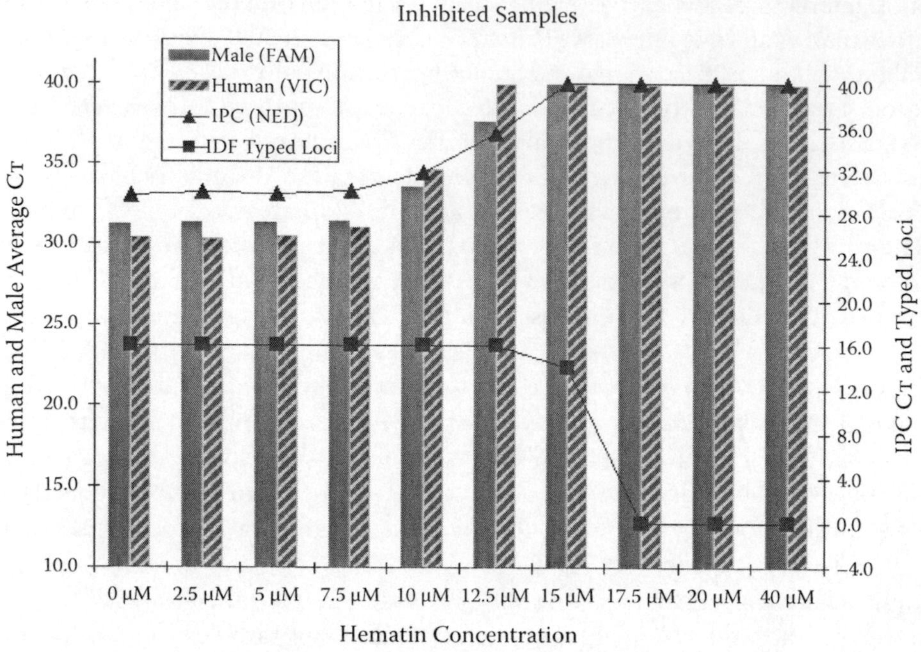

Figure 6.1 Inhibition study: Correlation between Quantifiler® Duo DNA Quantification Kit and AmpFℓSTR® Identifiler® PCR Amplification Kit performance in samples spiked with hematin.

STR kits like Identifiler Plus, NGM™, and Powerplex 16 HS kits use PCR chemistries that are vastly more robust to PCR inhibitors compared to earlier kits, including the quantification kits themselves. However, in general, monitoring the IPC C_T value can help the operator in making an informed decision as to whether to proceed with the STR typing (and choose the most appropriate chemistry among those commercially available) or reprocess the sample for extraction. Figure 6.1 illustrates the correlation between the IPC C_T values obtained from the Quantifiler Duo DNA Quantification kit reaction and the quality of the profile obtained using the AmpFℓSTR Identifiler PCR Amplification Kit. The curves that represent the IPC C_T and the number of loci correctly typed are almost specular images, demonstrating that, as the IPC C_T gradually increases, the quality and the completeness of the STR profile decreases. In the presence of higher quantities of PCR inhibitors, underestimation of the quantity of DNA is observed since amplification of both the IPC target and quantification target gets affected. For example, Seo et al. (2011) observed a decrease of 38.5% to 51.4% and ≥71.8% in the quantity of estimated DNA in three tested samples in the presence of 2.4 ng/μL and ≥ 3.2 ng/μL of humic acid, respectively, using the Quantifiler Human DNA Quantification Kit.

The mechanisms of inhibition can be determined by examining the effect of amplicon length, melting temperature, and sequence on amplicons generated from a single target (Funes-Huacca et al. 2011; Opel et al. 2010). A change in melt curve is a result of altered configuration of PCR products due to inhibitor binding. The different inhibition mechanisms include Taq inhibition, altered DNA template binding, and reduced reaction efficiency. The type of inhibition depends on the chemical nature of the inhibitor co-extracted (Opel et al. 2010). Nicklas and Buel (2003b, 2005) proposed to detect the presence of PCR inhibitor from the nature of amplification curves and raw data. In the presence of inhibitors, the shape of the amplification curve is altered: it is flatter and never reaches the plateau. The baseline value in the raw data is high when a high concentration of DNA is present, but no amplification occurs. Additionally, the samples that do not contain DNA should display a curve similar to NTC wells. Samples that do not amplify at all indicate the presence of an inhibitor.

Mutants of Taq DNA polymerase that are resistant to PCR inhibitors have been developed, enabling amplification of whole blood samples (Kermekchiev et al. 2009). These types of developments hold promise for advancing the capabilities of real-time PCR assays.

Mixtures of human male and female DNA: In general, mixtures of male and female contributors are considered challenging samples and very often the male portion is the minor component. Generating a conclusive male profile using an autosomal STR system on such samples is rather difficult, especially if the sample contains more than 10-fold excess of female DNA. Thus, it is useful to quantify or at least detect the male component as well as the amount of female DNA in the sample to select a Y-STR chemistry that is known to be more sensitive than the autosomal kits. Based on the quantity of human and human male DNA obtained using any qPCR system, one can derive the mixture ratio using the simple formula (Barbisin et al. 2009):

Male DNA: Female DNA Ratio = 1 : (Human DNA – Male DNA)/Male DNA

[All quantities in the above equation are expressed with the same unit, e.g., ng/μL.]

Mixture ratios determined by the real-time PCR assays are fairly accurate when the quantity of male DNA is above the stochastic threshold. For samples containing very small quantities of male DNA, the detection of the presence of male DNA can be valuable information for Y-STR profiling. The sensitivity of the multiplex real-time PCR assays is typically so high that it is possible to detect small quantities of male DNA even in the presence of a large excess of female DNA. Some examples reported in the literature are the following: 25 pg/μL of male DNA in the presence of 10,000-fold excess of female DNA (Applied Biosystems 2008a), 25 pg of male DNA in the presence of 5,000-fold excess of female DNA (Horsman et al. 2006), 8 pg of male DNA in the presence of 27,000-fold excess of female DNA (Hudlow et al. 2008), and 0.8 ng and 6 pg of male DNA in the presence of 1,000- or 10,000-fold excess of female DNA, respectively (Krenke et al. 2008).

Table 6.2 shows an example of how the mixture ratio calculated for a swab quantified with the Quantifiler Duo DNA Quantification Kit can direct the operator toward choosing the appropriate STR system. The male:female ratio for this sample is about 1:16 and therefore it will be challenging to resolve the male minor component using an autosomal STR kit. In this case it is advisable and more efficient to proceed directly with Y-STR analysis to obtain the male profile. Figures 6.2 and 6.3 illustrate the profiles obtained with the autosomal AmpFℓSTR Identifiler PCR Amplification Kit and the Y-STR AmpFℓSTR Yfiler® PCR

Table 6.2 Mixture Sample: A Swab Analyzed with Quantifiler® Duo DNA Quantification Kit

Sample Type	Human DNA Quantity (ng/μL)	Human Male DNA Quantity (ng/μL)	IPC C_T	M:F Ratio[a]
Swab	0.250	0.015	29.380	1:15.7

[a] Male DNA:Female DNA ratio.

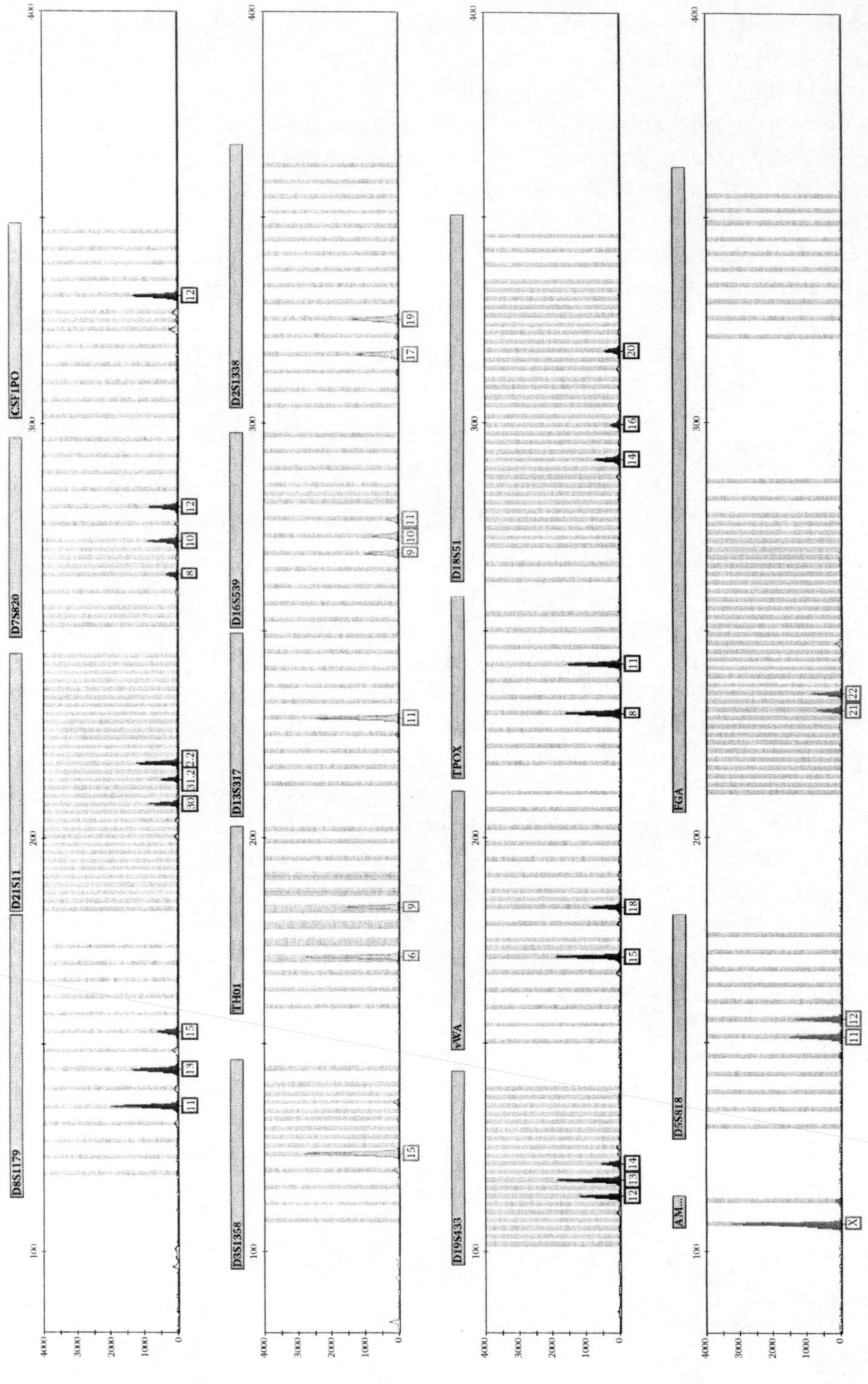

Figure 6.2 Mixture sample: Profile of the sample described in Table 6.2 obtained using AmpFℓSTR® Identifiler® PCR Amplification Kit.

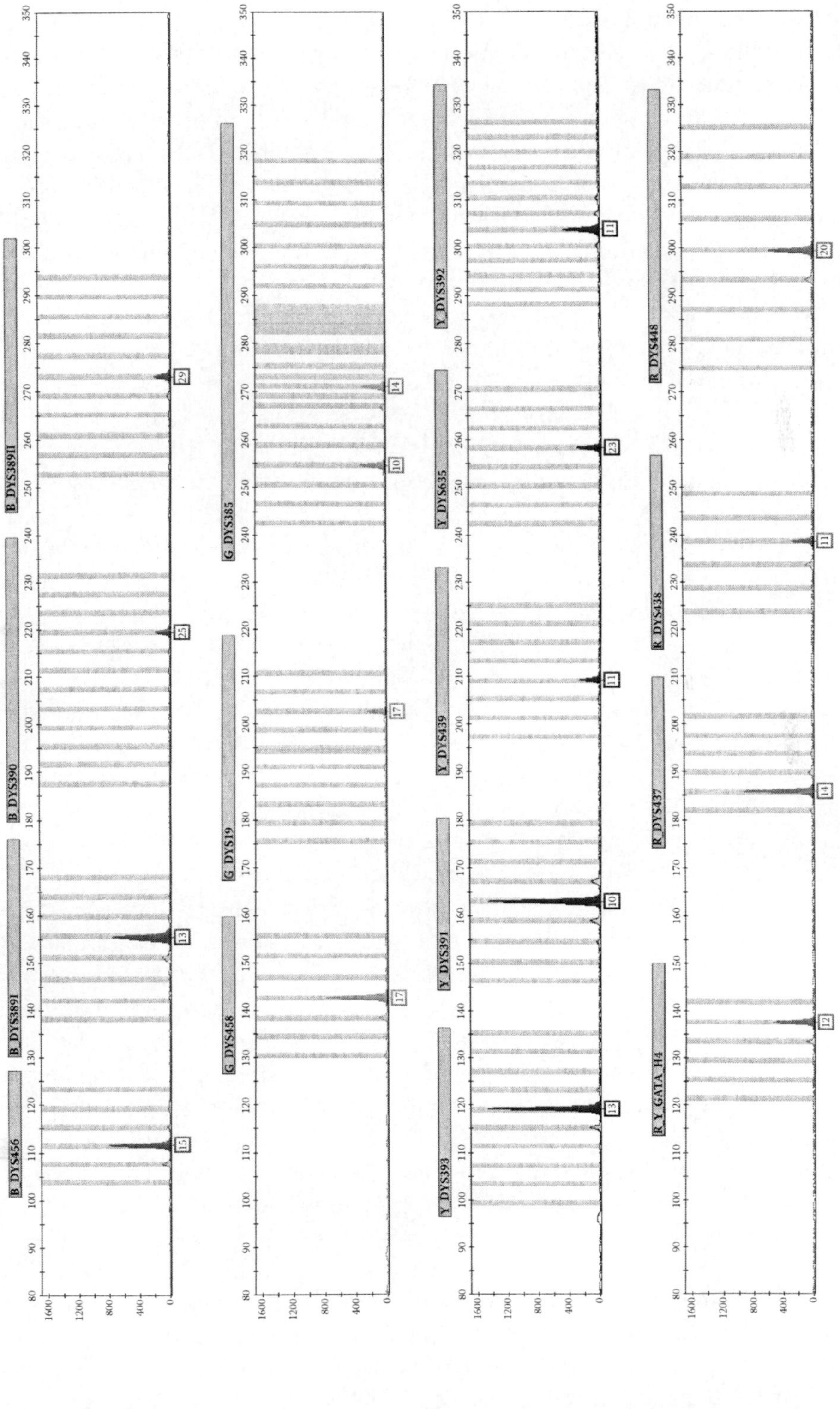

Figure 6.3 Mixture sample: Profile of the sample described in Table 6.2 obtained using AmpFℓSTR® Yfiler® PCR Amplification Kit.

Amplification Kit. They confirm that the qPCR information could reasonably predict the genotyping results, thereby increasing the success rate at first pass and decreasing sample consumption at the same time.

DNA Degradation: Biological samples, particularly evidence samples, are often exposed to uncontrolled environmental conditions prior to collection or receiving in the laboratory, leading to degradation of DNA to varying extents. When the extent of DNA degradation results in fragments smaller than the length of target loci, the results are inconclusive. STR genotyping systems that enable amplification of shorter fragments have been developed for genotyping of degraded samples (Butler et al. 2003; Mulero et al. 2008). SNP genotyping is also advocated for such samples (Kidd et al. 2006; Phillips et al. 2007). Thus, determining the extent of DNA degradation enables a scientist to use the appropriate genotyping system to obtain conclusive genotyping results. The degradation ratio is estimated by comparing the extent of amplification of a small amplicon and a large amplicon generated simultaneously in the real-time PCR assay. The real-time PCR assays described by Nicklas et al. (2012) and Hudlow et al. (2008) are targeted to determine the DNA degradation of nuclear DNA, whereas the one described by Alonso et al. (2004) targets mtDNA. As mentioned earlier in Section 6.3.4.2.2, Niederstätter et al. (2007) developed different modular duplex assays for determining DNA degradation for nuclear DNA and mtDNA. Sinha et al. (2013) developed a DNA-based qualitative/quantitative/inhibition assessment system that measures the extent of DNA degradation. The assay is designed to amplify two amplicons; 80 and 250 bp fragments from Yb8- and Ya5-lineage *Alu* insertion targets, respectively.

Correlation between assays: It is expected to see a certain discrepancy in the quantification results obtained with different methods, and many factors contribute to it. Some of these factors are the technology used in the assay, the accuracy of the predetermined quantity of the standard, the size of the utilized amplicon, the number of copies of the target gene, the variation in the amplification efficiencies, potential mutations at the primer binding sites, the susceptibility of the assay to the interference of molecules other than DNA present in the sample, and so forth. Nielsen et al. (2008) observed a wide variation in the quantity of DNA measured in commercially available DNA samples using five DNA quantification methods based on spectrophotometric, hybridization, and real-time PCR principles. The study concluded that a standard reference DNA material and a standard method for DNA quantification were needed. Timken et al. (2005), on the other hand, obtained a good agreement on the quantity of DNA measured in a sample containing ~6 ng/μL using slot-blot hybridization, nuTH01-mtND1 duplex qPCR, and Quantifiler Human DNA Quantification Kit. However, the quantity of DNA varied significantly across quantification methods for highly degraded DNA samples: the Quantifiler Human Kit provided a higher quantity of DNA compared to the nuTH01-mtND1 duplex assay. The difference is expected due to the difference in the size of the amplicons: 62 bp for Quantifiler Human Kit and about 180 bp for nuTH01. Slot-blot provided much lower quantities.

The performance of real-time assays depends on the genetic characteristics of the target gene(s). LaSalle et al. (2011) compared the performance of Quantifiler Duo and Plexor® HY kits that employ single- and multicopy targets, respectively, for quantification of human and human male DNA. It was observed that the single-copy target assay provided accurate measurement of the mixture ratio of human to human male DNA in a sample and the multicopy assay provided better detection of DNA at concentrations below 10 pg/μL.

However, both assays provided linear estimates for DNA over a broad range of input DNA (LaSalle et al. 2011).

6.4 Standard Reference Material 2372

The National Institute of Standards and Technology (NIST) provides several standard reference materials (SRMs) to many industries and organizations, including forensic laboratories. SRMs are certified reference materials (CRMs) that are issued under the NIST trademark and are well characterized using state-of-the-art measurement methods. In the year 2004, a study involving 82 forensic laboratories worldwide was organized to understand the measurement performance of different DNA quantification methods (Kline et al. 2005). The aims of the study were to examine the concentration effects and the performance at the lower DNA concentration levels that are frequently observed in forensic casework, to examine consistency across multiple laboratories with various methodologies, to examine single- versus multiple-source samples, and to study DNA stability over time and through shipping in two types of storage tubes (Kline et al. 2005). In October 2007, NIST released SRM 2372 Human DNA Quantitation Standard for use in forensic laboratories (Kline et al. 2009; Vallone et al. 2008). SRM 2372 consists of three DNA extracts: a single-source male, a multiple-source female, and a mixture of male and female sources (see NIST Web site for certificate and more details: http://www.cstl.nist.gov/biotech/strbase/srm2372.htm). All three components of SRM 2372 have very similar optical densities. These three components represent the nature of the samples routinely analyzed by forensic laboratories. SRM 2372 was assessed by 32 forensic laboratories in the United States, Canada, and Australia (Kline et al. 2009). NIST recommends that qPCR users calibrate their in-house or commercial standard with SRM 2372 using one or more dilution series, each series prepared from one SRM 2372 component (Kline et al. 2009). Component C, a mixture of male and female DNA from multiple sources, is not suitable for calibration of gender-specific assays. SRM 2372 can be used by the manufacturers of real-time assays to validate the values assigned to their own reference DNA provided in the kit. Individual forensic laboratories can use SRM 2372 to validate the concentration values assigned to their in-house or commercial DNA calibration standard (Vallone et al. 2008). The correct use of SRM 2372 would minimize the variation in the results obtained in different forensic laboratories that are using different quantification assays.

6.5 Conclusions

The different quantification methods developed for forensic DNA analysis over the years are based on spectrophotometry, fluorometry, hybridization, end-point PCR, or real-time PCR. Using these assays it is possible to detect DNA quantities in a sample at the picogram level. Fluorometric assays, such as PicoGreen dye-based assays, are generally used for single-source or reference samples. Real-time PCR methods have been widely adopted in the past few years, particularly for the analysis of evidence samples. Different principles employed in the real-time PCR assays to measure the quantity of DNA include the use of intercalating dyes that bind to the double-stranded DNA, the cleavage of TaqMan probes hybridized to the amplicons, hybridization of molecular beacons to the amplicons, Scorpions primers (a bifunctional molecule comprising primer and probe), and Plexor chemistry.

The multiplex capability of real-time PCR methods enables forensic scientists to obtain more information about the sample than merely the quantity of DNA. In addition to the quantification of DNA, it is highly recommended to perform a quality assessment of forensic evidence samples to enable the choice of the most appropriate genotyping system and to generate interpretable results on the first attempt. Assessment of DNA extract using the multiplexed real-time PCR assays aid in determining one or more of the following points of information: whether the sample contains sufficient human DNA, human male DNA, or mtDNA to proceed with STR, SNP, or mtDNA typing; the optimal amount of sample to use in the amplification reactions; the relative quantities of human male and female DNA in a sample that can assist in the selection of the applicable STR chemistry; whether PCR inhibitors are present in a sample that may require additional purification before proceeding to genotyping; and the extent of DNA degradation to select either miniSTR or SNP typing instead of traditional STR systems. This approach increases the success rate of DNA profiling, thereby reducing the number of samples that need reprocessing, which in turn reduces the cost of analysis and decreases the backlogs in forensic laboratories.

Life Technologies is currently developing Quantifiler HP and Quantifiler Trio kits (triplex and quadruplex real-time PCR assays, respectively) for the assessment of forensic evidence samples. The assays employ multicopy targets, allowing a subpicogram level of detection of human and human male DNA, fast PCR capability, enhanced resistance to PCR inhibitors, and the ability to determine the extent of DNA degradation. These assays are designed to achieve optimal performance with next-generation STR multiplex systems.

Acknowledgments

We thank Lisa Calandro, Heidi Kijenski, Robert Green, Lori Hennessy, and Manohar Furtado (Life Technologies) for useful discussions. We also thank the San Diego Police Department for providing the mixture sample data.

References

Afonina, I., M. Zivarts, I. Kutyavin, E. Lukhtanov, H. Gamper, and R. B. Meyer. 1997. Efficient priming of PCR with short oligonucleotides conjugated to a minor groove binder. *Nucleic Acids Res* 25 (13): 2657–60.

Allen, R. W. and V. M. Fuller. 2006. Quantitation of human genomic DNA through amplification of the amelogenin locus. *J Forensic Sci* 51 (1): 76–81.

Alonso, A., P. Martín, C. Albarrán et al. 2003. Specific quantification of human genomes from low copy number DNA samples in forensic and ancient DNA studies. *Croat Med J* 44 (3): 273–80.

Alonso, A., P. Martín, C. Albarrán et al. 2004. Real-time PCR designs to estimate nuclear and mitochondrial DNA copy number in forensic and ancient DNA studies. *Forensic Sci Int* 139 (2–3): 141–9.

Andréasson, H., U. Gyllensten, and M. Allen. 2002. Real-time DNA quantification of nuclear and mitochondrial DNA in forensic analysis. *BioTechniques* 33 (2): 402–11.

Andréasson, H., M. Nilsson, B. Budowle, H. Lundberg, and M. Allen. 2006. Nuclear and mitochondrial DNA quantification of various forensic materials. *Forensic Sci Int* 164 (1): 56–64.

Applied Biosystems. 2003. Quantifiler® Kits: Quantifiler® Human DNA Quantification Kit and Quantifiler® Y Human Male DNA Quantification Kits; User's Manual. Available from: http://www3.appliedbiosystems.com/cms/groups/applied_markets_support/documents/generaldocuments/cms_041395.pdf (accessed September 28, 2012).

Applied Biosystems. 2008a. Quantifiler® Duo DNA Quantification Kit; Product Bulletin. Available from: http://marketing.appliedbiosystems.com/images/Product_Microsites/QUANTDUO/pdf/53167_QuantDuo_PB_f3.pdf (accessed September 28, 2012).

Applied Biosystems. 2008b. Quantifiler® Duo DNA Quantification Kit; User Manual. Available from: http://www3.appliedbiosystems.com/cms/groups/applied_markets_support/documents/generaldocuments/cms_049050.pdf (accessed September 28, 2012).

Barbisin, M., R. Fang, C. E. O'Shea et al. 2008. A multiplexed system for quantification of human DNA and human male DNA and detection of PCR inhibitors in biological samples. *Forensic Sci Int Genet* Suppl Ser 1: 13–5.

Barbisin, M., R. Fang, C. E. O'Shea, L. M. Calandro, M. R. Furtado, and J. G. Shewale. 2009. Developmental validation of the Quantifiler Duo DNA Quantification kit for simultaneous quantification of total human and human male DNA and detection of PCR inhibitors in biological samples. *J Forensic Sci* 54 (2): 305–19.

Bataille, M., K. Crainic, M. Leterreux, M. Durigon, and P. de Mazancourt. 1999. Multiplex amplification of mitochondrial DNA for human and species identification in forensic evaluation. *Forensic Sci Int* 99 (3): 165–70.

Butler, J. M., Y. Shen, and B. R. McCord. 2003. The development of reduced size STR amplicons as tools for analysis of degraded DNA. *J Forensic Sci* 48 (5): 1054–64.

Chong, M. D., C. D. Calloway, S. B. Klein, C. Orrego, and M. R. Buoncristiani. 2005. Optimization of a duplex amplification and sequencing strategy for the HVI/HVII regions of human mitochondrial DNA for forensic casework. *Forensic Sci Int* 154 (2–3): 137–48.

Collins, P. J., L. K. Hennessy, C. S. Leibelt, R. K. Roby, D. J. Reeder, and P. A. Foxall. 2004. Developmental validation of a single-tube amplification of the 13 CODIS STR loci, D2S1338, D19S433, and amelogenin: The AmpFlSTR Identifiler PCR Amplification Kit. *J Forensic Sci* 49 (6): 1265–77.

Cupples, C. M., J. R. Champagne, K. E. Lewis, and T. D. Cruz. 2009. STR profiles from DNA samples with "undetected" or low quantifiler results. *J Forensic Sci* 54 (1): 103–7.

DNA Advisory Board. 2000. Quality Assurance Standards for DNA Testing and Databasing Laboratories (Forensic Science Communications, July 2000). Available from: http://www.fbi.gov/about-us/lab/forensic-science-communications/fsc/july2000/codispre.htm (accessed September 28, 2012).

Ellison, S. L. R., C. A. English, M. J. Burns, and J. T. Keer. 2006. Routes to improving the reliability of low level DNA analysis using real-time PCR. *BMC Biotechnol* 6: 33–43.

Funes-Huacca, M. E., K. Opel, R. Thompson, and B. R. McCord. 2011. A comparison of the effects of PCR inhibition in quantitative PCR and forensic STR analysis. *Electrophoresis* 32 (9): 1084–9.

Georgiou, C. D. and I. Papapostolou. 2006. Assay for the quantification of intact/fragmented genomic DNA. *Anal Biochem* 358 (2): 247–56.

Green, R. L., I. C. Roinestad, C. Boland, and L. K. Hennessy. 2005. Developmental validation of the quantifiler real-time PCR kits for the quantification of human nuclear DNA samples. *J Forensic Sci* 50 (4): 809–25.

Hiroshige, K., M. Soejima, T. Nishioka, S. Kamimura, and Y. Koda. 2009. Simple and sensitive method for identification of human DNA by allele-specific polymerase chain reaction of FOXP2. *J Forensic Sci* 54 (4): 857–61.

Holland, M. and T. Parson. 1999. Mitochondrial DNA sequence analysis-validation and use for forensic casework. *Forensic Sci Rev* 11: 21–50.

Holland, P. M., R. D. Abramson, R. Watson, and D. H. Gelfand. 1991. Detection of specific polymerase chain reaction product by utilizing the 5'-3' exonuclease activity of Thermus aquaticus DNA polymerase. *Proc Natl Acad Sci USA* 88 (16): 7276–80.

Horsman, K. M., J. A. Hickey, R. W. Cotton, J. P. Landers, and L. O. Maddox. 2006. Development of a human-specific real-time PCR assay for the simultaneous quantitation of total genomic and male DNA. *J Forensic Sci* 51 (4): 758–65.

Hudlow, W. R., M. D. Chong, K. L. Swango, M. D. Timken, and M. R. Buoncristiani. 2008. A quadruplex real-time qPCR assay for the simultaneous assessment of total human DNA, human male DNA, DNA degradation and the presence of PCR inhibitors in forensic samples: A diagnostic tool for STR typing. *Forensic Sci Int Genet* 2 (2): 108–25.

Jeffreys, A. J., V. Wilson, and S. L. Thein. 1985. Hypervariable "minisatellite" regions in human DNA. *Nature* 314 (6006): 67–73.

Johnson, S. C., C. B. Sherrill, D. J. Marshall, M. J. Moser, and J. R. Prudent. 2004. A third base pair for the polymerase chain reaction: Inserting isoC and isoG. *Nucleic Acids Res* 32 (6): 1937–41.

Kavlick, M. F., H. S. Lawrence, R. T. Merritt et al. 2011. Quantification of human mitochondrial DNA using synthesized DNA standards. *J Forensic Sci* 56 (6): 1457–63.

Kermekchiev, M. B., L. I. Kirilova, E. E. Vail, and W. M. Barnes. 2009. Mutants of Taq DNA polymerase resistant to PCR inhibitors allow DNA amplification from whole blood and crude soil samples. *Nucleic Acids Res* 37:(5): e40.

Kidd, K. K., A. J. Pakstis, W. C. Speed et al. 2006. Developing a SNP panel for forensic identification of individuals. *Forensic Sci Int* 164 (1): 20–32.

Kline, M. C., D. L. Duewer, J. W. Redman, and J. M. Butler. 2005. Results from the NIST 2004 DNA Quantitation Study. *J Forensic Sci* 50 (3): 570–8.

Kline, M. C., D. L. Duewer, J. C. Travis et al. 2009. Production and certification of NIST Standard Reference Material 2372 Human DNA Quantitation Standard. *Anal Bioanal Chem* 394 (4): 1183–92.

Krenke, B. E., N. Nassif, C. J. Sprecher, C. Knox, M. Schwandt, and D. R. Storts. 2008. Developmental validation of a real-time PCR assay for the simultaneous quantification of total human and male DNA. *Forensic Sci Int Genetics* 3 (1): 14–21.

Krenke, B. E., A. Tereba, S. J. Anderson et al. 2002. Validation of a 16-locus fluorescent multiplex system. *J Forensic Sci* 47 (4): 773–85.

Krenke, B. E., L. Viculis, M. L. Richard et al. 2005. Validation of a male-specific, 12-locus fluorescent short tandem repeat (STR) multiplex. *Forensic Sci Int* 148 (1): 1–14.

LaSalle, H. E., G. Duncan, and B. McCord. 2011. An analysis of single and multi-copy methods for DNA quantitation by real-time polymerase chain reaction. *Forensic Sci Int Genet* 5 (3): 185–93.

Mandrekar, M. N., A. M. Erickson, K. Kopp et al. 2001. Development of a human DNA quantitation system. *Croat Med J* 42 (3): 336–9.

Matsuda, H., Y. Seo, E. Kakizaki, S. Kozawa, E. Muraoka, and N. Yukawa. 2005. Identification of DNA of human origin based on amplification of human-specific mitochondrial cytochrome b region. *Forensic Sci Int* 152 (2-3): 109–14.

Moser, M. J. and J. R. Prudent. 2003. Enzymatic repair of an expanded genetic information system. *Nucleic Acids Res* 31 (17): 5048–53.

Mulero, J. J., C. W. Chang, L. M. Calandro et al. 2006. Development and validation of the AmpFlSTR Yfiler PCR amplification kit: A male specific, single amplification 17 Y-STR multiplex system. *J Forensic Sci* 51 (1): 64–75.

Mulero, J. J., C. W. Chang, R. E. Lagacé, et al. 2008. Development and validation of the AmpFlSTR MiniFiler PCR Amplification Kit: A MiniSTR multiplex for the analysis of degraded and/or PCR inhibited DNA. *J Forensic Sci* 53 (4): 838–52.

Nicklas, J. A. and E. Buel. 2003a. Development of an *Alu*-based, QSY 7-labeled primer PCR method for quantitation of human DNA in forensic samples. *J Forensic Sci* 48 (2): 282–91.

Nicklas, J. A. and E. Buel. 2003b. Development of an *Alu*-based, real-time PCR method for quantitation of human DNA in forensic samples. *J Forensic Sci* 48 (5): 936–44.

Nicklas, J. A. and E. Buel. 2003c. Quantification of DNA in forensic samples. *Anal Bioanal Chem* 376 (8): 1160–7.

Nicklas, J. A. and E. Buel. 2005. An *Alu*-based, MGB Eclipse real-time PCR method for quantitation of human DNA in forensic samples. *J Forensic Sci* 50 (5): 1081–90.

Nicklas, J. A. and E. Buel. 2006. Simultaneous determination of total human and male DNA using a duplex real-time PCR assay. *J Forensic Sci* 51 (5): 1005–15.

Nicklas, J. A., T. Noreault-Conti, and E. Buel. 2012. Development of a real-time method to detect DNA degradation in forensic samples. *J Forensic Sci* 57 (2): 466–71.

Niederstätter, H., S. Köchl, P. Grubwieser, M. Pavlic, M. Steinlechner, and W. Parson. 2007. A modular real-time PCR concept for determining the quantity and quality of human nuclear and mitochondrial DNA. *Forensic Sci Int Genet* 1 (1): 29–34.

Nielsen, K., H. S. Mogensen, J. Hedman, H. Niederstätter, W. Parson, and N. Morling. 2008. Comparison of five DNA quantification methods. *Forensic Sci Int Genet* 2 (3): 226–30.

Opel, K. L., D. Chung, and B. R. McCord. 2010. A study of PCR inhibition mechanisms using real-time PCR. *J Forensic Sci* 55 (1): 25–33.

Phillips, C., R. Fang, D. Ballard et al. 2007. Evaluation of the Genplex SNP typing system and a 49plex forensic marker panel. *Forensic Sci Int Genet* 1 (2): 180–5.

Promega Corporation. 2007. Plexor® HY System for the Applied Biosystems 7500 and 7500 FAST real-time PCR systems. Available from: http://www.promega.com/resources/protocols/technical-manuals/0/plexor-hy-system-protocols/ (accessed September 28, 2012).

Qiagen. 2011. Investigator® Quantiplex Handbook. Available from: http://www.qiagen.com/literature/default.aspx?Term = investigator+quantiplex&Language = EN&LiteratureType = 1+2+3+4+53&ProductCategory = 0 (accessed September 28, 2012).

Qiagen. 2012. Investigator® Quantiplex HYres Handbook. Available from: http://www.qiagen.com/literature/default.aspx?Term = investigator+quantiplex&Language = EN&LiteratureType = 1+2+3+4+53&ProductCategory = 0 (accessed September 28, 2012).

Sambrook, J. and D. W. Russell. 2001. *Molecular Cloning: A Laboratory Manual*, 3rd ed. Cold Spring Harbor Laboratory Press: Cold Spring Harbor, NY.

Scherer, M. and F. Di Pasquale. 2012. How to improve analysis of sexual assault samples using novel tools: Differential wash and human:male quantification. Presented at the *Green Mountain DNA Conference*, Burlington, VT.

Schmittgen, T. D. and K. J. Livak. 2008. Analyzing real-time PCR data by the comparative C_T method. *Nat Protoc* 3 (6): 1101–8.

Seo, S. B., H. Y. Lee, A. H. Zhang, H. Y. Kim, D. H. Shin, and S. D. Lee. 2011. Effects of humic acid on DNA quantification with Quantifiler® Human DNA Quantification kit and short tandem repeat amplification efficiency. *Int J Legal Med* 10.1007/s00414-011-0616-z.

Sherrill, C. B., D. J. Marshall, M. J. Moser et al. 2004. Nucleic acid analysis using an expanded genetic alphabet to quench fluorescence. *J Am Chem Soc* 126 (14): 4550–6.

Shewale, J. G., H. Nasir, E. Schneida, A. M. Gross, B. Budowle, and S. K. Sinha. 2004. Y-chromosome STR system, Y-PLEX 12, for forensic casework: Development and validation. *J Forensic Sci* 49 (6): 1278–90.

Shewale, J. G., E. Schneida, J. Wilson, J. A. Walker, M. A. Batzer, and S. K. Sinha. 2007. Human genomic DNA quantitation system, H-Quant: Development and validation for use in forensic casework. *J Forensic Sci* 52 (2): 364–70.

Sifis, M. E., K. Both, and L. A. Burgoyne. 2002. A more sensitive method for the quantitation of genomic DNA by *Alu* amplification. *J Forensic Sci* 47 (3): 589–92.

Sinha, S. K., A. H. Montgomery, G. M. Pineda, and R. Thompson. 2013. Development of a highly sensitive quantification system for assessing DNA quality in forensic samples. *Presentation—65th American Academy of Forensic Sciences Annual Meeting*. Washington, DC.

Swango, K. L., W. R. Hudlow, M. D. Timken, and M. R. Buoncristiani. 2007. Developmental validation of a multiplex qPCR assay for assessing the quantity and quality of nuclear DNA in forensic samples. *Forensic Sci Int* 170 (1): 35–45.

Swango, K. L., M. D. Timken, M. D. Chong, and M. R. Buoncristiani. 2006. A quantitative PCR assay for the assessment of DNA degradation in forensic samples. *Forensic Sci Int* 158 (1): 14–26.

Tak, Y. K., W. Y. Kim, M. J. Kim et al. 2012. Highly sensitive polymerase chain reaction-free quantum dot-based quantification of forensic genomic DNA. *Anal Chim Acta* 721: 85–91.

Timken, M. D., K. L. Swango, C. Orrego, and M. R. Buoncristiani. 2005. A duplex real-time qPCR assay for the quantification of human nuclear and mitochondrial DNA in forensic samples: Implications for quantifying DNA in degraded samples. *J Forensic Sci* 50 (5): 1044–60.

Tyagi, S., D. P. Bratu, and F. R. Kramer. 1998. Multicolor molecular beacons for allele discrimination. *Nat Biotechnol* 16 (1): 49–53.

Vallone, P., M. C. Kline, D. L. Duewer et al. 2008. Development and usage of a NIST standard reference material for real-time PCR quantitation of human DNA. *Forensic Sci Int Genet* Suppl Ser 1: 80–2.

Von Wurmb-Schwark, N., R. Higuchi, A. P. Fenech, C. Elfstroem, C. Meissner, M. Oehmichen, and G. A. Cortopassi. 2002. Quantification of human mitochondrial DNA in a real time PCR. *Forensic Sci Int* 126 (1): 34–9.

Von Wurmb-Schwark, N., T. Schwark, M. Harbeck, and M. Oehmichen. 2004. A simple Duplex-PCR to evaluate the DNA quality of anthropological and forensic samples prior short tandem repeat typing. *Legal Med (Tokyo, Japan)* 6 (2): 80–8.

Walker, J. A., D. J. Hedges, B. P. Perodeau et al. 2005. Multiplex polymerase chain reaction for simultaneous quantitation of human nuclear, mitochondrial, and male Y-chromosome DNA: Application in human identification. *Anal Biochem* 337 (1): 89–97.

Walker, J. A., G. E. Kilroy, J. Xing, J. Shewale, S. K. Sinha, and M. A. Batzer. 2003. Human DNA quantitation using *Alu* element-based polymerase chain reaction. *Anal Biochem* 315 (1): 122–8.

Walsh, P. S., J. Varlaro, and R. Reynolds. 1992. A rapid chemiluminescent method for quantitation of human DNA. *Nucleic Acids Res* 20 (19): 5061–5.

Whitcombe, D., J. Theaker, S. P. Guy, T. Brown, and S. Little. 1999. Detection of PCR products using self-probing amplicons and fluorescence. *Nat Biotechnol* 17 (8): 804–7.

Wilson, J., V. Fuller, G. Benson et al. 2010. Molecular assay for screening and quantifying DNA in biological evidence: The modified Q-TAT assay. *J Forensic Sci* 55 (4): 1050–7.

STRs—Proven Genotyping Markers

III

Principles, Practice, and Evolution of Capillary Electrophoresis as a Tool for Forensic DNA Analysis

7

JAIPRAKASH G. SHEWALE
LIWEI QI
LISA M. CALANDRO

Contents

7.1 Introduction 132
7.2 Forensic DNA Analysis 133
  7.2.1 Fragment Analysis and STR Typing 133
  7.2.2 DNA Sequencing 136
  7.2.3 SNP Typing 137
7.3 Requirements of CE for STR Analysis 137
7.4 Evolution of CE Systems and Utility in Forensic DNA Analysis 138
  7.4.1 Development of Polymer and Capillaries 140
  7.4.2 Injection and Run Parameters 142
  7.4.3 Detection 145
7.5 CE Instruments for Forensic DNA Analysis 147
  7.5.1 Sizing Precision 149
  7.5.2 Key Features of 3500 Genetic Analyzers 149
  7.5.3 Normalization and Consistency of 3500 Genetic Analyzers 152
7.6 Electrophoretic Microdevices and Integrated Systems 155
7.7 Other Applications of CE 155
7.8 Conclusions 155
Acknowledgments 156
References 156

Abstract: Capillary electrophoresis (CE) is a versatile and widely used analysis platform with application in diverse areas such as analytical chemistry, chiral separations, clinical, forensics, molecular biology, natural products, organic chemistry, and the pharmaceutical industry. Forensic applications of CE include fragment analysis, DNA sequencing, single nucleotide polymorphism (SNP) typing, and analysis of gunshot residues, explosive residues, and drugs. Fragment analysis is a widely used method for short tandem repeat (STR) profiling for human identification (HID) due to the single-base resolution capability of CE. This approach circumvents the tedious and expensive approach of DNA sequencing for STR typing. The high sizing precision, ability to detect fluorescence emitted from multiple dyes, automated electrophoretic runs, and data collection software are key factors in the worldwide adoption of CE as the preferred platform for forensic DNA analysis. The most common CE systems used in forensic DNA analysis include the ABI PRISM® 310, 3100, 3100 Avant, 3130, 3130*xl*, 3500, and 3500xL Genetic Analyzers (GAs). The 3500 series GAs are developed with features useful for forensic scientists, including a normalization feature for analysis of the data designed to reduce the variation in peak height from instrument to instrument and injection to injection. Other hardware and software features include improved temperature control, radio frequency identification (RFID) tags for monitoring instrument consumables, HID-focused software features, and security and maintenance.

7.1 Introduction

Capillary electrophoresis (CE) instrumentation is one of the most notable developments in the forensic DNA typing workflow, perhaps second only to the invention of the polymerase chain reaction (PCR). The occurrence of regions containing repeat sequences within the DNA molecule and their polymorphic nature was discovered in the early 1980s. Within a span of a few years, Jeffreys et al. demonstrated the use of regions containing a variable number of tandem repeats (VNTR) for human identification (Jeffreys et al. 1985a and 1985b). Subsequently, in the 1990s, highly polymorphic short tandem repeat (STR) markers replaced VNTRs. Initially amplified STR fragments were separated using slab-gel electrophoresis and detected by silver staining. Inventions in fluorescent dyes, fluorescence detection, and multiplex PCR paved the way to the currently used STR typing protocols (Fregeau and Fourney 1993; Micka et al. 1996; Ricci et al. 2000; Sullivan et al. 1992; Wang et al. 1995). A CCT triplex system was successfully developed in the early 1990s for simultaneous amplification of CSF1PO, TPOX, and THO1 loci (Micka et al. 1996). Multiplexing allowed simultaneous separation and detection of multiple amplified fragments in one lane and now in one capillary. Polyacrylamide gel-based sequencers like the ABI PRISM 373 DNA Sequencer and ABI PRISM 377 DNA Sequencer were routinely used for STR analysis for more than a decade, even after development of a single capillary instrument, as the 96 lanes conferred high-throughput capability (Frazier et al. 1996; Kimpton et al. 1996). CE instruments with capillary arrays have since replaced these gel-based sequencers in almost all forensic laboratories. The term CE in this chapter refers to both capillary electrophoresis and capillary-array electrophoresis. CE results have been admissible in courts of law since 1996 (Kuffner et al. 1996).

The uniqueness of CE is its versatility; inorganic ions, organic molecules, and macromolecules can be separated on the same instrument—and in most cases the same capillary—while changing only the composition of the running buffer and separation medium. Thus, CE is the most widely used analytical method replacing gel electrophoresis, high-performance liquid chromatography (HPLC), gas chromatography, and other separation methods (Issaq 2000; Thormann et al. 1999). CE possesses extremely high resolving capability due to its plug flow and minimal diffusion. CE offers numerous advantages over slab-gel electrophoresis in forensic DNA analysis applications (Buel et al. 1998; Butler et al. 1998, 2004; Dovichi 1997; Isenberg et al. 1996; Mansfield et al. 1998; McCord et al. 1995):

- Injection, separation, and detection of the amplified DNA fragments are fully automated.
- High voltage and efficient heat dissipation in the capillary achieves rapid separation, leading to faster run times.

- Software products for automated collection and handling of the data are available.
- Real-time data viewing is possible.
- Quantitative information can be derived.
- Precision is increased.
- High-resolution separation is provided.
- Results are highly reproducible.
- Chances of cross-contamination/carryover are minimal since the capillary gets rinsed after each run and samples are contained within the capillary.
- Only a small portion of a sample is consumed, so it can be retested/reinjected if needed.
- Generation of hazardous waste is drastically reduced.
- Labor-intensive manual procedures of pouring gels and sample loading are eliminated.

It is important to note that quantitative differences observed in the CE data may not exactly correlate to the quantitative differences in the original samples. Preferential amplification may affect the proportions of amplified DNA detected on CE, specifically with regard to the minor contributor proportion. The CE results—for example, peak height and peak balance—are utilized for optimization of multiplex PCR reactions (Butler et al. 2001).

7.2 Forensic DNA Analysis

Commonly used methodologies in forensic DNA analysis that employ CE are STR analysis (Butler et al. 2004; Collins et al. 2004; Ensenberger et al. 2010; Mulero et al. 2006, 2008, 2009), mitochondrial DNA (mtDNA) sequencing (Chong et al. 2005; Gabriel et al. 2001; Holland and Parson 1999; Ruiz-Martinez et al. 1993), and typing of single nucleotide polymorphisms (SNPs) (Borsting et al. 2008; Fang et al. 2009; Sanchez et al. 2006; Thomas et al. 2008). In addition, other methods like mRNA marker analysis for the identification of body fluids (Juusola and Ballantyne 2005), DNA methylation-based forensic identification (Frumkin et al. 2011), and chimerism analysis (Nagy et al. 2006) are being developed. Of these, STR profiling is the most widely used method. The discussion in this chapter, therefore, is centered on STR profiling although the general principles of CE are applicable to other DNA analysis methodologies as well. Readers may refer to previous review articles for more information on the principles and applications of CE (Butler et al. 2004; Carey and Mitnik 2002; Carrilho 2000; Dovichi 1997; Heller 2001; Issaq 2000; Swinney and Bornhop 2000; Thormann et al. 1999).

7.2.1 Fragment Analysis and STR Typing

STR profiling is the most commonly used method for human identification (HID) applications like paternity testing, profiling for generating databases, forensic casework sample analysis, identification of missing persons, identification of victims of mass disasters, and so forth. The approach is to amplify the DNA from biological sources for multiple human-specific highly polymorphic STR loci in a single PCR using fluorescently labeled primers. The amplified fragments are then separated by electrophoresis, the raw data is analyzed using software that determines the size of each amplified fragment, and genotypes by comparison to alleles in an allelic ladder that are run on the same plate. The primers for amplification of STR loci in the multiplex systems are designed so that the alleles of the loci amplified with the primers labeled with the same dye do not overlap each other. Using primers labeled with different dyes, it is possible to analyze loci generating amplified alleles with similar fragment sizes. This concept has been used to develop multiplexes containing 20–24 loci (Butler et al. 2002; Wang et al. 2012). The capability of the CE instrument in the resolution of spectral overlap of the dyes, physicochemical properties of the dyes, and sensitivity and robustness of the multiplex PCR determine the number of loci that can be analyzed simultaneously.

Typically, one would determine the exact number of repeat units at an STR locus by sequencing the amplified fragment. However, this exercise is laborious and expensive, and needs specialized expertise. Alternatively, the number of repeat units can be computed from the length of the amplified fragment following strict conditions for amplification and primer sequences. A key factor in accurate STR typing by fragment analysis is to amplify STR loci in samples, controls, and allelic ladders using identical primers residing outside the STR region with the same PCR conditions so that the variation in the length of the amplified fragment is directly related to the length of the STR region and hence genotype. During STR typing by CE, the size of the amplified product is determined from the calibration curve generated using size-standard fragments run simultaneously with the sample in the same injection (i.e., well). The calibration curve is created by plotting the size of the fragments present in the size standard versus the migration data point. The allele designation is then assigned by comparing the size of the amplified allele in the sample with the size of the same allele in the allelic ladder, which contains common alleles from major human population groups that have previously been sequenced to confirm the number of repeat units (Lazaruk et al. 1998; Smith 1995). Fragment analysis is a preferred method over sequencing for obtaining the STR genotype of a sample due to its simplicity, lower cost, ease of interpretation, and high throughput. Although most of the STR alleles differ in size by the length of the repeat unit (e.g., 4 bases when the repeat unit comprises 4 bases), some rare alleles do exist that differ in size by an incomplete repeat unit; for example, alleles 9.3 and 10 at the locus TH01 (Butler et al. 2003; Puers et al. 1993; Urquhart et al. 1994). For accurate typing of alleles that vary in length by 1 base, it is desired that the size of an allele be no more than 0.5 base pairs apart from the measured allele in the allelic ladder. Determining the size of amplified products and ascertaining the allele are critical steps in genotyping.

Accurate size determination requires co-injection of the size standards with each sample. This is because, unlike slab-gel electrophoresis, each injection in a capillary is an independent electrophoresis run. The nature of the capillary walls, internal diameter, temperature, current, sieving matrix, and sample composition contribute to migration variation among the capillaries. The size of the DNA fragment is established relative to the size standard using one of several methods for creating a size-calling curve such as Global Southern or Local Southern. The Global Southern method generates a best-fit sizing curve through all size standards selected and then calculates the size of the unknown fragments. The Local Southern method, on the other hand, selects four size-standard points that are nearest in size to the unknown fragment to derive the best-fit curve (Elder and Southern 1983). It is postulated that the Local Southern method provides a better estimate of the size since there is not a linear relationship between size and mobility over a wide range of sizes. Hartzell et al. (2003) studied the effect of varying the temperature between 35°C and 70°C on the migration and size determination of DNA fragments. The results indicated that the

Global Southern method was preferable to Local Southern in situations where temperature fluctuation can occur.

As mentioned earlier, the allelic ladder needs to be generated using the same primers that are used in the amplification of samples. Commercial kits provide the reagents for multiplex amplification together with the amplified allelic ladder. Applied Biosystems, part of Life Technologies (Foster City, California), Promega Corporation (Madison, Wisconsin), and Qiagen (Hilden, Germany) are major providers of multiplex STR kits worldwide. The primers used for amplification of a given STR locus may vary from one manufacturer to the other and also between two STR amplification kits offered by the same manufacturer. Thus, the size of an allele may vary from one STR amplification kit to the other. It is, therefore, important to use the PCR reagents and allelic ladder from the same kit for accurate genotyping. Multiple allelic ladders are typically run on a 96-well plate since minor variations in experimental parameters like electrical current, temperature, buffer, and separating medium can alter the relative mobility and contribute to differences across the plate in the sizing of alleles in the allelic ladder relative to amplified alleles from the sample and may ultimately impact allele typing. For automated computerized analysis, one can fix the bin for the size of known and well-characterized alleles in the allelic ladder. Another approach is to assign a variable bin based on the size determined on the electrophoresis run on which both the allelic ladder and sample are run. The fixed-bin method provides consistent results only when electrophoretic conditions are consistent from one injection to the other, which is difficult to achieve on CE. The variable-bin method minimizes the impact of minor fluctuations in electrophoresis conditions on genotyping (Laclair et al. 2004). The variable-bin method, therefore, is the method of choice for comparing different instrument platforms and processing the data from CE instruments wherein each injection on a capillary and each capillary in a capillary array is an independent electrophoresis run exhibiting fluctuations in the electrophoresis conditions.

The ability to resolve DNA fragments that differ in length by 1 base pair is a critical factor in the selection of a separation system for fragment analysis. It is also important to reliably identify actual DNA peaks as opposed to PCR and electrophoresis artifacts. PCR artifacts include stutter, incomplete adenylation (-A peak), unequal amplification of heterozygous loci (peak imbalance), higher amounts of amplified product (off-scale data), stochastic effects, degraded dye products, and nonspecific amplification. Electrophoresis artifacts include electrical spike peaks, poor resolution, spectral bleed-through, air bubbles, and so forth. The ultimate goal is to differentiate between a true allele peak and a peak resulting from PCR and/or electrophoresis artifacts that can be mistaken as a true allele. A threshold for detection of fluorescence for each electrophoresis instrument is determined during the validation studies conducted by a laboratory.

The size of a DNA fragment obtained from different instrument platforms—for example, 310 GA and 377 DNA Sequencers—may vary, even using different in-lane size standards on the same platform (Lazaruk et al. 1998; Mansfield et al. 1996). For example, a D3S1358 allele 12 was sized as 114.3 base pairs (bp) on a 377 DNA Sequencer using a 36-cm gel and 111.0 bp on 310 GA using POP-4® (Lazaruk et al. 1998). Nevertheless, the alleles were correctly genotyped since an allelic ladder was run simultaneously during the respective electrophoresis runs. Thus, sizing precision across the array is more important than the accuracy of sizing with respect to genotyping. Sizing precision is defined as the ability to obtain reproducible DNA fragment sizes from injection to injection on a CE instrument or from lane to lane on a slab gel. Accuracy, on the other hand, is defined by how close the calculated size of a DNA fragment is to the actual length in nucleotides as determined by sequencing. The variation in sizing is minimized, to a great extent, by using an allelic ladder on the same electrophoresis run on a slab gel or in multiple wells within a set of samples being run on CE and by using the variable-bin method for allele designation. The set of samples are defined as one tray of samples on a CE or one gel run, both containing any number of samples. Sizing precision is highly important when comparing different samples in determining whether the samples are consistent with having originated from the same source—for example, evidence and reference samples in forensic casework. Alleles at an STR locus that are of the same size can occasionally differ in their sequence in the repeat region or flanking region resulting from SNPs (Barber et al. 1995; Walsh et al. 1996). Lazaruk et al. (1998) injected five times each allele 14 and 14′ that exhibited sequence variability on a 310 GA and observed mean sizes of 167.04 and 167.20 bases, respectively. This variability did not affect the allele designation of 14 since the variation in the size was within the window of 0.5 base used for allele designation. Sizing precision of <0.15 bp standard deviation means 99.7% of all alleles that are of the same length in nucleotides should fall within ± 0.45 nucleotide of the allelic ladder's allele size (Butler et al. 2004; Smith 1995). Further, the occurrence of sample alleles falling outside the allelic ladder bin due to measurement error will be rare, and this error should not be reproducible when the sample is rerun. Thus, reinjection of a sample is a common practice to confirm microvariant alleles or to eliminate migration anomalies whenever observed (Mansfield et al. 1998).

7.2.2 DNA Sequencing

DNA sequencing is widely used to identify genes and mutations to confirm successful site-directed mutagenesis. Smith et al. (1986) reported the first automated sequencer wherein four sequencing reactions were run using labeled primers with different fluorescent dyes; the reaction products were pooled and separated in a single lane of a sequencing gel. The pace of DNA sequencing is growing exponentially with the developments in next-generation sequencing. Sequencing of variable regions in mtDNA is being performed in forensic laboratories. mtDNA is a circular genome of 16,569 bp comprising coding and noncoding regions. mtDNA analysis gained importance in HID because of its maternal inheritance, noncoding, or D-loop, sequence variability between humans, and its high copy number (>5,000 copies per cell). mtDNA, therefore, is an excellent tool for analysis of degraded samples or those containing minute quantities of genomic DNA. Hair, bone, and tooth are preferred samples for mtDNA analysis. In general, hypervariable regions 1 and 2 (HV1 and HV2) of the D-loop containing the highest degree of sequence polymorphism are sequenced (Chong et al. 2005; Gabriel et al. 2001; Grzybowski 2000; Holland and Parson 1999; Levin et al. 1999). The mtDNA genome first sequenced by Anderson et al. (1981) serves as a reference sequence. The occurrence of heteroplasmy, or the presence of more than one mtDNA type in an individual, is disadvantageous since it can complicate results interpretation (Bender et al. 2000) or advantageous as it can increase the discrimination power (Grzybowski 2000). In addition to detecting variability in HV1 and HV2 regions, sequencing methods may be used for analysis of SNPs in the coding region (Alvarez-Iglesias et al. 2007). This approach increases the discrimination power of mtDNA typing and facilitates haplogroup assignment of mtDNA profiles in both human population studies and medical research.

A typical workflow for mtDNA sequencing comprises isolation of DNA, amplification of fragments for HV1 and HV2 regions in fragments of 400 bp or less based on the extent of degradation of the sample, cycle sequencing of the amplified products, removal of unincorporated fluorescent terminators from completed cycle-sequencing reactions, and separation of products of cycle sequencing on CE using POP-6 polymer (Chong et al. 2005; Holland and Parson 1999; Stewart et al. 2003). Stewart et al. (2003) compared 376 sequences generated using the gel-based 377 DNA Sequencer, single-capillary system 310 GA, and 16-capillary array system 3100 GA; the same sequencing results were obtained on each platform. The sample types included blood, saliva, bone, and hair.

7.2.3 SNP Typing

SNPs are base substitutions, insertions, or deletions that occur at single positions in the genome. It is interesting to know that 85% of human variation is derived from SNPs (Budowle and van Dall 2008). SNPs are relatively stable markers that can be used for lineage-based case studies like identification of missing persons, inheritance cases, and cases where no direct reference sample is available. SNP typing is advantageous over STR profiling in certain types of samples such as highly degraded samples and when more genetic information is desired. A major disadvantage of SNPs is the biallelic nature, which means only three possible genotypes (e.g., homozygous A, heterozygous AB, and homozygous B) are observed among different population groups at a given SNP locus, providing a much lower power of discrimination per locus than highly polymorphic STR loci. It is estimated that 50–100 SNPs are required to achieve a power of discrimination similar to 13 STR loci. The SNPs for forensic analysis are grouped into four categories: (1) identity testing SNPs, (2) lineage informative SNPs, (3) ancestry informative SNPs, and (4) phenotype informative SNPs (Budowle and van Dall 2008).

The SNP*for*ID consortium selected 52 biallelic SNPs with a high level of polymorphism in the major population groups (Sanchez et al. 2006, http://www.snpforid.org). The random match probability using these 52 SNPs was 5.0×10^{-19}. This set of markers is useful for forensic casework, paternity cases, and immigration cases (Borsting et al. 2008; Sanchez et al. 2006; Thomas et al. 2008). Kidd and his group identified a panel of 92 SNPs for human identification after extensive population studies (Fang et al. 2009; Kidd et al. 2006; Pakstis et al. 2007). These SNPs were selected such that they have average heterozygosities of >0.4 and Fst values <0.06 on the 44 human population groups representing the major regions of the world. The GenPlex™ system used for forensic applications involves amplification of the SNP loci followed by an oligo ligation assay using allele and locus-specific oligonucleotides, immobilization of ligation products on a microtiter plate, hybridization of reporter oligonucleotides (ZipChute® probes), elution of hybridized ZipChute probes, and detection on the CE (Fang et al. 2009; Thomas et al. 2008).

7.3 Requirements of CE for STR Analysis

Precision between CE runs is critical since the samples and allelic ladders are processed in a sequential manner in a single-capillary instrument or in separate capillaries in a capillary-array instrument. In both cases, each electrophoresis run is independent since the electrophoresis parameters like voltage, current, and temperature can vary marginally. In contrast, the samples and allelic ladders loaded in different lanes on a gel are subjected to electrophoretic conditions that are similar for all lanes on a gel. Over the years, many forensic scientists have put forth desired characteristics for a separation and detection system that can be employed in STR profiling (Butler et al. 2004; Carey and Mitnik 2002; Isenberg et al. 1998):

- Accurate sizing of fragments over the fractionation range of the STR multiplex amplification systems (in general 75–500 bp, the fractionation range of commercially available STR kits).
- Single-base resolution over the fractionation range for identification and typing of microvariant alleles.
- High run-to-run precision to enable comparison of mobility of alleles in allelic ladders to the alleles in the samples.
- Effective color separation of the fluorescent dyes used for labeling the primers to avoid bleed-through of different colors and minimize the occurrence of pull-up peaks.
- Good temperature control for a higher degree of run-to-run precision.
- Better sensitivity and automation.
- Higher throughput and faster methods without compromising data quality.
- Integration of data collection software and analysis software for detection of CE, PCR, and dye artifacts.
- User-friendly data analysis software.

7.4 Evolution of CE Systems and Utility in Forensic DNA Analysis

Tiselius (1930) introduced electrophoresis as an analytical technique by demonstrating the separation of blood plasma proteins (albumin from α-, β-, and γ-globulins), for which he received the Nobel Prize in 1948. Since this discovery, several advancements have been made in the field of paper and gel electrophoresis. A milestone advancement in gel electrophoresis was separation of proteins on polyacrylamide gels under denaturing conditions using urea or sodium dodecyl sulfate (Swank and Munkers 1971). The path toward CE originated when Hjerten (1967) demonstrated that electrophoretic separations can be carried out in a 300-μM glass tube with separated compounds detected by UV absorption. The term "free zone electrophoresis" was used for this technique. However, more earnest development of CE systems began after the studies on electrophoresis in small-diameter tubes by Jorgenson and Lukacs (1981) and Mikkers et al. (1979). The simple and efficient instrument setup described by Jorgenson and Lukacs (1981) is the basis of most commercial CE platforms. The setup was composed of a narrow (internal diameter <100 μM) fused-silica capillary, high voltage (30 kV), and on-column UV detection for separation of ionic species. Terabe et al. (1984) introduced micellar electrokinetic chromatography for the separation of neutral compounds by adding micelle and SDS to the buffer solution. The first commercial CE instrument was marketed in 1988 (Issaq 2000).

CE has become a powerful method for DNA analysis; it enabled completion of the human genome project faster than expected and is routinely used for genotyping (Marshall and Pennisi 1998). Major breakthroughs in the area of CE were the development of replaceable polymer as a separation medium and availability of laser-induced fluorescence (LIF) detection. Some important studies reported in the literature in the development of CE for forensic DNA analysis are summarized here. A major advantage of capillary-array systems over single-capillary systems is increased throughput. The use of capillary-array electrophoresis in DNA analysis to increase the speed and throughput was first demonstrated by Mathies's group (Huang et al. 1992a,b; Mathies and Huang 1992). Later, DNA fragments generated from a fourplex STR system (VWFA, TH01, TPO, and CSF) were separated on arrays of hollow fused-silica capillaries filled with denaturing and replaceable hydroxyethylcellulose (HEC) sieving matrices; plasmid M13mp18 was used as an internal standard and the separated fragments were detected by a two-color confocal fluorescence scanner (Wang et al. 1996). Several other investigators described the use of 48-capillary-array electrophoresis with a confocal fluorescence scanner for STR analysis (Mansfield et al. 1998; Vainer et al. 1997). Earlier versions of CE that hosted UV detection or LIF detection of a single color required bracketing of the allelic ladder or amplified alleles with dual internal size standard for accurate typing of STR alleles (Butler et al. 1994, 1995; McCord et al. 1993a,b). Infrared (IR) fluorescence detectors were developed in the year 2000 (Ricci et al. 2000). Development of GAs with multicolor fluorescence detection expanded the ability to genotype multiple loci simultaneously (Lazaruk et al. 1998). Butler et al. (1994) demonstrated high resolution of a HUMTHO1 (human THO1) allelic ladder on a CE instrument using a 1% solution of HEC as a separation matrix, YOPRO-1 as an intercalating dye, and LIF detection. Isenberg et al. (1996) developed a protocol for DNA typing of a twoplex PCR (D1S80 and Amelogenin) on a Beckman P/ACE 2050 CE instrument with a Laser Module 488 argon ion laser (Beckman Instruments, Fullerton, California). The results were compared with slab-gel electrophoresis; CE systems proved useful in detection of DNA mixtures up to 16:1 due to the increased sensitivity and ability to quantitate the extent of amplification. Mansfield et al. (1996, 1997) demonstrated the use of a capillary-array electrophoresis prototype developed by Molecular Dynamics for STR analysis. Gao et al. (1999) described genotyping of a fourplex STR system (vWF, TH01, TPOX, and CSF1PO loci) on a 96-capillary-array prototype employing polyvinylpyrrolidone (PVP) as a separation matrix. The capillary array was used for over 27 runs. The crosstalk between adjacent capillaries was <0.89%.

Lazaruk et al. (1998) demonstrated the utility of the ABI PRISM 310 GA for STR profiling of nine loci using the Profiler Plus® Kit. The 310 GA was readily accepted worldwide by the forensic community. The key features of the 310 GA include analysis of up to 96 samples sequentially without interference by the user, detection of amplified products as they move past a laser detection window near the end of the capillary, LIF detection on a charge-coupled device (CCD) camera that stimulates all wavelengths from 525 to 680 nm, reload of fresh polymer into the capillary between injections, and capillary life of ≥100 injections. Isenberg et al. (1998) investigated the effects of different temperatures, injection times, running voltages, and sieving matrix compositions on the separation of STR fragments on 310 GA. The multiplex systems used were the AmpFℓSTR Blue triplex (D3S1358, vWA, and FGA loci) and CTTv quadruplex (CSF1PO, TPOX, TH01, and vWA). The conclusions for the study were: 3% HEC (M_r 40 K Daltons) with 7.1 M urea provided the highest resolution of 1–2 bases for alleles up to 300 bases in length, sizing precision was better at 60°C compared to 30°C, higher temperature provided longer capillary lifetime (>100 runs at 60°C compared to >40 runs at 30°C), 15kV voltage (field strength of 375 V/cm) was optimal for separation of the allelic ladders in 20 min, and precision in sizing of 0.19 bp was achieved. The investigators observed a concordance of genotypes when 100 unrelated individuals were profiled using slab-gel electrophoresis and 310 GA. However, using this protocol the resolution values were greater than 2 bases for 300 base fragments, which is not optimal for forensic applications. It is important to note that the POP-4® polymer was not used in this study as a separation matrix. An interlaboratory comparison study using slab gel and 310 GA for 80 samples exhibited concordance of the allele calls using allelic ladders and internal size standard, an important validation study for the then-new CE platform (Buel et al. 1998). Gill et al. (2001) used the 96-capillary-array electrophoresis system ABI PRISM® 3700 GA for analysis of 10 STR loci and Amelogenin using the AmpFℓSTR SGM Plus kit. POP-6 polymer was used as a separation matrix. The mean size of each allele in the allelic ladder was determined by analysis of at least five different allelic ladders across the array. DNA profiles of 1,112 samples were generated using the 3700 GA and 377 DNA Sequencer. Complete concordance of the allele designation was observed. Moretti et al. (2001) conducted extensive validation studies comparing the 310 GA and 377 DNA Sequencer with the FMBIO II Fluorescent Imaging device and FluorImager systems for typing the Combined DNA Index System (CODIS) 13 STR loci using a variety of sample types. The sample types included pristine DNA, blood and semen deposited on carpet, denim, leather, nylon, wallboard, and wood extracted after 1 day and 20 weeks, environmentally insulted samples, nonprobative casework samples, and mixture samples (mixtures of blood/saliva, blood/semen, and semen/saliva). The STR kits investigated were Profiler Plus, COfiler®, and GenePrint PowerPlex, and subsets of these kits. The investigators concluded that the examined electrophoresis-based DNA typing methods are sufficiently robust for DNA analysis in a forensic laboratory for the vast majority of human biological sample types. Similar studies, though less extensive, were performed by Schubbert (2001) using the 377 DNA Sequencer, 3700 GA, and 3100 GA following the European DNA Profiling Group (EDNAP) requirements.

7.4.1 Development of Polymer and Capillaries

The mobility of molecules during electrophoresis in a free solution is primarily determined by net charge. However, the macromolecules are separated based on the net charge and size (rather, stoke radius) when electrophoresis is performed on a gel; size plays a role since the molecules need to migrate through the gel (or medium). Unlike a mixture of proteins (or amphoteric molecules), DNA molecules have constant charge/mass ratio and their electrophoretic mobility in free solution is equal to or independent of chain length. A sieving matrix, therefore, is introduced into the capillary to achieve size-based separation. Understanding the relationship between the polymer type, size, viscosity, and concentration and peak resolution are key aspects to optimizing DNA separation on CE. In the early 1990s, polyacrylamide gel-filled capillaries were used for high-resolution separation (Heiger et al. 1990; Williams et al. 1994). However, such capillaries were not robust in nature and provided limited life in terms of reuse. During the same period, the use of open capillaries filled with soluble sieving polymer that provides hundreds of repeated uses were developed and used in DNA separation (Butler 1998; McCord et al. 1993a,b). Development of entangled polymer buffers is a major landmark contribution for the automated and routine analysis of DNA fragments by CE. Entangled polymers, though characterized by intermolecular interactions and effective separation media for electrophoresis, are not gels since they can be held only in a container-like capillary. Polyacrylamide or agarose gels, on the other hand, bear a physical state. Use of polymer solutions in electrophoresis was demonstrated in 1977 by Bode (1977a,b). The concept was further explored with advancements in the development of CE systems. Linear polyacrylamide (LPA), developed in 1993, was the most widely used polymer for DNA sequencing applications (Ruiz-Martinez et al. 1993). However, this polymer had high viscosity and capillaries had to be precoated to suppress the electroosmotic flow (EOF). PVP, which is effective in reducing the EOF, was explored as a sieving matrix for DNA analysis on CE (Gao and Yeung 1998). PVP, being hydrophobic, exhibited excellent self-coating effects in bare capillaries, thereby enhancing the performance in separation of DNA (Madabhushi et al. 1996). However, PVP was not accepted for routine use due to extremely large mobility shifts of the dye-labeled fragments. Some investigators demonstrated the use of HEC as a separating medium for CE (Barron et al. 1994; Butler et al. 1994; Isenberg et al. 1998). The physicochemical properties of the polymer used for CE must meet separation/resolution expectations: provide noncovalent coating to the capillary surface to reduce the EOF, contain manageable viscosity for holding in the capillary as well as for easy replacement after every electrophoresis run, allow separation of DNA fragments differing in size by a single base, result in sizing precision between alleles of the same length of <0.15 nucleotide standard deviation, provide a denaturing environment, have analysis time less than 30 min per sample/run, and allow longer capillary life. Poly(N,N-dimethylacrylamide) (PDMA) polymers and the performance-optimized polymers POP-4 and POP-6 were developed to meet these requirements (Lazaruk et al. 1998; Madabhushi 1998; Rosenblum et al. 1997). In addition to reduction in EOF, coating of the capillary surface reduces the interactions of DNA with the silica surface. Performance-optimized polymers have a defined ionic strength and uniform distribution of polymer chain length and pore size, and are well characterized for pH and temperature effects, separation efficiency, and denaturing conditions (Hahn et al. 2001). Thus, POP-4 and POP-6 polymers are the most commonly used polymers in the forensic DNA laboratory. In general, POP-4 polymer is used for fragment analysis and POP-6 polymer is used for sequencing applications. POP-4 polymer is composed of 4% linear dimethylacrylamide, 8 M urea, 5% 2-pyrrolidinone, and 1 mM EDTA (Levin et al. 1999; Rosenblum et al. 1997; Wenz et al. 1998). The major difference between POP-4 and POP-6 polymers is the concentration of dimethylacrylamide. The advantage of using a self-coating polymer compared to permanent coating is the ability to recoat the capillary surface after each run and to reduce the interactions between the constituents of the polymer and the coating material. These factors improve the reproducibility and extend the life of capillaries. The presence of methyl groups in the PDMA polymer makes it more hydrophobic compared to LPA. Increased hydrophobic nature is prone to hydrophobic interaction of certain DNA fragments. Boulos et al. (2008) developed a formulation comprising PVP and HEC for separation of single-stranded DNA ranging from 50 to 500 bp on CE. The approach takes advantage of the coating capability of PVP and the high efficiency of separation of HEC. Incorporating HEC probably alters the nature of interactions of DNA fragments with PVP, thereby providing consistent separation of dye-labeled DNA fragments. The PVP-HEC formulation has a low viscosity (388 centipoise, cP), which facilitates the filling of the capillaries.

Two hypotheses are proposed to explain the migration and separation of DNA in the linear polymers: transient entanglement mechanism (Barron et al. 1994) and nontangling collisions between the DNA and polymer (Sunada and Blanch 1998). The assumption in both hypotheses is generation of a drag force on the DNA molecule. Further, the extent of the drag force is proportional to the size of the molecule. The transient entanglement coupling mechanism was modified to the nontangling collisions mechanism after studies with videomicroscopy of DNA electrophoresing through polymer solutions. In addition to the size, the mobility of DNA is sequence dependent. For example, AT-rich sequences exhibit anomalous migration rates relative to internal size standards due to differences in DNA-gel interactions affected by secondary structure formed by DNA bending (Hagerman 1990; Sullivan et al. 1992). However, one would expect similar migration effects on the corresponding alleles in the allelic ladder because of sequence homology, thereby preventing the sequence-related migration differences and consequently the accuracy of allele designations. In addition to the length of the DNA fragment and sequence characteristics, physicochemical properties of the fluorophore dye also contribute to the mobility during CE (Hahn et al. 2001; Shewale et al. 2003). Hahn et al. (2001) have shown that dye-specific mobility anomalies exist for 5′-fluorophor–labeled, single-stranded DNA fragments on CE; fragments labeled with fluorescein dye derivatives have a higher mobility than the corresponding rhodamine-dye–labeled fragments. For accurate genotyping, therefore, it is important to co-separate a defined allelic ladder labeled with the same dye as the unknown sample fragments, which is a common practice in commercial STR kits. However, SNP variation of the same length allele may have different migration.

Separation of DNA fragments takes place during migration through a capillary that holds the polymer. The diameter, length, construction material, coating of internal surface, and detection window of the capillary are important factors. The diameter and length of the capillary contribute to the resistance during electrophoresis. The rate of mobility of DNA fragments may change from run to run due to EOF. The internal surface of the capillary made from silica should be coated or chemically modified to reduce the EOF since residual charges on the silica surface induce a flow of the bulk solution toward the cathode (Fung and Yeung 1995; Hjerten 1985; Madabhushi et al. 1996; Rosenblum et al. 1997). The internal diameter of capillaries ranges from 50 to 100 uM. The narrow capillary enables high electrical fields without the overheating problems associated with the use of high voltages. This is due to enhanced heat dissipation resulting from the small cross-sectional area of the capillary. During repeated use of a capillary, a shift in the migration of DNA fragments (i.e., precision of separation), may occur due to multiple factors like ionic strength, temperature, uniformity of separation medium, and fluctuations in the electrical current. Such shifts can affect the accuracy of sizing and hence the genotype. Co-migration of an internal size standard, to some extent, corrects for such microshifts. However, it is important to monitor the runs to detect such shifts. The recommendation, therefore, is to replace the capillaries before their useful lifetime or whenever a drift in migration of DNA fragments is observed to eliminate possible errors in allele sizing or loss of desired resolution. A typical capillary array is schematically represented in Figure 7.1. The capillaries are mounted in such a way that one end is aligned with the cathode for sample injection, the majority of the capillaries contact an oven for maintaining the temperature, the detection windows of the capillaries are aligned for detection of the migrating molecules, and the other end of the capillaries is aligned with the anode.

7.4.2 Injection and Run Parameters

A sample can be injected into a capillary either by hydrodynamic (pressure) injection, siphoning, or electrokinetic injection. Of these, electrokinetic injection is virtually the only method used in forensic laboratories. DNA fragments are injected into the capillary via

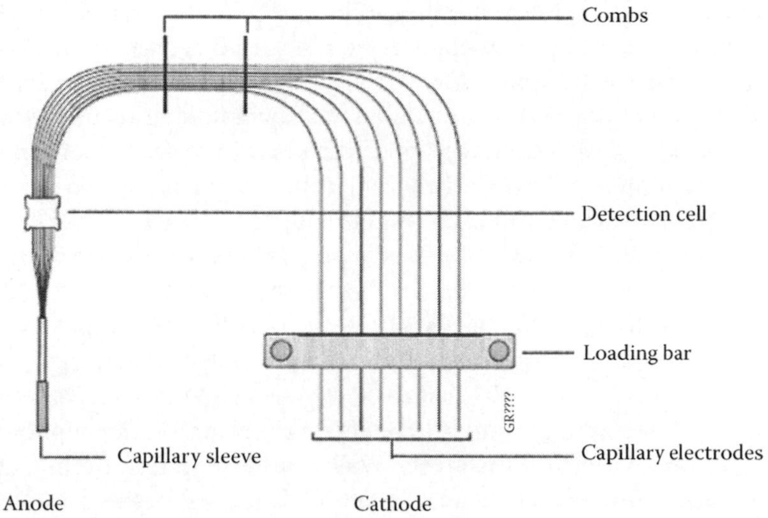

Figure 7.1 Schematic representation of a capillary array used in CE. The capillary sleeve provides a seal for alignment with the polymer block. Capillary electrodes hold the capillary ends in position. The loading bar supports the capillaries and provides a high-voltage connection to the capillary electrodes. Combs separate the capillaries to maintain consistent position and heat distribution in the oven. The detection cell aligns and holds the capillaries in place for laser detection.

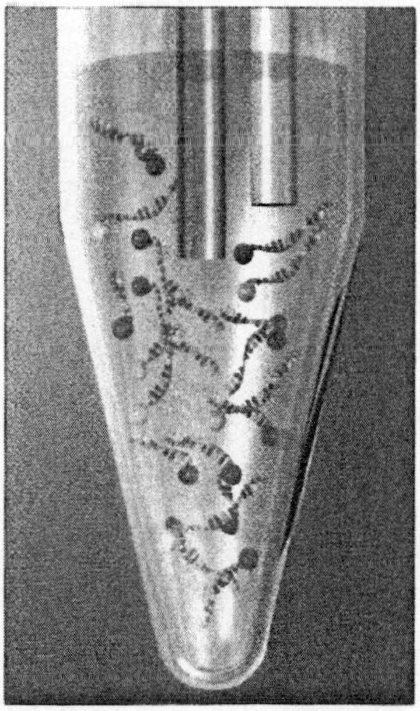

Figure 7.2 (See color insert.) Mechanism of electrokinetic injection during CE. The negatively charged molecules enter the capillary, aligned with the cathode, as they migrate toward the anode at the other end of the capillary. (Adapted from Applied Biosystems, 2002b, ABI Prism® 310 Genetic Analyzer—The measure of enabling technology, Product Bulletin, Publication #106PB07–01.)

electrokinetic injection, wherein the negatively charged particles migrate into the capillary toward the anode when a voltage is applied (Figure 7.2). In this process, negatively charged ions are transferred via electromotive force. DNA and other negatively charged molecules are introduced at the cathode end of the capillary as current flows from the cathode to the anode. Sample "stacking," wherein DNA molecules stack in a narrow zone at the interface of the capillary and sample solution, results when the voltage is applied during electrokinetic injection. Stacking enables minimal band broadening, larger sample loading into the capillary, and better resolution. One may classify electrokinetic injection as the onset of the electrophoresis run since DNA molecules are within the separation matrix (polymer). The quantity of DNA injected into a capillary is determined by the electric field, injection time, concentration of DNA in the sample, diameter of capillary, ionic strength of the buffer, and other interfering ions such as Cl^- that compete with DNA for electrokinetic injection (Butler et al. 2004; Rose and Jorgenson 1988). Smaller molecules with higher charge/mass ratio like Cl^- and residual primers from the PCR would migrate into the capillary faster than larger amplified products of interest. The high ionic strength of amplified samples can result in low amounts of DNA being injected because salts like KCl and $MgCl_2$ present in the PCR compete with the DNA during the electrokinetic injection. Thus, the amount of DNA fragments that get injected into the capillary is inversely proportional to the salt concentration. Different approaches are proposed to control the salt concentration in the sample prepared for CE and the methods that remove these interfering molecules enhance the sensitivity of the system (Butler 1998; Forster et al. 2008; Smith and Ballantyne 2007). In general, laboratories routinely control the volume of the amplified sample used in the preparation for the CE run. Peak heights obtained after multiple injections of the same sample may vary because of changes in the ionic strength and concentration of DNA in the sample with each successive injection (Isenberg et al. 1998). DNA forms complexes with divalent metal ions like Zn^{2+}, Cu^{2+}, and Ni^{2+}. Hartzell and McCord (2005) observed that these metal ion-DNA complexes exhibit altered mobility or peak broadening due to conformational change. The extent of peak broadening or shift in the mobility is determined by the type of metal ion, its concentration, fragment length and sequence, and the level of complex retained at the time of injection. It is important to correlate such occurrences to the sample history while interpreting evidence samples that were exposed to environmental insults or contaminated at the crime scene.

The pH and ionic strength of buffer play an important role in electrophoresis. The buffer keeps the DNA fragments in solution and provides charge carriers for the electrophoretic current. In simple terms, buffer is the most inexpensive yet critical component of the electrophoresis. High concentration (strength or molarity) of buffer decreases the resistance, resulting in overheating of the capillary, band broadening, and poor resolution. The composition of anode and cathode buffer changes, to varying degrees, when the same buffer is used for multiple electrophoresis runs. Further, altered pH changes the fluorescence emission of some dyes (Singer and Johnson 1997). Thus, it is a good practice to replace the buffer as frequently as possible during repeated use of CE.

Samples for STR analysis on CE are generally prepared by diluting a portion of the amplified product with deionized formamide. Formamide, a strong denaturing agent, is used in preparation of single-stranded DNA fragments. Denaturation is further ensured by rapid heating to 95°C for a few minutes followed by snap cooling. Formic acid, a product of formamide degradation, being negatively charged, competes with the DNA fragments for injection. Thus, it is advisable to aliquot and store high-purity formamide (conductivity <80 microsiemens). In addition to the formamide treatment, denaturing conditions are maintained by performing the electrophoresis in the presence of urea and at elevated temperature, normally 60°C. The goal is to keep the DNA fragments denatured since conformation of DNA fragments contributes to their mobility. Single-stranded DNA (ssDNA) molecules exhibit better relation of size to length, are more flexible, and interact more efficiently with the polymer. Denaturants prevent formation of secondary structures like loops or hairpins that can result in anomalous migration. In spite of these measures, some investigators have reported postinjection hybridization in certain alleles (McLaren et al. 2008).

Factors that affect the migration of DNA fragments through the capillary during electrophoresis include polymer, capillary diameter, length of capillary, electrophoresis buffer, field strength (or voltage), and temperature. Temperature also alters the viscosity of the polymer and hence separation of DNA fragments. It takes a few hours on capillary-array systems or 2 days on a single-capillary system for analysis of 96 samples. Thus, it is advisable to run allelic ladders in multiple wells spread over the plate to monitor or to average out the migration differences due to microchanges in the electrophoretic conditions among different injections. An amplified sample can be injected at different injection parameters to obtain a profile meeting the acceptance criteria of a laboratory. For example, a sample exhibiting a lower relative fluorescence unit (RFU) value than the threshold value may generate an acceptable and interpretable result by increasing the injection time. Varying the quantity of amplified product during preparation of the sample for CE is yet another approach. However, a laboratory needs to generate a dataset during validation studies to understand the limits in variation of electrophoretic parameters to support this approach. In general, it is observed that (a) the quantity of amplified product that gets injected into a capillary as measured by the peak height is increased, though not proportional, to increasing the injection time and voltage, (b) broader peaks are observed at longer injection times and at lower voltage for electrophoresis runs, (c) better resolution is achieved with longer run times (or capillary length) and at higher temperature, and (d) there is a correlation between the length of amplicon being separated and the migration capability of polymer (Buel et al. 2001, 2003). The overall performance of an electrophoresis run can be judged from the resolution, shape, and breadth of peaks in the electropherogram. Loss of resolution calls for evaluation of reagents as well as instrument issues. Electrophoresis conditions such as current, voltage, and temperature are available in the data file for each capillary for troubleshooting purposes. It is important to remember that the peak height, peak width, and resolution can be controlled by varying the electrophoretic conditions, but the stochastic issues resulting from PCR cannot be corrected or camouflaged.

7.4.3 Detection

The different methods used in detection of ions, small molecules, and macromolecules during CE are grouped into four modalities; absorbance, fluorescence, electrochemical, and refractive index (Swinney and Bornhop 2000). The most sensitive online detection system for CE is LIF. It is capable of detecting fluorescently derivatized single molecules. Readers may refer to the article by Swinney and Bornhop (2000) for detailed discussion on the different detectors developed for CE. Typically, 488-nm argon is used to detect the DNA fragment during CE. The laser is focused on the detection window located at the anode end of the capillary. The laser excites the dyes covalently attached to the primers and hence incorporated in the amplified fragments. Excited dyes emit fluorescent light at the

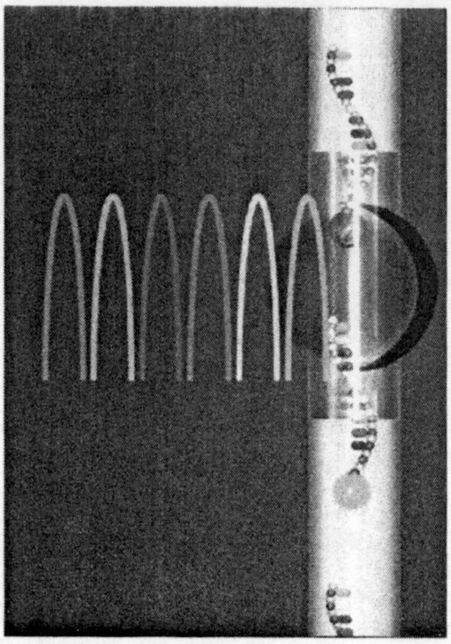

Figure 7.3 (See color insert.) Detection of the DNA fragments during CE. DNA fragments labeled with fluorescent dyes pass through the detection window during migration toward the anode. The laser excites the dyes, causing them to emit light at wavelengths larger than that of the laser. Emitted light is collected by a CCD camera. Finally, the software converts the pattern of emissions into colored peaks. (Adapted from Applied Biosystems, 2002b, ABI Prism® 310 Genetic Analyzer—The measure of enabling technology, Product Bulletin, Publication #106PB07–01.)

different wavelengths (approximately 400–600 nm) determined by the chemical structure of the dye. The emitted fluorescence is separated into its constituent wavelengths and captured by a photoarray detector (Figure 7.3). The 310 GA uses virtual filters to collect the light striking the CCD camera at a given wavelength. A silicon chip is used in the 3100, 3130, and 3500 series GA instruments for converting the incident fluorescence into digital information. The response from the detector is measured in RFUs. The software translates the output as peaks termed as an electropherogram. The RFU or extent of fluorescence is proportional to the quantity of labeled DNA that was injected into the capillary.

Some regions of dye fluorescence used in the multiplex system overlap. This spectral overlap is eliminated by studying the spectra of each dye individually and ignoring the overlapping regions by use of computer software algorithms. The exercise is known as spectral calibration or matrix generation. The spectral-calibration file or matrix file is generated from a separate electrophoretic run using the single-dye labels, and contains information about how much of the collected light falling on a filter is due to the intended light emission and how much is contaminating light. One spectral-calibration file or matrix file can be used for all runs sharing the same conditions and dyes. If run conditions are changed, a new matrix file or spectral calibration must be generated using the new run conditions. The data-collection software defines certain areas on the CCD array for the collection of the fluorescent emissions from the dye labels in the dye set. These areas are called virtual filters. There can be any number of virtual filters, since the filter is simply a software-designated site on the CCD array. Virtual filters are grouped into sets and referred to by a letter, such as Virtual Filter Set E5, G5, E, D, C, or F. Fluorescent dye labels generally come in sets of four or five. There are several different dye sets for use in different types of experiments. The fluorescence from each dye set must be collected using the correct virtual filter set. One selects the virtual filter set that corresponds to the experiment's dye set by choosing a module file when setting up the run. Some portion of a dye's emission profile may fall on a virtual filter other than the one intended to collect its emission maximum. The dyes in each dye set are selected to have widely spaced emission maximums to minimize overlap of the emission profiles on the CCD array. However, overlap still occurs to some extent. Changes in run conditions affect dye fluorescence and the amount of spectral overlap. In cases where the concentration of the amplified fragment is higher than the limits of the calibration algorithm, there could be bleed-through of fluorescence of a dye into the detection window of other dye(s). Bleed-through of fluorescence can also occur when the CCD camera alignment changes or the laser degenerates. It is important to follow the recommended maintenance guidelines for the instrument.

CE-based multiplex STR typing systems currently used in forensic laboratories amplify 16 to 22 STR loci simultaneously. This is achieved by generating amplicons of similar size for two or more loci using primers labeled with fluorescent dyes with minimal spectral overlap. Similarly, the analysis of sequencing reactions on CE instruments requires detection of four bases. The selection of dyes is made so that a single argon-ion laser excites all dyes but exhibits emission of fluorescence in different regions of the spectra that can be quantified by employing a CCD camera. The detection window is located at the flag end (anode end) of the capillary, an appropriate location for detection of fragments after separation.

7.5 CE Instruments for Forensic DNA Analysis

In the late 1990s and early 2000s, CE instruments were commercially available from several manufacturers like Life Technologies (Foster City, California), Amersham Biosciences (Piscataway, New Jersey), SpectruMedix Corporation (State College, Pennsylvania), Beckman Coulter (Fullerton, California), and Molecular Dynamics (Sunnyvale, California) (Butler et al. 2004; Carrilho 2000). Protocols for STR analysis on the Beckman P/ACE 2050 with a Laser Module 488 argon ion laser were developed (Butler 1998; McCord et al. 1995). Butler (1998) used Beckman P/ACE with Windows® software (version 3) to control the CE instrument and data were collected on Waters Millennium 2010 software version 2.0 (Waters Chromatography, Bedford, Massachusetts). The most commonly used CE systems in human identification laboratories in the past decade have been the 310, 3100, and 3130 series of GAs. Features of these GAs are summarized in Table 7.1. The single-capillary 310 GA was rapidly adopted by forensic laboratories soon after its commercialization, particularly in casework applications. It is still being used in low-throughput laboratories and for reinjection of samples in high-throughput laboratories because of its simplicity. The 3130*xl* GA has been the preferred platform for high-throughput laboratories for applications like paternity testing, databasing, casework, and so forth. A few laboratories have adopted the 3730 GA for STR profiling of single-source samples. The next-generation 3500 series of GAs have been recently commercialized by Applied Biosystems (Applied Biosystems 2010c). Most of the hardware, operations, and capabilities of the 3500 GA and 3500xL GA are identical except for the number of capillaries in the array: an 8-capillary array in the 3500 GA and a 24-capillary array in the 3500xL GA. The interior parts of the 3500 GA are shown in Figure 7.4.

Table 7.1 Features of Most Commonly Used Capillary Electrophoresis Platforms in Forensic DNA Analysis

Feature	310 GA[a]	3100 Series GA	3130 Series GA	3730 Series DNA Analyzer	3500 Series GA
Number of capillaries	1	4 (3100 Avant GA) or 16 (3100 GA)	4 (3130 GA) or 16 (3130*xl* GA)	48 (3730 DNA Analyzer) or 96 (3730xL DNA Analyzer)	8 (3500 GA) or 24 (3500xL GA)
Capillary array length (cm) for fragment analysis	47 (36 separation length)	36	36	36	36
Capillary array length (cm) for standard sequence applications	61 (50 separation length)	50	50	50	50
Polymer for fragment analysis	POP-4®	POP-4®	POP-4®	POP-4®	POP-4®
Polymer for standard sequence applications	POP-6®	POP-6®	POP-6® POP-7®	POP-7®	POP-6® POP-7®
DNA size performance	Single-base resolution up to 250 bases, precision ≤ 0.15 for 50–400 base pairs	Single-base resolution up to 400 bases, precision ≤ 0.15 for 50–400 base pairs	Single-base resolution up to 400 bases, precision ≤ 0.15 for 50–400 base pairs	Single-base resolution up to 400 bases, precision ≤ 0.15 for 50–400 base pairs	Single-base resolution up to 400 bases, precision ≤ 0.15 for 50–400 base pairs
Laser	10 mW argon-ion multi-line laser	25 mW argon-ion multi-line laser	25 mW argon-ion multi-line laser	25 mW argon-ion multi-line laser	10–25 mW solid-state single-line laser
Excitation lines	488 nm and 514.5 nm	488 nm and 514.5 nm	488 nm and 514.5 nm	488 nm and 514.5 nm	505 nm
Detection unit	CCD[b] camera, 525–650 nm, virtual filters	CCD camera, 525–650 nm, virtual filters	CCD camera, 525–650 nm, virtual filters	CCD camera, 525–650 nm, virtual filters	CCD camera, 525–650 nm, virtual filters
Dye channels	5 Dyes	5 Dyes	5 Dyes	5 Dyes	6 Dyes
Power supply	Normal	High voltage	High voltage	High voltage	Normal
Electrophoresis voltage	100 V–15 kV	Up to 20kV	Up to 20kV	Up to 20kV	Up to 20kV
Maximum power dissipation	1,440 W	2,500 W	2,500 W	2,500 W	320 W
Oven design	Plate, heating by electric resistance	Air convection, Peltier heat exchange	Air convection, Peltier heat exchange	Air convection, Peltier heat exchange	Air convection, Peltier heat exchange
Oven temperature range	Ambient to 60°C	18–65°C	18–65°C	18–70°C	18–70°C
Operating environment	15–30°C; 20–80% humidity	15–30°C; 20–80% humidity	15–30°C; 20–80% humidity	15–30°C; 20–80% humidity	15–30°C; 20–80% humidity

(continued)

Table 7.1 Features of Most Commonly Used Capillary Electrophoresis Platforms in Forensic DNA Analysis (Continued)

Feature	310 GA[a]	3100 Series GA	3130 Series GA	3730 Series DNA Analyzer	3500 Series GA
Radio frequency ID (RFID) tags	No	No	No	No	Yes
Peak height normalization	No	No	No	No	Yes
Commercialization year	1995	2000	2003	2005	2010
Reference	AB[c] 2010a	AB 2002a	AB 2010b	AB 2007	AB 2010c

[a] Genetic analyzer.
[b] Charge-coupled device.
[c] Applied Biosystems.

7.5.1 Sizing Precision

As discussed earlier (Section 7.2.1), sizing precision is important for STR fragment analysis since some alleles differ by only 1 base in size. Accurate sizing of fragments over the fractionation range of the STR multiplex amplification systems (in general 75–500 bp), single-base resolution over the fractionation range, and high run-to-run precision to enable comparison of mobility of alleles in the allelic ladder to the alleles in the samples are key performance requirements for the use of CE for separation and detection of amplified fragments. The sizing precision for all alleles in the allelic ladder from the Identifiler kit was determined for the 310, 3130, 3130*xl*, 3500, 3500xL, and 3730 GAs. The results are summarized in Figure 7.5. The sizing precision for all alleles in the Identifiler allelic ladder in the present study was less than 0.1 bp on all investigated CE platforms. Further, we did not observe statistically significant differences in the sizing precision among different loci, demonstrating the accuracy of sizing over the entire fractionation range for genotyping. This high level of precision provides a measure of reliability of CE for forensic DNA analysis. Although the sizing precision for all investigated CE platforms overlaps, a close relationship between the platforms from the same series of GA (e.g., 3130 GA and 3130*xl* GA as well as 3500 GA and 3500xL GA) is observed. The sizing-precision values for the 3730 GA were slightly higher than the other GAs.

7.5.2 Key Features of 3500 Genetic Analyzers

The 3500 and 3500xL GAs are designed to support the performance needs of forensic DNA analysis and to support compliance with certain ISO 17025 requirements. The basic principles and operations of the 3500 and 3500xL GAs are similar to the 310, 3100, and 3130 series of GAs. However, incorporation of innovative hardware design and software enhanced the performance of the 3500 and 3500xL GAs to attain a higher level of data quality, reliability, reduced run time, and increased throughput. Some salient features of the 3500 and 3500xL GAs are summarized here.

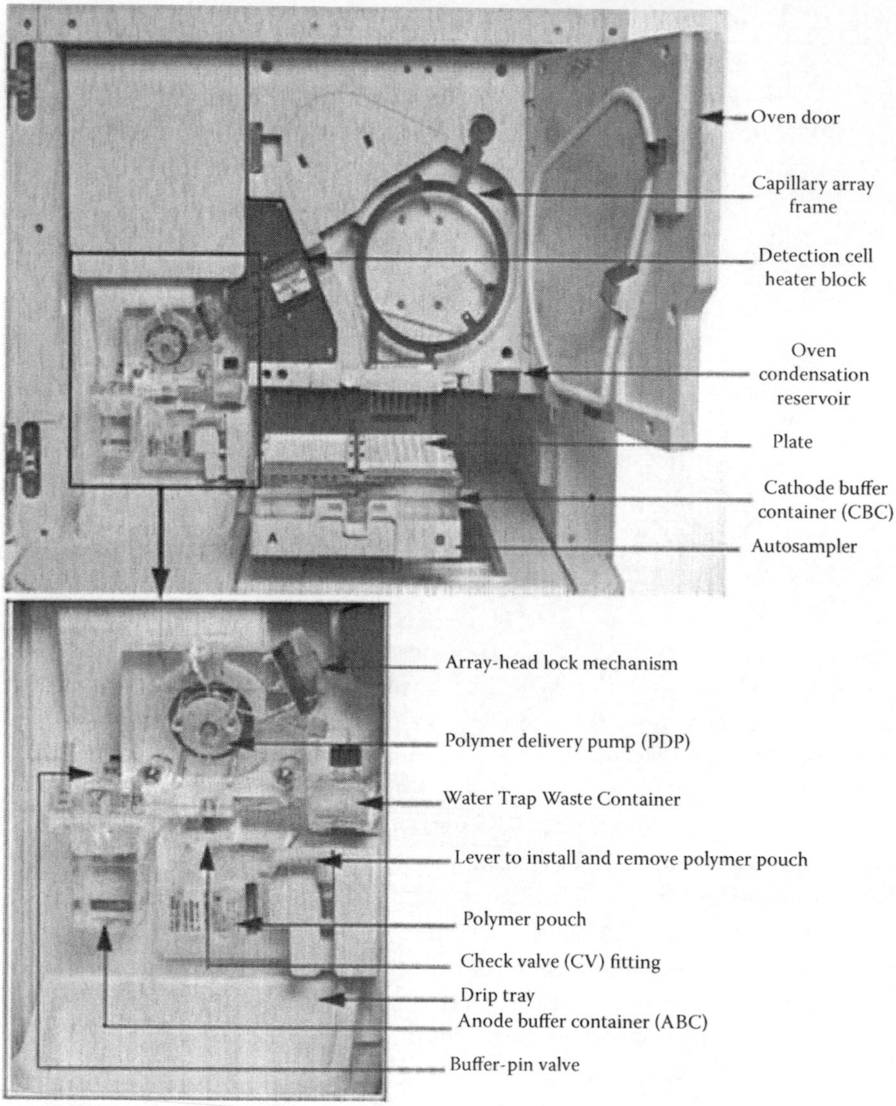

Figure 7.4 Interior components of the 3500 GA that are accessible to the operator. The autosampler holds the sample plates and cathode buffer container. The oven maintains uniform capillary-array temperature. Condensation from the oven gets collected in an oven condensation container. A capillary-array frame holds the capillaries in position. The detection-cell heater block holds the detection cell in place and maintains the detection cell temperature of 50°C. The polymer delivery pump delivers polymer into the capillary array and allows automated maintenance procedures. The pump-block unit comprises a displacement pump chamber, polymer chambers, piston water seal, array port, lower pump block, and check-valve fitting. The lower pump block contains the buffer valve and anode electrode buffer gasket and holds the anode buffer container.

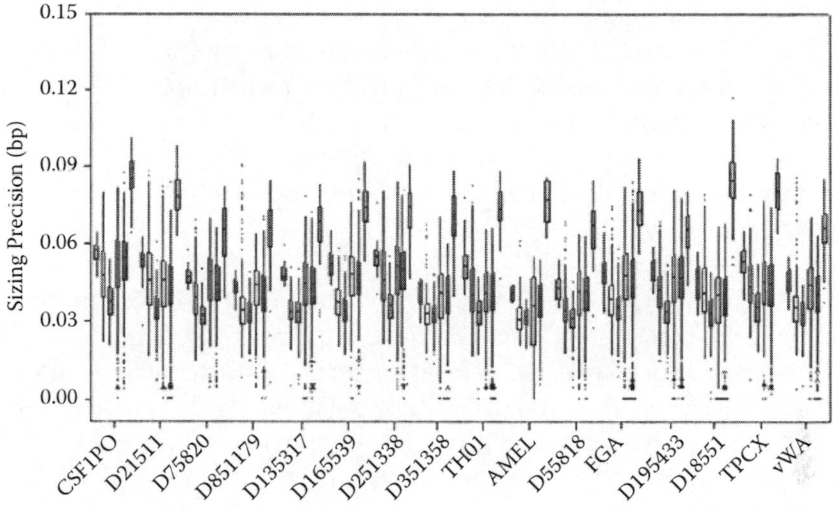

Figure 7.5 Sizing precision of all alleles in the Identifiler® allelic ladder injected on different CE platforms. Sizing precision presented from left to right are per 48 injections on the 310 GA, per 16 injections on the 3130 GA, per 16 injections on the 3130*xl* GA, per 8 injections on the 3500 GA, per 24 injections on the 3500xL GA, and per 48 injections on the 3730 GA. The box represents the interquartile range of the data points, the bar within the box represents the median data point, lower whisker extends to the lowest data value within the lower limit, upper whisker extends to the highest data value within the upper limit, and asterisks are outliers.

The changes in the hardware redesign were focused to achieve effective temperature control, reduce wastage of polymer and chances of contamination, and minimize energy consumption. Specific improvements in the hardware are:

1. The capillary-array oven is more compact with an improved door seal and locking mechanism, Peltier heat unit, and fan-circulated air. These features provide precise control of the temperature during the electrophoresis, thereby giving consistent performance and reduced run time.
2. The effects of fluctuation in the room temperature on the separation of DNA molecules are further reduced due to the shorter length of capillaries at the cathode end.
3. The detection-cell heater block holds the detection cell firmly and maintains the detection cell temperature at 50°C independent of electrophoresis run temperature.
4. Electrophoresis reagents such as POP polymers, anode buffer, cathode buffer, and conditioning reagents are available in prefilled containers, which minimize handling variations as well as the chances for introduction of contaminants.
5. A compact polymer pump design reduces polymer waste, potential for bubble formation, and instrument setup time.
6. A new single-line 505 nm, solid-state, long-life laser utilizes standard power supply, reducing the power consumption and heat exhaust.

7. Radio frequency identification (RFID) technology is used to track the lot numbers, serial numbers, expiration dates, on-instrument lifetime, and usage of the consumables like POP polymer, anode buffer, cathode buffer, and capillary array. The RFID tags are carefully designed. They are passive, cannot be read by the computer, read in stationary position, have limited read range (within a few centimeters of the reader), and cannot execute code. The tracked data are retrievable for quality-control records and audit purposes.

HID-focused software is developed to reduce the time for instrument setup and management of the data:

1. Data collection and QC analysis software that defines peak detection, sizing, and quality values are integrated. Thus, real-time assessment of data quality is possible.
2. Electrophoresis data is collected in HID file format, which contains information on samples, electrophoresis run conditions, consumables (lot number, expiration date, and time on the instrument), reinjection information whenever applicable, analysis parameters, and sample normalization factor when selected. This data can be transferred to secondary analysis software like GeneMapper® ID-X v1.2 or newer versions for genotyping.
3. Data collection software is preconfigured for AmpFℓSTR kits, allowing for simplified run setup. Additional features of the software include security and maintenance:
    1. Data chain of custody is maintained by controlling access through Security, Audit, and eSignatures that can be tracked. These features can be configured to meet the needs of the individual laboratory.
    2. Calibration and maintenance scheduling calendar functionality helps timely completion of such activities. The reports on security and maintenance are useful for audit, quality control, and troubleshooting purposes.

7.5.3 Normalization and Consistency of 3500 Genetic Analyzers

The sensitivity of a CE system is characterized by the signal-to-noise ratio and not necessarily by the absolute peak height. An instrument showing low peak intensity with low noise may be as sensitive as another instrument showing high peak intensity and high noise. The sensitivity of the CE can vary from one instrument to the other partly due to the varying intensity of the laser and capabilities of the CCD camera. The sensitivity of an instrument may gradually decline due to aging of the laser, a dirty capillary detection window, misalignment of the CCD camera, or degradation of dyes. Gao et al. (1999) described the normalization concept during the development of an algorithm for analysis of data collected from a 96-capillary array prototype to identify unknown alleles. The peaks for each locus in the allelic ladder are normalized to a selected peak in that capillary. According to this algorithm, the peak selected for normalization will always show a residual value of zero (residuals were calculated as the fractional changes in peak heights between the unknown sample and pooled ladder). Most of the other allele peaks will give residuals close to zero since their relative intensities should remain constant in the absence of unknown alleles. In mixtures of allelic ladder and unknown sample, residuals significantly higher than zero indicate the presence of an unknown allele.

In high-throughput laboratories, it is desired that the variation in the observed peak height of amplified fragments is minimal among multiple injections or on different instruments. Variation in the signal intensity among the instruments and capillaries is controlled in the 3500 and 3500xL GAs by factory standardization and an analysis software normalization feature that normalizes the signal intensity based on the intensity of size-standard peaks. Factory standardization involves calibration of the laser intensity and resolution of the CCD camera. The normalization feature in 3500 and 3500xL GAs attenuates signal variations associated with instrument, capillary array, sample salt load, and injection variability among capillaries and instruments. Normalization can be applied during primary analysis of the data. An example of raw data without and with application of the normalization feature is presented in Figure 7.6. The normalization feature in the 3500 Series Data Collection Software is useful for normalizing the observed intensity of amplified fragments with reference to the amplitude of their co-injected size standards using GeneScan™ 600 LIZ® size standard v2.0 (GeneScan 600 LIZ v2). Thus, GeneScan 600 LIZ v2 functions as a size standard for peak sizing as well as an internal standard for signal-height normalization. The quantity of GeneScan 600 LIZ v2 co-injected with samples is constant and would provide similar peak height when injected into different capillaries and on different instruments. Thus, the observed intensity of internal size-standard peaks from different injections or instruments can be normalized and the extent of variation can

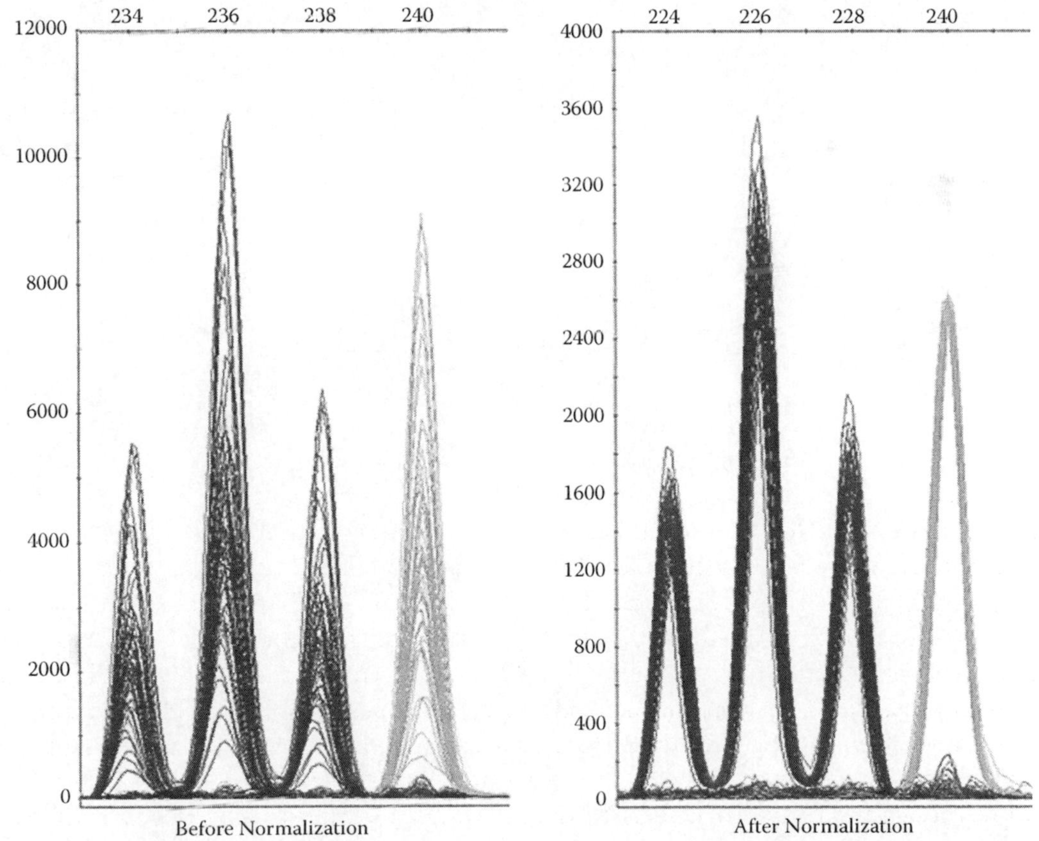

Figure 7.6 Example of raw data before and after normalization from different injections run on a 3500 GA.

be applied (or corrected) to the peak height of the amplified fragments from the sample. A normalization factor is derived for each sample on a particular instrument in the laboratory. The average peak height of the size standard is determined by comparing the height of the 11 most stable size-standard peaks (200, 220, 240, 260, 280, 300, 314, 320, 340, 360, and 400 bp) present in the GeneScan 600Liz v2 internal size standard from all instruments in the same laboratory. Normalization by the internal size standard in each injection is performed by calculating the normalization factor as the ratio between a preset peak height of the size standard (the normalization target) and the average peak height of the size standard, and multiplying the height of each sample peak by the normalization factor within the injection. The normalization feature needs to be selected in the 3500 Data Collection Software prior to the run in order for normalization of the data during the analysis to occur. However, raw data thus collected during the electrophoresis can be analyzed using the normalization feature "on" or "off" in the analysis software. The effect of normalization in minimizing the observed variation in the peak height between injections can be judged from the data presented in Figure 7.7. The genomic DNA from three human donors and the 007 control DNA were amplified in four replicates using 0.062, 0.125, 0.5, 1.0, and 1.5 ng of template DNA and the Identifiler Plus Kit. The samples were placed in the 96-well plate such that each sample would be injected into the capillaries spreading across the entire array. The samples were run with normalization activated in the 3500 Data Collection Software. The data were analyzed with both normalization options ("on" and "off") in the GeneMapper ID-X v1.2 software. The normalization performance was evaluated by the degree of peak-height variation. The coefficient of variance (CV) was calculated as the

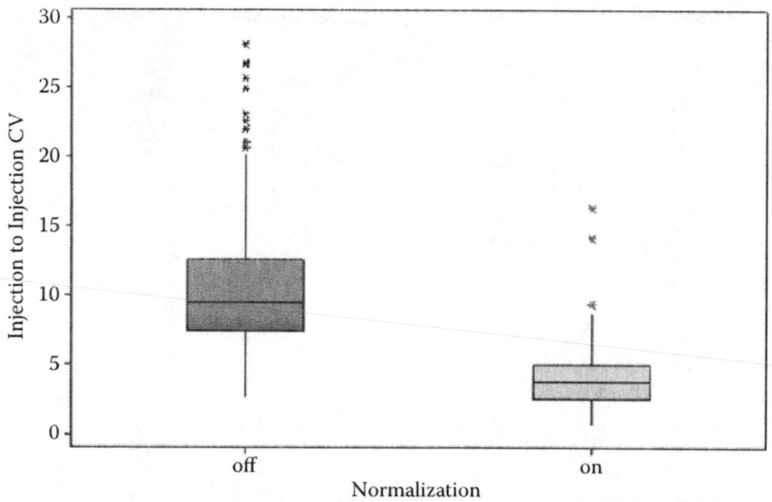

Figure 7.7 Coefficient-of-variance values for the peak heights observed in 15 injections on a 3500xL GA. Genomic DNA from three human donors and the 007 control DNA were amplified in four replicates using 0.062, 0.125, 0.5, 1.0, and 1.5 ng of template DNA and the Identifiler® Plus Kit. The data was analyzed with the normalization feature "on" and "off" as indicated on the X-axis. The box represents the interquartile range of the data points, the bar within the box represents the median data point, lower whisker extends to the lowest data value within the lower limit, upper whisker extends to the highest data value within the upper limit, and asterisks are outliers.

percent ratio of the standard deviation of the peak heights to the mean peak height. It was used to evaluate the peak-height variations among injections. The average CV in peak height from one injection to the other when normalization by the internal standard was not enabled was 11%. With the normalization by the internal standard turned "on," the average injection to injection CV was reduced significantly to 4%.

7.6 Electrophoretic Microdevices and Integrated Systems

The work on developing microdevices containing a microchannel structure in glass, fused silica, or plastic materials for electrophoresis and its application to STR analysis was initiated by some investigators in early 2000 (Carey and Mitnik 2002). Simultaneous analysis of CCTv, a fourplex STR multiplex system comprising CSF1PO, TPOX, TH01, and vWA loci, in 2 min on a 2.6-cm chip was demonstrated in 1997 by Schmalzing et al. (1997). Liu et al. (2011) described an integrated sample cleanup and 12-lane radial microfabricated capillary-array electrophoresis system for STR analysis. Integrated systems for STR profiling are now in an advanced state of development (Easley et al. 2006; Giese et al. 2009; Hurth et al. 2010; Jovanovich 2010; Li et al. 2010). These systems aim at a sample-to-answer approach wherein all steps of STR analysis including DNA extraction, amplification of STR loci, and electrophoretic separation of amplified fragments are performed on an integrated platform.

7.7 Other Applications of CE

CE has proved to be a highly versatile technique for separation of molecules, small and large, such as inorganic ions, organic acids, amino acids, peptides, proteins, natural products, illicit drugs, chiral separations, and nucleic acids (Issaq 2000). Thus, several applications have been developed on CE spanning diverse areas such as analytical chemistry, chiral separations, clinical research, forensic studies, molecular biology, natural products, organic chemistry, the pharmaceutical industry, and others (Issaq 2000; Thormann et al. 1999). Examples of forensic applications besides DNA analysis include analysis of gunshot and explosive residues, and drug analysis (Northrop et al. 1994). Clinical applications include: (a) analysis of small molecules and ions, drug seizures, drugs and small exogenous compounds in biofluids and tissues, and endogenous small molecules and ions in biofluids and tissues; (b) analysis of proteins like serum proteins, lipoproteins, hemoglobins, transferring urinary proteins, and proteins in cerebrospinal fluid; (c) diagnosis of certain genetic diseases like cystic fibrosis, chronic myeloid leukemia, Prader–Willi syndrome, Angelman syndrome, Fragile-X syndrome, and so forth.

7.8 Conclusions

Innovations in CE played a key role in bringing forensic DNA analysis to the forefront. Single-base resolution capabilities with high precision allow genotyping of STR markers by fragment analysis, thereby eliminating the need for cumbersome DNA sequencing. Development of fluorescent dyes and optical systems for resolving the fluorescence emitted from multiple dyes enabled analysis of amplified products from multiplex STR amplification systems. Thus, it is possible to amplify, separate, and detect amplified fragments from 16 to 20 STR loci, simultaneously. Capillary arrays and walk-away automated operations for sample injection, electrophoretic separation, and detection increased the throughput of forensic laboratories, thereby saving valuable time of forensic analysts. The 3500 series GAs are new additions to the CE family. Incorporation of innovative hardware design and software have enhanced the performance capabilities of the 3500 and 3500xL GAs to attain higher levels of data quality, reliability, reduced run time, and increased throughput. Normalization functionality has been developed to reduce variation in the observance of data (peak height in particular) between multiple instruments and injections. In addition, the 3500 and 3500xL GAs offer several unmet hardware and software features including temperature controls, RFID tags for the primary consumables, HID-focused software features, and security and maintenance. Besides STR analysis, CE platforms are used for mtDNA sequencing, SNP typing, tissue identification, and other fragment-analysis-based applications in forensic DNA analysis.

Acknowledgments

We thank Ariana Wheaton, Ben Johnson, and Jeff Sailus from Life Technologies for useful discussions.

References

Alvarez-Iglesias, V., J. C. Jaime, A. Carracedo, and A. Salas. 2007. Coding region mitochondrial DNA SNPs: Targeting East Asian and Native American haplogroups. *Forensic Sci Int Genet* 1: 44–55.

Anderson, S., A. T. Bankier, B. G. Barell et al. 1981. Sequence and organization of the human mitochondrial genome. *Nature* 290: 457–65.

Applied Biosystems. 2002a. ABI Prism® 3100 Genetic Analyzer and ABI Prism® 3100-*Avant* Genetic Analyzer User Reference Guide. Part Number 4335393 Rev. A. Available at http://www3.appliedbiosystems.com/cms/groups/mcb_support/documents/generaldocuments/cms_041300.pdf.

Applied Biosystems. 2002b. ABI Prism® 310 Genetic Analyzer—The measure of enabling technology. Product Bulletin, Publication # 106PB07-01.

Applied Biosystems. 2007. Applied Biosystems 3730 DNA Analyzer human identification validation report. Publication No. 112UB18-01.

Applied Biosystems. 2010a. ABI Prism® 310 Genetic Analyzer User Guide. Part Number 4317588 Rev. B. Available at http://www3.appliedbiosystems.com/cms/groups/mcb_support/documents/generaldocuments/cms_041158.pdf.

Applied Biosystems. 2010b. Applied Biosystems 3130/3130*xl* Genetic Analyzers Getting Started Guide. Part Number 4352715 Rev. D. Available at http://www3.appliedbiosystems.com/cms/groups/mcb_support/documents/generaldocuments/cms_041468.pdf.

Applied Biosystems. 2010c. Applied Biosystems 3500/3500xL Genetic Analyzer User Guide. Part Number 4401661 Rev. C. Available at http://www3.appliedbiosystems.com/cms/groups/mcb_support/documents/generaldocuments/cms_069856.pdf.

Barber, M. D., R. C. Piercy, J. F. Andersen, and B. H. Parkin. 1995. Structural variation of novel alleles at the Hum vWA and Hum FES/FPS short tandem repeat loci. *Int J Leg Med* 108: 31–5.

Barron, A. E., H. W. Blanch, and D. S. Soane. 1994. A transient entanglement coupling mechanism for DNA separation by capillary electrophoresis in ultra-dilute polymer solutions. *Electrophoresis* 15: 597–615.

Bender, K., P. M. Schneider, and C. Rittner. 2000. Application of mtDNA sequence analysis in forensic casework for the identification of human remains. *Forensic Sci Int* 113: 103–7.

Bode, H. J. 1977a. The use of liquid polyacrylamide in electrophoresis. I. Mixed gels composed of agar-agar and liquid polyacrylamide. *Anal Biochem* 83: 204–10.

Bode, H. J. 1977b. The use of liquid polyacrylamide in electrophoresis. II. Relationship between gel viscosity and molecular sieving. *Anal Biochem* 83: 364–71.

Borsting, C., J. J. Sanchez, H. E. Hansen, A. J. Hansen, H. Q. Brun, and N. Morling. 2008. Performance of the SNP*for*ID 52 SNP-plex assay in paternity testing. *Forensic Sci Int* 2: 292–300.

Boulos, S., O. Cabrices, M. Blas, and B. McCord. 2008. Development of an entangled polymer solution for improved resolution on DNA typing by CE. *Electrophoresis* 29: 4695–703.

Budowle, B. and A. van Dall. 2008. Forensically relevant SNP classes. *BioTechniques* 44: 603–10.

Buel, E., M. LaFountain, and M. Schwartz. 2003. Using resolution calculation to access changes in capillary electrophoresis run parameters. *J Forensic Sci* 48: 77–9.

Buel, E., M. LaFountain, M. Schwartz, and M. Walkinshaw. 2001. Evaluation of capillary electrophoresis performance through resolution measurements. *J Forensic Sci* 46: 341–5.

Buel, E., M. B. Schwartz, and M. J. LaFountain. 1998. Capillary electrophoresis STR analysis: Comparison to gel-based systems. *J Forensic Sci* 43: 164–70.

Butler, J. M. 1998. The use of capillary electrophoresis in genotyping STR loci. *Method Mol Biol* 98: 279–89.

Butler, J. M., E. Buel, F. Crivellente, and B. R. McCord. 2004. Forensic typing by capillary electrophoresis using the ABI Prism 310 and 3100 genetic analyzers for STR analysis. *Electrophoresis* 25: 1397–412.

Butler, J. M., B. R. McCord, J. M. Jung, and R. O. Allen. 1994. Rapid analysis of the short tandem repeat HUMTHO1 by capillary electrophoresis. *BioTechniques* 17: 1062–4.

Butler, J. M., B. R. McCord, J. M. Jung, J. A. Lee, B. Budowle, and R. O. Allen. 1995. Application of dual internal standards for precise sizing of polymerase chain reaction products using capillary electrophoresis. *Electrophoresis* 16: 974–80.

Butler, J. M., C. M. Ruitberg, and P. M. Vallone. 2001. Capillary electrophoresis as a tool for optimization of multiplex PCR reactions. *Fresenius J Anal Chem* 369: 200–5.

Butler, J. M., R. Schoske, P. M. Vallone, M. C. Kline, A. J. Redd, and M. F. Hammer. 2002. A novel multiplex for simultaneous amplification of 20 Y chromosome STR markers. *Forensic Sci Int* 129: 10–24.

Butler, J. M., R. Schoske, P. M. Vallone, J. W. Redman, and M. C. Kline. 2003. Allele frequencies for 15 autosomal STR loci on U.S. Caucasian, African American, and Hispanic populations. *J Forensic Sci* 48: 908–11.

Carey, L., and L. Mitnik. 2002. Trends in DNA forensic analysis. *Electrophoresis* 23: 1386–97.

Carrilho, E. 2000. DNA sequencing by capillary array electrophoresis and microfabricated array systems. *Electrophoresis* 21: 55–65.

Chong, M. D., C. D. Calloway, S. B. Klein, C. Orrego, and M. R. Buoncristiani. 2005. Optimization of a duplex amplification and sequence strategy for the HVI/HVII regions of human mitochondrial DNA for forensic casework. *Forensic Sci Int* 154: 137–48.

Clark, P., A. E. Stark, R. J. Walsh, R. Jardine, and N. G. Martin. 1981. A twin study of skin reflectance. *Ann Hum Biol* 8: 529–41.

Collins, P. J., L. K. Hennessy, C. S. Leibelt, R. K. Roby, D. J. Redder, and P. A. Foxall. 2004. Developmental validation of a single-tube amplification of the 13 CODIS STR loci, D2S1338, D19S433 and amelogenin: The AmpFlSTR® Identifiler PCR amplification kit. *J Forensic Sci* 49: 1265–77.

Dovichi, N. J. 1997. DNA sequencing by capillary electrophoresis. *Electrophoresis* 18: 2393–99.

Easley, C. J., J. M. Karlinsey, J. M. Bienvenue et al. 2006. A fully integrated microfluidic genetic analysis system with sample-in–answer-out capability. *Proc Natl Acad Sci USA* 103: 19272–77.

Elder, J. K. and E. M. Southern. 1983. Measurement of DNA length by gel electrophoresis. Part II: Comparison of methods for relating mobility to fragment length. *Anal Biochem* 128: 227–31.

Ensenberger, M. G., J. Thompson, B. Hill et al. 2010. Developmental validation of the PowerPlex® 16 HS system: An improved 16-locus fluorescent STR multiplex. *Forensic Sci Int Genet* 4: 257–64.

Fang, R., A. J. Pakstis, F. Hyland et al. 2009. Multiplexed SNP detection panels for human identification. *Forensic Sci Int Genet* Suppl Ser 2: 538–9.

Forster, L., J. Thomson, and S. Kutranov. 2008. Direct comparison of post-28-cycle PCR purification and modified capillary electrophoresis methods with the 34-cycle "low copy number" (LCN) method for analysis of trace forensic DNA samples. *Forensic Sci Int Genet* 2: 318–28.

Frazier, R. R. E., E. S. Millican, S. K. Watson et al. 1996. Validation of the Applied Biosystems Prism™ 377 automated sequencer for forensic short tandem repeat analysis. *Electrophoresis* 17: 1550–2.

Fregeau, C. J. and R. M. Fourney. 1993. DNA typing with fluorescently tagged short tandem repeats: A sensitive and accurate approach to human identification. *BioTechniques* 15: 100–19.

Frumkin, D., A. Wasserstrom, B. Budowle, and A. Davidson. 2011. DNA methylation-based forensic tissue identification. *Forensic Sci Int Genet* 5: 517–24.

Fung, E. N. and E. S. Yeung. 1995. High-speed DNA sequencing by using mixed poly(ethylene oxide) solutions in uncoated capillary columns. *Anal Chem* 67: 1913–9.

Gabriel, M. N., E. F. Huffine, J. H. Ryan, M. M. Holland, and T. J. Parsons. 2001. Improved MtDNA sequence analysis of forensic remains using a "mini-primer set" amplification. *J Forensic Sci* 46: 247–53.

Gao, Q., H. Pang, and E. S. Yeung. 1999. Simultaneous genetic typing from multiple short tandem repeat loci using a 96-capillary array electrophoresis system. *Electrophoresis* 20: 1518–26.

Gao, Q. and E. S. Yeung. 1998. A matrix for DNA separation: Genotyping and sequencing using ply(vinylpyrrolidone) solution in uncoated capillaries. *Anal Chem* 70: 1382–88.

Giese, H., R. Lam, R. Selden, and E. Tan. 2009. Fast multiplexed polymerase chain reaction for conventional and microfluidic short tandem repeat analysis. *J Forensic Sci* 54: 1287–96.

Gill, P., P. Koumi, and H. Allen. 2001. Sizing short tandem repeat alleles in capillary array gel electrophoresis instruments. *Electrophoresis* 22: 2670–8.

Grzybowski, T. 2000. Extremely high levels of human mitochondrial DNA heteroplasmy in single hair roots. *Electrophoresis* 21: 548–53.

Hagerman, P. J. 1990. Sequence-directed curvature of DNA. *Annu Rev Biochem* 59: 755–81.

Hahn, M., J. Wilhelm, and A. Pingoud. 2001. Influence of fluorophor dye labels on the migration behavior of polymerase chain reaction amplified short tandem repeats during denaturing capillary electrophoresis. *Electrophoresis* 22: 2691–700.

Hartzell, B., K. Graham, and B. McCord. 2003. Response of short tandem repeat systems to temperature and sizing methods. *Forensic Sci Int* 133: 228–34.

Hartzell, B. and B. McCord. 2005. Effect of divalent metal ions on DNA studied by capillary electrophoresis. *Electrophoresis* 26: 1046–56.

Heiger, D. N., A. S. Cohen, and B. L. Krager. 1990. Separation of DNA restriction fragments by high performace capillary electrophoresis with low and zero cross-linked polyacrylamide using continuous and pulsed electric fields. *J Chromatogr* 516: 33–48.

Heller, C. 2001. Principles of DNA separation with capillary electrophoresis. *Electrophoresis* 22: 629–43.

Hjerten, S. 1967. Free zone electrophoresis. *Chromatogr Rev* 9: 122–219.

Hjerten, S. 1985. High-performance electrophoresis: Elimination of electroendosmosis and solute adsorption. *J Chromatogr* 347: 191–8.

Holland, M. M. and T. J. Parson. 1999. Mitochondrial DNA sequence analysis—Validation and use for forensic casework. *Forensic Sci Rev* 11: 21–50.

Huang, X. C., M. A. Quesada, and R. A. Mathies. 1992a. Capillary array electrophoresis using laser-excited confocal fluorescence detection. *Anal Chem* 64: 967–72.

Huang, X. C., M. A. Quesada, and R. A. Mathies. 1992b. DNA sequencing using capillary array electrophoresis. *Anal Chem* 64: 2149–54.

Hurth, C., S. D. Smith, A. R. Nordquist et al. 2010. An automated instrument for human STR identification: Design, characterization, and experimental validation. *Electrophoresis* 31: 3510–7.

Isenberg, A. R., R. O. Allen, K. M. Keys, J. B. Smerick, B. Budowle, and B. R. McCord. 1998. Analysis of two multiplexed short tandem repeat systems using capillary electrophoresis with multiwavelength fluorescence detection. *Electrophoresis* 19: 94–100.

Isenberg, A. R., B. R. McCord, B. W. Koons, B. Budowle, and R. O. Allen. 1996. DNA typing of a polymerase chain reaction amplified D1S80/amelogenin multiplex using capillary electrophoresis and a mixed entangled polymer matrix. *Electrophoresis* 17: 1505–11.

Issaq, H. J. 2000. A decade of capillary electrophoresis. *Electrophoresis* 21: 1921–39.

Jeffreys, A. J., V. Wilson, and S. L. Theis. 1985a. Hypervariable "minisatellite" regions in human DNA. *Nature* 314: 67–73.

Jeffreys, A. J., V. Wilson, and S. L. Theis. 1985b. Individual-specific "fingerprints" of human DNA. *Nature* 316: 76–79.

Jorgenson, J. W. and K. D. Lukacs. 1981. Zone electrophoresis in open-tubular glass capillaries. *Anal Chem* 53: 1298–302.

Jovanovich, J. 2010. Apollo 200™ human identity system—The first fully integrated, sample-to-profile STR-based system for forensics. *Proc 21st Int Symp Human Identification*, San Antonio, TX.

Juusola, J. and J. Ballantyne. 2005. Multiplex mRNA profiling for the identification of body fluids. *Forensic Sci Int* 152: 1–12.

Kidd, K. K., A. J. Pakstis, W. C. Speed et al. 2006. Developing a SNP panel for forensic identification of individuals. *Forensic Sci Int* 164: 20–32.

Kimpton, C. P., N. J. Oldroyd, S. K. Watson et al. 1996. Validation of highly discriminating multiplex short tandem repeat amplification systems for individual identification. *Electrophoresis* 17: 1283–93.

Kuffner, C. A. Jr., E. Marchi, J. M. Morgado, and C. R. Rubio. 1996. Capillary electrophoresis and Daubert: Time for admission. *Anal Chem* 68: 241A–46A.

Laclair, B., C. J. Fregeau, K. L. Bowen, and R. M. Fourney. 2004. Precision and accuracy in fluorescent short tandem repeat DNA typing: Assessment of benefits imparted by the use of allelic ladders with the AmpFℓSTR® Profiler Plus™ Kit. *Electrophoresis* 25: 790–6.

Lazaruk, K., P. S. Walsh, F. Oaks et al. 1998. Genotyping of forensic short tandem repeat (STR) systems based on sizing precision in a capillary electrophoresis instrument. *Electrophoresis* 19: 86–93.

Levin, B. C., H. Cheng, and D. J. Reeder. 1999. A human mitochondrial DNA standard reference material for quality control in forensic identification, medical diagnosis, and mutation detection. *Genomics* 55: 135–46.

Li, J., B. Strachan, J. Lounsbury et al. 2010. Defining new microfluidic methodologies for forensic DNA sample preparation. *Proc 21st Int Symp Human Identification*, San Antonio, TX.

Liu, P., J. R. Scherer, S. A. Greenspoon, T. N. Chiesl, and R. A. Mathies. 2011. Integrated sample cleanup and capillary array electrophoresis microchip for forensic short tandem repeat analysis. *Forensic Sci Int Genet* 5: 484–92.

Madabhushi, R. S. 1998. Separation of 4-color DNA sequencing extension products in noncovalently coated capillaries using low viscosity polymer solutions. *Electrophoresis* 19: 224–30.

Madabhushi, R. S., S. M. Menchen, J. W. Efcavitch, and P. D. Grossman. 1996. Polymers for separation of biomolecules by capillary electrophoresis. U.S. Patent No. 5,567,292.

Mansfield, E. S., J. M. Robertson, M. Vainer et al. 1998. Analysis of multiplexed short tandem repeat (STR) systems using capillary array electrophoresis. *Electrophoresis* 19: 101–7.

Mansfield, E. S., M. Vainer, S. Enad et al. 1996. Sensitivity, reproducibility, and accuracy in short tandem repeat genotyping using capillary array electrophoresis. *Genome Res* 6: 893–903.

Mansfield, E. S., M. Vainer, S. Enad et al. 1997. Rapid sizing of polymorphic microsatellite markers by capillary array electrophoresis. *J Chromatogr* A 781: 295–305.

Marshall, E. and E. Pennisi. 1998. Hubris and the human genome. *Science* 280: 994–5.

Mathies, R. A. and X. C. Huang. 1992. Capillary array electrophoresis: An approach to high-speed, high-throughput DNA sequencing. *Nature* 359: 167–9.

McCord, B. R., A. R. Isenberg, J. M. Butler, and R. O. Allen. 1995. Rapid and automated genetic typing using capillary electrophoresis for the analysis of STRs and D1S80. *Proc 6th Int Symp Human Genetics* pp. 81–7.

McCord, B. R., J. M. Jung, and E. A. Holleran. 1993a. High resolution capillary electrophoresis of forensic DNA using a non-gel sieving buffer. *J Liq Chromatogr* 16: 1963–81.

McCord, B. R., D. M. McClure, and J. M. Jung. 1993b. Capillary electrophoresis of PCR-amplified DNA using fluorescence detection with an intercalating dye. *J Chromatogr* 652: 75–82.

McLaren, R. S., M. G. Ensenberger, B. Budowle et al. 2008. Post-injection hybridization of complementary DNA strands on capillary electrophoresis platforms: A novel solution for dsDNA artifacts. *Forensic Sci Int Genet* 2: 257–73.

Micka, K. A., C. J. Sprecher, A. M. Lins et al. 1996. Validation of multiplex polymorphic STR amplification sets developed for personal identification applications. *J Forensic Sci* 41: 582–90.

Mikkers, F. E. P., F. M. Everaerts, and T. P. E. M. Verheggen. 1979. High-performance zone electrophoresis. *J Chromatogr* 169: 11–20.

Mitnik, L., L. Carey, R. Burger et al. 2002. High-speed analysis of multiplexed short tandem repeats with an electrophoretic microdevice. *Electrophoresis* 23: 719–26.

Moretti, T. R., A. L. Baumstark, D. A. Defenbaugh, K. M. Keys, J. B. Smerick, and B. Budowle. 2001. Validation of short tandem repeats (STRs) for forensic usage: Performance testing of fluorescent multiplex STR systems and analysis of authentic and simulated forensic samples. *J Forensic Sci* 46: 647–60.

Mulero, J. J., C. W. Chang, L. M. Calandro et al. 2006. Development and validation of the AmpFlSTR® Yfiler® PCR amplification kit: a male specific, single amplification 17 Y-STR multiplex system. *J Forensic Sci* 51: 64–76.

Mulero, J. J., C. W. Chang, R. E. Lagacé et al. 2008. Development and validation of the AmpFlSTR® MiniFiler™ PCR amplification kit: A MiniSTR multiplex for the analysis of degraded and/or PCR inhibited DNA. *J Forensic Sci* 53: 838–52.

Mulero, J., R. Green, N. Oldroyd, and L. Hennessy. 2009. Development of a new forensic STR multiplex with enhanced performance for degraded and inhibited samples. *23rd World Congr Int Soc Forensic Genet.*

Nagy, M., J. Rascon, G. Massenkeil, W. Ebell, and L. Toewer. 2006. Evaluation of whole-genome amplification of low-copy-number DNA in chimerism analysis after allogeneic stem cell transplantation using STR marker typing. *Electrophoresis* 27: 3028–37.

Northrop, D. M., B. R. McCord, and J. M. Butler. 1994. Forensic applications of capillary electrophoresis. *J Capillary Electrop* 1: 158–68.

Pakstis, A. J., W. C. Speed, J. R. Kidd, and K. K. Kidd. 2007. Candidate SNPs for a universal individual identification panel. *Hum Genet* 121: 305–17.

Puers, C., H. A. Hammond, L. Jin, C. T. Caskey, and J. W. Schumm. 1993. Identification of repeat sequence heterogeneity at the polymorphic short tandem repeat locus HUMTH01[AATG]n and reassignment of alleles in population analysis by using a locus-specific allelic ladder. *Am J Hum Genet* 53: 953–8.

Ricci, U., I. Sani, S. Guarducci et al. 2000. Infrared fluorescent automated detection of thirteen short tandem repeat polymorphisms and one gender-determining system of the CODIS core system. *Electrophoresis* 21: 3564–70.

Rose, D. J. and J. W. Jorgenson. 1988. Characterization and automation of sample introduction methods for capillary zone electrophoresis. *Anal Chem* 60: 642–8.

Rosenblum, B. B., F. Oaks, S. Menchen, and B. Johnson. 1997. Improved single-strand DNA sizing accuracy in capillary electrophoresis. *Nucleic Acid Res* 25: 3925–9.

Ruiz-Martinez, M. C., J. Berka, A. Belenkii, F. Foret, A. W. Miller, and B. L. Karger. 1993. DNA sequencing by capillary electrophoresis with replaceable linear polyacrylamide and laser-induced fluorescence detection. *Anal Chem* 65: 2851–8.

Sanchez, J. J., C. Phillips, C. Borsting et al. 2006. A multiplex assay with 52 single nucleotide polymorphisms for human identification. *Electrophoresis* 27: 1713–24.

Schmalzing, D., L. Koutny, A. Adourian, P. Belgrader, P. Matsudaira, and D. Ehrlich. 1997. DNA typing in thirty seconds with a microfabricated device. *Proc Natl Acad Sci USA* 94: 10273–8.

Schubbert, R. 2001. DNA-Profiling using Promega's PowerPlex® 16 kit on an ABI 3100 System. *Proc 12th Int Symp Human Identification*, Biloxi, MI.

Shewale, J. G., H. Nasir, and S. K. Sinha. 2003. Variation in migration of the DNA fragments labeled with fluorescent dyes on the 310 Genetic Analyzer and its implication in the genotyping. *J Assoc Genet Technol* 29: 60–3.

Singer, V. L. and I. D. Johnson. 1997. Fluorophore characteristics: Making intelligent choices in application-specific dye selection. *Proc 8th Int Symp Human Identification,* Madison, WI.

Smith, L. M., J. Z. Sanders, R. J. Kaiser et al. 1986. Fluorescence detection in automated DNA sequence analysis. *Nature* 321: 674–9.

Smith, P. J. and J. Ballantyne. 2007. Simplified low-copy-number DNA analysis by post-PCR purification. *J Forensic Sci* 52: 820–9.

Smith, R. N. 1995. Accurate size comparison of short tandem repeat alleles amplified by PCR. *BioTechniques* 18: 122–8.

Stewart, J. E. B., P. J. Aagaard, E. G. Pokorak, D. Polanskey, and B. Budowle. 2003. Evaluation of a multicapillary electrophoresis instrument for mitochondrial DNA typing. *J Forensic Sci* 48: 571–80.

Sullivan, K. M., S. Pope, P. Gill, and J. M. Robertson. 1992. Automated DNA profiling by fluorescent labeling of PCR products. *PCR Methods Appl* 2: 34–40.

Sunada, W. M. and H. W. Blanch. 1998. A theory for the electrophoretic separation of DNA in polymer solutions. *Electrophoresis* 19: 3128–36.

Swank, R. T. and K. D. Munkers. 1971. Molecular weight analysis of oligopeptides by electrophoresis in polyacrylamide gel with sodium dodecyl sulphate. *Anal Biochem* 39: 462–77.

Swinney, K. and D. J. Bornhop. 2000. Detection in capillary electrophoresis. *Electrophoresis* 21. 1239–50.

Terabe, S., K. Otsuka, K. Ichikawa, A. Tsuchuya, and T. Ando. 1984. Electrokinetic separations with micellar solutions and open-tubular capillaries. *Anal Chem* 56: 111–3.

Thomas, C., M. Stangegaard, C. Borsting, A. J. Hansen, and N. Morling. 2008. Typing of 48 autosomal SNPs and amelogenin with GenPlex SNP genotyping system in forensic genetics. *Forensic Sci Int Genet* 2: 1–6.

Thormann, W., A. B. Wey, I. S. Lurie et al. 1999. Capillary electrophoresis in clinical and forensic analysis: Recent advances and breakthrough to routine applications. *Electrophoresis* 20: 3203–36.

Tiselius, A. 1930. Thesis; *Nova Acta Regiae Sociates Scientiarum Upsallensis* Ser. IV, 7, Number 4.

Urquhart, A., C. P. Kimpton, T. J. Downes, and P. Gill. 1994. Variation in short tandem repeat sequences—A survey of twelve microsatellite loci for use as forensic identification markers. *Int J Legal Med* 107: 13.

Vainer, M., S. Enad, V. Dolnik et al. 1997. Short tandem repeat typing by capillary array electrophoresis: Comparison of sizing accuracy and precision using different buffer systems. *Genomics* 41: 1–9.

Walsh, P. S., N. J. Fildes, and R. Reynolds. 1996. Sequence analysis and characterization of stutter products at the tetranucleotide repeat locus vWA. *Nucleic Acids Res* 24: 2807–12.

Wang, D., J. Mulero, S. Gopinath, W. Norona, L. Calandro, and L. Hennessy. 2012. Development of a "Global" STR multiplex for human identification analysis. *21st Int Symp Forensic Sci,* Hobart, Tasmania.

Wang,Y., J. Ju, B. A. Carpenter, J. M. Atherton, G. F. Sensabaugh, and R. A. Mathies. 1995. Rapid sizing of short tandem repeat alleles using capillary array electrophoresis and energy transfer fluorescent. *Anal Chem* 67: 1197–203.

Wang, Y., J. M. Wallin, J. Ju, G. F. Sensabaugh, and R. A. Mathies. 1996. High-resolution capillary array electrophoretic sizing of multiplexed short tandem repeat loci using energy-transfer fluorescent primers. *Electrophoresis* 15: 1485–90.

Wenz, H. M., J. M. Robertson, S. Menchen et al. 1998. High-precision genotyping by denaturing capillary electrophoresis. *Genome Res* 8: 69–80.

Williams, P. E., M. A. Marino, S. A. Del Rio, L. A. Turni, and J. M. Devaney. 1994. Analysis of DNA restriction fragments and polymerase chain reaction products by capillary electrophoresis. *J Chromatogr* A680: 525–40.

Next-Generation STR Genotyping Kits for Forensic Applications

8

JULIO J. MULERO
LORI K. HENNESSY

Contents

8.1 Introduction 164
8.2 Contents of a Commercial STR Kit 164
8.3 STR Kits in the United States 165
8.4 STR Kits in Europe 169
8.5 Next-Generation International STR Kits 170
8.6 MiniSTR Kits 170
8.7 Kits That Amplify the SE33 Marker 171
8.8 Nonstandard STR Markers in Forensics 174
8.9 Sex-Chromosome STR Kits 174
8.10 Genotypic Concordance 176
8.11 Conclusions 176
Acknowledgments 177
References 177

Abstract: Forensic DNA typing has been a constantly evolving field driven by innovations from academic laboratories as well as kit manufacturers. Central to these technological advances has been the transition from multilocus-probe restriction fragment length polymorphism (RFLP) methods to short tandem repeat (STR) PCR-based assays. STRs are now the markers of choice for forensic DNA typing and a wide variety of commercial STR kits have been designed to meet the various needs of a forensic lab. This review provides an overview of the commercial STR kits made available since the year 2000 and explains the rationale for creating these kits. Substantial progress has been made in key areas such as sample throughput, speed, and sensitivity. For example, a significant advancement for databasing labs was the capability of direct amplification from a blood or buccal sample without need for DNA extraction or purification, enabling increased throughput. Other key improvements are greater tolerance for inhibitors (e.g., humic acid, hematin, and tannic acid) present in evidence samples, polymerase chain reaction (PCR) cycling times decreased by 1–1.5 h, and greater sensitivity with improved buffer components and thermal cycling conditions. These improvements that have been made over the last 11 years have enhanced the ability of forensic laboratories to obtain a DNA profile from more challenging samples. However, with the proliferation of kits from different vendors, the primer binding sequences of the loci vary, which could result in discordant events that would need to be resolved either via a database-driven software solution or simply by evaluating discordant samples with multiple kits.

8.1 Introduction

DNA genotyping has become the tool of choice in forensic analysis of biological samples (Butler 2005). This technology was developed more than 25 years ago and has evolved from using restriction fragment length polymorphism (RFLP) to short tandem repeats (STR), mitochondrial DNA sequencing, and single nucleotide polymorphisms (SNP). Among these, STR reigns as the DNA marker of choice for the vast majority of human identification applications (Butler 2006). The widespread adoption of STR genotyping kits for forensic applications started in the late 1990s and coincided with the introduction of standardized markers by the Forensic Science Service (FSS) in the United Kingdom (Lygo et al. 1994). In the early 1990s, the FSS developed small multiplexes for forensic use (Kimpton et al. 1996) but it was not until 1996 when they developed a second-generation multiplex (SGM) (Sparkes et al. 1996a,b), which provided an improved set of markers for large-scale human identification. SGM was made up of six STRs (TH01, VWA, FGA, D8S1179, D18S51, and D21S11) and Amelogenin as a sex identification marker and had a modest random match probability of approximately 1 in 50 million. In 1999, Applied Biosystems (now part of Life Technologies, Foster City, California) launched the AmpFℓSTR SGM Plus kit, an 11-plex that, in addition to the six STR markers used in the original SGM kit, contained D3S1358, D16S539, D2S1338, and D19S433 (Cotton et al. 2000).

In 1997, the United States followed suit with the establishment of a core set of 13 STR loci known as the Combined DNA Index System (CODIS) loci. This set allowed for an average random-match probability of one in a quadrillion (1×10^{-15}) (Budowle et al. 1999; Moretti et al. 2001). Once a demand for STR multiplexes was established in the United States, two vendors, the Promega Corporation (Madison, Wisconsin) and Life Technologies, established themselves as the primary purveyors of STR kits for the forensic community. Initially, two PCR reactions were needed for the amplification of all 13 CODIS markers in kit combinations such as the AmpFℓSTR Profiler Plus and AmpFℓSTR COfiler (Life Technologies) (Holt et al. 2002) or PowerPlex 1.1 and PowerPlex 2.1 (Promega) (Levadokou et al. 2001). These kits also had overlapping markers as an internal control to minimize the possibility of a sample mixup.

Starting in 2000, kits such as the PowerPlex 16 System (Promega) (Krenke et al. 2002) and the AmpFℓSTR Identifiler PCR Reaction Kit (Life Technologies) (Collins et al. 2004) were capable of amplifying all 13 CODIS markers in a single PCR reaction along with the Amelogenin sex identification marker and two additional STR markers, Penta D and E in the PowerPlex 16 kit and D2S1338 and D19S433 in the Identifiler kit. This chapter will focus on the new developments in STR kits that have taken place after the release of the aforementioned 16-STR multiplexes, which have now been on the marketplace for more than 10 years.

8.2 Contents of a Commercial STR Kit

A commercial STR kit contains reagents to efficiently amplify and genotype a DNA sample. Depending on the kit, the DNA sample may or may not need to be purified (i.e., direct amplification kits). The typical kit consists of a primer mix, a master mix, an allelic ladder, and a positive control DNA. The amplification step is carried out by a polymerase chain reaction (PCR) event that simultaneously amplifies multiple STR loci. The amplification step is carried out with a hot-start thermostable DNA polymerase in a master mix containing reagents needed for the optimal performance of the enzyme. The primer mix contains a set of forward and reverse primers for each locus where one of the locus-specific primers is fluorescently labeled for subsequent detection during the separation of the fragments via capillary electrophoresis. The allelic ladder, which is essential for genotyping, is an artificial mixture of the common alleles present in the human population for each particular STR marker (Sajantila et al. 1992). The allelic ladder is added during the electrophoresis step and is run in parallel with the amplified samples as a standard. The positive control DNA serves as an internal control when following the recommended amplification conditions. In addition to the kit components, a manufacturer will provide evidence that the kit has been extensively validated and that it meets the criteria needed for its intended use.

8.3 STR Kits in the United States

The AmpFℓSTR Identifiler PCR Amplification Kit and the PowerPlex 16 System have been used widely by both forensic databasing and casework laboratories to genotype STRs. This has resulted in the addition of millions of STR profiles to DNA databases, leading to the resolution of countless cases by the criminal justice system. The proven success of DNA genotyping during the investigative process has driven the passage of new legislation to collect more DNA samples from offenders and/or arrestees (Berson 2009) and resulted in the use of DNA evidence for an expanded number and wider variety of criminal cases. In response to the increased demand for DNA analysis, laboratories are continuously seeking to implement enhanced technologies that allow them to process database and casework samples more efficiently and effectively.

Casework laboratories face the challenge of having to amplify DNA extracted from a wide variety of sample types. For instance, crimes committed outdoors present challenges when body fluids are deposited on soil, sand, and other surfaces that contain substances that may co-extract with the perpetrators' DNA and prevent PCR amplification (McNally et al. 1989; Sutlovic et al. 2005). Indoors, inhibitors of the PCR might be encountered on surfaces such as fabric containing textile dyes, leather, and wood furniture. The amplification of a DNA sample containing an inhibitor of the PCR may result in loss of the alleles from the larger-sized STR loci or even complete amplification failure at all loci. While new and improved methods for sample extraction can remove most inhibitors of the PCR (Brevnov et al. 2009), there are still cases where sample extraction may not be an option due to the small quantities available or because the laboratory itself has not adopted these new sample-extraction technologies.

The AmpFℓSTR Identifiler® Plus PCR Amplification Kit (Life Technologies) was developed to address the desire for greater sensitivity and improved performance on inhibited samples (Figure 8.1). The Identifiler Plus kit employs the same primers as the widely used Identifiler Kit (Table 8.1) but employs a much more robust master mix and modified thermal cycling parameters that help it overcome most inhibitors of the PCR. The PowerPlex 16 HS System is an updated version of the PowerPlex 16 System. The primers and dyes remain unchanged, and similarly to the Identifiler Plus kit, it introduces an enhanced buffer system with improved robust performance when compared to the original version of the kit.

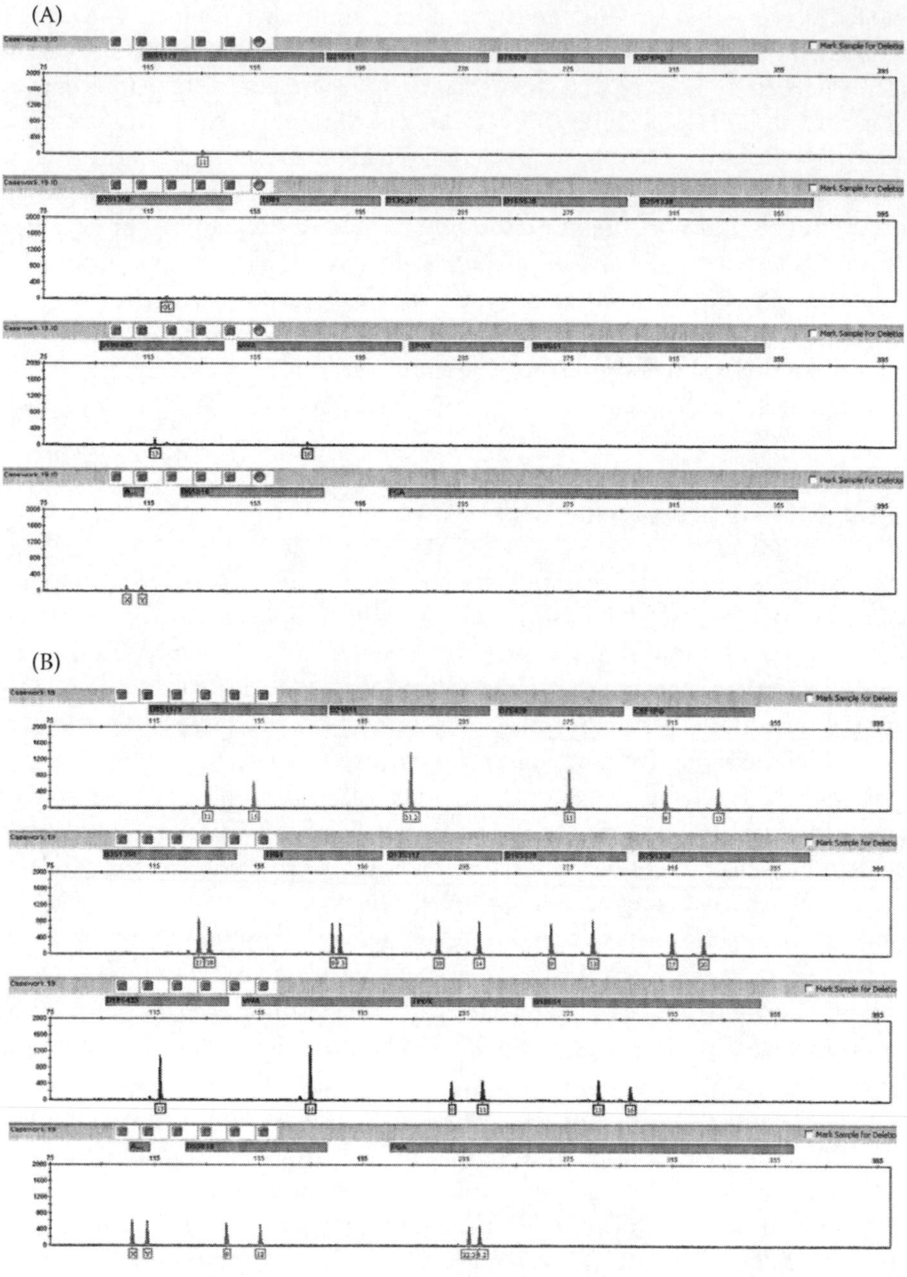

Figure 8.1 Cigarette butt (data from customer) amplified with the Identifiler™ (A) and Identifiler Plus™ kits (B). Six alleles were detected using the Identifiler kit and a full profile was obtained using the Identifiler Plus kit.

Databasing laboratories face the challenge of amplifying DNA from blood or buccal cells. Crude biological samples such as blood and buccal cells contain many substances that can inhibit the PCR (Akane et al. 1994; Al-Soud and Radstrom 2001; De Franchis et al. 1998). Furthermore, many database laboratories use FTA cards (Whatman, Inc., Piscataway, New Jersey) for archiving blood or buccal samples at room temperature to allow re-interrogation of the DNA profiles at any time. FTA cards are treated with proprietary chemicals

Table 8.1 Autosomal STR Kits Produced after the Year 2000

Powerplex® 16 Powerplex® 16HS	Identifiler™ Identifiler Plus™ Identifiler Direct™	Powerplex® 18D	NGM™	NGM SElect™ NGM SElect™ Express	Powerplex® ESX-16 Powerplex® ESI-16	Powerplex® ESX-17 Powerplex® ESI-17	ESSplex SE	Powerplex® ES	SEfiler™ SEfiler Plus™	Powerplex® 21	Globalfiler™ Globalfiler™ Express	PowerPlex® Fusion
Amelogenin	Amelogenin	Amelogenin	Amelogenin	Amelogenin	Amelogenin	Amelogenin	Amelogenin	Amelogenin	Amelogenin	Amelogenin	Amelogenin	Amelogenin
D3S1358	D3S1358	D3S1358	D3S1358	D3S1358	D3S1358	D3S1358	D3S1358	D3S1358	D3S1358	D3S1358	D3S1358	D3S1358
D8S1179	D8S1179	D8S1179	D8S1179	D8S1179	D8S1179	D8S1179	D8S1179	D8S1179	D8S1179	D8S1179	D8S1179	D8S1179
D18S51	D18S51	D18S51	D18S51	D18S51	D18S51	D18S51	D18S51	D18S51	D18S51	D18S51	D18S51	D18S51
D21S11	D21S11	D21S11	D21S11	D21S11	D21S11	D21S11	D21S11	D21S11	D21S11	D21S11	D21S11	D21S11
FGA	FGA	FGA	FGA	FGA	FGA	FGA	FGA	FGA	FGA	FGA	FGA	FGA
TH01	TH01	TH01	TH01	TH01	TH01	TH01	TH01	TH01	TH01	TH01	TH01	TH01
vWA	vWA	vWA	vWA	vWA	vWA	vWA	vWA	vWA	vWA	vWA	vWA	vWA
D16S539	D16S539	D16S539	D16S539	D16S539	D16S539	D16S539	D16S539		D16S539	D16S539	D16S539	D16S539
CSF1PO	CSF1PO	CSF1PO								CSF1PO	CSF1PO	CSF1PO
D5S818	D5S818	D5S818								D5S818	D5S818	D5S818
D7S820	D7S820	D7S820								D7S820	D7S820	D7S820
D13S317	D13S317	D13S317								D13S317	D13S317	D13S317
TPOX	TPOX	TPOX								TPOX	TPOX	TPOX
	D2S1338	D2S1338	D2S1338	D2S1338	D2S1338	D2S1338	D2S1338		D2S1338	D2S1338	D2S1338	D2S1338
	D19S433	D19S433	D19S433	D19S433	D19S433	D19S433	D19S433		D19S433	D19S433	D19S433	D19S433
Penta E		Penta E								Penta E		Penta E
Penta D		Penta D								Penta D		Penta D
				SE33		SE33	SE33	SE33	SE33		SE33	
										D6S1043		
			D1S1656	D1S1656	D1S1656	D1S1656	D1S1656			D1S1656	D1S1656	D1S1656
			D2S441	D2S441	D2S441	D2S441	D2S441				D2S441	D2S441
			D10S1248	D10S1248	D10S1248	D10S1248	D10S1248				D10S1248	D10S1248
			D12S391	D12S391	D12S391	D12S391	D12S391			D12S391	D12S391	D12S391
			D22S1045	D22S1045	D22S1045	D22S1045					D22S1045	D22S1045
											DYS391	DYS391
											Y Indel	

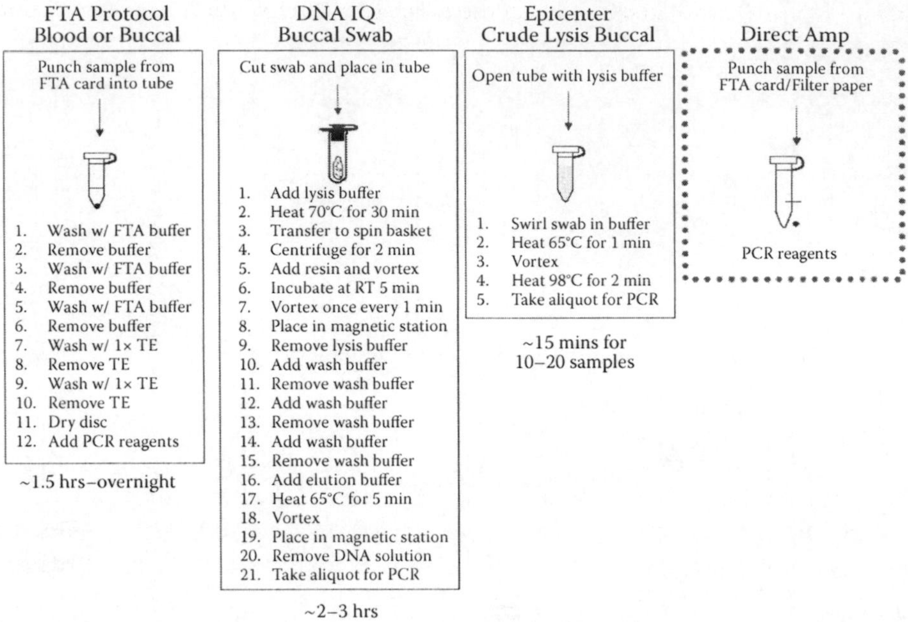
FTA Protocol
Blood or Buccal
Punch sample from FTA card into tube
1. Wash w/ FTA buffer
2. Remove buffer
3. Wash w/ FTA buffer
4. Remove buffer
5. Wash w/ FTA buffer
6. Remove buffer
7. Wash w/ 1× TE
8. Remove TE
9. Wash w/ 1× TE
10. Remove TE
11. Dry disc
12. Add PCR reagents
~1.5 hrs–overnight
DNA IQ
Buccal Swab
Cut swab and place in tube
1. Add lysis buffer
2. Heat 70°C for 30 min
3. Transfer to spin basket
4. Centrifuge for 2 min
5. Add resin and vortex
6. Incubate at RT 5 min
7. Vortex once every 1 min
8. Place in magnetic station
9. Remove lysis buffer
10. Add wash buffer
11. Remove wash buffer
12. Add wash buffer
13. Remove wash buffer
14. Add wash buffer
15. Remove wash buffer
16. Add elution buffer
17. Heat 65°C for 5 min
18. Vortex
19. Place in magnetic station
20. Remove DNA solution
21. Take aliquot for PCR
~2–3 hrs
Epicenter
Crude Lysis Buccal
Open tube with lysis buffer
1. Swirl swab in buffer
2. Heat 65°C for 1 min
3. Vortex
4. Heat 98°C for 2 min
5. Take aliquot for PCR
~15 mins for
10–20 samples
Direct Amp
Punch sample from
FTA card/Filter paper
PCR reagents

Figure 8.2 Workflow comparison of traditional versus direct-amplification approaches.

that lyse cells, inactivate pathogens, and inhibit the growth of bacteria (Burgoyne 1996). As a result, a wash-and-dry procedure is required to remove these PCR inhibitors from the cards. The wash-and-dry procedure takes approximately 1–3 h or more depending on the protocol used (manual versus automated) and the number of samples being processed.

The AmpFℓSTR Identifiler Direct PCR Amplification Kit (Life Technologies) (Wang et al. 2011) introduced in 2010 and the PowerPlex 18D (Promega) (Oostdik et al. 2011) introduced in 2011 represent a new class of STR kits optimized to allow direct amplification of single-source blood and buccal samples on an FTA card without the need for sample extraction and purification. These kits employ optimized PCR buffer formulations and an improved PCR cycling protocol designed specifically to overcome the PCR inhibitors present within the crude sample and contained within the FTA card (Figure 8.2). The Identifiler Direct Kit amplifies the same loci and employs the same primer sequences used in the Identifiler kit (Table 8.1). The PowerPlex 18D kit amplifies all of the markers in the original PowerPlex 16 kit plus the markers D2S1338 and D19S433 (Table 8.1). It also has an improved allelic ladder featuring many additional alleles not found in the original PowerPlex 16 kit. Both of these direct amplification kits have now been approved for use in laboratories generating DNA records for the National DNA Index System (NDIS) database. The PowerPlex 21 System (Ensenberger et al. 2012) also released in 2011 is designed to work with a variety of sample types, including casework samples, while also being compatible with direct amplification from FTA card punches as well as nontreated paper.

Recently, the utilization of these direct amplification strategies has been expanded to non-FTA-based collection devices such as the Bode Buccal DNA Collector™ (Bode Technology, Lorton, Virginia). These collectors do not contain embedded chemicals that can efficiently lyse the collected buccal cells and thus require a cell-lysis step prior to the addition of STR assay reagents. The Bode Punchprep (Bode Technology) is a reagent added to a 1.2-mm punch and then incubated at 70°C for about 20 min in order to efficiently lyse the cells.

The Prep-n-Go buffer (Life Technologies) introduced in 2011 is another reagent containing chemicals that can efficiently lyse the buccal cells but without the need for a heating step.

8.4 STR Kits in Europe

In 2005, the European Network of Forensic Science Institutes (ENFSI) and the European DNA Profiling Group (EDNAP) published recommendations on new multiplex developments in order to encourage standardization within Europe (Gill et al. 2006). The recommendations included the adoption of three new miniSTR loci (D2S441, D10S1248, and D22S1045) to improve the success rate of degraded DNA markers. The inclusion of two midi-STR loci (D1S1656 and D12S391) also helped improve the discrimination power of the STR multiplex. These recommendations encouraged manufacturers to develop amplification kits capable of producing more discriminating DNA profiles on a wider number of the more challenging samples increasingly encountered in forensic laboratories.

As a direct response to the ENFSI/EDNAP recommendations, three manufacturers developed kits that meet both the requirements stipulated by the ENFSI/EDNAP groups for a chemistry that can deliver improved data quality with casework samples and a powerful level of discrimination to support European data-sharing initiatives.

Promega Corporation developed a suite of four DNA-profiling kits including Powerplex® ESI-16 (European Standard Investigator), PowerPlex ESX-16 (European Standard Extended), and two kits containing the SE33 marker (discussed in section 8.7 below) (Tucker et al. 2011, 2012). The ESI-16 and ESX-16 kits share primer sequences for only three STR loci (D3S1358, D21S11, and vWA). The ESI-16 and ESX-16 use a five-dye detection chemistry to enable the amplification of the five new ENFSI loci (D1S1656, D2S441, D10S1248, D12S391, and D22S1045) in conjunction with the existing loci currently in use in the United Kingdom and Europe. In the ESX system, the "ENFSI loci" have been designed as mini- and midi-STRs, while the original STR amplicons have been designed to fit in as standard-size amplicons. This configuration provides a complementary approach to that offered by the PowerPlex ESI System, such that using both systems together enables one to amplify eight loci under 200 bases in the ESX-16 kit (D3S1358, TH01, D10S1248, D1S1656, D22S1045, vWA, D2S441, D12S391) and an additional five loci under 200 bases in the ESI-16 kit (D19S433, D16S539, D18S51, D8S1179, and the most frequently observed FGA alleles).

Life Technologies developed the AmpFℓSTR NGM Kit to address the new standards from the European forensic community (Budowle et al. 2011). The NGM Kit features improvements in robustness, data quality, and level of discrimination to meet the challenge of the European data-sharing initiatives. The NGM Kit simultaneously amplifies the 10 SGM Plus Kit loci (D3S1358, vWA, D16S539, D2S1338, D8S1179, D21S11, D18S51, D19S433, TH01, and FGA) together with the five additional loci approved for inclusion in the expanded European Standard Set of Loci (D10S1248, D22S1045, D2S441, D1S1656, and D12S391). This kit uses the same five-dye detection system utilized in previous Life Technologies kits while keeping the same primer sequences that were shared with the SGM Plus kit, with the exception of the D8S1179 locus where an additional primer was added to recover null alleles from Chamorro and Filipino populations as in the Identifiler kit (Leibelt et al. 2003). Qiagen (Hilden, Germany) introduced the Investigator ESSplex Kit (Qiagen 2010) featuring the same loci as the NGM Kit. It also features a five-dye detection system.

The new kits from all vendors have demonstrated the ability to amplify a diverse range of sample types, representative of the types of evidence submitted to operational forensic laboratories, and have shown to produce better-quality results in models of DNA degradation and common models of inhibition when compared to the SGM Plus kit (Tucker et al. 2011, 2012).

8.5 Next-Generation International STR Kits

In 2011, the CODIS Core Loci Working Group announced a plan to increase the number of CODIS loci (Hares 2012). The reasons for expanding the number of loci are to reduce the likelihood of adventitious matches in the national database, to increase international compatibility, and to increase discrimination power to aid in missing persons cases. In their communication the loci were divided into the categories of required (Section A) and recommended (Section B). Among the required loci are two gender identification markers (Amelogenin and DYS391) as well as four markers unique to the European standard set of loci (D1S1656, D12S391, D2S441, and D10S1248). Responding to this recent announcement, manufacturers are developing truly international STR multiplexes that amplify in a single PCR reaction the European as well as the U.S. loci.

Life Technologies has responded to this challenge by developing two application-specific kits: the GlobalFiler™ and GlobalFiler™ Express PCR Amplification kits. Both kits share the same marker configuration and incorporate all of the 23 required and recommended loci announced by the CODIS core loci working group (Table 8.1). The kits also have an additional gender identification marker (Y indel) in the smaller range of the multiplex to enable efficient amplification from degraded DNA samples. The GlobalFiler kit has been designed with casework type samples in mind. It is optimized for the efficient amplification of low levels of DNA and to overcome common inhibitors of the PCR. The GlobalFiler Express kit is designed for the amplification of single-source samples such as those originating from blood or buccal samples deposited on treated paper, untreated paper, and swab substrates.

Promega Corporation is developing its own version of an international STR multiplex. The PowerPlex Fusion System consists of 24 markers (Table 8.1) (Oostdik et al. 2012). This assay differs from the GlobalFiler kits in two significant ways: (1) It amplifies all of the new required loci but only two of the three recommended loci (TPOX and D22S1045). In lieu of the recommended marker SE33, the PowerPlex Fusion system amplifies the Penta D and Penta E loci to aid in paternity and forensic cases and to allow searching of databases that already include profiles with these Penta loci. (2) The PowerPlex Fusion System is a dual-purpose kit in that it can be used for common casework samples as well as the direct amplification of reference sample types stored on paper with only minor changes to the PCR amplification conditions.

At the time of this writing these kits are not yet available for purchase.

8.6 MiniSTR Kits

Environmental exposure degrades DNA molecules by breaking them into smaller pieces (Lindahl 1993). In order for PCR amplification to occur, the DNA template must be intact not only where the two primers bind but also between the primers so that full extension can occur. If the DNA template were broken, the PCR would be unsuccessful because primer extension will halt at the break in the template. The observation that small PCR amplification products are preferentially amplified in degraded DNA samples was first reported after typing victims from the 1993 Branch Davidian fire disaster in Waco, Texas (Whitaker et al. 1995).

Several publications have demonstrated improvements in genotyping degraded DNA samples by repositioning primers as close as possible to the STR repeat region (Butler et al. 2003; Coble and Butler 2005; Tsukada et al. 2002; Wiegand and Kleiber 2001). These primer changes result in smaller PCR products termed "miniSTRs" and increase the potential number of template molecules available for the PCR. In 2007, the AmpFℓSTR MiniFiler PCR Amplification Kit (Life Technologies) became the first commercial kit designed to amplify miniSTRs (Hill et al. 2007; Mulero et al. 2008). It features eight of the largest-sized loci in the Identifiler kit as miniSTRs (Table 8.2). Five of these loci (D16S539, D21S11, D2S1338, D18S51, and FGA) are also five of the largest loci in the AmpFℓSTR NGM and NGM SElect kits. Together with the gender-identification locus Amelogenin, this nine-locus multiplex enables simultaneous amplification of the loci that often fail detection during the amplification of severely degraded DNA samples (Figure 8.3). Seven of these loci (D7S820, D13S317, D16S539, D21S11, D18S51, CSF1PO, FGA) are part of the core loci required by CODIS (Table 8.2).

The PowerPlex S5 System (Promega) is primarily a screening tool, but it can also be used as a miniSTR kit (Promega 2008) (Table 8.2). It amplifies four CODIS STR loci (D18S51, D8S1179, TH01, and FGA) plus Amelogenin. The amplicons for all loci are smaller than 260 bp.

The Investigator Hexaplex ESS Kit (Qiagen 2010) was designed for prescreening strategies, follow-up analysis of already genotyped samples, or analysis of highly degraded DNA (maximum amplicon size of 225 bp). This kit coamplifies six STR markers (Table 8.2) and the sex-identification marker Amelogenin. This kit complements the SGM Plus set of loci, or a CODIS set including D2S1338 and D19S433, to complete the full European Standard Set.

8.7 Kits That Amplify the SE33 Marker

The SE33 locus is the most informative tetranucleotide STR used for human identification (Butler et al. 2008) The SE33 locus is highly polymorphic and exhibits structural variation as well as complex length and sequence polymorphisms, with some microvariants differing by 1 bp (Polymeropoulos et al. 1992; Rolf et al. 1997). SE33 is also known for having the highest mutation rate of any STR in forensic use, which may limit its use in relationship testing (Dauber et al. 2012; Müller et al. 2010). The Federal Criminal Police Office of Germany (BKA) included the SE33 locus as one of the eight core genetic loci to be included in their national database. The seven other required loci were D3S1358, D8S1179, D18S51, TH01, vWA, FGA, and D21S11. In 2002, two commercial kits were introduced in Germany, both containing the SE33 marker. PowerPlex ES (Promega) contained eight STR loci plus Amelogenin (Table 8.1). The AmpFℓSTR SEfiler kit (Life Technologies) (Coticone et al. 2004) contained all of the markers in the PowerPlex ES kit plus three additional ones (D16S539, D2S1338, and D19S433) (Table 8.1).

In 2007, the AmpFℓSTR SEfiler Plus kit was introduced to the German market (Mulero et al. 2008). It contained the same loci and primer sequences as the SEfiler™ kit but used

Table 8.2 Specialty Autosomal STR Kits Produced after the Year 2000

Minifiler™	Powerplex® S5	Hexaplex ESS	HDplex	Powerplex® CS7
Amelogenin	Amelogenin	Amelogenin	Amelogenin	
D18S51	D18S51		D18S51	
FGA	FGA			
	TH01	TH01		
D2S1338				
D7S820				
D13S317				
D16S539				
D21S11				
CSF1PO				
	D8S1179			
		D1S1656		
		D2S441		
		D10S1248		
		D12S391		
		D22S1045		
			D1S2325	
			D2S1360	
			D3S1744	
			D4S2366	
			D5S2500	
			D6S474	
			D7S1517	
			D8S1132	
			D10S2325	
			D12S391	
			D21S2055	
			SE33	
				F13B
				F13A01
				FESFPS
				LPL
				Penta C
				Penta D
				Penta E

improved synthesis and purification processes to minimize the presence of dye-labeled artifacts. Other improvements included modified PCR cycling conditions for enhanced sensitivity and a new buffer formulation that improved performance with inhibited samples when compared to the original SEfiler kit. In 2009, yet another generation of SE33-containing kits was introduced. Following the European recommendations for new STR markers described in the previous section, the PowerPlex ESI-17 and PowerPlex ESX-17 (Table 8.1) were introduced in countries needing the SE33 marker (Hill et al. 2011). Similarly, in 2010, Life Technologies released the AmpFℓSTR NGM SElect kit (Green et al. 2012) and in 2012 the direct amplification AmpFℓSTR NGM SElect Express kit, both of which combine the SE33 marker with the

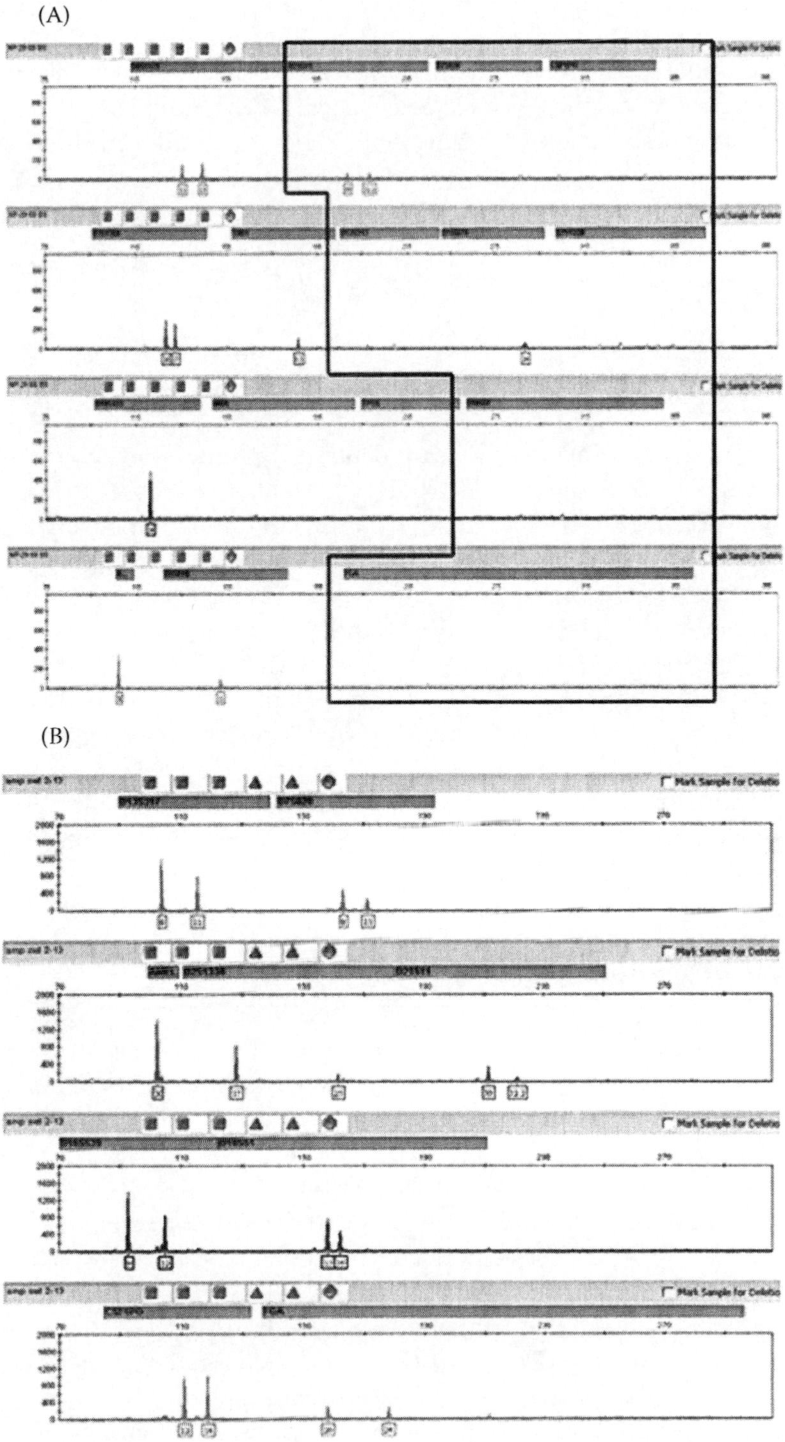

Figure 8.3 Bone sample (data from customer) amplified with the Identifiler Plus™ kit (A) illustrating recovery of 11 alleles. The area outlined in black highlights loci contained in the MiniFiler™ kit that yielded a full profile for this sample (B).

markers in the NGM kit (Table 8.1). These kits also share (with the exception of SE33 primer sequences in the NGM kit) all of the other primer sequences.

The Investigator ESSplex SE Kit (Qiagen) amplifies the same 16 STRs as the NGM SElect kit including Amelogenin (Table 8.1) (Qiagen 2010). According to the manufacturer, the kit was developed specifically for blood, buccal swabs, and forensic stains and can be used for forensic applications, identification of tissue culture cell lines, evaluation of population-level genetic structure, and paternity testing.

It is likely that the inclusion of the SE33 marker in these next-generation, high-performing kits coupled with the increased discrimination capacity that this marker offers will lead to additional countries outside of Germany using this marker for routine forensic applications.

8.8 Nonstandard STR Markers in Forensics

In some very special cases involving kinship deficiency testing, the use of additional nonstandard STR markers could prove valuable. The PowerPlex CS7 System, Custom (Promega 2010), is a multiplex STR assay for relationship testing and human identification. The PowerPlex CS7 System allows coamplification and three-color detection of seven STR loci (Table 8.2). The PowerPlex CS7 System contains two loci, Penta D and Penta E, which overlap with the loci included in the PowerPlex 16 and PowerPlex 16 HS Systems. This feature allows the PowerPlex CS7 System to be used as a confirmatory kit in paternity applications using the five unshared STR loci to supplement the genotype and increase the available information.

The Investigator HDplex Kit (Qiagen 2010) includes several highly polymorphic STR markers (Table 8.2) (Kuzniar et al. 2006; Schmid et al. 2005) that are not included in commonly used marker standards and were developed specifically to discriminate between closely related individuals. According to the manufacturer, the kit was designed for difficult forensic and paternity cases.

8.9 Sex-Chromosome STR Kits

As was the case with autosomal STRs, the widespread use of Y-STR kits occurred in the late 1990s and coincided with the acceptance of standardized markers for forensic use. The selected loci became known as the "minimal haplotype" loci in Europe and consisted of nine Y-STR markers: DYS19, DYS389I, DYS389II, DYS390, DYS391, DYS392, DYS393, and DYS385 a/b. In 2003, the U.S. Scientific Working Group on DNA Analysis Methods (SWGDAM) recommended the use of DYS438 and DYS439 in addition to the minimal haplotype loci. In 2003, the first commercially available Y-STR kit capable of amplifying the 11 recommended loci plus the Amelogenin marker was developed by ReliaGene Technologies (New Orleans, Louisiana) and was named Y-PLEX™ 12 (Shewale et al. 2004). Several months later the PowerPlex Y kit (Promega) (Krenke et al. 2005) was launched. This kit amplified the same loci as the Reliagene kit but replaced the Amelogenin marker with the DYS437 locus (Table 8.3). The following year, Life Technologies released the 17-plex AmpFℓSTR Yfiler PCR amplification kit (Table 8.3) (Mulero et al. 2006). This kit not only amplified the same loci as the previously released PowerPlex Y kit but it added six additional markers with very good diversity values when compared to the historical minimal haplotype markers. The additional loci DYS456, DYS458, DYS635, GATA H4,

Table 8.3 Current Commercial Y-STR Kits

Yfiler™	Powerplex® Y	Argus Y-12 QS	Powerplex® Y23
DYS19	DYS19	DYS19	DYS19
DYS385 a/b	DYS385a/b	DYS385a/b	DYS385a/b
DYS389 I/II	DYS389 I/II	DYS389 I/II	DYS389 I/II
DYS390	DYS390	DYS390	DYS390
DYS391	DYS391	DYS391	DYS391
DYS392	DYS392	DYS392	DYS392
DYS393	DYS393	DYS393	DYS393
DYS437	DYS437	DYS437	DYS437
DYS438	DYS438	DYS438	DYS438
DYS439	DYS439	DYS439	DYS439
DYS448			DYS448
DYS456			DYS456
DYS458			DYS458
Y GATA H4			Y GATA H4
DYS635			DYS635
			DYS576
			DYS481
			DYS549
			DYS533
			DYS570
			DYS643

DYS437, and DYS448 have since been adopted by the forensic community and thousands of samples from around the world have been genotyped with this multiplex. Qiagen offers the Investigator™ Argus Y-12 QS Kit (Qiagen 2010), which amplifies the same 12 Y-STRs found in the Promega PowerPlex Y kit (Table 8.3). The Argus Y-12 includes an internal control (Quality Sensor) that provides helpful information about PCR efficiency and about the presence of PCR inhibitors in tested samples.

In 2012, Promega Corporation launched the PowerPlex Y23 System (Thompson et al. 2012). This multiplex includes all 17 loci from the Y-Filer Kit, as well as six new informative Y-STR loci (Table 8.3). The PowerPlex Y23 System allows Y-STR analysis of both human forensic samples and database samples. It features a faster amplification time and better tolerance to inhibitors of the PCR when compared to previous generations of Y-STR multiplexes.

The widespread use of Y-STR kits has been facilitated by the creation of government- and academic-sponsored Y-STR databases around the world. The U.S. Y-STR Database (http://usystrdatabase.org/) is a searchable listing of 11- to 23-locus Y-STR haplotypes. This project is funded by the National Institute of Justice and managed by the National Center for Forensic Science (NCFS) in conjunction with the University of Central Florida. This database does not function like the more commonly used CODIS database. This is a population database only and is intended for use in estimating Y-STR haplotype population frequencies for forensic case work purposes. As of 2011, the U.S. Y-STR database version 2.5 contained 18,658 samples typed with the minimal haplotype plus SWGDAM-recommended loci and 8,487 samples typed with the 17 loci in the Yfiler kit. In Europe, Lutz Roewer and others have developed the largest Y-STR database in the world (http://www.yhrd.org) with sample contributions from more than 100 countries (Willuweit and

Roewer 2007). As of 2011, more than 97,500 samples have been typed with the European minimal haplotype and 36,400 samples with the 17 loci from the Yfiler kit.

Recently, a new set of Y-STR markers known as rapid mutating Y-STRs offer the potential of being able to discriminate within family members due to their high mutation rate when compared to the currently used markers (Ballantyne et al. 2010, 2012). These Y-STR markers could be very useful in forensic individual identification but they may have limited use in paternity testing or genealogical applications due to their high mutation rates. The inclusion of these markers in a commercial kit may have to wait until additional information is gathered regarding allele frequencies, gene diversity values, and possible recognition from a forensic organization as loci worth adding to already existing Y-STR databases.

Several publications have addressed X-chromosomal STR markers in forensic cases that involve kinship testing (Luo et al. 2011; Tillmar et al. 2011). These cases are rare but it appears that in some special deficiency cases there is an advantage to using X-STR over autosomal or even Y-STRs. To date, only one commercial kit exists for this very specific application. The Investigator Argus X-12 Kit (Qiagen 2010) enables simultaneous amplification of 12 X-chromosomal markers for kinship and paternity testing, as well as population genetics and anthropological studies. In addition, this kit is also suited for analysis of forensic stains, such as female traces on a male background. The markers are clustered into four linkage groups (L1 to L4) consisting of three markers per group, and thus each set of three markers is handled as a haplotype for genotyping (Qiagen 2010).

8.10 Genotypic Concordance

Although many of the kits from different manufacturers share the same loci, the primer sequences differ among the kits and thus raise a potential problem with genotypic concordance. Discordance between results generated using different STR primer sequences is a well-characterized phenomenon and most often results from primer binding-site mutations affecting at least one of the different primers, leading to the occurrence of null alleles (Heinrich et al. 2004; Hill et al. 2007, 2011; Kline et al. 2011). Discordance could also involve indel polymorphisms between the repeat region of the STR locus and the primer binding site (Rolf et al. 2011). Another instance of discordance was recently characterized as a region of secondary structure adjacent to the SE33 repeat region, which, if amplified, can affect the electrophoretic mobility of the amplification product and creates discordance among commercial STR kits (Wang et al. 2012). National DNA databases manage discordances by utilizing error-tolerant searching algorithms in their searches so that differences of one allele due to results obtained with different STR kits do not result in false exclusions. A simple strategy for countries that maintain the reference DNA would be to confirm the discrepancy by retesting the sample with the appropriate alternative kit. However, kit manufacturers should verify that their new kits remain as concordant as possible with previous generation kits to minimize these occurrences.

8.11 Conclusions

The new next-generation commercial kits discussed in this chapter have focused on improving performance with inhibited samples, increasing sensitivity, improving throughput by direct amplification, and to a lesser extent, faster thermal cycling time. In some of the newer kits the current PCR cycling times are now 1–1.5 h shorter than first-generation kits. Further investigations to decreasing cycling time can increase the throughput of a laboratory and potentially expand the DNA typing into other applications such as analysis of individuals at the airport or borders. Recent studies have shown that using alternative enzymes can decrease cycling time to ~17 min (Giese et al. 2009; Vallone et al. 2008; Wang et al. 2009). However, it is necessary to balance the decrease in time to result against the need to address factors that can impact interpretation of a DNA profile such as: generation of stutter products, nontemplate addition, intralocus balance, accuracy, and species specificity.

The development of new kits and the resulting competition among kit manufacturers provide a "driving force" for the progress of forensic genetics as it encourages continued innovation. The forensic community in turn responds by demanding products with better performance, ease of use, and increased discrimination potential. Recently, the CODIS Core Loci Working Group issued a communication outlining a plan to increase the number of CODIS Core loci (Hares 2012). The main reasons for expanding the CODIS core loci are to reduce the likelihood of adventitious matches in the national database, to increase international compatibility, and to increase discrimination power to aid in missing persons cases. Manufacturers are currently developing new kits that will meet the new requirements for additional markers. It is expected that these new products will deliver on all of the product features that kit users have grown to expect based on the current product offerings.

Recently, Life Technologies announced the development of Yfiler Plus, a new 27 Y-STR multiplex that coamplifies the 17 markers from the Yfiler kit with 10 additional highly polymorphic Y-STR markers (DYS576, DYS627, DYS460, DYS518, DYS570, DYS449, DYS481, DYF387S1a/b, and DYS533). These 10 new loci include 7 rapidly mutating Y-STR loci which allow for improved discrimination of related individuals.

Acknowledgments

We thank Robert Green of Life Technologies for useful discussions.

References

Akane, A., K. Matsubara, H. Nakamura, S. Takahashi, and K. Kimura. 1994. Identification of the heme compound copurified with deoxyribonucleic acid (DNA) from bloodstains, a major inhibitor of polymerase chain reaction (PCR) amplification. *J Forensic Sci* 39:362–72.

Al-Soud, W. A. and P. Radstrom. 2001. Purification and characterization of PCR-inhibitory components in blood cells. *J Clin Microbiol* 39:485–93.

Ballantyne, K. N., M. Goedbloed, R. Fang et al. 2010. Mutability of Y-chromosomal microsatellites: Rates, characteristics, molecular bases, and forensic implications. *Am J Hum Genet* 87:341–53.

Ballantyne, K. N., V. Keerl, A. Wollstein et al. 2012. A new future of forensic Y-chromosome analysis: Rapidly mutating Y-STRs for differentiating male relatives and paternal lineages. *Forensic Sci Int Genet* 6:208–18.

Berson, S. B. 2009. Debating DNA collection. 2011. *NIJ Journal No. 264*. http://www.ojp.usdoj.gov/nij/journals/264/debating-DNA.htm (accessed September 2012).

Brevnov, M. G., H. S. Pawar, J. Mundt, L. M. Calandro, M. R. Furtado, and J. G. Shewale. 2009. Developmental validation of the PrepFiler Forensic DNA Extraction Kit for extraction of genomic DNA from biological samples. *J Forensic Sci* 54:599–607.

Budowle, B., J. Ge, R. Chakraborty et al. 2011. Population genetic analyses of the NGM STR loci. *Int J Legal Med* 125(1):101–09.

Budowle, B., T. R. Moretti, A. L. Baumstark, D. A. Defenbaugh, and K. M. Keys. 1999. Population data on the thirteen CODIS core short tandem repeat loci in African Americans, U.S. Caucasians, Hispanics, Bahamians, Jamaicans, and Trinidadians. *J Forensic Sci* 44:1277–86.

Burgoyne, L. A. 1996. Solid medium and method for DNA storage. U.S. Patent 5,496,562.

Butler, J. M. 2005. *Forensic DNA Typing: Biology, Technology, and Genetics of STR Markers*, 2nd ed. Elsevier: New York, NY.

Butler, J. M. 2006. Genetics and genomics of core short tandem repeat loci used in human identity testing. *J Forensic Sci* 51:253–65.

Butler, J. M., C. R. Hill, M. C. Kline et al. 2008. The single most polymorphic STR Locus: SE33 performance in US populations. *Forensic Sci Int Genet* Suppl Ser 1:23–5.

Butler, J. M, Y. Shen, and B. R. McCord. 2003. The development of reduced size STR amplicons as tools for analysis of degraded DNA. *J Forensic Sci* 48:1054–64.

Coble, M. D. and J. M. Butler. 2005. Characterization of new miniSTR loci to aid analysis of degraded DNA. *J Forensic Sci* 50:43–53.

Collins, P. J., L. K. Hennessy, C. S. Leibelt, R. K. Roby, D. J. Reeder, and P. A. Foxall. 2004. Developmental validation of a single-tube amplification of the 13 CODIS STR loci, D2S1338, D19S433, and Amelogenin: The AmpFℓSTR® Identifiler® PCR Amplification Kit. *J Forensic Sci* 49:1265–77.

Coticone, S. R., N. Oldroyd, H. Philips, and P. Foxall. 2004. Development of the AmpFℓSTR SEfiler PCR amplification kit: A new multiplex containing the highly discriminating ACTBP2 (SE33) locus. *Int J Legal Med* 118:224–34.

Cotton, E. A., R. F. Allsop, J. L. Guest et al. 2000. Validation of the AMPFℓSTR SGM plus system for use in forensic casework. *Forensic Sci Int* 112:151–61.

Dauber, E. M., A. Kratzer, F. Neuhuber et al. 2012. Germline mutations of STR-alleles include multi-step mutations as defined by sequencing of repeat and flanking regions. *Forensic Sci Int Genet* 6:381–86.

De Franchis, R., N. C. Cross, N. S. Foulkes, and T. M. Cox. 1998. A potent inhibitor of Taq polymerase copurifies with human genomic DNA. *Nucleic Acids Res* 16:10355.

Ensenberger, M. G., P. M. Fulmer, B. E. Krenke, R. S. McLaren, C. J. Sprecher, and D. R. Storts. 2012. Development of the PowerPlex® 21 System. *Profiles in DNA 2012.*

Giese, H., R. Lam, R. Selden, and E. Tan. 2009. Fast multiplexed polymerase chain reaction for conventional and microfluidic short tandem repeat analysis. *J Forensic Sci* 54:1287–96.

Gill, P., L. Fereday, N. Morling, and P. M. Schneider. 2006. New multiplexes for Europe—Amendments and clarification of strategic development. Forensic Sci Int 163:155–57.

Green, R. L., R. E. Lagacé, N. J. Oldroyd, L. K. Hennessy, and J. J. Mulero. 2012. Developmental validation of the AmpFℓSTR® NGM SElect™ PCR Amplification Kit: A next-generation STR multiplex with the SE33 locus. *Forensic Sci Int Genet* [Epub ahead of print] PubMed PMID: 22742953.

Hares, D. R. 2012. Expanding the CODIS core loci in the United States. *Forensic Sci Int Genet* 6:e52–54.

Heinrich, M., M. Muller, S. Rand, B. Brinkman, and C. Hohoff. 2004. Allelic drop-out in the STR system ACTBP2 (SE33) as a result of mutations in the primer binding region. *Int J Legal Med* 118:361–63.

Hill, C. R., D. L. Duewer, M. C. Kline et al. 2011. Concordance and population studies along with stutter and peak height ratio analysis for the PowerPlex® ESX 17 and ESI 17 systems. *Forensic Sci Int Genet* 5:269–75.

Hill, C. R., M. C. Kline, J. J. Mulero et al. 2007. Concordance study between the AmpFℓSTR MiniFiler PCR amplification kit and conventional STR typing kits. *J Forensic Sci* 52:870–3.

Holt, C. L., M. Buoncristiani, J. M. Wallin, T. Nguyen, K. D. Lazaruk, and P. S. Walsh. 2002. TWGDAM validation of AmpFℓSTR PCR amplification kits for forensic DNA casework. *J Forensic Sci* 47:66–96.

Kimpton, C. P., N. J. Oldroyd, S. K. Watson et al. 1996. Validation of highly discriminating multiplex short tandem repeat amplification systems for individual identification. *Electrophoresis* 17:1283–93.

Kline, M. C., C. R. Hill, A. E. Decker, and J. M. Butler. 2011. STR sequence analysis for characterizing normal, variant, and null alleles. *Forensic Sci Int Genet* 5:329–32.

Krenke, B. E., A. Tereba, S. J. Anderson et al. 2002. Validation of a 16-locus fluorescent multiplex system. *J Forensic Sci* 47:773–85.

Krenke, B. E., L. Viculis, M. L. Richard et al. 2005. Validation of a male-specific, 12-locus fluorescent short tandem repeat (STR) multiplex. *Forensic Sci Int* 148:1–14. Erratum in: *Forensic Sci Int* 151:109. Corrected and republished in: *Forensic Sci Int* 151:111–24.

Kuzniar, P., E. Jastrzebska, and R. Ploski. 2006. Validation of nine non-CODIS STR loci for forensic use in a population from Central Poland. *Forensic Sci Int* 159:258–60.

Leibelt, C., B. Budowle, P. Collins et al. 2003. Identification of a D8S1179 primer binding site mutation and the validation of a primer designed to recover null alleles. *Forensic Sci Int* 133:220–7.

Levadokou, E. N., D. A. Freeman, M. J. Budzynski et al. 2001. Allele frequencies for fourteen STR loci of the PowerPlex 1.1 and 2.1 multiplex systems and Penta D locus in Caucasians, African-Americans, Hispanics, and other populations of the United States of America and Brazil. *J Forensic Sci* 46:736–61. Erratum in: *J Forensic Sci* 46:1533.

Lindahl, T. 1993. Instability and decay of the primary structure of DNA. *Nature* 362:709–15.

Luo, H. B., Y. Ye, Y. Y. Wang et al. 2011. Characteristics of eight X-STR loci for forensic purposes in the Chinese population. *Int J Legal Med* 125:127–31.

Lygo, J. E., P. E. Johnson, D. J. Holdaway et al. 1994. The validation of short tandem repeat (STR) loci for use in forensic casework. *Int J Legal Med* 107:77–89.

McNally, L., R. C. Shaler, M. B. Baird, P. De Forest, and L. Kobilinsky. 1989. Evaluation of deoxyribonucleic acid (DNA) isolated from human bloodstains exposed to ultraviolet light, heat, humidity, and soil contamination. *J Forensic Sci* 34:1059–69.

Moretti, T. R., A. L. Baumstark, D. A. Defenbaugh, K. M. Keys, J. B. Smerick, and B. Budowle. 2001. Validation of short tandem repeats (STRs) for forensic usage: Performance testing of fluorescent multiplex STR systems and analysis of authentic and simulated forensic samples. *J Forensic Sci* 46:647–60.

Mulero, J. J., C. W. Chang, L. M. Calandro et al. 2006. Development and validation of the AmpFℓSTR Yfiler PCR amplification kit: A male specific, single amplification 17 Y-STR multiplex system. *J Forensic Sci* 51:64–75.

Mulero, J. J., C. W. Chang, R. E. Lagacé et al. 2008. Development and validation of the AmpFℓSTR MiniFiler PCR Amplification Kit: A MiniSTR multiplex for the analysis of degraded and/or PCR inhibited DNA. *J Forensic Sci* 53:838–52.

Mulero, J. J., N. J. Oldroyd, M. T. Malicdem, and L. K. Hennessy. 2008. Developmental validation of the AmpFℓSTR SEfiler Plus PCR amplification kit: An improved multiplex with enhanced performance for inhibited samples. *Forensic Sci Int Genet* Suppl Ser 1:121–3.

Müller, M., U. Sibbing, C. Hohoff, and B. Brinkmann. 2010. Haplotype-assisted characterization of germline mutations at short tandem repeat loci. *Int J Legal Med* 124:177–82.

Oostdik, K., M. Ensenberger, B. Krenke, C. Sprecher, and D. Storts. 2011. The PowerPlex® 18D System: A direct amplification STR system with reduced thermal cycling time. *Profiles in DNA 2011*.

Oostdik, K., M. Ensenberger, C. Sprecher, J. Bourdeau-Heller, B. Krenke, and D. Storts. 2012. Bridging databases for today and tomorrow: The Powerplex® Fusion System. *Profiles in DNA 2012*.

Polymeropoulos, M. H., D. S. Rath, H. Xiao, and C. R. Merrill. 1992. Tetranucleotide repeat polymorphism at the human beta-actin related pseudogene H-beta-Ac-psi-2 (ACTBP2). *Nucleic Acids Res* 20:1432.

Promega. 2010. Powerplex CS7 System, http://www.promega.com/products/genetic-identity/str-analysis-for-forensic-and-paternity-testing/ (accessed May 2013).

Qiagen. 2010. Investigator ESSplex Kit, http://www.qiagen.com/Products/Lab-Focus/Human-Identity-and-Forensics/STR-Technology/ (accessed May 2013).

Rolf, B., N. Bulander, and P. Wiegand. 2011. Insertion-/deletion polymorphisms close to the repeat region of STR loci can cause discordant genotypes with different STR kits. *Forensic Sci Int Genet* 5:339–41.

Rolf, B., M. Schurenkamp, A. Junge, and B. Brinkmann. 1997. Sequence polymorphism at the tetranucleotide repeat of the human beta-actin related pseudogene H-beta-Acpsi-2 (ACTBP2) locus. *Int J Legal Med* 110:69–72.

Sajantila, A., S. Puomilahti, V. Johnsson, and C. Ehnholm. 1992. Amplification of reproducible allele markers for amplified fragment length polymorphism analysis. *BioTechniques* 12:16–22.

Schmid, D., K. Anslinger, and B. Rolf. 2005. Allele frequencies of the ACTBP2 (= SE33), D18S51, D8S1132, D12S391, D2S1360, D3S1744, D5S2500, D7S1517, D10S2325 and D21S2055 loci in a German population sample. *Forensic Sci Int* 16.151:303–5.

Shewale, J. G., H. Nasir, E. Schneida, A. M. Gross, B. Budowle, and S. K. Sinha. 2004. Y-chromosome STR system, Y-PLEX 12, for forensic casework: Development and validation. *J Forensic Sci* 49:1278–90.

Sparkes, R., C. Kimpton, S. Gilbard et al. 1996. The validation of a 7-locus multiplex STR test for use in forensic casework. II. Artefacts, casework studies and success rates. *Int J Legal Med* 109(4):195–204.

Sparkes, R., C. Kimpton, S. Watson et al. 1996. The validation of a 7-locus multiplex STR test for use in forensic casework. I. Mixtures, ageing, degradation and species studies. *Int J Legal Med* 109:186–94.

Sutlovic, D., M. Definis Gojanovic, S. Andelinovic, D. Gugic, and D. Primorac. 2005. Taq polymerase reverses inhibition of quantitative real time polymerase chain reaction by humic acid. *Croat Med J* 46:556–62.

Thompson, J. M., M. M. Ewing, D. R. Rabbach, P. M. Fulmer, C. J. Sprecher, and D. R. Storts. 2012. The PowerPlex® Y23 System: A new Y-STR multiplex for casework and database applications. *Profiles in DNA 2012*.

Tillmar, A. O., T. Egeland, B. Lindblom, G. Holmlund, and P. Mostad. 2011. Using X-chromosomal markers in relationship testing: Calculation of likelihood ratios taking both linkage and linkage disequilibrium into account. *Forensic Sci Int Genet* 5:506–11.

Tsukada, K., K. Takayanagi, H. Asamura, M. Ota, and H. Fukushima. 2002. Multiplex short tandem repeat typing in degraded samples using newly designed primers for the TH01, TPOX, CSF1PO, and vWA loci. *Legal Med* 4:239–45.

Tucker, V. C., A. J. Hopwood, and C. J. Sprecher. 2011. Developmental validation of the PowerPlex® ESI 16 and PowerPlex® ESI 17 Systems: STR multiplexes for the new European standard. *Forensic Sci Int Genet* 5:436–48.

Tucker, V. C., A. J. Hopwood, C. J. Sprecher et al. 2012. Developmental validation of the PowerPlex® ESX 16 and PowerPlex® ESX 17 Systems. *Forensic Sci Int Genet* 6:124–31.

Vallone, P. M., C. R. Hill, and J. M. Butler. 2008. Demonstration of rapid multiplex PCR amplification involving 16 genetic loci. *Forensic Sci Int Genet* 3: 42–5.

Wang, D. Y., C. W. Chang, R. E. Lagacé, N. J. Oldroyd, and L. K. Hennessy. 2011. Development and validation of the AmpFℓSTR® Identifiler® Direct PCR Amplification Kit: A multiplex assay for the direct amplification of single-source samples. *J Forensic Sci* 56:835–45.

Wang, D. Y., C. C. Chang, N. J. Oldroyd, and L. K. Hennessy. 2009. Direct amplification of STRs from blood or buccal cell samples. *Forensic Sci Int Genet* Suppl Ser 2:113–5.

Wang, D. Y., R. L. Green, R. E. Lagacé, N. J. Oldroyd, L. K. Hennessy, and J. J. Mulero. 2012. Identification and secondary structure analysis of a region affecting electrophoretic mobility of the STR locus SE33. *Forensic Sci Int Genet* 6:310–16.

Whitaker, J. P., T. M. Clayton, A. J. Urquhart et al. 1995. Short tandem repeat typing of bodies from a mass disaster: High success rate and characteristic amplification patterns in highly degraded samples. *BioTechniques* 18:670–7.

Wiegand, P. and M. Kleiber. 2001. Less is more—Length reduction of STR amplicons using redesigned primers. *Int J Legal Med* 114:285–7.

Willuweit, S. and L. Roewer, 2007. International Forensic Y Chromosome User Group. Y chromosome haplotype reference database (YHRD): Update. *Forensic Sci Int Genet* 1:83–7.

Biology and Genetics of New Autosomal STR Loci Useful for Forensic DNA Analysis

9

JOHN M. BUTLER
CAROLYN R. HILL

Contents

9.1 Introduction 181
9.2 Core Loci Used in the United States and Europe 182
9.3 Commercial Kits 183
9.4 Autosomal STR Loci in Current Commercial Kits 185
9.5 Additional Autosomal STR Loci 191
9.6 Relative Variability of Autosomal STR Loci 191
9.7 Expansion of U.S. Core Loci 194
9.8 Conclusions 195
Acknowledgments 195
References 196

Abstract: Short tandem repeats (STRs) are regions of tandemly repeated DNA segments found throughout the human genome that vary in length (through insertion, deletion, or mutation) with a core repeated DNA sequence. Forensic laboratories commonly use tetranucleotide repeats, containing a four base pair (4 bp) repeat structure such as GATA. In 1997, the FBI Laboratory selected 13 STR loci that form the backbone of the U.S. national DNA database. Building on the European expansion in 2009, the FBI announced plans in April 2011 to expand the U.S. core loci to as many as 20 STRs to enable more global DNA data sharing. Commercial STR kits enable consistency in marker use and allele nomenclature between laboratories and help improve quality control. The STRBase Web site, maintained by the U.S. National Institute of Standards and Technology, contains helpful information on STR markers used in human identity testing.

9.1 Introduction

Eukaryotic genomes are full of repeated DNA sequences (Ellegren 2004). These repeated DNA sequences come in all sizes and are typically designated by the length of the core repeat unit and the number of contiguous repeat units or the overall length of the repeat region. Long repeat units may contain several hundred to several thousand bases in the core repeat.

- DNA regions with repeat units that are 2 to 7 bp in length are called microsatellites, simple sequence repeats (SSRs), or most commonly short tandem repeats (STRs). STRs have become popular DNA repeat markers because they are easily amplified by the polymerase chain reaction (PCR) without the problems of differential amplification. This is because both alleles from a heterozygous individual are similar in size since the repeat size is small. The number of repeats in STR markers can be highly variable among individuals, which make these STRs effective for human identification purposes.
- In the past two decades, a number of tetranucleotide STRs have been explored for application to human identification. The types of STR markers that have been sought have included short STRs for typing degraded DNA materials (Coble and Butler 2005), STRs with low stuttering characteristics for analyzing mixtures (Bacher and Schumm 1998), and male-specific Y chromosome STRs for analyzing male-female mixtures from sexual crimes (Carracedo and Lareu 1998; Kayser et al. 2004). In order to take advantage of the product rule and be able to combine the genetic information across multiple loci, autosomal STR markers used in forensic DNA typing are typically chosen from separate chromosomes or are widely spaced on the same chromosome to avoid any problems with linkage between the markers.
- STR sequences vary not only in the length of the repeat unit and the number of repeats but also in the rigor with which they conform to an incremental repeat pattern. STRs are often divided into several categories based on the repeat pattern. Simple repeats contain units of identical length and sequence, compound repeats comprise two or more adjacent simple repeats, and complex repeats may contain several repeat blocks of variable unit length as well as variable intervening sequences (Urquhart et al. 1994). Not all alleles for an STR locus contain complete repeat units. Even simple repeats can contain nonconsensus alleles that fall in between alleles with full repeat units. Microvariants are alleles that contain incomplete repeat units (e.g., the allele 9.3 at the TH01 locus [Puers et al. 1993]).

9.2 Core Loci Used in the United States and Europe

For DNA typing markers to be effective across a wide number of jurisdictions, a common set of standardized markers must be used. Since their selection by the FBI Laboratory in November 1997 (Budowle et al. 1998; Butler 2006), a core set of 13 STR loci have been required within the United States to upload DNA profiles to the national DNA database (NDIS, National DNA Index System). The 13 core U.S. loci currently used by the Combined DNA Index System (CODIS) software to enable DNA matches within the United States are CSF1PO, FGA, TH01, TPOX, vWA, D3S1358, D5S818, D7S820, D8S1179, D13S317, D16S539, D18S51, and D21S11. Only eight loci overlap with STR data gathered in the United Kingdom and most other European nations.

While there are over 20,000 tetranucleotide STR loci now known to exist in the human genome due to the Human Genome Project efforts completed after the 1997 selection of the 13 CODIS STRs, not many of these loci have extensive population data collected to date—nor have these markers been tested in multiplex PCR assays or with forensic DNA samples. The need to maintain connection to the legacy data for the original STRs or at least a significant subset of previous data limits the loci that can be considered for expanding marker sets (Butler 2012; Gill et al. 2006).

In April 2009, the European Network of Forensic Science Institutes (ENFSI) voted to adopt five additional STR loci (D12S391, D1S1656, D2S441, D10S1248, and D22S1045) to their already existing European Standard Set (ESS) of seven STRs (TH01, vWA, FGA, D8S1179, D18S51, D21S11, and D3S1358). These 12 EU core loci are typically accompanied by D16S539, D2S1338, and D19S433 when amplified with commercially available STR kits (Figure 9.1).

In April 2011, the FBI Laboratory proposed an expanded set of core STR loci for the United States in order to (1) reduce the likelihood of adventitious matches as the number of profiles stored in the U.S. national DNA database increases, (2) increase international compatibility to assist law enforcement data-sharing efforts, and (3) increase the discrimination power to assist missing persons cases (Hares 2012a; FBI 2012). As noted in Figure 9.1, the proposed required loci for an expanded U.S. core are CSF1PO, FGA, TH01, vWA, D3S1358, D5S818, D7S820, D8S1179, D13S317, D16S539, D18S51, D21S11, D12S391, D1S1656, D2S441, D10S1248, D2S1338, D19S433, and DYS391 (Hares 2012a,b). In addition, the sex-typing marker amelogenin is included as a required locus.

9.3 Commercial Kits

Forensic DNA scientists rarely (and in the United States essentially never) use their own STR assays, opting to purchase quality controlled commercial kits instead. These kits come with allelic ladders, positive control samples, and premixed reagents including the fluorescently labeled oligonucleotides (primers) that target the specific locations in the human genome to be PCR amplified. Figure 9.1 includes the commonly used kits in the United States and Europe. Many of these kits enable simultaneous, multicolor fluorescence detection of 15 STRs and the sex-typing marker amelogenin in a single PCR reaction. For example, both Identifiler and PowerPlex 16 kits enable typing of the U.S. core 13 STRs plus two additional loci—Identifiler provides D2S1338 and D19S433 and PowerPlex 16 provides Penta D and Penta E (Figure 9.1).

To meet the needs of the European community, new STR kits have been released since 2009 by Promega (PowerPlex ESX and ESI 16 or 17), Life Technologies/Applied Biosystems (AmpFℓSTR NGM and NGM SElect), and Qiagen (ESSplex and ESSplex SE). These kits enable amplification of 15 or 16 STR loci with amelogenin using five-dye chemistry. More recently, Promega has released the PowerPlex Fusion kit (Promega 2012) and Life Technologies has produced the GlobalFiler kit (Life Technologies 2012). The loci included in these 24plex kits, which are designed to meet the expanded U.S. core loci requirements, are listed in Figure 9.1.

Our group at NIST has been involved in numerous concordance studies to help locate primer binding site mutations that lead to discordance results between commercial kits using different PCR primers (Hill et al. 2010, 2011a). Across all kits and samples examined we have observed a concordance rate of over 99.8% (Hill et al. 2011b).

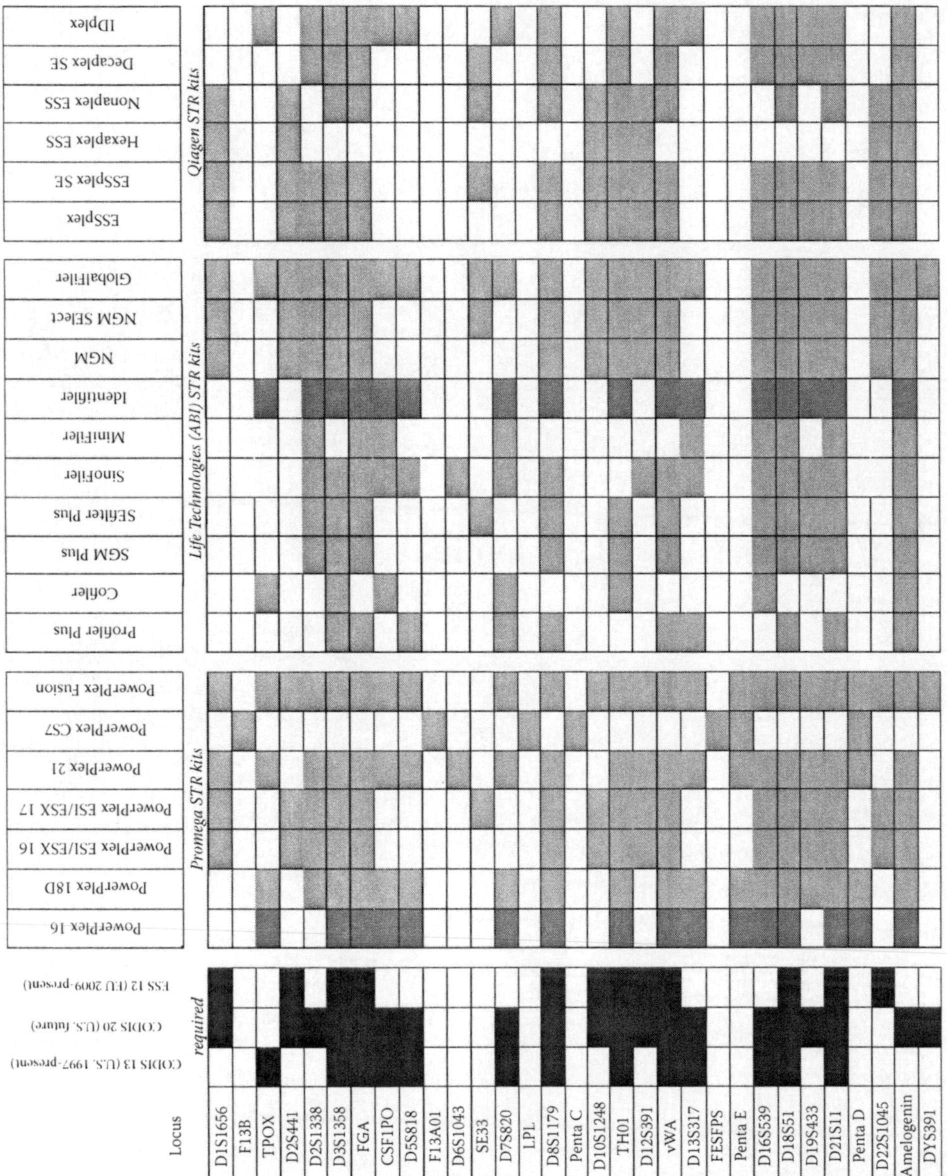

Figure 9.1 STR loci present in various sets of U.S. or European required loci and commercial kits used worldwide. Loci are listed by chromosomal position (Chr) with locations of syntenic pairs boxed in bold font. CODIS stands for the FBI's Combined DNA Index System and ESS refers to the European Standard Set. The GlobalFiler™ kit also includes a Y indel to confirm AMEL Y.

9.4 Autosomal STR Loci in Current Commercial Kits

Each of the 29 autosomal STR loci used in common commercial kits has unique characteristics, either in terms of the number of alleles present, the type of repeat sequence, or the kinds of microvariants that have been observed. This section reviews some of the basic details on each of the core and commonly used STR loci that are present in commercial STR kits (Table 9.1).

- **CSF1PO** is a simple tetranucleotide repeat found in the sixth intron of the c-fms proto-oncogene for the CSF-1 receptor on the long arm of chromosome 5. Common alleles contain an AGAT core repeat and range in size from 5 to 17 repeats. The PowerPlex 16 PCR products for CSF1PO are 41 bp larger than those generated with AmpFℓSTR kits. Mobility modifiers are included with CSF1PO in the Identifiler kit to increase the apparent PCR product size by around 25 bp.
- **FGA** is a compound tetranucleotide repeat found in the third intron of the human alpha fibrinogen locus on the long arm of chromosome 4. FGA has also been referred to in the literature as FIBRA or HUMFIBRA. The locus contains a CTTT repeat flanked on either side by degenerate repeats. Reported alleles range in size from 12.2 repeats to 51.2 repeats, spanning over 35 repeats. A 2-bp deletion, from the loss of a CT, in the region just prior to the core repeat motif is responsible for the x.2 microvariant alleles that are very prevalent in this STR system. PCR products from Promega's PowerPlex 16 STR kit are 112 bp larger than those generated with Applied Biosystems AmpFℓSTR kits for equivalent alleles.
- **TH01** is a simple tetranucleotide repeat found in intron 1 of the tyrosine hydroxylase gene on the short arm of chromosome 11. The locus name arises from the initials for tyrosine hydroxylase and intron 1 (i.e., 01). The locus is sometimes incorrectly referred to as THO1, with a letter O instead of a zero. In the literature, TH01 has also been referred to as TC11 and HUMTH01. TH01 has a simple tetranucleotide sequence with a repeat motif of TCAT on the upper strand in the GenBank reference sequence with alleles ranging in size from 3 to 14 repeats. The repeat motif is commonly referenced as AATG, which is correct for the complementary (bottom) strand to the GenBank reference sequence. A common microvariant allele that exists in Caucasians contains a single base deletion from allele 10 (Puers et al. 1993) and is designated allele 9.3 (Bar et al. 1994). PCR products from Promega's PowerPlex 16 STR kit are 8 bp smaller than those generated with Applied Biosystems AmpFℓSTR kits for equivalent alleles.
- **TPOX** is a simple tetranucleotide repeat found in intron 10 of the human thyroid peroxidase gene near the very end of the short arm of chromosome 2. This STR locus possesses a simple AATG repeat with alleles ranging in size from 4 to 16 repeats. It is the least polymorphic of the 29 commonly used STR loci (see Table 9.3). PCR products from Promega's PowerPlex 16 STR kit are 45 bp larger than those generated with Applied Biosystems AmpFℓSTR kits for equivalent alleles. Tri-allelic patterns are more prevalent in TPOX than in most other forensic STR markers (see STRBase, http://www.cstl.nist.gov/biotech/strbase/tri_tab.htm).
- **VWA** is a compound tetranucleotide repeat found in intron 40 of the von Willebrand Factor gene on the short arm of chromosome 12. VWA has also been

Table 9.1 Information on 29 Autosomal STR Loci Present in Commercial Kits

Locus (UniSTS)*	Chromosomal Location	Physical Position	GenBank Accession (Allele Repeat #)	Category and Repeat Motif	Allele Range**
F13B	1q31 factor XIIIb subunit	Chr 1 197.008 Mb	M64554 (10)	simple AAAT	6 to 11
D1S1656 (58809)	1q42	Chr 1 230.905 Mb	G07820 (15.3)	compound TAGA	10 to 19.3
TPOX (240638)	2p25.3 thyroid peroxidase, 10th intron	Chr 2 1.493 Mb	M68651 (11)	simple AATG	5 to 13
D2S441 (71306)	2p14	Chr 2 68.239 Mb	AC079112 (12)	compound TCTA/TCAA	8 to 17
D2S1338 (30509)	2q35	Chr 2 218.879 Mb	AC010136 (23)	compound TGCC/TTCC	15 to 27
D3S1358 (148226)	3p21.31	Chr 3 45.582 Mb	AC099539 (16)	compound TCTA/TCTG	11 to 20
FGA (240635)	4q31.3 alpha fibrinogen, 3rd intron	Chr 4 155.509 Mb	M64982 (21)	compound CTTT/TTCC	16.2 to 43.2
D5S818 (54700)	5q23.2	Chr 5 123.111 Mb	AC008512 (11)	simple AGAT	7 to 15
CSF1PO (156169)	5q33.1 c-fms proto-oncogene, 6th intron	Chr 5 149.455 Mb	X14720 (12)	simple AGAT	7 to 15
F13A01	6p24	Chr 6 6.321 Mb	M21986 (7)	simple AAAG	3.2 to 17
SE33 (ACTBP2)	6q14 beta-actin related pseudogene	Chr 6 88.987 Mb	V00481 (26.2)	complex AAAG	6.3 to 36
D6S1043 (23182)	6q15	Chr 6 92.450 Mb	G08539 (11)	compound AGAT/AGAC	8 to 26
D7S820 (74895)	7q21.11	Chr 7 83.789 Mb	AC004848 (13)	simple GATA	6 to 14
LPL	8p22 lipoprotein lipase gene, 6th intron	Chr 8 19.815 Mb	D83550 (8)	simple AAAT	7 to 15
D8S1179 (83408)	8q24.13	Chr 8 125.907 Mb	AF216671 (13)	compound TCTA/TCTG	8 to 18
Penta C	9p13	Chr 9 37.920 Mb	NT_008413 (12)	simple AAAAC	5 to 16
D10S1248 (51457)	10q26.3	Chr 10 131.093 Mb	AL391869 (13)	simple GGAA	8 to 19
TH01 (240639)	11p15.5 tyrosine hydroxylase, 1st intron	Chr 11 2.192 Mb	D00269 (9)	simple TCAT	5 to 11

(*continued*)

Table 9.1 Information on 29 Autosomal STR Loci Present in Commercial Kits (Continued)

Locus (UniSTS)*	Chromosomal Location	Physical Position	GenBank Accession (Allele Repeat #)	Category and Repeat Motif	Allele Range**
VWA (240640)	12p13.31 von Willebrand Factor, 40th intron	Chr 12 6.093 Mb	M25858 (18)	compound TCTA/TCTG	11 to 21
D12S391 (2703)	12p13.2	Chr 12 12.450 Mb	G08921(20)	compound AGAT/AGAC	14 to 27
D13S317 (7734)	13q31.1	Chr 13 82.692 Mb	AL353628 (11)	simple TATC	8 to 15
FESFPS	15q25 c-fes/fps proto-oncogene	Chr 15 91.432 Mb	X06292 (11)	simple ATTT	5 to 14
Penta E	15q26.2	Chr 15 97.374 Mb	AC027004 (5)	simple AAAGA	5 to 25
D16S539 (45590)	16q24.1	Chr. 16 86.386 Mb	AC024591 (11)	simple GATA	5 to 15
D18S51 (44409)	18q21.33	Chr 18 60.949 Mb	AP001534 (18)	simple AGAA	9 to 28
D19S433 (33588)	19q12	Chr 19 30.416 Mb	AC008507 (14)	compound AAGG/TAGG	9 to 18.2
D21S11 (240642)	21q21.1	Chr 21 20.554 Mb	AP000433 (29.1)	complex TCTA/TCTG	24.2 to 39
Penta D	21q22.3	Chr 21 45.056 Mb	AP001752 (13)	simple AAAGA	2.2 to 17
D22S1045 (49680)	22q12.3	Chr 22 37.536	AL022314 (17)	simple ATT	8 to 19

Note: The 13 CODIS core loci are highlighted in bold font. (Adapted from Butler, J. M., 2006, Genetics and genomics of core short tandem repeat loci used in human identity testing, *J Forensic Sci* 51: 253–65; Butler, J. M., 2012, *Advanced Topics in Forensic DNA Typing: Methodology*, Elsevier Academic Press: San Diego.) Physical positions are from the February 2009 Human Genome reference GRCh37 assembly. (Adapted from Thanakiatkrai, P. and L. Welch, 2011, Evaluation of nucleosome forming potentials (NFPs) of forensically important STRs, *Forensic Sci Int Genet* 5: 285–90.)

* UniSTS is a comprehensive database of sequence tagged sites (STSs) available on the NCBI Web site: http://www.ncbi.nlm.nih.gov/entrez/query.fcgi?db=unists.

** From Butler, J. M., C. R. Hill, and M. D. Coble, 2012, Variability of new STR loci and kits in U.S. population groups. *Profiles in DNA*, available at http://www.promega.com/resources/articles/profiles-in-dna/.

referred to in the literature as vWF and vWA. It possesses a TCTA repeat interspersed with a TCTG repeat with alleles ranging in size from 10 to 25 repeats. PCR products from Promega's PowerPlex 16 kit are 29 bp smaller than those generated with Applied Biosystems AmpFℓSTR kits for equivalent alleles.

- **D3S1358** is a compound tetranucleotide repeat found on the short arm of chromosome 3. This locus possesses both TCTA and TCTG repeat units with alleles ranging in size from 6 to 26 repeats. PCR products from Promega's PowerPlex 16 kit are 2 bp larger than those generated with Applied Biosystems kits for equivalent alleles.

- **D5S818** is a simple tetranucleotide repeat found on the long arm of chromosome 5. The locus possesses AGAT repeat units with alleles ranging in size from 4 to 29 repeats. In both Promega and Applied Biosystems STR kits, D5S818 is one of the smaller-sized loci and as such should appear more often than some of the other loci in degraded DNA samples. Only a few rare microvariants have been reported at this STR marker and a G→T mutation has been reported 55 bp downstream of the repeat (Jiang et al. 2011). PCR products from Promega's PowerPlex 16 kit are 15 bp smaller than those generated with Applied Biosystems kits for equivalent alleles.
- **D7S820** is a simple tetranucleotide repeat found on the long arm of chromosome 7. The locus possesses primarily a GATA repeat with alleles ranging in size from 5 to 16 repeats. However, a number of microvariant alleles have been reported. In some cases, these x.1 and x.3 alleles likely result from a variation in the number of T nucleotides found in a poly(T) stretch that occurs 13 bases downstream of the core GATA repeat. Sequencing has revealed that "on-ladder" alleles contain nine tandem T's while x.3 alleles contain eight T's and x.1 alleles contain 10 T's (Egyed et al. 2000). PCR products from Promega's PowerPlex 16 STR kit are 42 bp smaller than those generated with Applied Biosystems kits for equivalent alleles.
- **D8S1179** is a compound tetranucleotide repeat found on chromosome 8. In early publications by the Forensic Science Service, D8S1179 is listed as D6S502 because of a labeling error in the Cooperative Human Linkage Center database from which this STR was chosen (Barber and Parkin 1996; Oldroyd et al. 1995). The locus consists primarily of alleles containing TCTA although a TCTG repeat unit enters the motif for all alleles larger than 13 repeats, usually at the second or third position from the 5'-end of the repeat region (Barber and Parkin 1996). Alleles range in size from 6 to 20 repeats. PCR products from Promega's PowerPlex 16 kit are 80 bp larger than those generated with Applied Biosystems kits for equivalent alleles. AmpFℓSTR Identifiler and Profiler Plus ID kits possess an extra, unlabeled D8S1179 reverse primer to prevent allele dropout in a small portion of Asian populations due to a mutation in the middle of the primer-binding site (Leibelt et al. 2003).
- **D13S317** is a simple TATC tetranucleotide repeat found on the long arm of chromosome 13. Common alleles contain between 5 to 17 repeat units. PCR products from Promega's PowerPlex 16 STR kit are 36 bp smaller than those generated with Applied Biosystems AmpFℓSTR kits for equivalent alleles. A 4 bp deletion has been reported 24 bases downstream from the core TATC repeat that can impact allele designations with different primer sets (Butler et al. 2003; Drabek et al. 2004).
- **D16S539** is a simple tetranucleotide repeat found on the long arm of chromosome 16. This locus possesses a core repeat unit of GATA ranging from 4 to 17 repeat units in length. PCR products from Promega STR kits are 12 bp larger than those generated with Applied Biosystems kits for equivalent alleles. In the Identifiler kit, mobility modifiers are used to adjust observed sizes with D16S539. A point mutation (T→A) 38 bp downstream of the STR repeat impacts the reverse primers for both Applied Biosystems and Promega primer sets. Applied Biosystems added a redundant unlabeled primer in their COfiler, SGM Plus, and Identifiler kits so that both possible alleles could be amplified (Wallin et al. 2002). On the other hand, Promega altered their D16S539 reverse primer sequence between kits but kept the overall amplicon size the same (Butler et al. 2001; Krenke et al. 2002; Masibay et al. 2000).

- **D18S51** is a simple tetranucleotide AGAA repeat found on the long arm of chromosome 18. Alleles range in size from 5.3 to 40 repeats. A number of x.2 allele variants exist due to a 2 bp deletion from a loss of AG in the 3'-flanking region (Barber and Parkin 1996). More than 70 alleles have been reported for D18S51 (Butler 2012), making it one of the more polymorphic of the commonly used loci. PCR products from Promega's PowerPlex 16 kit are 22 bp larger than those generated with Applied Biosystems AmpFℓSTR kits for equivalent alleles.
- **D21S11** is a complex tetranucleotide repeat found on the long arm of chromosome 21. A variable number of TCTA and TCTG repeat blocks surround a constant 43 bp section made up of the sequence {$[TCTA]_3$ TA $[TCTA]_3$ TCA $[TCTA]_2$ TCCA TA}. Alleles range in size from 12 to 43.2 repeats. The x.2 microvariant alleles arise primarily from a 2 bp (TA) insertion on the 3'-end of the repeat region (Brinkmann et al. 1996). PCR products from Promega's PowerPlex 16 kit are 17 bp larger than those generated with Applied Biosystems AmpFℓSTR kits for equivalent alleles. D21S11 is far more polymorphic than can be easily detected with sized-based length separations. Fine differences in the D21S11 allele structures can only be determined by DNA sequencing since so many of the alleles have the same length but different internal sequence structure because some of the repeat units are switched around. For example, there are four different alleles designated as 30 repeats, which are indistinguishable by size-based methods alone (Butler 2012).
- **D2S1338** is a compound tetranucleotide repeat found on the long arm of chromosome 2. Alleles ranging from 10 to 31 repeats have been observed. D2S1338 has a high heterozygosity and is present in the SGM Plus, Identifiler, SEfiler Plus, MiniFiler, NGM, and NGM SElect kits from Applied Biosystems and the PowerPlex ESI 16/17, PowerPlex ESX 16/17, and PowerPlex 18D kits from Promega.
- **D19S433** is a compound tetranucleotide repeat located on chromosome 19 with observed alleles ranging from 5.2 to 20 repeats. The x.2 alleles are due to an AG deletion prior to the core AAGG repeat (Heinrich et al. 2005). D19S433 is present in the SGM Plus, Identifiler, SEfiler Plus, NGM, and NGM SElect kits from Applied Biosystems and the PowerPlex ESI 16/17, PowerPlex ESX 16/17, and PowerPlex 18D kits from Promega.
- **Penta D** is a pentanucleotide repeat found on chromosome 21 about 25 Mb from D21S11. Alleles ranging from 1.1 to 19 repeats have been observed although some of the shorter alleles are likely due to flanking region deletions (Kline et al. 2011). Penta D is present in the PowerPlex 16 and PowerPlex 18D kits from Promega.
- **Penta E** is a pentanucleotide repeat with very low stutter product formation that is located on the long arm of chromosome 15 with alleles ranging from 5 to 32 repeats. Penta E is highly polymorphic (see Table 9.3) and is present in the PowerPlex 16 and PowerPlex 18D kits from Promega.
- **D1S1656** is a tetranucleotide repeat found on the long arm of chromosome 1 with alleles ranging from 8 to 20.3 repeats. The x.3 alleles arise from a TGA insertion typically after four full TAGA repeats. It is part of the extended European Standard Set and is present in NGM and NGM SElect kits from Applied Biosystems, the PowerPlex ESI and ESX Systems from Promega, and the Qiagen ESSplex and ESSplex SE kits.
- **D2S441** is a tetranucleotide repeat located on the short arm of chromosome 2 more than 60 Mb from TPOX. It can be amplified as a miniSTR and works well on

degraded DNA samples (Coble and Butler 2005). A compound repeat motif with TCTA and TCAA sequences can range from 8 to 17 repeats. Some x.3 alleles have been observed as well as same-size, different-sequence alleles (Phillips et al. 2011). It is part of the extended European Standard Set and is present in NGM and NGM SElect kits from Applied Biosystems, the PowerPlex ESI and ESX Systems from Promega, and the Qiagen ESSplex and ESSplex SE kits.

- **D10S1248** is a simple tetranucleotide repeat found on the long arm of chromosome 10 and possesses 7 to 19 GGAA repeats. It can be amplified as a miniSTR and works well on degraded DNA samples (Coble and Butler 2005). It is part of the extended European Standard Set and is present in NGM and NGM SElect kits from Applied Biosystems, the PowerPlex ESI and ESX Systems from Promega, and the Qiagen ESSplex and ESSplex SE kits.
- **D12S391** is a highly polymorphic compound tetranucleotide found on the short arm of chromosome 12 only 6.3 megabases from VWA. It possesses over 50 different alleles ranging from 13 to 27.2 repeats in length. A number of same-size, different-sequence alleles have been identified through sequence analysis (Lareu et al. 1996; Phillips et al. 2011). It is part of the extended European Standard Set and is present in NGM and NGM SElect kits from Applied Biosystems, the PowerPlex ESI and ESX Systems from Promega, and the Qiagen ESSplex and ESSplex SE kits.
- **D22S1045** is a simple trinucleotide repeat found on chromosome 22 with alleles ranging from 7 to 20 ATT repeats. While it is not as polymorphic as most of the other 29 core and common STR loci (see Table 9.3), it can be amplified as a miniSTR and works well on degraded DNA samples (Coble and Butler 2005). It is part of the extended European Standard Set and is present in NGM and NGM SElect kits from Applied Biosystems, the PowerPlex ESI and ESX Systems from Promega, and the Qiagen ESSplex and ESSplex SE kits.
- **D6S1043** is a compound tetranucleotide repeat with alleles ranging from 8 to 25 AGAT or AGAC repeats. Some x.2 and x.3 alleles have been reported in population studies. D6S1043 is part of the Sinofiler kit from Applied Biosystems and has been used to date almost exclusively in Chinese and other Asian population studies. D6S1043 is located less than 4 Mb from SE33 on the long arm of chromosome 6.
- **SE33** is the most variable STR locus studied to date. It is located on the long arm of chromosome 6 and contains a core AAAG repeat structure. Appendix 1 in *Advanced Topics in Forensic DNA Typing: Methodology* (Butler 2012) describes 178 observed alleles ranging from 3 to 49 repeats. For example, sequence analysis has revealed 15 different 29.2 alleles possessing a variety of internal sequence combinations. SE33 is a core locus for the German national DNA database and with its growing availability in the NGM SElect and PowerPlex ESI 17 and ESX 17 Systems is being adopted by other laboratories around Europe.
- **F13B** is a simple tetranucleotide repeat with alleles ranging from 6 to 12 AAAT repeats. It is located on the long arm of chromosome 1. F13B is one of the least polymorphic loci (see Table 9.3) and not commonly used in the forensic community today. It is present in the Promega FFFL and PowerPlex CS7 kits that are sometimes used by relationship-testing laboratories.
- **F13A01** is located on the short arm of chromosome 6. It is a simple tetranucleotide repeat with alleles ranging from 3 to 17 AAAG repeats. A 3.2 allele, which is fairly common in some populations, is due to a 2-base deletion of the GT nucleotides

from a AGTAAAA sequence immediately adjacent to the 3'-repeat flanking reigon. F13A01 was present in the early UK Forensic Science Service quadruplex (Kimpton et al. 1994), is part of the Promega FFFL, and is now included in the PowerPlex CS7 kit used primarily by relationship testing laboratories.

- **FESFPS** is located on the long arm of chromosome 15 near Penta E. FESFPS is a simple tetranucleotide repeat with alleles ranging from 5 to 14 ATTT repeats. A 5'-flanking region sequence variant involving an A to C transversion has been reported that occurs in multiple alleles (Barber et al. 1995). FESFPS was present in the early UK Forensic Science Service quadruplex (Kimpton et al. 1994), is part of the Promega FFFL, and is now included in the PowerPlex CS7 kit used primarily by relationship testing laboratories.
- **LPL** or **HUMLIPOL** is located within intron 6 of the lipoprotein lipase gene on the short arm of chromosome 8. LPL contains 7 to 15 AAAT repeats and exhibits a similar level of variability as TPOX (see Table 9.3). This simple tetranucleotide repeat is part of the Promega FFFL kit and is included in the PowerPlex CS7 kit used primarily by relationship-testing laboratories.
- **Penta C** is a simple pentanucleotide repeat located on the short arm of chromosome 9. In a set of 1036 unrelated U.S. population samples (Butler et al. 2012), Penta C displays alleles in the 5 to 16 AAAAC repeat range and exhibits a medium-level variability similar to D13S317 (see Table 9.3). Penta C is part of the PowerPlex CS7 kit used primarily by relationship-testing laboratories.

9.5 Additional Autosomal STR Loci

To improve results with challenging DNA samples, a set of 26 autosomal STR loci were characterized in our laboratory at NIST (Hill et al. 2008). These loci have also been developed into a 26plex multiplex amplification for rapid examination of reference samples (Hill et al. 2009). Table 9.2 contains information on these 26 STR loci, which were also selected to be genetically well-spaced from most of the 29 commercially available loci listed in Table 9.1.

9.6 Relative Variability of Autosomal STR Loci

Table 9.3 includes a summary of U.S. population results from the 52 autosomal STR loci discussed in this chapter. The loci are listed in rank order based on their probability of identity and heterozygosity. These results were obtained on a set of 1,036 unrelated NIST U.S. population samples (Butler et al. 2012) involving 361 Caucasian, 342 African American, 236 Hispanic, and 97 Asian samples. A number of the newly available STR loci, such as D12S391 and D1S1656, provide a better probability of identity than widely used loci such as D18S51 and FGA. For each locus in Table 9.3, calculations are provided for observed heterozygosity (H), polymorphism information content (PIC), and the probability of identity across the combined U.S. population dataset. The "genotypes observed" results column can also be a useful metric to locus performance. For example, a comparison of D1S1656 and Penta D in Table 9.3 is instructive. These two loci both have 15 alleles observed, yet D1S1656 has 82 genotypes observed while Penta D only

Table 9.2 Information on NIST 26 MiniSTR Loci*

STR Locus	Chromosomal Location	Physical Position (GRCh37 Assembly)	GenBank Accession (Allele Repeat #)	Category and Repeat Motif	Allele Range
D1GATA113	1p36.23	Chr 1 7.443 Mb	Z97987 (11)	simple GATA	7 to 13
D1S1627	1p21.1	Chr 1 106.964 Mb	AC093119 (13)	simple ATT	10 to 16
D1S1677	1q23.3	Chr 1 163.560 Mb	AL513307 (15)	simple TTCC	9 to 19
D2S441	2p14	Chr 2 68.239 Mb	AC079112 (12)	compound TCTA/TCAA	8 to 17
D2S1776	2q24.3	Chr 2 169.145 Mb	AC009475 (11)	simple AGAT	7 to 15
D3S4529	3p12.1	Chr 3 85.852 Mb	AC117452 (13)	simple ATCT	11 to 18
D3S3053	3q26.31	Chr 3 171.751 Mb	AC069259 (9)	simple TATC	7 to 13
D4S2408	4p15.1	Chr 4 31.304 Mb	AC110763 (9)	simple ATCT	7 to 13
D4S2364	4q22.3	Chr 4 93.517 Mb	AC022317 (9)	simple ATTC	7 to 11
D5S2500	5q11.2	Chr 5 58.699 Mb	AC008791 (17)	compound GATA/GATT	14 to 24
D6S1017	6p21.1	Chr 6 41.677 Mb	AL035588 (10)	simple ATCC	7 to 14
D6S474	6q21	Chr 6 112.879 Mb	AL357514 (17)	complex GATA/GACA	13 to 20
D8S1115	8p11.21	Chr 8 42.536 Mb	AC090739 (9)	simple ATT	9 to 20
D9S1122	9q21.2	Chr 9 79.689 Mb	AL161789 (12)	simple TAGA	9 to 17
D9S2157	9q34.2	Chr 9 136.035 Mb	AL162417 (10)	simple ATA	7 to 19
D10S1435	10p15.3	Chr 10 2.243 Mb	AL354747 (11)	simple TATC	5 to 19
D10S1248	10q26.3	Chr 10 131.093 Mb	AL391869 (13)	simple GGAA	7 to 19
D11S4463	11q25	Chr 11 130.872 Mb	AP002806 (14)	simple TATC	10 to 17
D12ATA63	12q23.3	Chr 12 108.322 Mb	AC009771 (13)	compound TAA/CAA	10 to 20
D14S1434	14q32.13	Chr 14 95.308 Mb	AL121612 (13)	complex CTGT/CTAT	9 to 17
D17S974	17p13.1	Chr 17 10.519 Mb	AC034303 (11)	simple CTAT	5 to 12
D17S1301	17q25.1	Chr 17 72.681 Mb	AC016888 (12)	simple AGAT	9 to 15
D18S853	18p11.31	Chr 18 3.990 Mb	AP005130 (14)	simple ATA	9 to 16
D20S482	20p13	Chr 20 4.506 Mb	AL121781 (14)	simple AGAT	9 to 19
D20S1082	20q13.2	Chr 20 53.866 Mb	AL158015 (14)	simple ATA	8 to 17
D22S1045	22q12.3	Chr 22 37.536 Mb	AL022314 (17)	simple ATT	7 to 20

Note: The three loci in bold font (D2S441, D10S1248, and D22S1045) are part of the European Standard Set, ESS (Figure 9.1). (Physical positions are from Thanakiatkrai, P. and L. Welch, 2011, Evaluation of nucleosome forming potentials (NFPs) of forensically important STRs, *Forensic Sci Int Genet* 5: 285–90.)

* From Hill, C. R. et al., 2008, Characterization of 26 miniSTR loci for improved analysis of degraded DNA samples, *J Forensic Sci* 53: 73–80.

has 68 genotypes observed. The greater number of genotypes formed with the different combinations of alleles in D1S1656 leads to better heterozygosity (0.8804 compared to 0.8786) and probability of identity (0.0229 vs. 0.0336) values. Furthermore, additional genotype combinations mean that D1S1656 will likely be more useful than Penta D for detecting contributors in DNA mixtures. Further analyses with these samples are available in previous publications (Hill et al. 2008, 2011a).

Table 9.3 Results for 52 Autosomal STRs and DYS391 across 1,036 U.S. Population Samples**

STR Locus	Alleles Observed	Genotypes Observed	Heterozygosity	Polymorphism Information Content	Probability of Identity
SE33	52	304	0.9353	0.9418	0.0066
Penta E	23	138	0.8996	0.9079	0.0147
D2S1338	13	68	0.8793	0.8827	0.0220
D1S1656	15	93	0.8890	0.8833	0.0224
D18S51	22	93	0.8687	0.8713	0.0258
D12S391	24	113	0.8813	0.8683	0.0271
FGA	27	96	0.8745	0.8586	0.0308
D6S1043	27	109	0.8494	0.8529	0.0321
Penta D	16	74	0.8552	0.8413	0.0382
D9S2157*	12	56	0.8424	0.8327	0.0395
D21S11	27	86	0.8330	0.8285	0.0403
D12ATA63*	11	40	0.8261	0.8142	0.0489
D8S1179	11	46	0.7992	0.7971	0.0558
D19S433	16	78	0.8118	0.7952	0.0559
vWA	11	39	0.8060	0.7861	0.0611
F13A01	16	56	0.7809	0.7719	0.0678
D7S820	11	32	0.7944	0.7667	0.0726
D16S539	9	28	0.7761	0.7604	0.0749
D3S4529*	7	22	0.7663	0.7583	0.0753
D13S317	8	29	0.7674	0.7573	0.0765
TH01	8	24	0.7471	0.7550	0.0766
Penta C	12	49	0.7732	0.7575	0.0769
D6S474*	9	27	0.7899	0.7480	0.0837
D2S441	15	43	0.7828	0.7446	0.0841
D1S1677*	11	33	0.7554	0.7428	0.0841
D10S1248	12	39	0.7819	0.7440	0.0845
D4S2408*	7	18	0.7101	0.7348	0.0850
D5S2505*	10	31	0.7464	0.7409	0.0853
D2S1776*	9	29	0.7500	0.7357	0.0858
D18S853*	8	26	0.7192	0.7263	0.0901
D6S1017*	8	28	0.7518	0.7308	0.0907
D3S1358	11	30	0.7519	0.7292	0.0915
D22S1045	11	44	0.7606	0.7351	0.0921
D11S4463*	8	27	0.7409	0.7187	0.0949
D8S1115*	11	43	0.6649	0.7021	0.0966
F13B	7	20	0.6911	0.7213	0.0973
D10S1435*	17	41	0.7591	0.7154	0.0990
CSF1PO	9	31	0.7558	0.7045	0.1054
D17S974*	8	21	0.7355	0.7046	0.1065
D1S1627*	7	23	0.7373	0.6977	0.1100
D5S818	9	34	0.7297	0.6986	0.1104
FESFPS	12	36	0.7230	0.6916	0.1128
D3S3053*	7	20	0.7174	0.6882	0.1143

(*continued*)

Table 9.3 Results for 52 Autosomal STRs and DYS391 across 1,036 U.S. Population Samples (Continued)**

STR Locus	Alleles Observed	Genotypes Observed	Heterozygosity	Polymorphism Information Content	Probability of Identity
D20S1082*	9	28	0.7174	0.6865	0.1190
D14S1434*	8	21	0.6902	0.6701	0.1261
LPL	9	27	0.7027	0.6638	0.1336
TPOX	9	28	0.6902	0.6555	0.1358
D20S482*	10	27	0.6703	0.6389	0.1439
D9S1122*	9	26	0.7301	0.6560	0.1446
D17S1301*	7	19	0.6377	0.6175	0.1590
D1GATA113*	7	22	0.6612	0.5999	0.1697
D4S2364*	5	9	0.5000	0.4733	0.2791
DYS391	7	7	—	0.4390	0.4758

Note: Loci with one asterisk (*) were only examined in a subset of 552 samples. The 13 CODIS STR loci are highlighted for comparison purposes.

** From Butler, J. M. et al., 2012, Variability of new STR loci and kits in U.S. population groups. *Profiles in DNA*, available at http://www.promega.com/resources/articles/profiles-in-dna/.

9.7 Expansion of U.S. Core Loci

As noted earlier and illustrated in Figure 9.1, the FBI-sponsored CODIS Core Loci Working Group (Hares 2012a,b) has proposed expanding the required U.S. core loci from the current 13 to 18 autosomal STRs, of which 14 overlap with those commonly used in European STR typing kits, such as NGM and PowerPlex ESX 16 (see Figure 9.1). In addition, the sex-typing marker amelogenin has been included as a required locus and a Y-chromosome marker DYS391 has been proposed to help confirm sex-typing results in the event of an amelogenin Y deletion. The primary advantage of using DYS391 is that it is located more than 7 Mb from amelogenin. Moreover, DYS391 is a fairly stable locus possessing few duplications or null alleles with a relatively narrow allele range (Butler 2011, 2012).

In the proposed expanded U.S. core loci (Hares 2012a,b), three loci were listed as "recommended" (Section B) rather than "required" (Section A). The Section B loci include TPOX, D22S1045, and SE33. TPOX is the least variable of the 13 CODIS core loci (see Table 9.3) and thus is first to be considered for removal from the required list without significant impact to overall DNA profile information. D22S1045 is a trinucleotide repeat with accompanying higher stutter products (both N-3 and N+3 stutter) and one of the least variable loci of the expanded European Standard Set (see Table 9.3). SE33 is a valuable locus in terms of variability but requires a significant amount of "electrophoretic real estate" due to its wide allele range. SE33 also has a much higher mutation rate than other autosomal STR loci and could therefore adversely impact kinship associations in missing persons cases.

While it would be nice to have future STR kits that include all 29 commonly used loci shown in Table 9.1, there is limited electrophoretic space in creating STR kits with nonoverlapping PCR products that are less than 500 bp in size using the five-dye channels available in current instrumentation. Perhaps with future kits and six-dye instrument capabilities, such as available with the ABI 3500 and 3500xL Genetic Analyzers, all desired STR loci can be incorporated into a single kit.

In the past, INTERPOL has adopted the European Standard Set as its core set of loci. If future STR typing kits are created that are capable of analyzing all of the European loci as a subset of the expanded U.S. core, then there may very well be the possibility of a global autosomal STR panel. Only time will tell if robust kits can be developed to accomplish this feat. Since this chapter was originally written, Promega has released PowerPlex Fusion and Life Technologies has released GlobalFiler, both of which are 24plex STR kits covering the new expanded U.S. core loci requirements.

U.S. efforts to expand the number of core loci will likely follow patterns established and lessons learned with the expansion of the European Standard Set (Butler 2011). Following initial announcements of new proposed European loci (Gill et al. 2006), companies produced prototype kits for evaluation. The final expanded set of loci was decided after kit availability and data review. It is important that following the final selection of expanded loci, population data be gathered (Butler et al. 2012) and software implemented before required compliance with the new loci. In the ongoing U.S. locus expansion effort, as with the 2005–2011 European effort to go from 7 to 12 core loci, data-driven decisions will be made with improved casework capabilities being a priority (Federal Bureau of Investigation 2012).

9.8 Conclusions

The growing list of publications describing the application of STR loci to forensic DNA typing exceeds 3,500 references (see NIST STRBase Web site, http://www.cstl.nist.gov/biotech/strbase). STR markers have become important tools for human identity testing. Commercially available STR kits are now widely used in forensic and paternity-testing laboratories. The adoption and expansion of core STR loci in national DNA databases around the world ensures that these STR markers will be used for many years to come.

Acknowledgments

Support of the NIST Human Identity Project Team, including Margaret Kline, Peter Vallone, David Duewer, Mike Coble, Erica Butts, and Kristen O'Connor, is greatly appreciated. Collaborations with Applied Biosystems, Promega Corporation, and Qiagen have enabled the collection of concordance data between many of the commercial kits. Forensic DNA research conducted at NIST is supported by an interagency agreement between the National Institute of Justice and the NIST Office of Law Enforcement Standards. Points of view in this document are those of the authors and do not necessarily represent the official position or policies of the U.S. Department of Justice. Certain commercial equipment, instruments, and materials are identified in order to specify experimental procedures as completely as possible. In no case does such identification imply a recommendation or endorsement by NIST, nor does it imply that any of the materials, instruments, or equipment identified are necessarily the best available for the purpose.

References

Bacher, J. and J. W. Schumm. 1998. Development of highly polymorphic pentanucleotide tandem repeat loci with low stutter. *Profiles in DNA* 2: 3–6.

Bar, W., B. Brinkmann, P. Lincoln, W. R. Mayr, and U. Rossi. 1994. DNA recommendations—1994 report concerning further recommendations of the DNA Commission of the ISFH regarding PCR-based polymorphisms in STR (short tandem repeat) systems. *Int J Leg Med* 107: 159–60.

Barber, M. D. and B. H. Parkin. 1996. Sequence analysis and allelic designation of the two short tandem repeat loci D18S51 and D8S1179. *Int J Leg Med* 109: 62–5.

Barber, M. D., R. C. Piercy, J. F. Andersen, and B. H. Parkin. 1995. Structural variation of novel alleles at the Hum vWA and Hum FES/FPS short tandem repeat loci. *Int J Legal Med* 108: 31–5.

Brinkmann, B., E. Meyer, and A. Junge. 1996. Complex mutational events at the HumD21S11 locus. *Hum Genet* 98: 60–4.

Budowle, B., J. Ge, R. Chakraborty et al. 2011. Population genetic analyses of the NGM STR loci. *Int J Legal Med* 125: 101–9.

Budowle, B., T. R. Moretti, S. J. Niezgoda, and B. L. Brown. 1998. CODIS and PCR-based short tandem repeat loci: Law enforcement tools. *Proc 2nd Eur Symp Human Identification*, pp. 73–88. Available at http://www.promega.com/products/pm/genetic-identity/ishi-conference-proceedings/2nd-eshi-oral-presentations/.

Butler, J. M. 2012. *Advanced Topics in Forensic DNA Typing: Methodology*. Elsevier Academic Press: San Diego, CA.

Butler, J. M. 2006. Genetics and genomics of core short tandem repeat loci used in human identity testing. *J Forensic Sci* 51: 253–65.

Butler, J. M. 2011. NIST update. Presentation made at the National CODIS Conference (November 14, 2011); http://www.cstl.nist.gov/strbase/pub_pres/NIST-Update-CODIS2011.pdf.

Butler, J. M., J. M. Devaney, M. A. Marino, and P. M. Vallone. 2001. Quality control of PCR primers used in multiplex STR amplification reactions. *Forensic Sci Int* 119: 87–96.

Butler, J. M. and C. R. Hill. 2012. Biology and genetics of new autosomal STR loci useful for forensic DNA analysis. *Forensic Sci Rev* 24: 15–26.

Butler, J. M., C. R. Hill, and M. D. Coble. 2012. Variability of new STR loci and kits in U.S. population groups. *Profiles in DNA*. Available at http://www.promega.com/resources/articles/profiles-in-dna/.

Butler, J. M., R. Schoske, P. M. Vallone, J. W. Redman, and M. C. Kline. 2003. Allele frequencies for 15 autosomal STR loci on U.S. Caucasian, African American, and Hispanic populations. *J Forensic Sci* 48: 908–11.

Carracedo, A. and M. V. Lareu. 1998. Development of new STRs for forensic casework: Criteria for selection, sequencing and population data and forensic validation. *Proc 9th Int Symp Human Identification*, pp. 89–107. Available at http://www.promega.com/products/pm/genetic-identity/ishi-conference-proceedings/9th-ishi-oral-presentations/.

Coble, M. D. and J. M. Butler. 2005. Characterization of new miniSTR loci to aid analysis of degraded DNA. *J Forensic Sci* 50: 43–53.

Drabek, J., D. T. Chung, J. M. Butler, and B. R. McCord. 2004. Concordance study between Miniplex assays and a commercial STR typing kit. *J Forensic Sci* 49: 859–60.

Egyed, B., S. Furedi, M. Angyal et al. 2000. Analysis of eight STR loci in two Hungarian populations. *Forensic Sci Int* 113: 25–7.

Ellegren, H. 2004. Microsatellites: Simple sequences with complex evolution. *Nat Rev Genet* 5: 435–45.

Federal Bureau of Investigation. 2012. Planned process and timeline for implementation of additional CODIS core loci. Available at http://www.fbi.gov/about-us/lab/codis/planned-process-and-timeline-for-implementation-of-additional-codis-core-loci.

Ge, J., Eisenberg, A., and B. Budowle. 2012. Developing criteria and data to determine best options for expanding the core CODIS loci. *Investig Genet* 3: 1.

Gill, P., L. Fereday, N. Morling, and P. M. Schneider. 2006. The evolution of DNA databases—Recommendations for new European STR loci. *Forensic Sci Int* 156: 242–4.

Hares, D. R. 2012a. Expanding the CODIS core loci in the United States. *Forensic Sci Int Genet* 6(1): e52–4.

Hares, D. R. 2012b. Addendum to expanding the CODIS core loci in the United States. *Forensic Sci Int Genet* 6(5): e135.

Heinrich, M., H. Felske-Zech, B. Brinkmann, and C. Hohoff. 2005. Characterisation of variant alleles in the STR systems D2S1338, D3S1358 and D19S433. *Int J Legal Med* 119: 310–3.

Hill, C. R., J. M. Butler, and P. M. Vallone. 2009. A 26plex autosomal STR assay to aid human identity testing. *J Forensic Sci* 54: 1008–15.

Hill, C. R., D. L. Duewer, M. C. Kline et al. 2011a. Concordance and population studies along with stutter and peak height ratio analysis for the PowerPlex (R) ESX 17 and ESI 17 Systems. *Forensic Sci Int Genet* 5: 269–75.

Hill, C. R., M. C. Kline, M. D. Coble, and J. M. Butler. 2008. Characterization of 26 miniSTR loci for improved analysis of degraded DNA samples. *J Forensic Sci* 53: 73–80.

Hill, C. R., M. C. Kline, D. L. Duewer, and J. M. Butler. 2010. Strategies for concordance testing. *Profiles in DNA.* Available at http://www.promega.com/resources/articles/profiles-in-dna/2010/strategies-for-concordance-testing/.

Hill, C. R., M. C. Kline, D. L. Duewer, and J. M. Butler. 2011b. Concordance testing comparing STR multiplex kits with a standard data set. *Forensic Sci Int Genet* Suppl Ser 3: e188–9.

Jiang, W., M. Kline, P. Hu, and Y. Wang. 2011. Identification of dual false indirect exclusions on the D5S818 and FGA loci. *Legal Med* 13: 30–4.

Katsanis, S. H. and J. K. Wagner. 2013. Characterization of the standard and recommended CODIS markers. *J Forensic Sci* 58(Suppl 1): S169–72.

Kayser, M., R. Kittler, A. Ralf et al. 2004. A comprehensive survey of human Y-chromosomal microsatellites. *Am J Hum Genet* 74: 1183–97.

Kimpton, C. P., D. Fisher, S. Watson et al. 1994. Evaluation of an automated DNA profiling system employing multiplex amplification of four tetrameric STR loci. *Int J Legal Med* 106: 302–11.

Kline, M. C., C. R. Hill, A. E. Decker, and J. M. Butler. 2011. STR sequence analysis for characterizing normal, variant, and null alleles. *Forensic Sci Int Genet* 5: 329–32.

Krenke, B. E., A. Tereba, S. J. Anderson et al. 2002. Validation of a 16-locus fluorescent multiplex system. *J Forensic Sci* 47: 773–85.

Lareu, M. V., M. C. Pestoni, F. Barros, A. Salas, and A. Carracedo. 1996. Sequence variation of a hypervariable short tandem repeat at the D12S391 locus. *Gene* 182: 151–3.

Leibelt, C., B. Budowle, P. Collins, Y. Daoudi, T. Moretti, G. Nunn, D. Reeder, and R. Roby. 2003. Identification of a D8S1179 primer binding site mutation and the validation of a primer designed to recover null alleles. *Forensic Sci Int* 133: 220–7.

Life Technologies. 2012. Development of a next generation global multiplex. Slides available at http://www.invitrogen.com/site/us/en/home/Products-and-Services/Applications/Human-Identification/globalfiler_str_kit.html.

Masibay, A., T. J. Mozer, and C. Sprecher. 2000. Promega Corporation reveals primer sequences in its testing kits [letter]. *J Forensic Sci* 45: 1360–2.

O'Connor, K. L., C. R. Hill, P. M. Vallone, and J. M. Butler. 2011. Linkage disequilibrium analysis of D12S391 and vWA in U.S. population and paternity samples. *Forensic Sci Int Genet* 5(5): 538–40. Erratum in *Forensic Sci Int Genet* 5(5): 541–2.

O'Connor, K. L. and A. O. Tillmar. 2012. Effect of linkage between vWA and D12S391 in kinship analysis. *Forensic Sci Int Genet* 6(60): 840–4.

Oldroyd, N. J., A. J. Urquhart, C. P. Kimpton et al. 1995. A highly discriminating octoplex short tandem repeat polymerase chain reaction system suitable for human individual identification. *Electrophoresis* 16: 334–7.

Phillips, C., D. Ballard, P. Gill, D. Syndercombe-Court, A. Carracedo, and M. V. Lareu. 2012. The recombination landscape around forensic STRs: Accurate measurement of genetic distances between syntenic STR pairs using HapMap high density SNP data. *Forensic Sci Int Genet* 6: 354–65.

Phillips, C., L. Fernandez-Formoso, M. Garcia-Magarinos et al. 2011. Analysis of global variability in 15 established and 5 new European Standard Set (ESS) STRs using the CEPH human genome diversity panel. *Forensic Sci Int Genet* 5: 155–69.

Promega. 2012. PowerPlex Fusion System. http://www.promega.com/products/pm/genetic-identity/powerplex-fusion.

Puers, C., H. A. Hammond, L. Jin, C. T. Caskey, and J. W. Schumm. 1993. Identification of repeat sequence heterogeneity at the polymorphic short tandem repeat locus HUMTH01[AATG]n and reassignment of alleles in population analysis by using a locus-specific allelic ladder. *Am J Hum Genet* 53: 953–8.

Schneider, P. M. 2009. Expansion of the European Standard Set of DNA database loci—The current situation. *Profiles in DNA* 12(1): 6–7. Available from http://www.promega.com/resources/articles/profiles-in-dna/2009/expansion-of-the-european-standard-set/.

Sensabaugh, G. F. 1982. Biochemical markers of individuality. In Saferstein, R. (ed.) *Forensic Science Handbook,* Prentice-Hall: New York, pp. 338–415.

Thanakiatkrai, P. and L. Welch. 2011. Evaluation of nucleosome forming potentials (NFPs) of forensically important STRs. *Forensic Sci Int Genet* 5: 285–90.

Urquhart, A., C. P. Kimpton, T. J. Downes, and P. Gill. 1994. Variation in short tandem repeat sequences—A survey of twelve microsatellite loci for use as forensic identification markers. *Int J Leg Med* 107: 13–20.

Wallin, J. M., C. L. Holt, K. D. Lazaruk, T. H. Nguyen, and P. S. Walsh. 2002. Constructing universal multiplex PCR systems for comparative genotyping. *J Forensic Sci* 47: 52–65.

Hidden Variation in Microsatellite Loci
Utility and Implications for Forensic DNA

10

JOHN V. PLANZ
THOMAS A. HALL

Contents

10.1 Introduction 199
10.2 Molecular Biology of Interrupted Repeats 201
10.3 Mutation Modeling for STRs 203
10.4 Population Observations 207
10.5 Value to Forensic Investigations 212
10.6 Conclusions 215
References 215

Abstract: Short tandem repeat (STR) analysis has been the standard in forensic DNA examinations for almost 15 years. The purpose of this chapter is to provide some perspective on the biological nature of STR alleles themselves, underlying distributions of alleles in the STR loci that are routinely used and to discuss features of these alleles that are not observable with the currently employed methods. Many of the internationally standardized STR loci contain variations of their interrupted repeat structures, either due to the compound or complex nature of the locus or due to nucleotide variations within the simple repeat motif, which inevitably leads them to become more stratified at the population level. Current STR typing procedures utilizing polymerase chain reaction (PCR) amplification followed by fragment analysis via capillary or gel electrophoresis does not provide the resolution to discern these polymorphisms. Thus, current designation of alleles is operationally and not biologically defined. Although in the comparison of an evidentiary STR profile to that of a potential contributor, the biological nature of the allele may not be of consequence, when comparisons require assumptions of relatedness between individuals, the biological nature of shared alleles becomes an underlying focus. Herein we will discuss the nature of these additional allelic polymorphisms, what is known of their distribution among the STR loci utilized in forensic testing and within populations, and the advantages this level of allelic discrimination has in forensic and relationship testing.

10.1 Introduction

Short tandem repeat (STR) loci have become a standard genetic tool for population studies, conservation genetics, evolutionary modeling, disease association studies, and human identity testing analyses worldwide (Budowle and Chakraborty 2001; Crosby et al. 2009; Jorde et al. 1997; Wang et al. 2005; Weber and Broman 2001). Suites of loci, such as the 13-locus Combined DNA Index System (CODIS) STR panel in the United States (Butler 2006) and the recently expanded European STR panel (Ge et al. 2009; Gill et al. 2006a,b), often provide sufficient information for evaluating population structure, genetic diversity, and allele attribution in parent-offspring relationships (Jorde et al. 1997; Kruckenhauser et al. 2009; Peacock et al. 2002; Presciuttini et al. 2003). STR loci are abundant within genomes and in combination they offer the high levels of individual differentiation necessary for many analyses. The identification and implementation of STR loci is adaptable to applications in conservation genetics, phylogenetics, and population studies. Microsatellite-enriched genomic libraries can be quickly constructed for unrepresented species, targeting specific STR sequence motifs (Hamilton et al. 1999; Julian and King 2003; Trujillo and Amelon 2009). More recently, human population studies have utilized the wealth of data available on microsatellite markers revealed in the various components of the Human Genome Project (Mammalian Genotyping Service 2011; McIver et al. 2011; NCBI Human Reference Genome Build 2009).

The standard approaches for STR typing rely on PCR amplification followed by gel or capillary electrophoresis (CE). Allele designations are made by measurements of amplicon size through comparison to allelic ladders of common variants. Some loci reflect changes due to insertion or deletion of partial repeat motifs as microvariants of the nominal allele. Although initially the presence of these polymorphisms raised concern regarding interpretation in forensic analyses (Crouse et al. 1999), expanded allelic ladders and more precise fragment sizing in CE have facilitated utilization of the expanded allelic diversity and discrimination provided by microvariants that have become valued in forensic, clinical, and population studies. Nucleotide variations have been observed within the repeat structure of many STR loci genome wide (Madsen et al. 2008; Oberacher et al. 2008; Pemberton et al. 2009; Planz et al. 2012). Most sequence-based analyses of STR loci have been conducted during the course of locus development and primer optimization or when the investigation of null or dropout alleles was warranted. Concordance studies between manufactured STR kits often revealed additional polymorphisms flanking the repeat region (Budowle et al. 2001a; Budowle and Sprecher 2001; Clayton et al. 2004; Cotton et al. 2000; Delamoye et al. 2004; Drabeck et al. 2004; Hill et al. 2007; Leibelt et al. 2003; Nelson et al. 2002; Vanderheyden et al. 2007). Although these polymorphisms often contained population-specific information, STR kit manufacturers generally redesigned primers or created degenerate primer cocktails that would mask the polymorphisms' presence to achieve consistent typing against standard allelic ladders (Leibelt et al. 2003; Nelson et al. 2002). Polymorphisms also exist within the repeating regions of STR loci, disrupting the standard repeat motif pattern in several of the forensically useful STR markers. These polymorphisms, often initiated by a single nucleotide change within an individual repeat block of the sequence motif, begin a complex process of evolution having direct impact on the allelic diversity and population distribution of a family of alleles. Patterns of mutation working on the locus over time have led to the compound and complex microsatellite structure observed in the genome (Table 10.1) (e.g., vWA, D21S11) (Moller et al. 1994; Oberacher et al. 2008; Walsh et al. 2003). Standard approaches to microsatellite analysis typically define alleles based solely on the variation in repeat motif number and, due to the methodology utilized in STR typing, consequently overlook the resolving power found in the hidden variation inherent at the locus.

Mass spectrometry approaches have been employed to analyze STRs and single nucleotide polymorphisms (SNPs), primarily through matrix-assisted laser desorption-ionization time-of-flight mass spectrometry (MALDI TOF-MS) (Butler and Becker 2001; Butler et al. 1999). The mass accuracy and resolution obtained with electrospray ionization (ESI) TOF-MS (Hall et al. 2009; Hofstadler et al. 2005; Oberacher et al. 2006, 2008) is significantly enhanced relative to previous MALDI TOF-MS approaches. A method combining ion-pair reversed-phase high-performance liquid chromatography and ESI quadrupole time-of-flight mass spectrometry (ICE-MS) was presented to simultaneously detect length and nucleotide variability in STRs (Oberacher et al. 2008; Pitterl et al. 2008). Although this method can be effectively used to evaluate PCR amplicons, it utilizes a complex HPLC approach for PCR product purification prior to the mass spectrometry analysis. A fully automated, high-throughput ESI-MS system has been developed by Ibis Biosciences, Inc., for mtDNA and STR typing (Ecker et al. 2006; Hall et al. 2009; Hofstader and Hall 2008) that offers a matched kit and instrument-based solution for forensic applications.

Numerous studies (Brinkmann et al. 1998; Kimmel and Chakraborty 1996; Kruglyak et al. 1998; Pemberton et al. 2009) have addressed the question of microsatellite diversity and the underlying molecular processes driving it, suggesting various models for microsatellite evolution. These studies have typically been based on fairly limited pedigrees or reported allele frequency data. Although allele length has often been suggested as a major factor affecting mutation rate in STR loci (Ge et al. 2009; Leopoldino and Pena 2003; Shinde et al. 2003; Xu et al. 2000), the effect motif interruptions have on the mutation process has only been addressed in a few studies (Brinkmann et al. 1998; Ellegren 2000; Kruglyak et al. 1998; Taylor et al. 1999). In this review we will discuss the nature of interrupted microsatellite repeats in the genome and the utility of capturing these variations for forensic DNA analyses.

10.2 Molecular Biology of Interrupted Repeats

Microsatellite or STR sequences represent some of the most polymorphic components in the genome (Jorde et al. 1997; Payseur et al. 2011). STR polymorphisms are principally characterized as length variations measured as differences in the occurrence of a defined number of repeat motif blocks at the chromosomal location. Although the microsatellite locus contains considerable variation at the population level, the sequence itself is considered simple due to its limited nucleotide diversity compared to the more complex nonrepetitive regions of the genome. Madsen et al. (2008) recently reported that approximately 4% of the human genome consists of STR sequences, with greater than 92% of the STR loci occurring within the exons of known human genes. The majority of these STR loci consists of repeat motifs of 3, 6, and 9 nucleotide blocks with overall repeat regions generally less than 25 nucleotides in length. Although this study focused on microsatellite loci outside of the selection criteria normally used for forensic marker systems, several features of the STR loci were revealed. Most notably, greater than 62% of the STRs evaluated in the genome were found to contain imperfect repeats, for example, the repeat series was interrupted by a motif that contained a polymorphism differing from the regular repeat motif, observable in the genomic sequence data. A pattern of repeated SNPs had previously been reported, indicating that not all SNPs observed in the human genome may have originated from independent mutational events, but may arise from a single polymorphism within an

Table 10.1 DNA Sequence Repeat Motifs of Commonly Used STR Loci

Repeat-Motif	Locus	Repeat Sequence
Simple	D2S441	$TCTA_n$
	D5S818	$AGAT_n$
	D7S820	$GATA_n$
	D10S1248	$GGAA_n$
	D16S539	$GATA_n$
	D18S51	$AGAA_n$
	D19S433	$AAGG_n$
	D22S1045	ATT_n
	CSF1PO	$TAGA_n$
	FGA	$CTTT_n$
	Penta D	$AAAGA_n$
	Penta E	$AAAGA_n$
	TH01	$AATG_n$ or $(AATG)_nATG(AATG)_3TPOXGAAT_n$
Compound	D1S1656	$(TAGA)_n(TGA)_{0\text{-}1}(TAGA)_n(TAGG)_{0\text{-}1}(TG)_5$
	D2S1338	$(TGCC)_{3\text{-}9}(TTCC)_n$
	D3S1358	$TCTA(TCTG)_{1\text{-}4}(TCTA)_n$
	D8S1179	$(TATC)_{2\text{-}3}(TGTC)_{1\text{-}3}(TATC)_n$ or $(TATC)_n$
	D12S391	$(AGAT)_{8\text{-}17}(AGAC)_{6\text{-}10}(AGAT)_{0\text{-}1}$
	D13S317	$TATC_n(AATC)_{0\text{-}1}$
	VWA	$TCTA(TCTG)_{1\text{-}4}(TCTA)_{0\text{-}1}(TGTC)_{0\text{-}4}(TCTA)_n$
Complex	D21S11	$(TCTA)_{4\text{-}6}(TCTG)_{5\text{-}7}(TCTA)_3$ TA $(TCTA)_3$ TCA $(TCTA)_2$TCCA TA $(TCTA)_n$
	SE33(ACTBP2)	$(AAAG)_3(AGAG)_3AAGA(GAAG)_3AA(AAAG)_2AA(AAAG)_3AG$ $(AAAG)_n$

STR being replicated under the mutational model observed for tandem repeats (Madsen et al. 2007). DNA sequence data from several STRs used in forensic analysis reveal these repeated patterns, defining the motif structure of the compound and complex repeat motifs (Table 10.1) (Oberacher et al. 2008; Pitterl et al. 2010). It can be noted that studies targeting the development of SNP panels for use in forensic identification applications have effectively avoided polymorphisms within repeat regions through the targeted selection of low F_{st} markers (Kidd et al. 2006; Pakstis et al. 2007, 2010). As higher F_{st} SNPs were culled in these studies, it can be surmised that the polymorphisms existing within the repeat structures of STRs may be of some value in population specificity.

STR loci were rapidly adopted by population geneticists and in forensic testing due to the considerable degree of variability that could be observed at the loci using a broad range of analytical methods (Budowle et al. 1998; Butler and Becker 2001; Jordan et al. 2009; Kruckenhauser et al. 2009; Peacock et al. 2002). The high level of polymorphism at microsatellite loci is largely attributable to the process of replication slippage (Pumpernik et al. 2008; Schlötterer and Tautz 1992; Weber and Wong 1993). Slippage occurs when the DNA polymerase pauses and dissociates from the DNA during synthesis. In this brief respite, the terminal section of the newly synthesized strand anneals to a repeat block proximal or distal to the one that served as its template during synthesis. Once synthesis resumes and is completed, the resulting DNA strand will be one or more repeat units longer or shorter than the original DNA template. This synthesis anomaly is regularly observed *in vitro* during PCR amplification, resulting in the generation of "stutter" alleles (Leclair et al. 2004; Walsh et al. 1996). Stutter peaks or bands are commonly observed in current STR typing systems with amplicons representing synthesis products 1, 2, and 3 repeat sizes larger and smaller than the nominal true allele, albeit at very low frequency/intensity. The distribution of the allelic spectrum observed within and among populations differs markedly from that which is observed in stutter products. Newly synthesized alleles may not reach observable frequency in a population for numerous generations, if at all, depending on the effect of processes such as genetic drift, selection, and population substructure. Additionally, while a newly synthesized allele may be different from its originating allele due to a mutational event, it may be masked completely in the population, as other alleles in the population of that particular repeat size may undoubtedly already exist. Thus, an *a priori* assumption that a high frequency of homozygosity in a population study using STRs is indicative of substructure or substantial inbreeding may be incorrect. Similarly, the presence of allele sharing at an STR locus between individuals under the premise of familial relationship may be more attributable to the presence of same-length alleles that are not identical by descent (IBD), especially when considering alleles that are present at higher frequencies in the population. In addition, the knowledge that a particular allele shared between two individuals or between an evidentiary profile and a known sample based on conventional typing may in fact be exclusionary brings a new challenge to interpretation of DNA mixtures and single-source comparisons alike.

Using methods different from classical gel or CE, interrupted motifs have been observed within the repeat structure of many of STR loci used in human identity testing (Fordyce et al. 2011; Madsen et al. 2008; Oberacher et al. 2008; Pitterl et al. 2010; Planz et al. 2012). The most notable effect of capturing both repeat number and motif variants simultaneously in STR analyses is an increase in the observed number of segregating alleles and the attendant locus-specific heterozygosity (Table 10.2). The increase in observed alleles stems largely from capturing polymorphisms that stratify the nominal alleles we observe with CE based solely on amplicon length. The largest increase appears in loci that contain compound and complex repeat motif structures, where the polymorphisms detected largely represent variations in the repeat number of internal interrupted repeats. Additionally, it has been observed that many individuals who were originally typed as homozygous with conventional methods are in actuality same-repeat-length heterozygotes (Figure 10.1). Homoplasy, in this manner, reduces the resolution of alleles at the STR loci when used in population studies, as shared alleles within populations may not truly be homologous (Brinkmann et al. 1998).

10.3 Mutation Modeling for STRs

Microsatellite loci are among the fastest-evolving regions of the genome with estimates of mutation rate on the order of 10^{-3}–10^{-4}/locus/generation (Ellegren 2000; Xu et al. 2000). Although used in population and gene-mapping studies for the past three decades, a complete understanding of the mutation process of microsatellite markers has lagged well behind their implementation. The majority of mutation information comes from limited-pedigree studies, usually associated with various genetic diseases in humans (Tran et al. 2004; Wooster et al. 1994; Xu et al. 2000). More recently, data compiled from standardized suites of loci utilized in parentage evaluation and identity testing have provided a wealth of information regarding mutation rates and their variability across loci, and possible

Table 10.2 Summary Statistics from Eight STR Loci Having Considerable Polymorphism Identified through Mass Spectrometry and/or DNA Sequence Analysis for European and U.S. Caucasian, African and African American, Asian, and U.S. Hispanic Populations

Population Group[a]	TP	N	AD	DP	H_o	PE	N	AD	DP	H_o	PE
				D3S1358					**D8S1179**		
Austrian[b]	CE	98	7	0.925	0.816	0.605	96	9	0.939	0.865	0.657
	MS	98	14	0.957	0.857	0.728	96	15	0.962	0.906	0.749
U.S. Caucasian[c]	CE	181	8	0.923	0.808	0.613	182	10	0.932	0.819	0.634
	MS	182	18	0.967	0.890	0.774	181	14	0.963	0.879	0.752
Khoisan (Africa)[b]	CE	108	5	0.830	0.639	0.437	108	8	0.935	0.731	0.625
	MS	108	13	0.940	0.787	0.675	108	14	0.942	0.741	0.652
African American[c]	CE	214	8	0.892	0.785	0.572	214	10	0.923	0.752	0.514
	MS	213	18	0.978	0.906	0.808	213	19	0.949	0.812	0.622
Yakut (Asia)[b]	CE	93	6	0.804	0.667	0.382	93	8	0.859	0.677	0.476
	MS	93	11	0.899	0.763	0.531	93	11	0.906	0.763	0.577
U.S. Hispanic[c]	CE	193	8	0.894	0.741	0.494	193	9	0.930	0.798	0.596
	MS	193	18	0.946	0.819	0.634	193	16	0.964	0.886	0.767
				D21S11					**vWA**		
Austrian[b]	CE	98	11	0.952	0.857	0.691	98	7	0.931	0.788	0.620
	MS	98	17	0.971	0.929	0.784	98	16	0.952	0.848	0.698
U.S. Caucasian[c]	CE	182	14	0.954	0.868	0.731	182	10	0.939	0.802	0.603
	MS	181	23	0.978	0.901	0.797	181	22	0.958	0.867	0.729
Khoisan (Africa)[b]	CE	108	20	0.977	0.898	0.810	108	9	0.933	0.824	0.624
	MS	108	40	0.983	0.917	0.873	108	20	0.967	0.852	0.756
African American[c]	CE	214	20	0.959	0.818	0.632	214	11	0.940	0.822	0.641
	MS	213	33	0.971	0.836	0.667	213	26	0.977	0.892	0.779
Yakut (Asia)[b]	CE	92	9	0.868	0.677	0.468	92	7	0.927	0.804	0.611
	MS	92	16	0.959	0.837	0.711	92	10	0.930	0.815	0.624
U.S. Hispanic[c]	CE	193	14	0.952	0.808	0.615	193	7	0.911	0.788	0.576
	MS	193	25	0.975	0.845	0.684	193	16	0.931	0.798	0.595
				D13S317					**D7S820**		
U.S. Caucasian[c]	CE	182	7	0.921	0.830	0.655	182	8	0.935	0.786	0.573
	MS	181	12	0.971	0.906	0.808	181	15	0.960	0.807	0.611
African American[c]	CE	214	7	0.861	0.706	0.437	214	8	0.927	0.771	0.546
	MS	213	12	0.953	0.798	0.596	213	12	0.938	0.789	0.578
U.S. Hispanic[c]	CE	193	7	0.945	0.819	0.635	193	9	0.924	0.736	0.486
	MS	193	12	0.975	0.902	0.799	193	14	0.948	0.788	0.576
				D5S818							
U.S. Caucasian[c]	CE	182	9	0.843	0.670	0.384					
	MS	181	15	0.926	0.818	0.632					

(*continued*)

Table 10.2 Summary Statistics from Eight STR Loci Having Considerable Polymorphism Identified through Mass Spectrometry and/or DNA Sequence Analysis for European and U.S. Caucasian, African and African American, Asian, and U.S. Hispanic Populations (Continued)

Population Group[a]	TP	N	AD	DP	H_o	PE	N	AD	DP	H_o	PE
African American[c]	CE	214	9	0.893	0.701	0.430					
	MS	213	17	0.910	0.831	0.658					
U.S. Hispanic[c]	CE	193	9	0.868	0.689	0.412					
	MS	193	13	0.955	0.762	0.530					
				D2S1338							
Austrian[b]	CE	95	11	0.956	0.916	0.743					
	MS	95	20	0.967	0.947	0.788					
Khoisan (Africa)[b]	CE	108	12	0.960	0.806	0.713					
	MS	108	33	0.986	0.935	0.889					
Yakut (Asia)[b]	CE	92	11	0.957	0.826	0.718					
	MS	92	19	0.972	0.88	0.789					

[a] TP: testing platform; N: sample size; AD: alleles detected; DP: discriminating power; H_o: observed heterozygosity; PE: power of exclusion; CE: capillary electrophoresis; MS: mass spectrometry.

[b] Data obtained from Pitterl, F. et al., 2010, Increasing the discrimination power of forensic STR testing by employing high-performance mass spectrometry, as illustrated in indigenous South African and Central Asian populations, *Int J Legal Med* 124:551–7.

[c] Data obtained from Planz, J. V. et al., 2012, Automated analysis of sequence polymorphism in STR alleles by PCR and direct electrospray ionization mass spectrometry, *Forensic Sci Int Genet* 6:594–606.

constraints imposed by repeat motif and allele size (American Association of Blood Banks 2005; Brinkmann et al. 1998; Deka et al. 1999; Ge et al. 2009). However, due to unaccounted variability in what constitutes an allele at an STR locus, for example, motif composition, undetected polymorphisms within repeat sequence, and so forth, the significant inconsistencies and broad variance reported in prior studies (Jorde et al. 1997; Sun et al. 2009; Zhivotovsky et al. 2001) are not unexpected.

A major concern in population and evolutionary studies involves linearity of calculated measures of genetic differentiation or distance (e.g., F_{st}, $(\delta\mu)^2$) with time (Sun et al. 2009; Zhivotovsky et al. 2001). Any study evaluating population substructure ultimately hopes to define a time of common ancestry for the populations/species under investigation. Genetic markers must accumulate stable polymorphisms at consistent and predictable rates to be effective in these types of investigations. Although most forensic investigations do not address the processes of population differentiation, similar biochemical processes can lead to "allele drop-in" when working at the stochastic sampling level of input DNA. For example, random stutter product formation in an early cycle of PCR, when initiated with very few copies of target template, could produce a product one repeat smaller or larger than the target template as the major product, with the actual target allele as a minority product or absent in rare cases.

The occurrence and position of interrupted motif blocks within the STR locus may play a significant role in how the mutation process occurs at forensic loci. Empirical data exist on dynamic mutations observed in disease-causing trinucleotide repeat loci where interruptions within repeat motifs slowed down mutation rate (Sinden et al. 2002). These data suggest that in some instances the altered repeat motif may actually stabilize portions

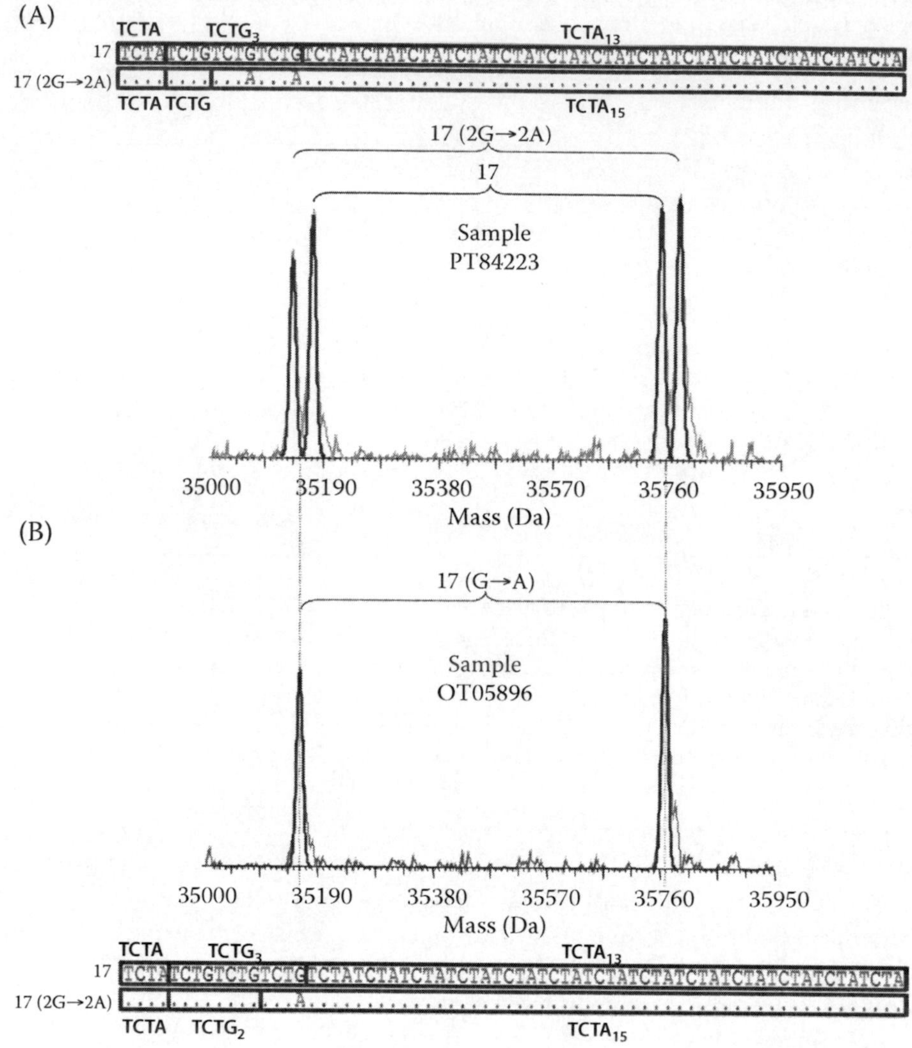

Figure 10.1 Distinct STR alleles can produce the same PCR product length at a given STR locus. (A) NIST population sample PT84223 is actually heterozygous at locus D3S1358, as observed with PCR and mass spectrometry, with a nominal allele 17 and an allele 17 with two G→A polymorphisms. The mass spectrum displays separate mass measurements for the forward and reverse DNA strands of each PCR product; the four discrete mass measurements in (A) correspond to two STR PCR products. The sequences of repeat regions for the two alleles for sample PT84223, as determined by Sanger sequencing, are shown above the mass spectrum, indicating the positions of the two G→A differences as tandem replacements of two TCTG motifs present in the nominal allele 17. (B) NIST population sample OT05896 displays a third allele, an allele 17 with a single G→A polymorphism, with the same length as the two alleles shown in (A). The sequence of one representative of a 17 (G→A) is shown below the mass spectrum in (B). (Samples provided by John Butler and Peter Vallone.)

of the overall repeat region, causing a reduction in mutation rate among the modified motif as compared to the unaltered repeat-motif region. Differential mutation rates within a locus would permit greater information content to be captured when comparing alleles of the same nominal length, but containing different numbers of motif categories.

Microsatellite alleles represent a character transformation series that is generally thought to develop through polymerase slippage during replication, causing an increase or decrease in repeat number at a locus as discussed previously (Pumpernik et al. 2008; Schlötterer and Tautz 1992; Weber and Wong 1993). Several models for the microsatellite mutation process have been developed stemming initially from the stepwise mutation model (SMM) originally proposed by Ohta and Kimura (1973). The SMM assumes an equal rate of increase or decrease in the repeat number by a single-motif block, with the rate being maintained across the allelic range of the locus. As work progressed on STR evolution, it became apparent that repeat loci did not expand indefinitely. Evidence suggested that there was selective pressure limiting allele size and favoring smaller alleles (Garza et al. 1995). Weber and Wong (1993) suggested a mechanism in which relatively rare, large deletions periodically stemmed the increasing allele size. Population studies suggest that STR loci that exhibit a small allelic range appear to adhere to a linear distance model, while loci with a larger allele-size variance exhibit no discernible pattern to explain deviations from linearity of distance measures (Jorde et al. 1997; Zhivotovsky et al. 2001). Taylor et al. (1999) suggested a balancing model in which motif interruptions would stabilize a microsatellite region followed by the deletion mechanism proposed by Weber and Wong (1993). Thus, the two primary causative agents for the instability of molecular clock modeling at microsatellite loci appear to be: (1) differing rates of mutation among alleles at a locus following a stepwise model, and (2) interruptions of the repeat motif that alter strand slippage. A model was introduced that incorporated both strand slippage and point mutation (Kruglyak et al. 1998); however, until recently methods needed to capture the appropriate empirical data to facilitate a thorough analysis required sequencing as polymorphisms other than insertions, or deletions are not captured by routine STR typing platforms. Allele distribution data generated by base composition analysis at several STR loci suggests that polymorphisms among the repeat structure often dichotomize the alleles in separately evolving classes (Figure 10.2) (Planz et al. 2012). The allele-frequency data at the D3S1358 locus shown in Figure 10.2 support the general model suggested by Taylor et al. (1999) and others (Brinkmann et al. 1998; Kruglyak et al. 1998), however, the allelic distributions between the African American and Caucasian datasets appear to be following alternate trends.

10.4 Population Observations

Several recent studies have reported on the enhanced allelic distributions observed when STR loci have been examined either by base composition analysis using mass spectrometry or through next-generation DNA sequencing (Fordyce et al. 2011; Oberacher et al. 2008; Pitterl et al. 2008, 2010; Planz et al. 2012). At present, the most practical method for a deeper analysis of the STR loci resides with amplification followed by electrospray ionization mass spectrometry (ESI-MS). Current deep-sequencing applications have limitations in read length that make them ineffective for all but the smallest microsatellite alleles (Fordyce et al. 2011), and as with traditional Sanger sequencing, the methods are cost prohibitive due to both chemistry and labor. Population samples have been examined for Caucasians (Austria and U.S.) (Oberacher et al. 2008; Planz et al. 2012), African Khoisan (Pitterl et al. 2010), African American (Planz et al. 2012), Asian Yakut (Pitterl et al. 2010), and U.S. Hispanic (Planz et al. 2012). Of the 16 total loci examined across these populations, 8 loci

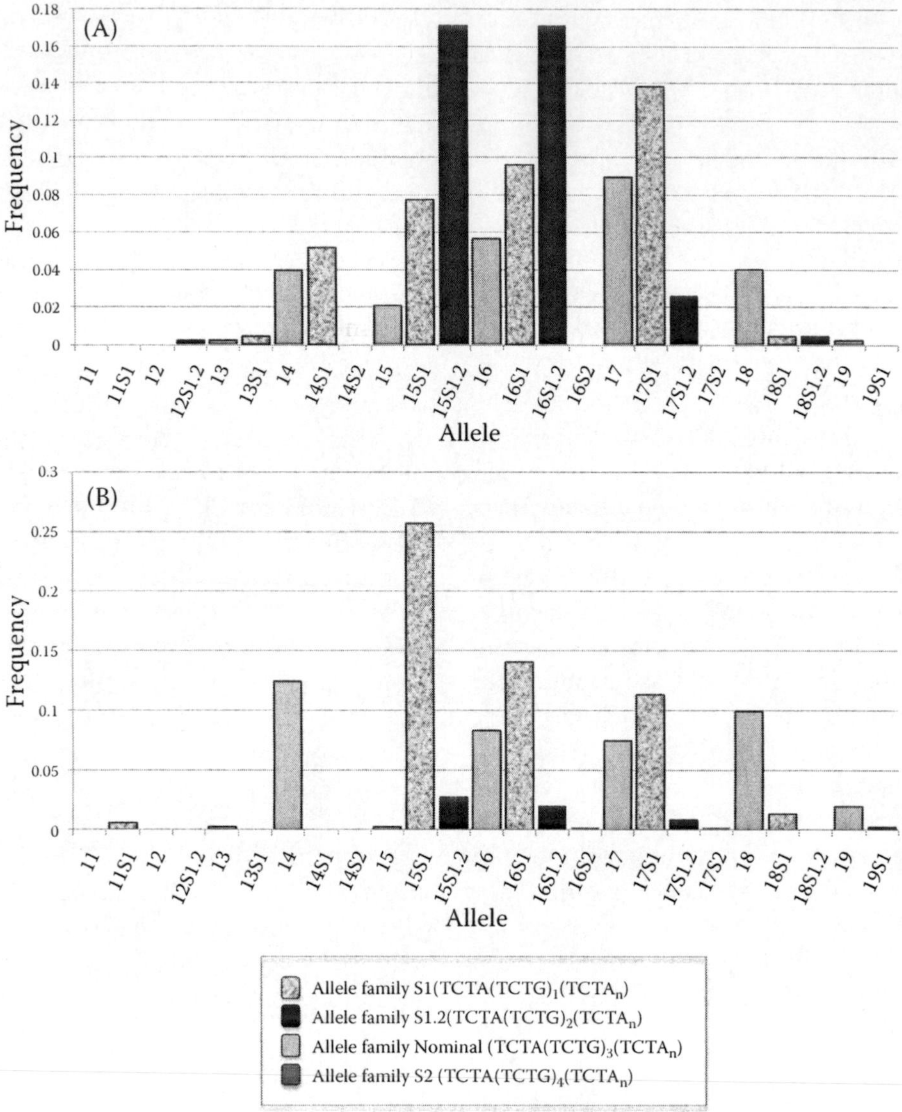

Figure 10.2 Allele distributions observed with mass spectrometry for D3S1358 in the U.S. African American population (A) and U.S. Caucasian population (B). Four allele families are observable based on variations in the number of repeats of the internal TCTG motif within the compound structure of the locus. Marked variations can be observed among the distributions for the different allelic forms, for example, although allele "15" is most common in both populations, the actual alleles based on sequence composition varies considerably with the S1.2 variant being prevalent in African Americans while the S1 variant is more common in Caucasians. Allele nomenclature addresses the overall tetranucleotide repeat number as well as nucleotide substitutions observed within the amplicon relative to a selected "nominal" allele. Nucleotide polymorphism suffixes S1-S12 specify various nucleotide substitutions within the amplicon; additional designations (e.g., S1.2, S1.3) reflect multiple changes of the same type from the "nominal" allele.

displayed considerable increases in the number of alleles detected when more advanced analytical methods were employed (Table 10.2). Comparable results were observed for 4 loci—D3S1358, D8S1179, D21S11, and vWA—between three population studies conducted using mass spectrometry (Oberacher et al. 2008; Pitterl et al. 2010; Planz et al. 2012).

Several loci—for example, CSF1P0, TH01, TPOX, D16S539, and D19S433—provided very little enhancement of the allelic range with the ESI-MS methods. Although the lack of greater variation at these loci minimizes their value as forensic identification markers, a more interesting question is: Why do these loci lack the variation observed at many of the other forensic STR loci? An initial observation might be that the first three loci—CSF1P0, TH01, and TPOX—share the characteristic of a location intronic to structural genes (Anker et al. 1992; Butler 2006; Hammond et al. 1994; Polymeropoulos et al. 1991). Although not a component of a gene product, it is unknown whether the intronic regions are responsible for any regulatory activities (Payseur et al. 2011). The limited allelic diversity and simple tetranucleotide motif structure suggest some level of selective pressure, acting across all populations, that would maintain low global F_{st} values observed at these loci (Taylor et al. 1999; Tishkoff and Kidd 2004). Conversely, other STR loci such as FGA, SE33, and vWA, which are also associated with structural gene loci (SE33 in the form of a pseudogene), display a great deal of polymorphism both at the nominal allele level, as well as in the presence of variations in repeat-motif structure among and within populations (Oberacher et al. 2008; Pitterl et al. 2010; Planz et al. 2012; Ruitberg et al. 2001). D16S539 and D19S433, however, lend no hint as to why additional polymorphisms are rare at these loci. D16S539 shares a repeat motif with CSF1PO (GATA for D16S539 and AGAT for CSF1PO, the difference based on naming convention, not DNA sequence) (Bar et al. 1997). This motif also exists within the repeat structure of D1S1656 and D12S391 recommended for the recently expanded European STR typing panel (Gill et al. 2006a,b). These two loci have been shown to be highly polymorphic due to the compound nature of their repeat-motif structure, however, extensive sequence analyses from population studies have not yet been reported (Gill et al. 1998).

Considerable attention is often given when drawing statistical conclusions on forensic evaluations surrounding population substructure and the application of appropriate F_{st} or θ value (Budowle et al. 2001b; Gill et al. 2003; National Research Council 1996). Tishkoff and Kidd (2004) discussed at length levels of population substructure as measured by F_{st} from SNPs, STRs, and protein polymorphisms. These authors suggest that the type of polymorphism examined may have an impact on the observed F_{st}. Measures of substructure of 3–5% are routinely observed with STRs contrasted to approximately 14% for SNPs (Tishkoff and Kidd 2004). The high heterozygosity in STRs fosters a lower F_{st} value. The routine typing methodology employed in genetic analysis can create artificial convergence in the recognition of alleles that are identical by state, but not by descent, due to masking of inherent polymorphisms. For example, the two alleles shown in Figure 10.1, panel A, are seen as identical when typed by CE, but are resolved as distinct when typed by mass spectrometry, demonstrating that by CE they could only be considered identical by state. Although convergence of polymorphic alleles will also occur (e.g., two D3S1358 allele 17 [2A→2G] can still be identical by state), the frequency of this observation will be reduced by the ability to resolve more of the distinct alleles that exist within populations.

An underlying assumption when evaluating sets of population data for substructure is that the alleles observed between population sets are homologous genetic entities. Several studies have evaluated population substructure using the forensic STR loci (Budowle and

Chakraborty 2001; Klintschar et al. 1999; Kracun et al. 2007; Sun et al. 2003) where loci-specific F_{st} have been reported to be consistently low. An evaluation of F_{st} across three U.S. population groups revealed higher than expected values for D3S1358 and vWA, indicating that some population-specific allele distributions may be present at these loci (Planz et al. 2012). A graphical depiction of the allele distributions derived from ESI-MS analysis on the U.S. Caucasian and African American populations clearly displays the layers of alternate allele distributions that exist at some of the STR loci (Figure 10.3). D8S1179 provides an example of a locus in which a near doubling of alleles was observed for the U.S. populations (Planz et al. 2012), as well as Austrian (Oberacher et al. 2008), Khoisan, and Yakut (Pitterl et al. 2010) populations. Two allele families are clearly represented at D8S1179 and a third allele family is also discernable in the African American dataset. One allele family consists of nominal alleles aligning with the standard GenBank reference sequence [G08710] motif of a simple TATC repeat. This allele family contains the alleles 8 through 14, with the 12 allele being the most frequent. This pattern is also observed in the Austrian dataset (Oberacher et al. 2008). However, the Khoisan and Yakut data do not clearly reflect this pattern, possibly due to their smaller sample sizes and/or isolated nature of those populations (Pitterl et al. 2010). The second allele distribution represents alleles that contain an interrupted motif structure in which a single TGTC repeat is present. This repeat motif is captured as an A→G polymorphism in mass spectrometry relative to the designated reference sequence. However, sequence analysis found that the position of this interrupting motif varies in its location within and among populations (Pitterl et al. 2010; Ruitberg et al. 2001). The modified repeat is positioned after the second or third unmodified repeat of the allele sequence. A third family of alleles is observed among African-origin populations that possess an additional A→G-bearing repeat motif. These alleles are observed only on the higher end of the overall allelic spectrum in alleles 13 and above. Sequence data suggests that this allele family is created by the slippage model described earlier (Section 10.2) originating from an allele already bearing an A→G-bearing repeat motif. Presence of a fourth allele family containing three copies of the TGTC repeat was observed in the African American population, reinforcing the evidence of further mutation through a slippage mechanism, with the possibility of different mutation rates effecting change in different regions of the allele. However, current sequence data is insufficient to thoroughly evaluate the evolutionary history of these alleles as they occur at lower frequencies in the populations. Thus, several underlying polymorphisms represented by the spectrum of alleles containing the interrupted repeats in D8S1179 remain hidden in mass spectrometry and can only be captured through sequence analysis.

However, resolution of repeat-motif polymorphism in D3S1358 is not masked as in D8S1179. Figure 10.2 provides a graphical depiction of the allele distribution at D3S1358 for the U.S. Caucasian and African American populations. Four distinct allele distributions are present at this locus, which has a repeat structure of $TCTA(TCTG)_2(TCTA)_{13}$ for the 16 allele in the reference sequence (AC099539). This reference sequence would actually correspond to a 16 (G→A) relative to a nominal reference allele of $TCTA(TCTG)_3(TCTA)_N$ selected in mass spectrometry studies (Oberacher et al. 2008; Pitterl et al. 2010; Planz et al. 2012). Allele family variations at this locus mostly consist of the loss or gain of a TCTG repeat localized at the beginning of the tandem repeat sequence (Figure 10.1), in addition to the expansion-contraction of the large repeat block of TCTA that regulates the majority of length variation at the locus. Allele results are comparable between the European and American studies, although sequence data was generated from alternate strands and referenced on different

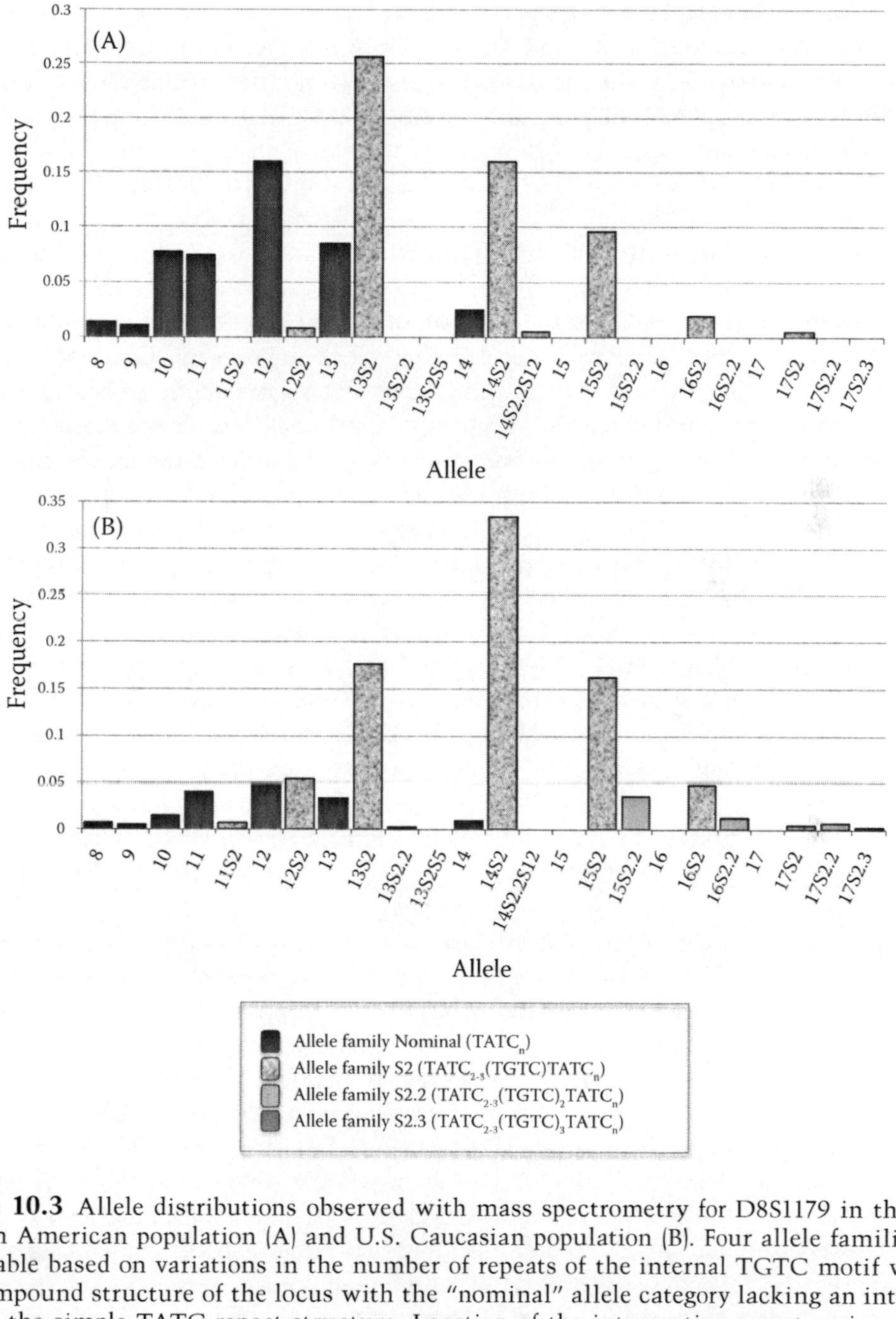

Figure 10.3 Allele distributions observed with mass spectrometry for D8S1179 in the U.S. African American population (A) and U.S. Caucasian population (B). Four allele families are observable based on variations in the number of repeats of the internal TGTC motif within the compound structure of the locus with the "nominal" allele category lacking an interruption to the simple TATC repeat structure. Location of the interrupting repeat varies within and among the populations, adding an additional level of complexity not observable with mass spectrometry. Allele nomenclature addresses the overall tetranucleotide repeat number as well as nucleotide substitutions observed within the amplicon relative to a selected "nominal" allele. Nucleotide polymorphism suffixes S1-S12 specify various nucleotide substitutions within the amplicon; additional designations (e.g., S2.2, S2.3) reflect multiple changes of the same type from the "nominal" allele.

known sequences. The four allele families that can be identified by mass spectrometry within D3S1358 are TCTA(TCTG)(TCTA)$_n$, TCTA(TCTG)$_2$(TCTA)$_n$, TCTA(TCTG)$_3$(TCTA)$_n$, representing the "nominal" allele, and TCTA(TCTG)$_4$(TCTA)$_n$. Presence of the increased diversity detectable at the locus increases its power of discrimination and power of exclusion markedly over what can be obtained using conventional CE analysis (Oberacher et al. 2008; Pitterl et al. 2010; Planz et al. 2012).

10.5 Value to Forensic Investigations

Ultimately, the most complete characterization of the variation within microsatellite loci will come through the continuing development of advanced sequencing technologies. Dephasing of multiple parallel reactions in current deep-sequencing applications relying on clonally expanded template clusters limit read lengths below what is required for all but the smallest microsatellite alleles. Although technologies are improving, the most commonly employed second-generation sequencing applications, Illumina and Life Sciences 454 platforms, have read lengths restricted to ~100 bp and ~350 bp, respectively. Data generated using these platforms have become the mainstay for deep coverage of high-complexity regions of the genome and *de novo* sequencing. However, when sequencing low-complexity regions, such as repeated motifs and homopolymeric stretches, depth of coverage is greatly reduced. Recently, McIver et al. (2011) reported that in a set of 376,695 microsatellites, between 48% and 96% of the loci were completely covered by at least a single read on the set of two family trios (mother, daughter, father) they studied, reflecting a global reduction in sequencing depth of two- to fivefold compared to nonrepeat regions of the genome. Recently, Fordyce et al. (2011) reported on the feasibility of applying high-throughput sequencing methods on the Roche FLX Genome Sequencer platform, specifically targeting commonly utilized forensic STR markers. A major hurdle addressed by the authors was the interrogation of the very large data file produced by deep-sequencing applications. The authors developed an algorithm that could parse the data file produced during the analysis that targeted sequence blocks containing the forensic STR sequences by locating primer sites in the sequence. The data were then further filtered to focus in on sequence regions that contained the entire repeat structure of the defined locus using specific sequence strings in both flanking regions of the locus. With sufficient depth of coverage, accurate sequence determinations can be differentiated from possible sequencing errors. Additionally, with sufficient data, stutter product percentages can be effectively assessed. Unlike most deep-sequencing studies, Fordyce et al. (2011) performed a site-directed study of five STR loci (CSF1PO, TH01, D5S818, D13S317, and D21S11) by the individual amplification of each locus across 10 human DNA samples. Of 6,488 sequence reads that advanced through screening by the sorting algorithm, 1,422 represented correct sequences for the total of 17 alleles evaluated in the study. Depth of coverage dropped to relatively low numbers for large loci, such as D21S11 (17 to 24 sequences per allele). Unfortunately, three of the five loci (CSF1PO, TH01, D5S818) selected by the authors for investigation represented loci that had been shown to be of limited polymorphism in previous studies (Oberacher et al. 2008; Pitterl et al. 2010; Planz et al. 2012; Ruitberg et al. 2001). Sequence analysis, however, would be of value in instances where the motif order is of interest, such as in population studies or familial analyses. A limited number of motif composition differences were observed by sequencing in D2S1338, D8S1179, D19S433, and D21S11 that would be unresolved with base composition analysis due to the canceling-out effect that dual changes have in the mass spectrometry analysis (Hall et al. 2009; Oberacher et al. 2008; Pitterl et al. 2010). The effect that these cryptic polymorphisms would play in routine forensic analyses would be expected to be minimal due to the presumed low frequency of these alleles in populations.

A considerable application of STR typing is in the area of relationship testing. Unlike typical forensic applications, relationship/kinship analyses must consider the possibility of mutational events occurring between potential relatives within the framework of the data analysis. To accurately assess motif structure and allele diversity for the purpose of understanding the mutation process, a STR locus must be sequenced through the entirety of its repeat-motif structure, including nonrepeating flanking sequences to accurately align consensus assemblies that would correctly locate polymorphisms among the repeat motifs as well as determine which portions of an interrupted allele is experiencing change. The importance of identifying the position of an altered repeat motif and how this site may serve as a regulator of the pattern and direction of allelic expansion or contraction at the locus will aid in developing better statistical models for supporting hypotheses involving mutation in relationship-testing cases. The actual process of mutation can be observed by evaluating family trios (mother, father, child) in which a Mendelian inconsistency is observed, especially when focusing on loci for which interrupted repeats are documented. However, the practicality of using deep-sequencing methods for routine casework is still limited by both throughput and read-length constraints of the technologies.

In many investigations, such as in relationship testing, missing persons identification, and forensic mixture analysis, the implementation of SNP assays and application of routine STR analyses are assumed to be effective. The presence of polymorphisms within the STR loci can have substantial impact of the concept of "allele sharing" as it is implied in the evaluation of relatedness (Figure 10.4). The ability to drill deeper into the composition of the STR allele would greatly enhance the reliability of familial/relatedness evaluations and reduce the number of uninformative associations considerably. Due to the relatively high rate of mutation that STR loci have (approximately $1–3 \times 10^{-3}$) (American Association of Blood Banks 2005; Ge et al. 2009; Weber and Wong 1993; Zhivotovsky et al. 2004), single inconsistencies in parentage evaluations are a common observation, being observed in approximately 2–3% of cases when the current STR kits are employed (Børsting and Morling 2011; Phillips et al. 2008; Poetsch et al. 2006). Although these events can be statistically addressed, the consequence is a greatly reduced parentage index, often supporting a hypothesis that favors a first-order relative as being the true parent over the tested individual. Børsting and Morling (2011) reported on the efficacy of utilizing a panel of 49 SNP markers to aid in the resolution of such cases. Albeit effective, SNP panels have not been readily implemented in many forensic laboratories. As a result, to address these concerns, additional genetic testing is typically performed using other STR loci, which can confound the matter. The independence of the STR markers, coupled with the high mutation rates, greatly supports the probability of observing a second mutational event when additional STR testing is performed in the case. These cases cannot be immediately relegated to being exclusionary. This observation is supported by the volume of parentage-testing evaluations, in which a suspected double mutation event is observed approximately every 5,000–6,000 parentage evaluations performed, a large proportion of which are observed when expanded panels of STR loci are used. Identifying hidden polymorphisms within the alleles at the STR locus may provide additional support for a particular nonmatching obligate allele to have been derived from a parental precursor. Similarly, the number of

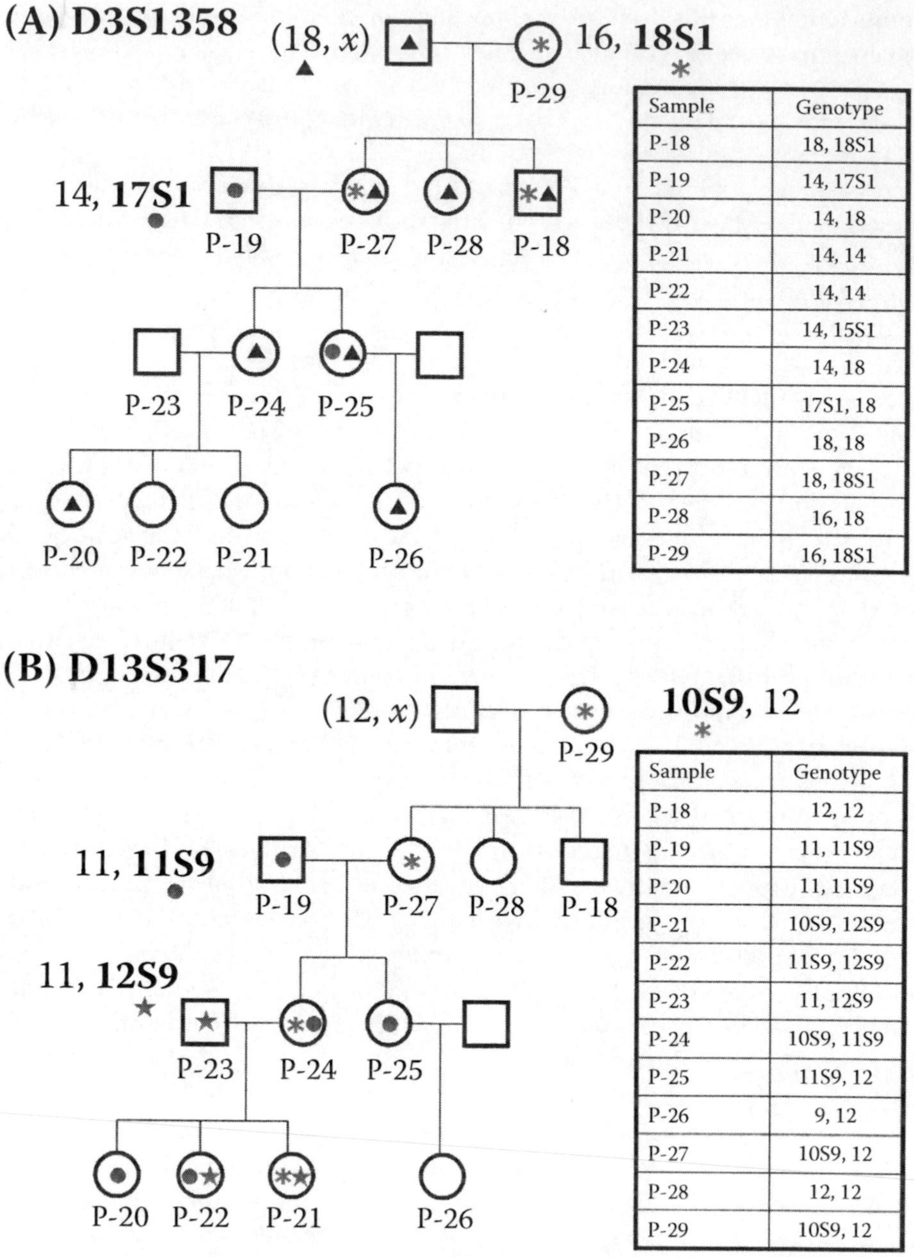

Sample	Genotype
P-18	18, 18S1
P-19	14, 17S1
P-20	14, 18
P-21	14, 14
P-22	14, 14
P-23	14, 15S1
P-24	14, 18
P-25	17S1, 18
P-26	18, 18
P-27	18, 18S1
P-28	16, 18
P-29	16, 18S1

Sample	Genotype
P-18	12, 12
P-19	11, 11S9
P-20	11, 11S9
P-21	10S9, 12S9
P-22	11S9, 12S9
P-23	11, 12S9
P-24	10S9, 11S9
P-25	11S9, 12
P-26	9, 12
P-27	10S9, 12
P-28	12, 12
P-29	10S9, 12

Figure 10.4 **(See color insert.)** Example of multigenerational pedigrees for which tracking of alleles is possible due to utilization of advanced typing methods. (A) Typing results for D3S1358 depicting resolution of allelic contribution of parents to offspring that represents same-length heterozygotes (P-18 and P-27). These two individuals have two "18" alleles that are different due to internal polymorphisms of the repeat motif. Origin of each allele can be traced to specific parents and the persistence of the allele originating from the founding male (who was not typed) can then be traced through the next three generations. (B) Typing results for D13S317 for the same pedigree also allowing discrimination of same-length heterozygotes for the "11" allele. Of note is the ability to distinguish the paternally derived 12S9 allele found in individuals P-21 and P-22 from the maternally inherited "12" allele in the cousin of these individuals, P-26. The "12" allele is ultimately inherited in this individual from the untyped founding male of the pedigree.

fortuitous associations in kinship evaluations driven by database searches should decrease sharply when the common alleles fostering the association are subdivided into two or more actual alleles.

10.6 Conclusions

Although the forensic community has been utilizing STR-typing technologies for almost 15 years, the suite of makers adopted for routine testing has not expanded substantially. To a large degree, the formation and growth of large national and international DNA databanks has fostered standardization of the loci examined. Recently, several proposals to expand the suite of markers to be included in databases have been fielded, primarily due to the statistical need to enhance discriminating power within large searchable databases. Very little attention has been placed on improving the resolving power inherent in the existing standard loci. Changes in technology in the sciences are inevitable. Where Sanger dideoxynucleotide sequencing with fragment analysis on CE instruments was the state of the art when the Human Genome Project was in process in the 1990s, culminating in the last decade with the publication of the genome (Lander et al. 2001; Venter et al. 2001), current approaches to DNA sequence analysis have far surpassed these traditional methods in accuracy, cost, and speed. It is inevitable that the forensic sciences must once again move forward in adopting new technologies as was done in the transition from allozyme electrophoresis to restriction fragment length polymorphism (RFLP) and then from RFLP to PCR-based systems, such as STRs. The two current technologies offering increased resolution of our existing STR loci represent a much shorter quantum leap in change than did the previous upgrades to the forensic toolkit just mentioned. Both the mass spectrometry and deep-sequencing approaches provide data that would be concordant with existing STR data based on fragment size since both technologies inherently capture this information. To fully utilize the increased power of the up-and-coming technologies, database systems need to adapt to the flexibility of advancing forms of genetic data. In a forensic comparison, the reality that further typing of a banked sample is inevitable must be acknowledged. The need for additional typing would hold true whether new technologies are adopted or whether additional loci are added to utilize the existing technologies. These considerations will be a focus of discussions for the entire suite of forensic practitioners, from analysts/examiners to administrators of laboratories, databases, and international consortiums. By far, STR markers will continue to provide the greatest power of discrimination in forensic examinations and genetic resolution in relationship testing available in the genome. As effective technologies continue to develop, the look and feel of forensic analyses of all forensic genetics applications will be changing and a clearer understanding of the nature of the STR allele that is more than skin deep will give analyses enhanced capabilities.

References

American Association of Blood Banks (AABB). 2005. Annual Report. Available at www.aabb.org.

Anker, R., T. Steinbrueck, and H. Donis-Keller. 1992. Tetranucleotide repeat polymorphism at the human thyroid peroxidase (hTPO) locus. *Hum Mol Genet* 1:137.

Bar,W., B. Brinkmann, B. Budowle et al. 1997. DNA recommendations. Further report of the DNA Commission of the ISFH regarding the use of short tandem repeat systems. International Society for Forensic Haemogenetics. *Int J Legal Med* 110:175–6.

Børsting, C. and N. Morling. 2011. Mutations and/or close relatives? Six case work examples where 49 autosomal SNPs were used as supplementary markers. *Forensic Sci Int Genet* 5:236–41.

Brinkmann, B., M. Klintschar, F. Neuhuber et al. 1998. Mutation rate in human microsatellites: Influence of the structure and length of the tandem repeat. *Am J Hum Genet* 62:1408–15.

Budowle, B. and R. Chakraborty. 2001. Population variation at the CODIS core short tandem repeat loci in Europeans. *Leg Med (Tokyo)* 3:29–33.

Budowle, B., A. Masibay, S. J. Anderson, et al. 2001a. STR primer concordance study. *Forensic Sci Int* 124:47–54.

Budowle, B., T. R. Moretti, S. J. Niezgoda et al. 1998. CODIS and PCR-based short tandem repeat loci: Law enforcement tools. in *Proc 2nd Eur Symp Human Identification*. Madison, WI: Promega Corporation.

Budowle, B., B. Shea, S. Niezgoda et al. 2001b. CODIS STR loci data from 41 sample populations. *J Forensic Sci* 46:453–89.

Budowle, B. and C. J. Sprecher. 2001. Concordance study on population database samples using the PowerPlex 16 kit and AmpFlSTR Profiler Plus kit and AmpFlSTR COfiler kit. *J Forensic Sci* 46:637–41.

Butler, J. M. 2006. Genetics and genomics of core short tandem repeat loci used in human identity testing. *J Forensic Sci* 51:253–65.

Butler, J. M. and C. H. Becker. 2001. Improved analysis of DNA short tandem repeats with time-of-flight mass spectrometry. *Science and Technology Research Report to NIJ* NCJ 188292.

Butler, J. M., J. Li, T. A. Shaler et al. 1999. Reliable genotyping of short tandem repeat loci without an allelic ladder using time-of-flight mass spectrometry. *Int J Legal Med* 112:45–49.

Clayton, T. M., S. M. Hill, L. A. Denton et al. 2004. Primer binding site mutations affecting the typing of STR loci contained within the AMPFlSTR SGM Plus kit. *Forensic Sci Int* 139:255–9.

Cotton, E. A., R. F. Allsop, J. L. Guest et al. 2000. Validation of the AMPFlSTR SGM plus system for use in forensic casework. *Forensic Sci Int* 112:151–61.

Crosby, M. K. A., L. E. Licht, and J. Fu. 2009. The effect of habitat fragmentation on finescale population structure of wood frogs (*Rana sylvatica*). *Conserv Genet* 10:1707–718.

Crouse, C. A., S. Rogers, E. Amiott et al. 1999. Analysis and interpretation of short tandem repeat microvariants and three-banded allele patterns using multiple allele detection systems. *J Forensic Sci* 44:87–94.

Deka, R., S. Guangyun, D. Smelser et al. 1999. Rate and directionality of mutations and effects of allele size constraints at anonymous, gene-associated, and disease-causing trinucleotide loci. *Mol Biol Evol* 16:1166–77.

Delamoye, M., C. Duverneuil, K. Riva et al. 2004. False homozygosities at various loci revealed by discrepancies between commercial kits: Implications for genetic databases. *Forensic Sci Int* 143:47–52.

Drabek, J., D. T. Chung, J. M. Butler et al. 2004. Concordance study between MiniPlex assays and a commercial STR typing kit. *J Forensic Sci* 49:859–60.

Ecker, D. J., J. Drader, J. Gutierrez et al. 2006. The Ibis T5000 Universal Biosensor—An automated platform for pathogen identification and strain typing. *JALA* 11:341–51.

Ellegren, H. 2000. Microsatellite mutations in the germline: Implications for evolutionary inference. *Trends Genet* 16:551–8.

Fordyce, S. L., M. C. Avila-Arcos, E. Rockenbauer et al. 2011. High-throughput sequencing of core STR loci for forensic genetic investigations using the Roche Genome Sequencer FLX platform. *Biotechniques* 51:127–33.

Garza, J. C., M. Slatkin, and N. B. Freimer. 1995. Microsatellite allele frequencies in humans and chimpanzees, with implications for constraints on allele size. *Mol Biol Evol* 12:594–603.

Ge, J., B. Budowle, X. G. Aranda et al. 2009. Mutation rates at Y chromosome short tandem repeats in Texas populations. *Forensic Sci Int Genet* 3:179–84.

Gill, P., E. d'Aloja, B. Dupuy et al. 1998. Report of the European DNA Profiling Group (EDNAP)—An investigation of the hypervariable STR loci ACTBP2, APOAI1 and D11S554 and the compound loci D12S391 and D1S1656. *Forensic Sci Int* 98:193–200.

Gill, P., L. Fereday, N. Morling et al. 2006a. The evolution of DNA databases—Recommendations for new European STR loci. *Forensic Sci Int* 156:242–4.

Gill, P., L. Fereday, N. Morling et al. 2006b. New multiplexes for Europe—Amendments and clarification of strategic development. *Forensic Sci Int* 163:155–7.

Gill, P., L. Foreman, J.S. Buckleton et al. 2003. A comparison of adjustment methods to test the robustness of an STR DNA database comprised of 24 European populations. *Forensic Sci Int* 131:184–96.

Hall, T. A., K. A. Sannes-Lowery, L. D. McCurdy et al. 2009. Base composition profiling of human mitochondrial DNA using polymerase chain reaction and direct automated electrospray ionization mass spectrometry. *Anal Chem* 81: 7515–26.

Hamilton, M. B., E. L. Pincus, A. Di Fiore et al. 1999. Universal linker and ligation procedures for construction of genomic DNA libraries enriched for microsatellites. *Biotechniques* 27:500–7.

Hammond, H. A., L. Jin, Y. Zhong et al. 1994. Evaluation of 13 short tandem repeat loci for use in personal identification applications. *Am J Hum Genet* 55:175–89.

Hill, C. R., M. C. Kline, J. J. Mulero et al. 2007. Concordance study between the AmpFlSTR MiniFiler PCR amplification kit and conventional STR typing kits. *J Forensic Sci* 52:870–3.

Hofstadler, S. A. and T. A. Hall. 2008. Analysis of DNA forensic markers using high-throughput mass spectrometry. *NIJ Final Report* Contract Number #2006 DN-BX-K011.

Hofstadler, S. A., R. Sampath, L. B. Blynm et al. 2005. TIGER: The universal biosensor. *Int J Mass Spectrom* 242:23–41.

Jordan, M. E., D. A. Morris, and S. E. Gibson. 2009. The influence of historical landscape change on genetic variation and population structure of a terrestrial salamander (*Plethodon cinereus*). *Conserv Genet* 10:1647–58.

Jorde, L. B., A. R. Rogers, M. Bamshad et al. 1997. Microsatellite diversity and the demographic history of modern humans. *Proc Natl Acad Sci USA* 94:3100–3.

Julian, S. E. and T. L. King. 2003. Novel tetranucleotide microsatellite DNA markers for the wood frog (*Rana sylvatica). Mol Ecol Notes* 3:256.

Kidd, K., A. Pakstis, W. Speed et al. 2006. Developing a SNP panel for forensic identification of individuals. *Forensic Sci Int* 164:20–32.

Kimmel, M. and R. Chakraborty. 1996. Measures of variation at DNA repeat loci under a general stepwise mutation model. *Theor Popul Biol* 50:345–67.

Klintschar, M., N. al-Hammadi, and B. Reichenpfader. 1999. Population genetic studies on the tetrameric short tandem repeat loci D3S1358, VWA, FGA, D8S1179, D21S11, D18S51, D5S818, D13S317 and D7S820 in Egypt. *Forensic Sci Int* 104:23–31.

Kracun, S., G. Curic, I. Birus et al. 2007. Population substructure can significantly affect reliability of a DNA-led process of identification of mass fatality victims. *J Forensic Sci* 52:874–8.

Kruckenhauser L., A. Bryant, S. Griffin et al. 2009. Patterns of within and between—Colony microsatellite variation in the endangered Vancouver Island marmot (*Marmota vancouverensis*). Implications for conservation. *Conserv Genet* 10:1759–72.

Kruglyak, S., R. Durrett, M. Schug et al. 1998. Equilibrium distributions of microsatellite repeat length resulting from a balance between slippage events and point mutations. *Proc Natl Acad Sci USA* 95:10774–78.

Lander, E., L. Linton, B. Birren et al. 2001. Initial sequencing and analysis of the human genome. *Nature* 409:860–921.

Leclair, B., C. Fregeau, K. Bowen et al. 2004. Systematic analysis of stutter percentages and allele peak height and peak area ratios at heterozygous STR loci for forensic casework and database samples. *J Forensic Sci* 49:968–80.

Leibelt, C., B. Budowle, P. Collins et al. 2003. Identification of a D8S1179 primer binding site mutation and the validation of a primer designed to recover null alleles. *Forensic Sci Int* 133:220–7.

Leopoldino, A. and S. Pena. 2003. The mutational spectrum of human autosomal tetranucleotide microsatellites. *Hum Mutat* 21:71–9.

Madsen, B., P. Villesen, and C. Wiuf. 2007. A periodic pattern of SNPs in the human genome. *Genome Res* 17:1414–19.

Madsen, B., P. Villesen, and C. Wiuf. 2008. Short tandem repeats in human exons: A target for disease mutations. *BMC Genomics* 9:410–18.

Mammalian Genotyping Service. 2011. http://research.marshfieldclinic.org/genetics/home/index.asp.

McIver, L., J. Fondon III, M. Skinner et al. 2011. Evaluation of microsatellite variation in the 1000 Genomes Project pilot studies is indicative of the quality and utility of the raw data and alignments. *Genomics* 97:193–9.

Moller, A., E. Meyer, and B. Brinkmann. 1994. Different types of structural variation in STRs: HumFES/FPS, HumVWA and HumD21S11. *Int J Legal Med* 106:319–23.

National Research Council. *The Evaluation of Forensic DNA Evidence*. Washington, DC: National Academy Press, 1996.

NCBI Human Reference Genome Build 36.1 March 1, 2009, ftp://ftp/ncbi.nih.gov/genomes/H_sapiens/ARCHIVE/BUILD.36.1/.

Nelson, M., E. Levedakou, J. Matthews et al. 2002. Detection of a primer-binding site polymorphism for the STR locus D16S539 using the Powerplex 1.1 system and validation of a degenerate primer to correct for the polymorphism. *J Forensic Sci* 47:345–9.

Oberacher, H., H. Niederstatter, F. Pitterl et al. 2006. Profiling 627 mitochondrial nucleotides via the analysis of a 23-plex polymerase chain reaction by liquid chromatography-electrospray ionization time-of-flight mass spectrometry. *Anal Chem* 78:7816–27.

Oberacher, H., F. Pitterl, G. Huber et al. 2008. Increased forensic efficiency of DNA fingerprints through simultaneous resolution of length and nucleotide variability by high-performance mass spectrometry. *Hum Mutat* 29:427–32.

Ohta, T. and M. Kimura. 1973. A model of mutation appropriate to estimate the number of electrophoretically detectable alleles in a finite population. *Genet Res* 22:201–4.

Pakstis, A., W. Speed, R. Fang et al. 2010. SNPs for a universal individual identification panel. *Hum Genet* 127:315–24.

Pakstis, A., W. Speed, J. Kidd et al. 2007. Candidate SNPs for a universal individual identification panel. *Hum Genet* 121:305–17.

Payseur, B., P. Jing, and R. Haasl. 2011. A genomic portrait of human microsatellite variation. *Mol Biol Evol* 28:303–12.

Peacock, M., V. Kirchoff, and S. Merideth. 2002. Identification and characterization of nine polymorphic microsatellite loci in the North American pika, *Ochotona princeps*. *Mol Ecol Notes* 2:360–2.

Pemberton, T., C. Sandefur, M. Jakobsson et al. 2009. Sequence determinants of human microsatellite variability. *BMC Genomics* 10:612–30.

Phillips, C., M. Fondevila, M. Garcia-Magarinos et al. 2008. Resolving relationship tests that show ambiguous STR results using autosomal SNPs as supplementary markers. *Forensic Sci Int Genet* 2:198–204.

Pitterl, F., H. Niederstatter, G. Huber et al. 2008. The next generation of DNA profiling—STR typing by multiplexed PCR—Ion-pair RP LC-ESI time-of-flight MS. *Electrophoresis* 29:4739–50.

Pitterl, F., K. Schmidt, G. Huber et al. 2010. Increasing the discrimination power of forensic STR testing by employing high-performance mass spectrometry, as illustrated in indigenous South African and Central Asian populations. *Int J Legal Med* 124:551–7.

Planz, J. V., K. A. Sannes-Lowery, D. D. Duncan et al. 2012. Automated analysis of sequence polymorphism in STR alleles by PCR and direct electrospray ionization mass spectrometry. *Forensic Sci Int Genet* 6:594–606.

Poetsch, M., C. Ludcke, A. Repenning et al. 2006. The problem of single parent/child paternity analysis—Practical results involving 336 children and 348 unrelated men. *Forensic Sci Int* 159:98–103.

Polymeropoulos, M., H. Xiao, D. Rath et al. 1991. Tetranucleotide repeat polymorphism at the human tyrosine hydroxylase gene (TH). *Nucleic Acids Res* 19:3753.

Figure 1.2 The luminescent reaction that occurs when BLUESTAR®, luminol, or leuco crystal violet are applied to a substrate. (Photograph from BLUESTAR Forensic Web site, http://www.bluestar-forensic.com/, accessed June 9, 2010.)

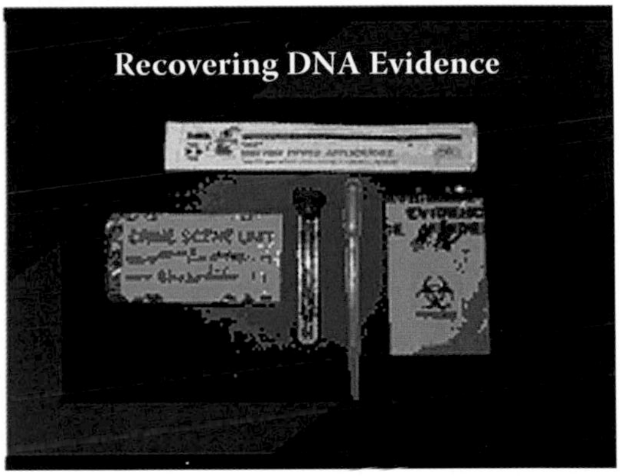

Figure 1.3 Swabs, distilled water, plastic pipettes, and paper envelopes are components of a basic DNA recovery kit. (Photograph by the author.)

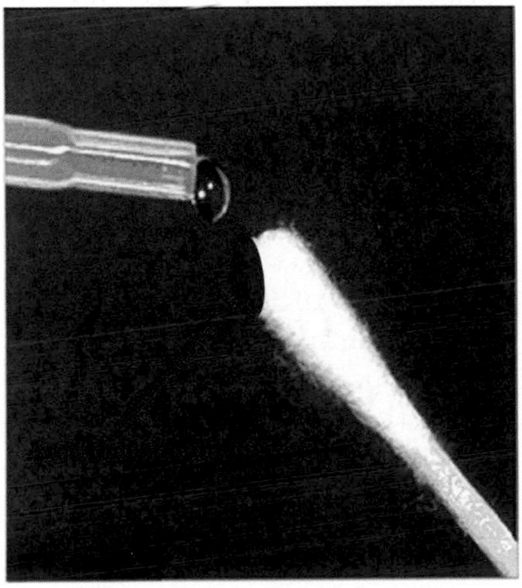

Figure 1.6 For proper hydration and swabbing, a single drop of distilled water should be applied to the side of a cotton-tipped swab. (Photograph from the New York City Office of the Chief Medical Examiner.)

Figure 1.7 The proper swabbing technique for the recovery of a dried blood sample from the side handle of a refrigerator door. Note the use of personal protective equipment. (Photograph by the author.)

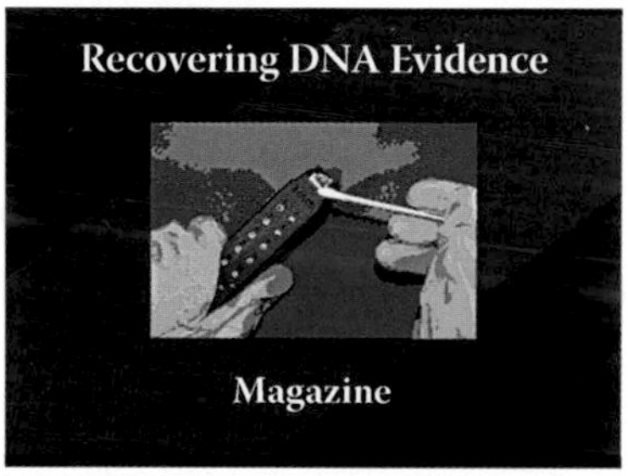

Figure 1.8 The lip area of a firearm's magazine is a potential source of DNA. (Photograph by the author.)

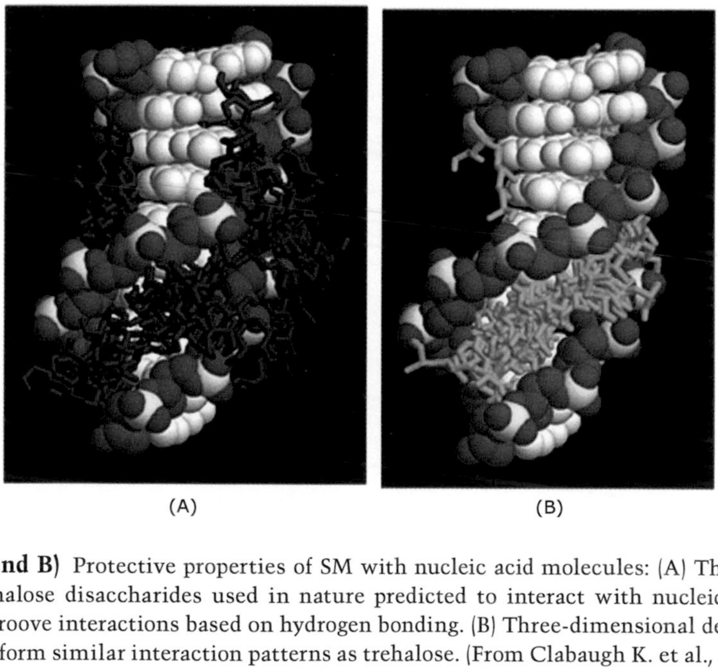

(A) (B)

Figure 2.1 (A and B) Protective properties of SM with nucleic acid molecules: (A) Three-dimensional depiction of trehalose disaccharides used in nature predicted to interact with nucleic acid molecules through minor groove interactions based on hydrogen bonding. (B) Three-dimensional depiction of SM as it is predicted to form similar interaction patterns as trehalose. (From Clabaugh K. et al., 2007, *18th Annu Meet Int Symp Human Identification*, Hollywood, CA, October 2007.)

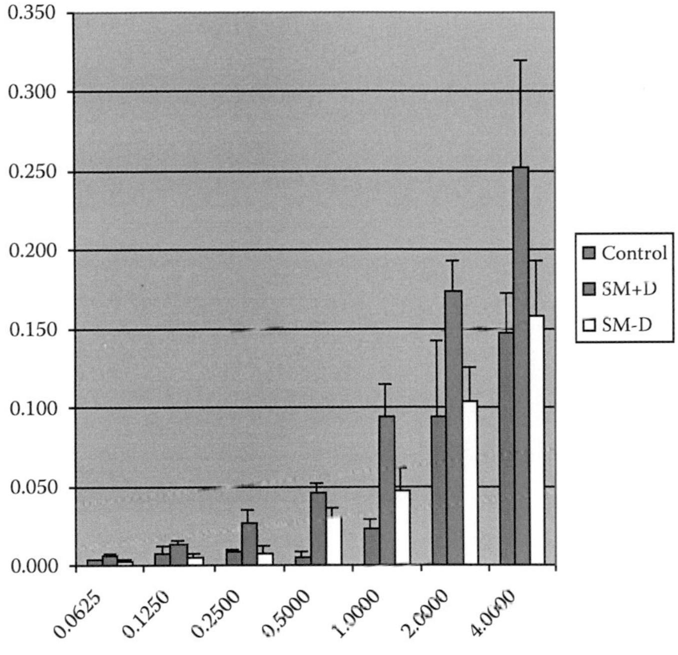

Figure 2.2A DNA recovery from samples stored in SampleMatrix versus –20°C: (A) Average DNA recovery of SampleMatrix-stored DNA versus –20°C frozen controls. Replicate DNA samples at seven different concentrations were stored at ambient room temperature in SampleMatrix with dessicant (SM+D), without dessicant (SM–D), or at –20°C as frozen liquid controls in polypropylene microfuge tubes. Quantification was performed utilizing the ABI Human Quantifiler kit as per manufacturer's recommendations. Recovery from SM+D-stored samples at room temperature was higher than that of frozen controls for every concentration. (From Ahmad T. et al., 2009, *Proc 2009 AAFS Annu Meet* 15:108–9; Lee S. B. et al., 2012, *Forensic Sci Int Genet* 6(1):31.40.)

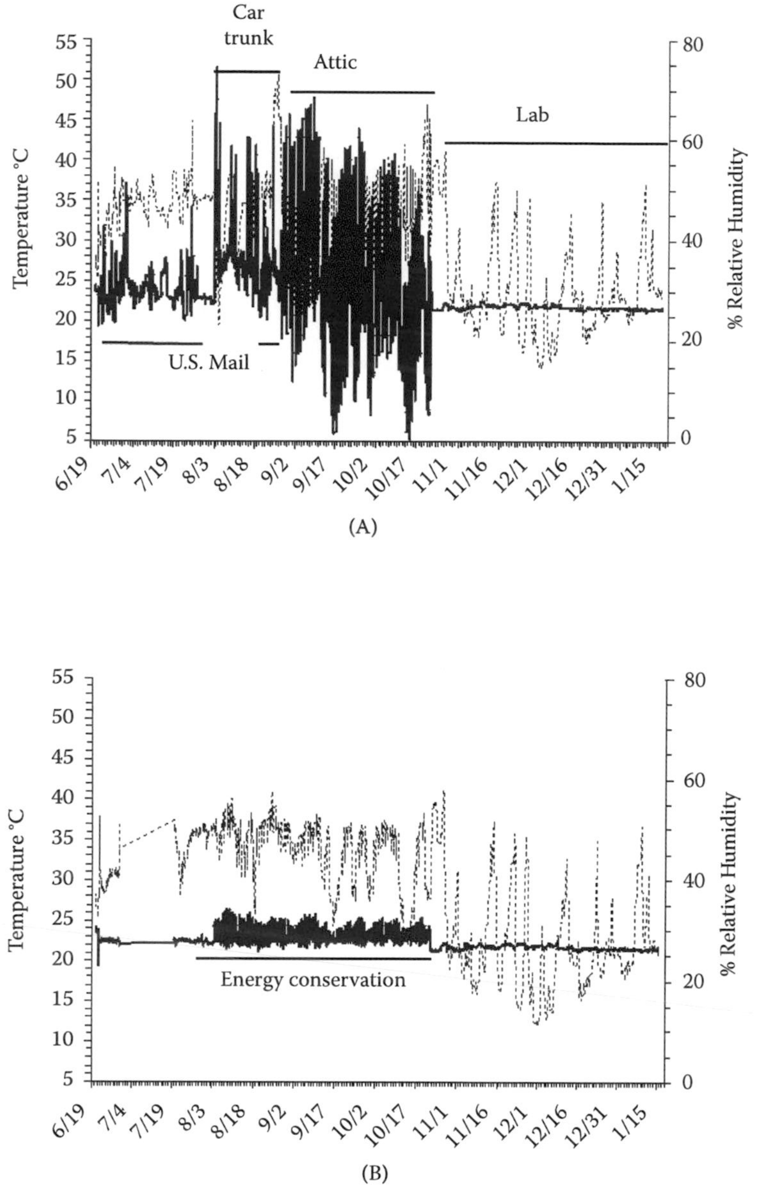

Figure 2.3 Temperature and humidity plots over 208 d of DNA storage and shipping at ambient temperature in the laboratory and in shipping containers: (A) temperature and humidity plots from shipped sample plate monitor and (B) temperature and humidity plots for room ambient temperature plate monitor. (From Clabaugh K. et al., 2007, *18th Annu Meet Int Symp Human Identification*, Hollywood, CA, October 2007.)

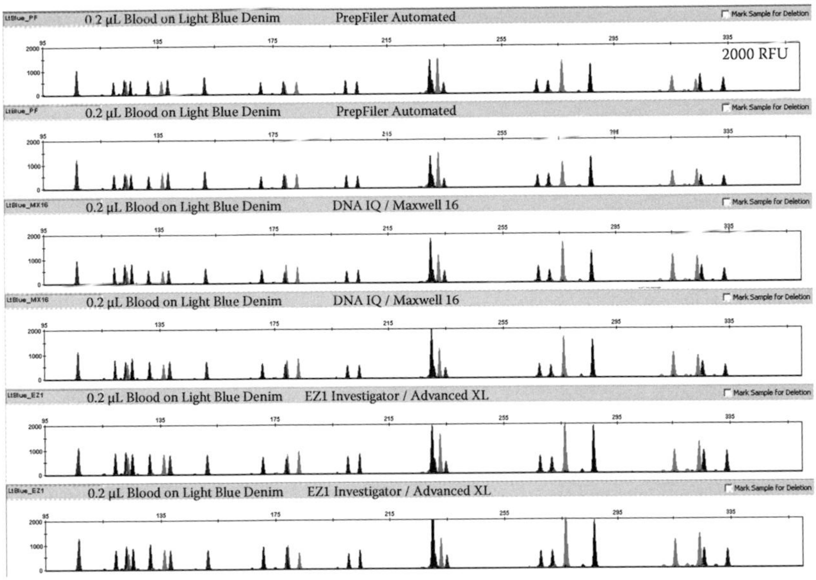

Figure 3.2 Representative STR profiles for 1 ng of DNA extracted from 0.2 μL blood spotted on light blue denim (4-mm punch) obtained using the Identifiler® PCR Amplification Kit. The DNA was extracted using either the PrepFiler® Automated Kit (top two panels), DNA IQ® Trace DNA Kit (middle two panels), or EZ1® Investigator kit (bottom two panels). Yield of DNA from respective kits is summarized in Figure 3.1.

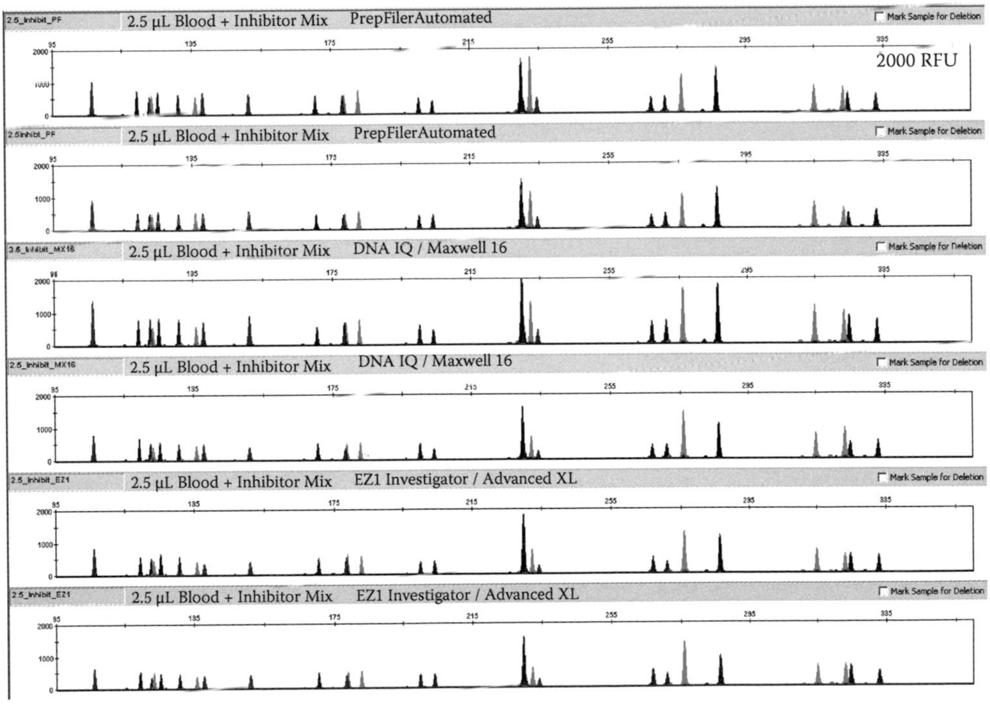

Figure 3.4 Representative STR profiles for 1 ng of DNA obtained from 2.5 μL blood spiked with an inhibitor mix (indigo, hematin, humic acid, and urban dust extract) using the Identifiler® PCR Amplification Kit. The DNA was extracted using either the PrepFiler® Automated Kit (top two panels), DNA IQ® Trace DNA Kit (middle two panels), or EZ1® Investigator kit (bottom two panels). Yield of DNA from respective kits is summarized in Figure 3.3.

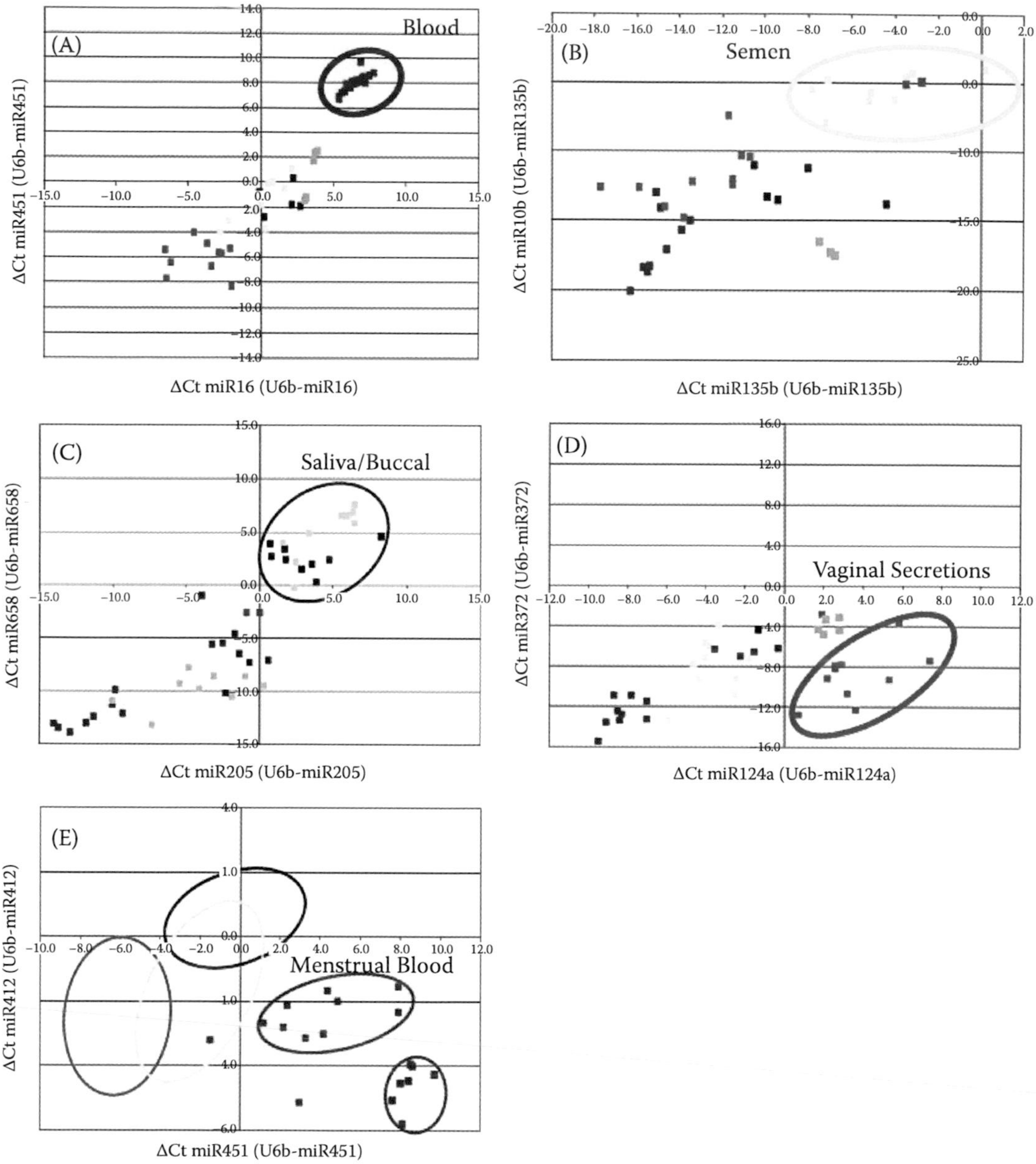

Figure 5.4 Evaluation of additional samples with miRNA body-fluid identification assays. To further demonstrate the ability of the miRNA assay to identify individual body fluids, additional samples were analyzed (blood, n = 19; semen, n = 11, including two vasectomized males; saliva, n = 18, including buccal and saliva samples; vaginal secretions, n = 11; menstrual blood, n = 11): (A) blood assay; (B) semen assay; (C) saliva assay; (D) vaginal secretions assay; (E) menstrual blood assay. Individual body-fluid data points are presented by colored squares: red, blood; yellow, semen; orange, semen from vasectomized males; dark blue, saliva; light blue, buccal swabs; green, vaginal secretions; pink, menstrual blood. (From Hanson, E. K., H. Lubenow, and J.Ballantyne, 2009, *Anal Biochem* 387:303. Reproduced with permission.)

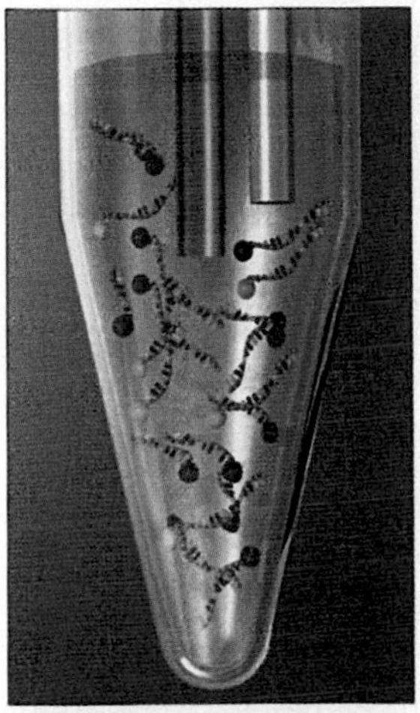

Figure 7.2 Mechanism of electrokinetic injection during CE. The negatively charged molecules enter the capillary, aligned with the cathode, as they migrate toward the anode at the other end of the capillary. (Adapted from Applied Biosystems, 2002b, ABI Prism® 310 Genetic Analyzer—The measure of enabling technology, Product Bulletin, Publication #106PB07–01.)

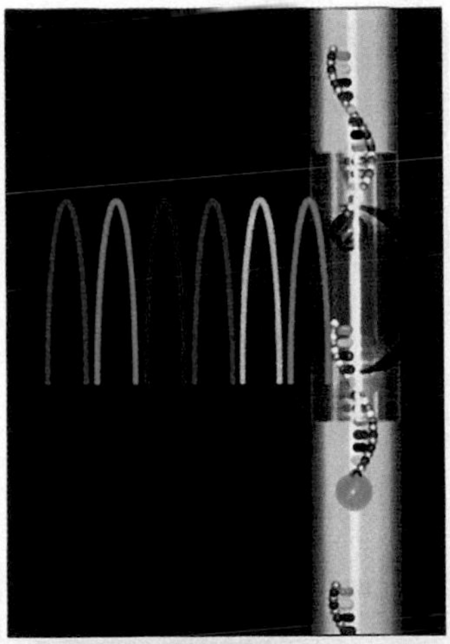

Figure 7.3 Detection of the DNA fragments during CE. DNA fragments labeled with fluorescent dyes pass through the detection window during migration toward the anode. The laser excites the dyes, causing them to emit light at wavelengths larger than that of the laser. Emitted light is collected by a CCD camera. Finally, the software converts the pattern of emissions into colored peaks. (Adapted from Applied Biosystems, 2002b, ABI Prism® 310 Genetic Analyzer—The measure of enabling technology, Product Bulletin, Publication #106PB07–01.)

(A) D3S1358

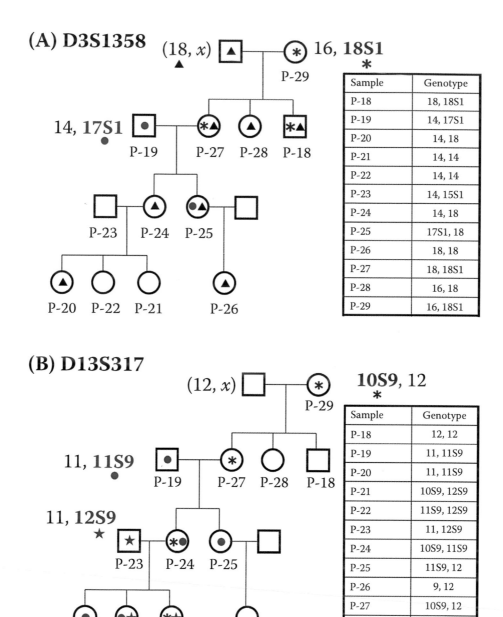

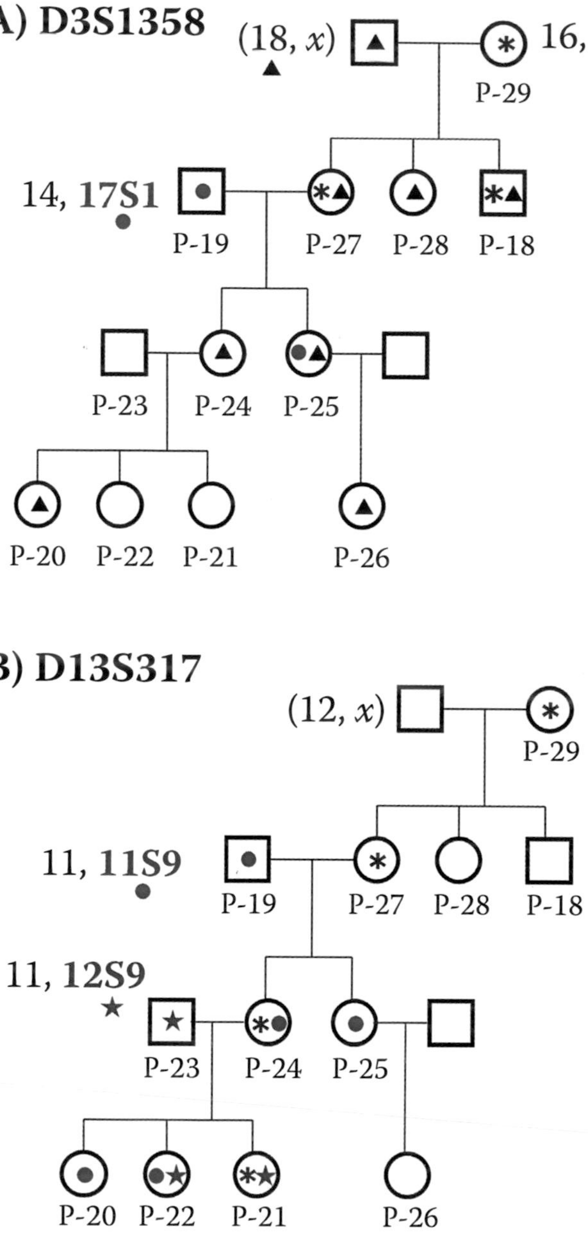

Sample	Genotype
P-18	18, 18S1
P-19	14, 17S1
P-20	14, 18
P-21	14, 14
P-22	14, 14
P-23	14, 15S1
P-24	14, 18
P-25	17S1, 18
P-26	18, 18
P-27	18, 18S1
P-28	16, 18
P-29	16, 18S1

Sample	Genotype
P-18	12, 12
P-19	11, 11S9
P-20	11, 11S9
P-21	10S9, 12S9
P-22	11S9, 12S9
P-23	11, 12S9
P-24	10S9, 11S9
P-25	11S9, 12
P-26	9, 12
P-27	10S9, 12
P-28	12, 12
P-29	10S9, 12

Figure 10.4 Example of multigenerational pedigrees for which tracking of alleles is possible due to utilization of advanced typing methods. (A) Typing results for D3S1358 depicting resolution of allelic contribution of parents to offspring that represents same-length heterozygotes (P-18 and P-27). These two individuals have two "18" alleles that are different due to internal polymorphisms of the repeat motif. Origin of each allele can be traced to specific parents and the persistence of the allele originating from the founding male (who was not typed) can then be traced through the next three generations. (B) Typing results for D13S317 for the same pedigree also allowing discrimination of same-length heterozygotes for the "11" allele. Of note is the ability to distinguish the paternally derived 12S9 allele found in individuals P-21 and P-22 from the maternally inherited "12" allele in the cousin of these individuals, P-26. The "12" allele is ultimately inherited in this individual from the untyped founding male of the pedigree.

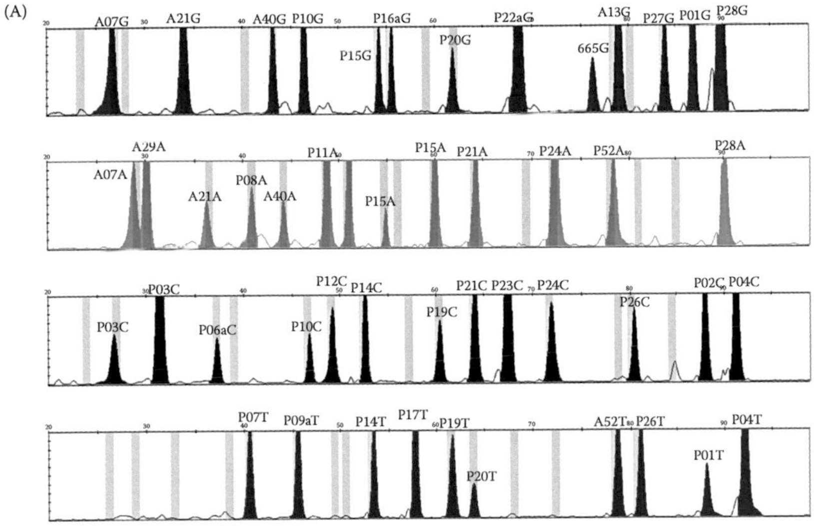

34 AIM-SNPs typed with SNaPshot

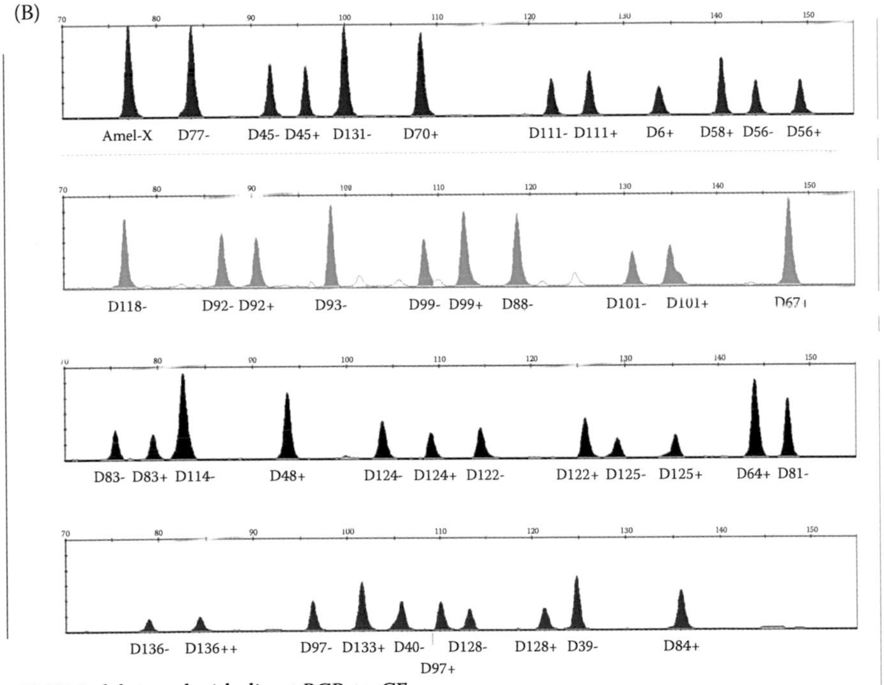

38 ID-Indels typed with direct PCR-to-CE

Figure 13.3 Two electropherograms representing typical forensic short-binary marker analysis: SNaPshot SNP genotyping patterns (upper pane) and PCR-to-CE indel genotyping patterns. The uppermost electropherogram (A) shows the SNP*for*ID 34-plex genotypes where heterozygotes have peaks labeled with different dyes depending on varying base-terminator combinations. The lower electropherogram (B) shows the 38-plex ID-Indel multiplex where heterozygotes have two successive peaks labeled with the same dye, separated by indel base motif size. The 34-plex component SNP codes used to identify each allele peak are outlined at: http://mathgene.usc.es/snipper/default3_34_new.html; the codes used for the 38-plex ID-Indels are outlined in Periera et al. (2009).

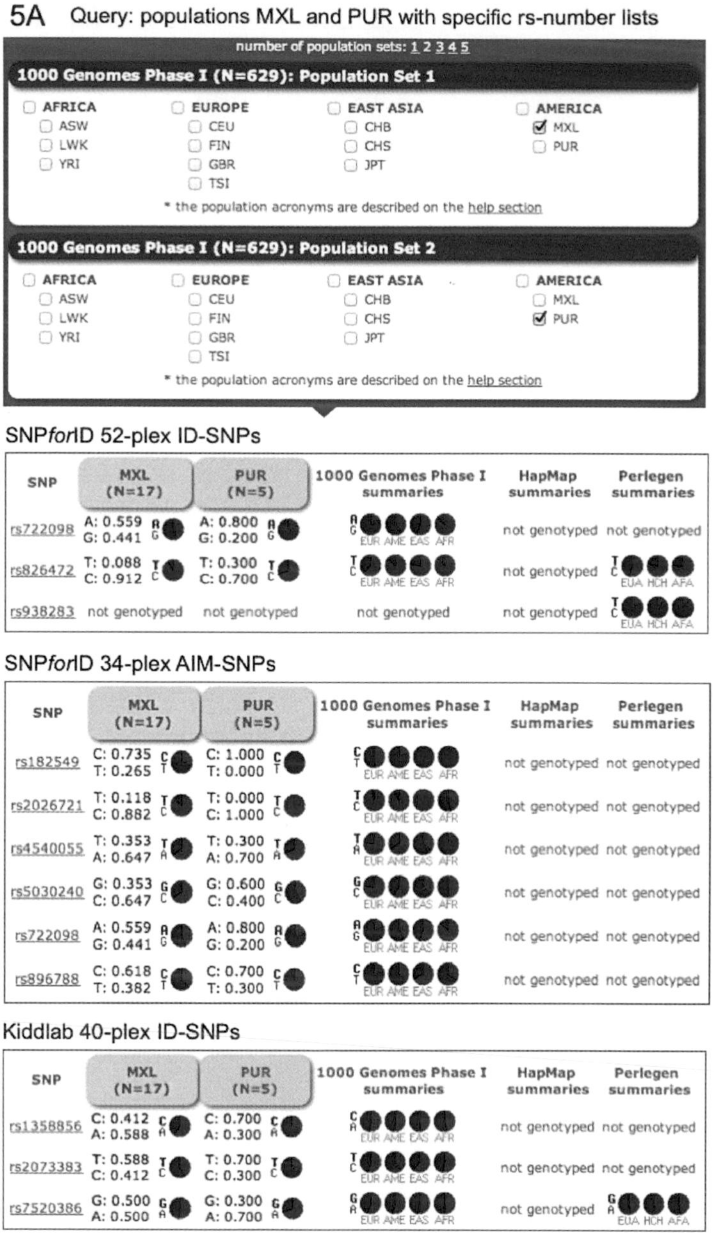

Figure 13.5 Five examples of studies using the SPSmart *ENGINES* SNP browser to explore 1000 Genomes data relevant to forensic analyses. (A) comparison of different database coverage of components of SNP*for*ID ID 52-plex, AIM 34-plex, and the Kiddlab 40-plex SNPs. (B) exploration of low-frequency SNPs in or around the PCR primer sequences of AIM-SNP rs2304925 (17:75551667). (C and D) Finding new SNP variation around the STR vWA. (E) Exploration of low-frequency coding region SNPs in the MC1R gene. Finally, (F) shows examples of tri-allelic SNPs now curated and listed in 1000 Genomes; data shows Japanese population frequencies found using the *ENGINES* SNP browser. (Continued)

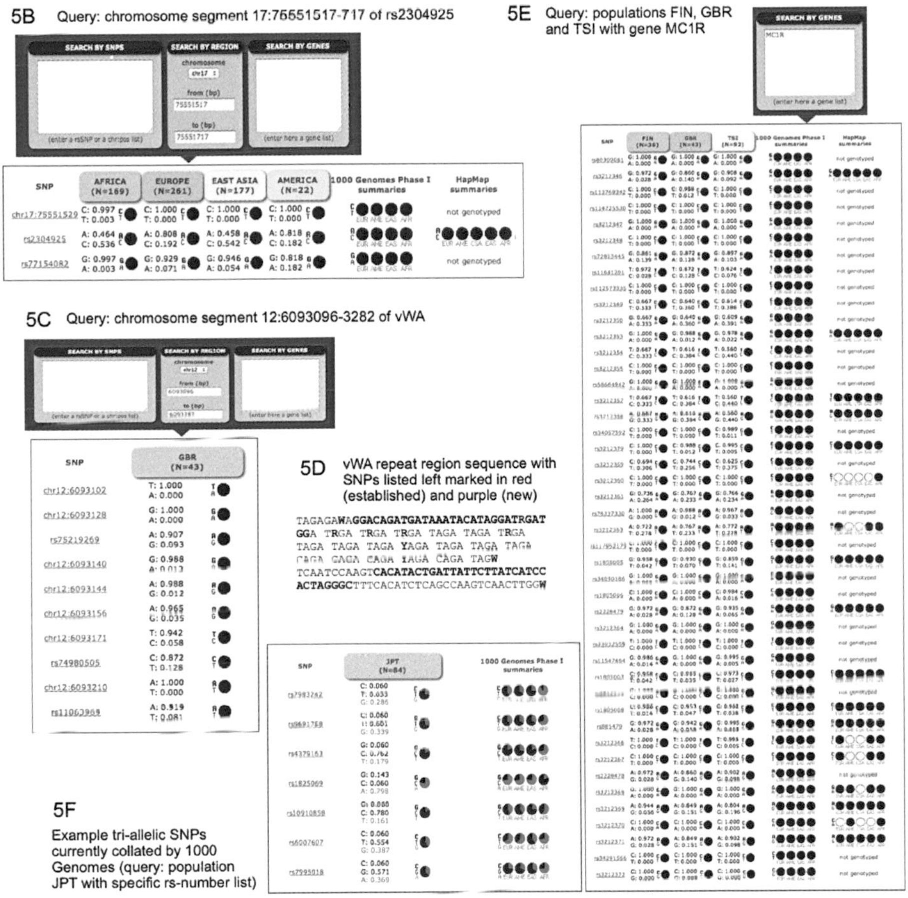

Figure 13.5 (Continued) Five examples of studies using the SPSmart *ENGINES* SNP browser to explore 1000 Genomes data relevant to forensic analyses. (A) comparison of different database coverage of components of SNP*for*ID ID 52-plex, AIM 34-plex, and the Kiddlab 40-plex SNPs. (B) exploration of low-frequency SNPs in or around the PCR primer sequences of AIM-SNP rs2304925 (17:75551667). (C and D) Finding new SNP variation around the STR vWA. (E) Exploration of low-frequency coding region SNPs in the MC1R gene. Finally, (F) shows examples of tri-allelic SNPs now curated and listed in 1000 Genomes; data shows Japanese population frequencies found using the *ENGINES* SNP browser.

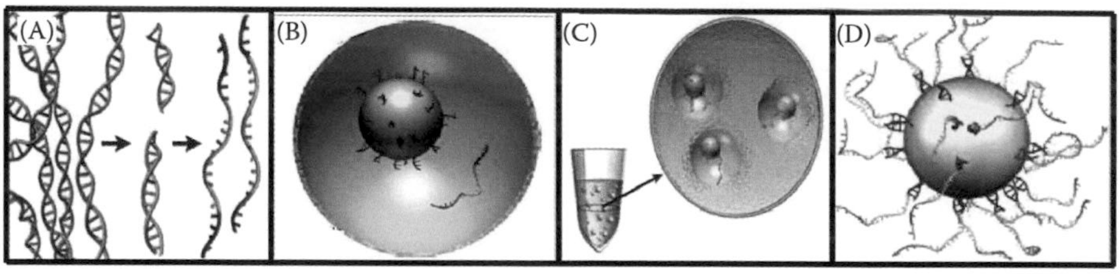

Figure 14.1 DNA polony formation via bead amplification. (A) The DNA is fragmented and denatured. Adaptors are ligated to the ends of the single-stranded fragments. (B) The adaptors hybridize to probes on the DNA capture beads. A single fragment attaching to a single bead. (C) Emulsion PCR. The DNA template–containing beads are emulsified with amplification reagents in a water-in-oil mixture. Each water droplet contains a bead. The droplets are amplified simultaneously. (D) A DNA polony. Millions of copies of the original DNA fragment attached to a bead. (Image reproduced with permission from 454 Sequencing, Roche Diagnostics, Indianapolis, Indiana.)

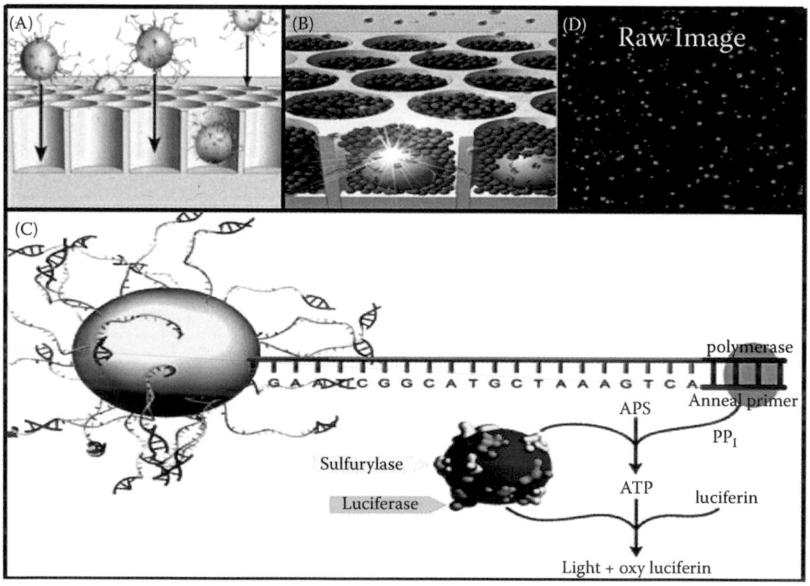

Figure 14.2 Principles of 454 sequencing. (A) A single bead enters each well of the PicoTiterPlate. (B) Simultaneous, independent sequencing of each bead by flooding the plate sequentially with each dNTP. (C) Pyrosequencing chemistry. Pyrophosphate (PPi) is released with the incorporation of a dNTP into the extending DNA strand. ATP sulfurylase forms ATP by joining the PPi with ADP. Luciferace utilizes the ATP to convert luciferin to oxyluciferin, which emits visible light. (D) The visible light is detected and recorded. (Image reproduced with permission from 454 Sequencing, Roche Diagnostics, Indianapolis, Indiana.)

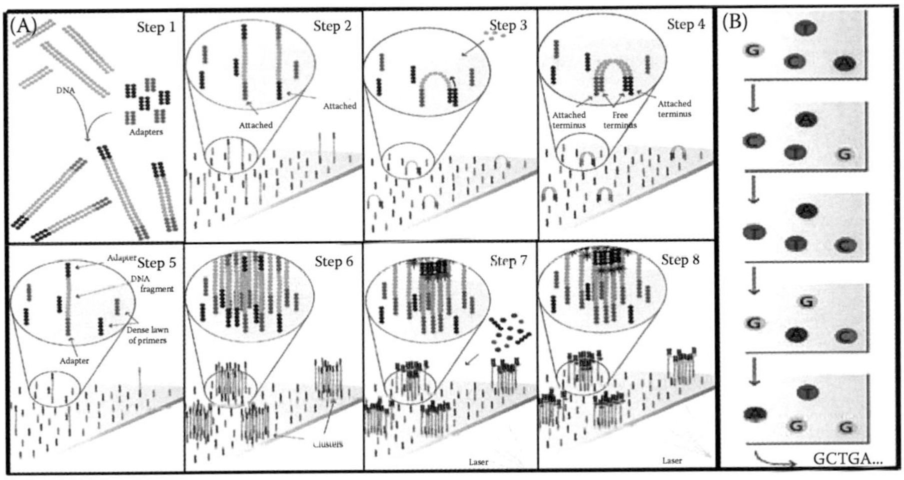

Figure 14.3 Principles of Illumina sequencing. (A) TruSeq chemistry: Step 1: DNA is fragmented and adapters are ligated to both ends of the fragments. Step 2: The adapters hybridize to the primers on the flow cell, immobilizing the template fragments. Steps 3–4: Bridge amplification. Step 3: Elongation occurs unidirectionally from the primer on the flow cell. The extended DNA molecule bends and attaches to the reverse PCR primer on the flow cell. Step 4: Elongation occurs from the reverse primer. Step 5: The bridged strands are denatured, forming single-stranded templates. Step 6: Steps 3–5 are repeated to form a cluster containing millions of copies of the template fragment. Step 7: The flow cell is flooded with fluorescently labeled terminator dNTPs; the incorporated dNTPs are recorded. Step 8: The fluorescent label and terminator are removed, and elongation/sequencing continues. (B) The incorporated dNTP is determined based on the color of the fluorescence detected in each position following each round of flow-cell nucleotide flooding. (Image reproduced with permission from Illumina, Inc., San Diego, California.)

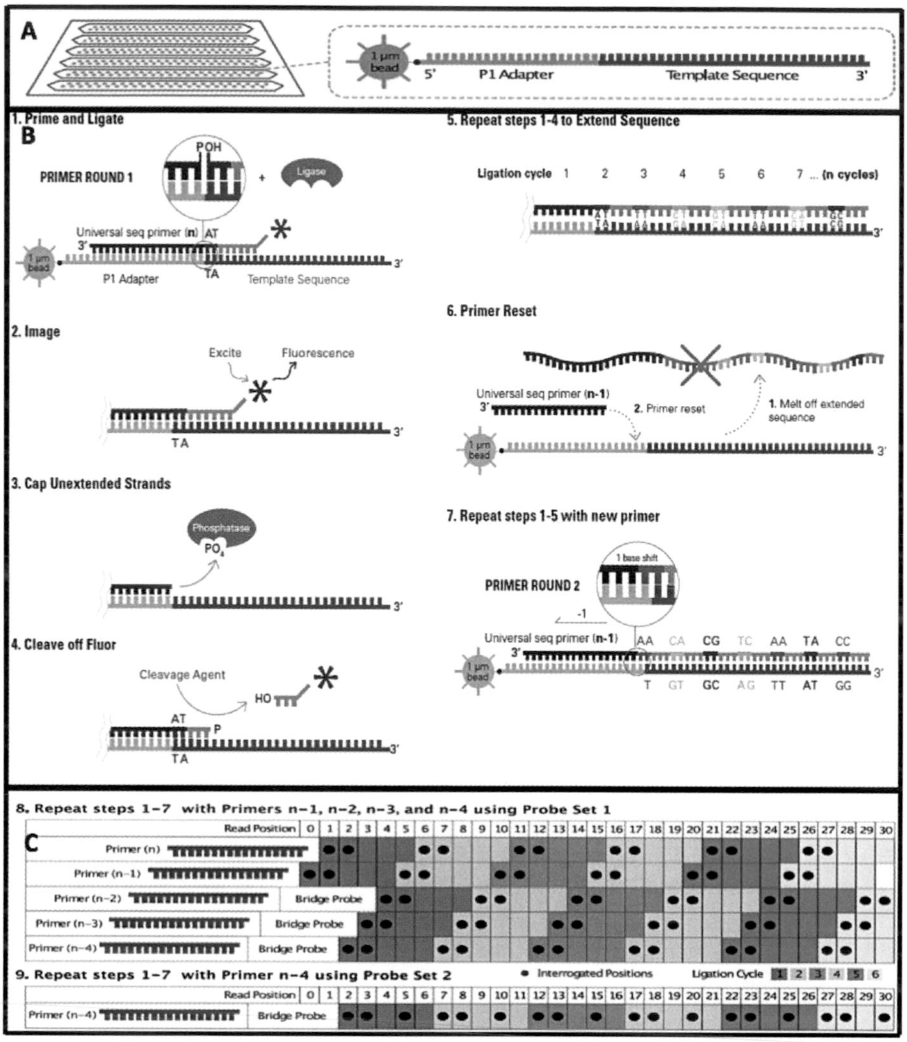

Figure 14.4 Principles of SOLiD sequencing. (A) A FlowChip with millions of DNA capture beads attached. A single bead contains a polony of a single DNA fragment, each with a 5′ P1 adapter sequence. (B) Sequence-by-ligation chemistry. Step 1: A universal primer binds to the P1 adapter. The chip is flooded with 8-mer probes. The probe corresponding to the first five nucleotides of the sequence ligates to the template. Step 2: A two-nucleotide dye system is used. The color of the fluorescence is specific to the first two nucleotides of the probe. The first two nucleotides in the sequence are identified. Steps 3–4: The fluorescent tag, along with the last three nucleotides of the probe are removed. Step 5: Steps 1–4 are repeated. The nucleotides of the sequence complimentary to the first two in each probe are identified along the template. Step 6: The primer sequence is melted off. A new universal primer binds to the P1 adapter at the n-1 position. Steps 7–8: Steps 1–5 are repeated for six sets of primers. Step 8: A schematic of the nucleotides identified with each primer set. The identified nucleotides are marked with a dot. Step 9: An additional primer set, the ECC sequencing round, is utilized for additional coverage and increased quality assurance. (Image reproduced with permission from Life Technologies Corporation, Foster City, California.)

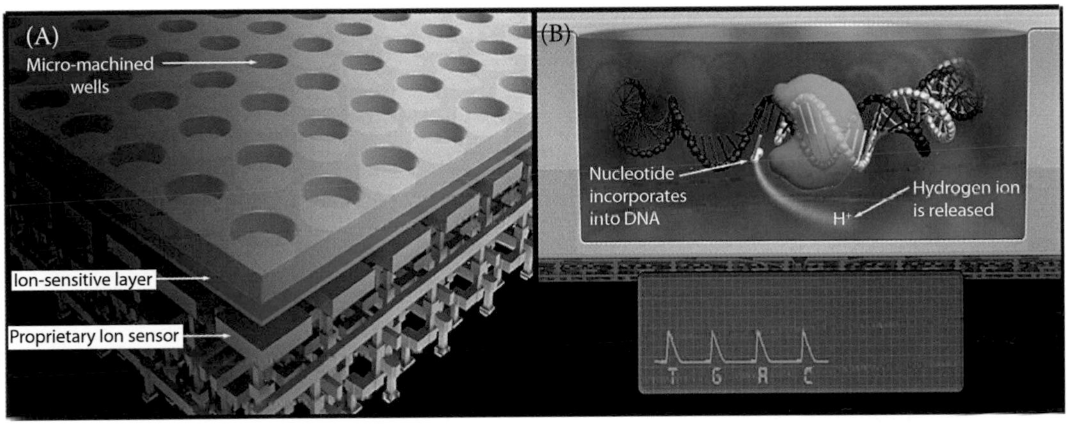

Figure 14.5 Principles of Ion Torrent sequencing. (A) Ion sequencing array chip. The top layer contains millions of micromachined wells. Underneath the wells is an ion-sensitive layer. The base of the chip contains a pH detector. (B) Semiconductor sequencing chemistry. Sequencing is performed by sequentially flooding the chip with each nucleotide. The incorporation of a dNTP into the growing DNA strand is accompanied by the release of a hydrogen ion. The pH of the solution within the ion-sensitive layer of the chip changes with the release of the hydrogen ion. The change in pH is sensed by the detector and recorded as nucleotide incorporation. (Image reproduced with permission from Ion Torrent Systems, Inc., South San Francisco, California.)

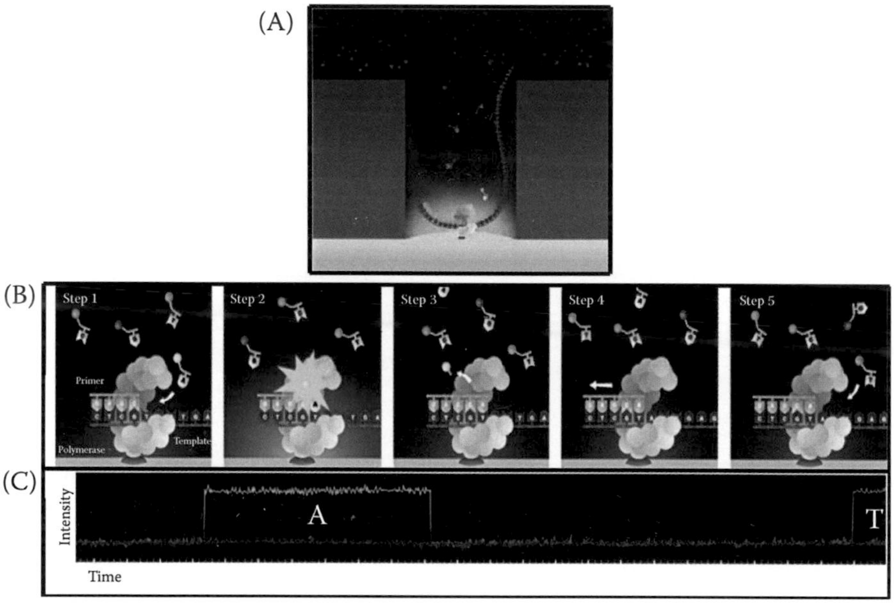

Figure 14.6 Principles of Pacific Biosciences sequencing. (A) A ZMW with a DNA polymerase attached to the slide at the bottom of the well. Sequencing of a single DNA molecule occurs by flooding the array with all four fluorescently labeled dNTPs. Only the bottom of the well illuminates due to the rapid diffusion and the slower incorporation of the dNTPs. (B) SMRT sequencing chemistry. Step 1: All four fluorescently labeled dNTPs surround the polymerase active site. Step 2: The complimentary base is incorporated into the growing DNA strand, and the fluorescence emitted is specific to that base. Step 3: The fluorescent tag is removed to allow for elongation to continue. Step 4: The DNA strand moves as the DNA polymerase translocates to the next position on the template strand. Step 5: Steps 1–4 are repeated. (C) Real-time detection of the fluorescence from the fluorophore tag of the incorporated dNTP. (Image reproduced with permission from Pacific Biosciences of California, Inc., Menlo Park, California.)

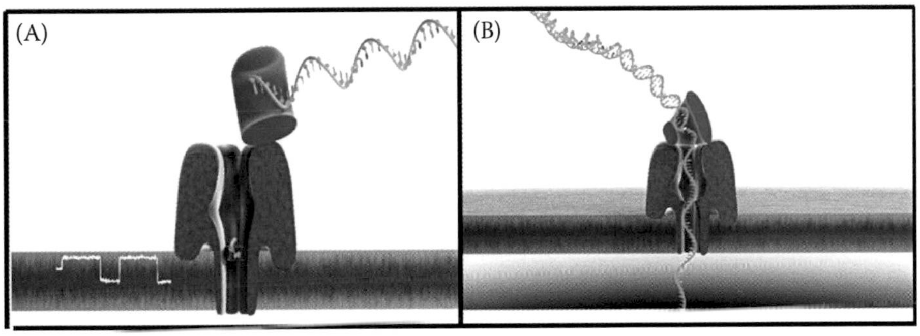

Figure 14.7 Principles of nanopore sequencing. (A) Exonuclease sequencing. The nanopore embedded in the membrane bilayer is shown in blue. The processive enzyme, shown in green, cleaves the nucleotide bases in a sequential order. The cleaved nucleotide goes into the pore and undergoes a binding event with cyclodextrin, shown in red. This causes a change in current that is recorded and used to determine the order of nucleotides passing through the pore. (B) Strand sequencing. An enzyme complex, shown in green, binds to a double-stranded DNA fragment. The enzyme unzips the duplex strand and feeds the single-stranded DNA through the pore one base at a time. As the different nucleotides are fed through the pore, the change in current is measured to determine the order of nucleotides that has passed through the pore. (Image reproduced with permission from Oxford Nanopore Technologies, Boston, Massachusetts.)

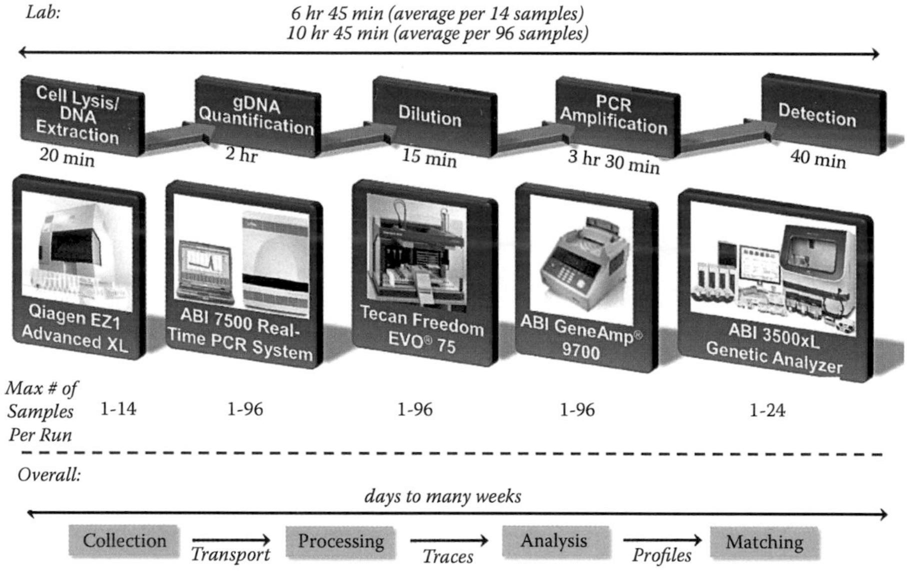

Figure 15.1 Instrumentation and process times associated with conventional forensic DNA analysis. Note: The times associated with each step are per run only and do not include any preparation or transfer times. (Adapted from presentation of Bienvenue, J. et al., Integrated microfluidics for forensic applications, *61st American Academy of Forensic Sciences Annual Meeting*, Denver, CO.)

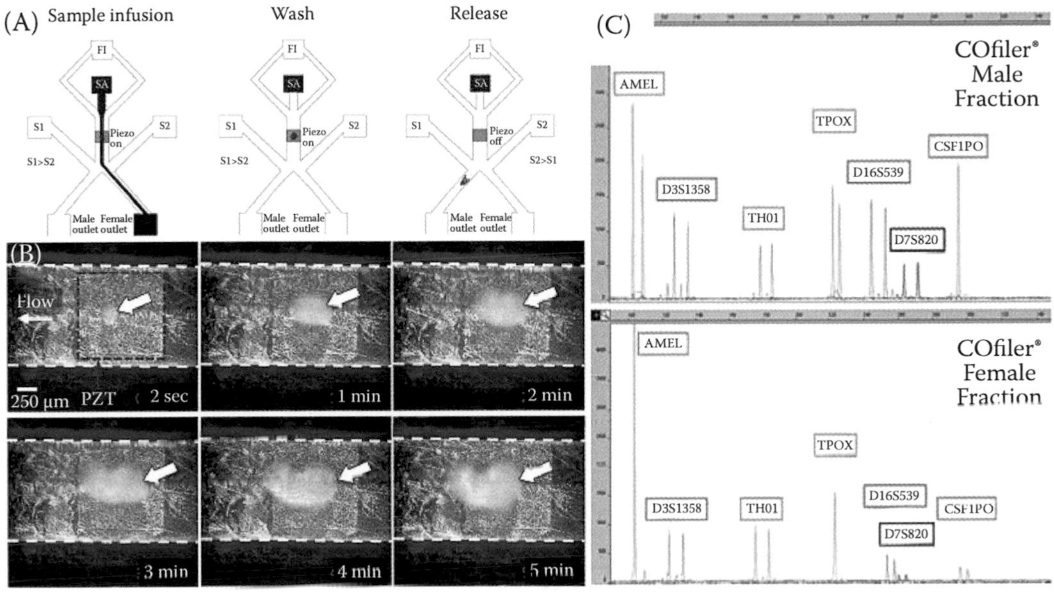

Figure 15.3 (A) Process flow for an acoustic differential extraction (ADE) microdevice. Intact sperm cells are trapped by acoustic forces, while lysed female DNA flows to one outlet. After rinsing any remaining female DNA away, the flow is switched to the second outlet and the trapped sperm cells are released. (B) Acoustic trapping of 335 µL of dilute semen over the course of 5 min. The microchannel walls are indicated with white dashed lines, the PZT transducer is outlined in the first frame and the white aggregate (indicated by block arrows) consists of the trapped sperm cells. (C) Male (top) and female (bottom) STR profiles obtained after (ADE). (From Norris, J. V. et al., 2009a, *Anal Chem* 81: 6089–95.)

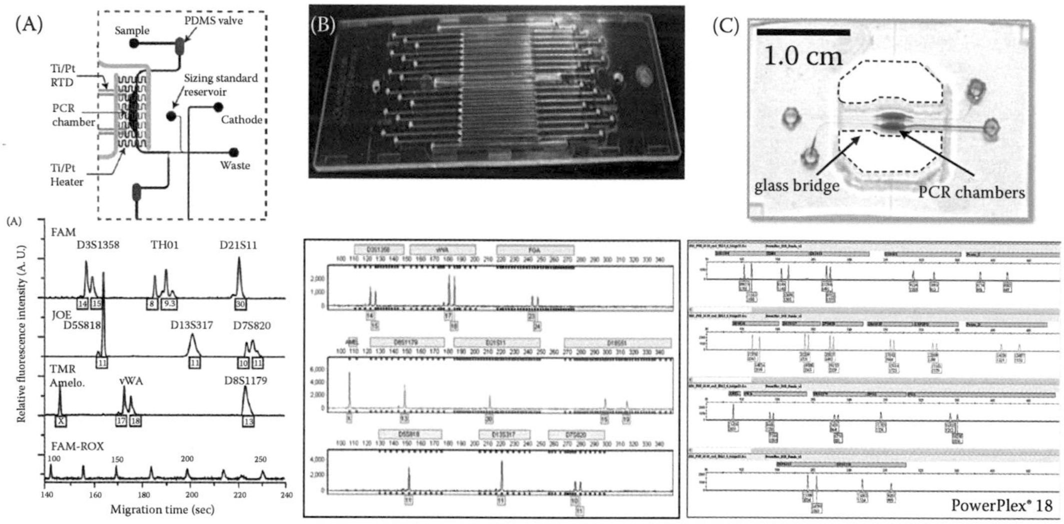

Figure 15.4 Comparison of different types of stationary PCR. (A) Glass microchip with microfabricated heaters and resistance temperature detectors included in the microdevice. This device was able to produce Amelogenin peaks in less than 15 min. (From Lagally, E. T., C. A. Emrich, and R. A. Mathies, 2001, *Lab Chip* 1: 102–7.) (B) Sixteen 7-μL PCR chambers are heated by a thermoelectric element and can yield full STR profiles (9 STR loci and Amelogenin) in 17.1 min. (From Giese, H. et al., 2009, *J Forensic Sci* 54: 1287–96.) (C) An IR-mediated PCR system uses 400-nL PCR chambers fabricated in glass and yields full STR profiles (17 STR loci and Amelogenin) in under 40 min. (From Easley, C. J., J. A. C. Humphrey, and J. P. Landers, 2007, *J Micromech Microeng* 17: 1758–66; Root, B. E., C. R. Reedy, K. A. Hagan et al., 2011, *Presentation—15th Int Conf Miniaturized Systems for Chemistry and Life Sciences*, Seattle, WA.)

(A) (B) (C)

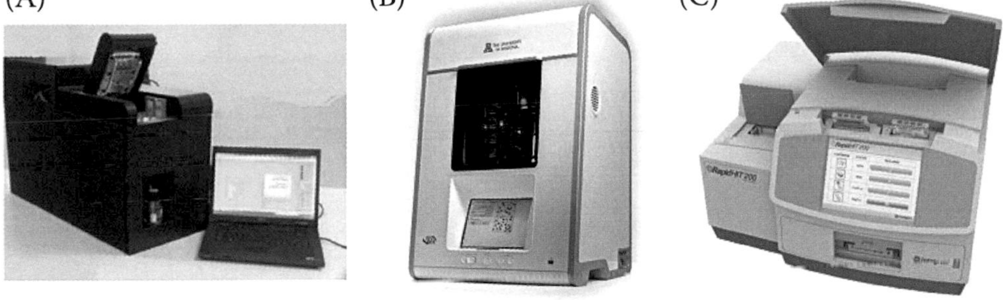

Figure 15.6 Prototype microfluidic systems: (A) Photograph of the RapID™ system under development by MicroLab Diagnostics, a subsidiary of ZyGEM, Inc. (From Root, B. E., C. R. Reedy, K. A. Hagan et al., 2011, *Presentation—15th Int Conf Miniaturized Systems for Chemistry and Life Sciences.* Seattle, WA.) (B) Photograph of the integrated, automated microfluidic system developed by the Forensic Science Service and the University of Arizona. (Center for ANBN [Applied Nanobioscience and Medicine], The University of Arizona, http://anbm.arizona.edu/research/current-projects-programs/rapid-dna-analysis.) (C) Photograph of the RapidHIT™ 200 Early Access Commercial System developed by IntegenX. (From El-Sissi, O. et al., 2010, RapidHIT 200 human DNA identification system, August 6, 2010, available online at: http://integenx.com/applications-2/dna-identity/.)

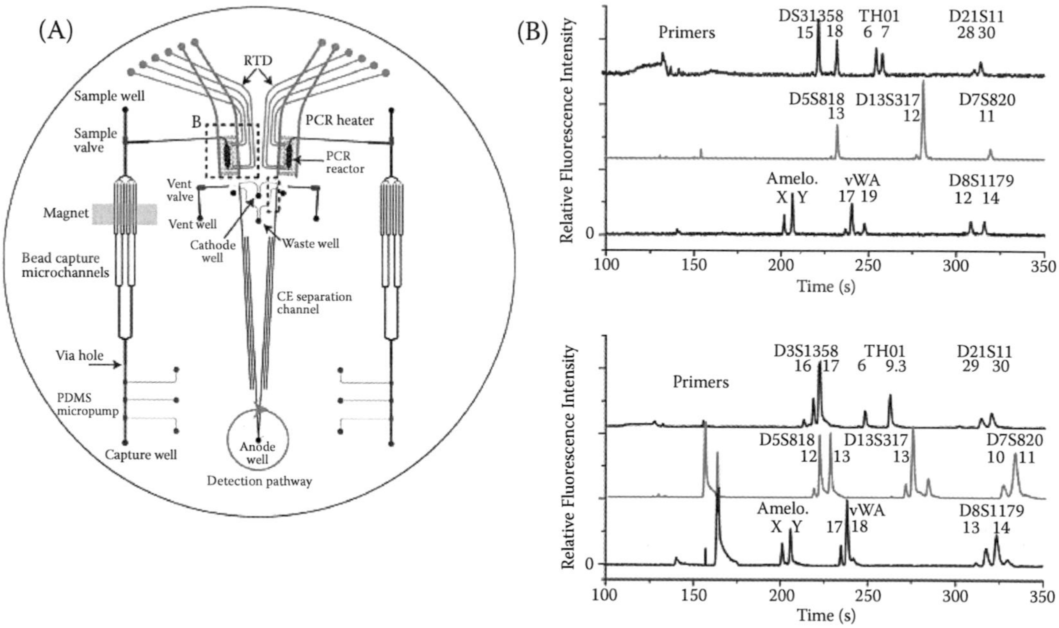

Figure 15.7 A fully integrated device for DNA extraction, STR PCR, and microchip electrophoresis developed by the University of California at Berkley and the Virginia Department of Forensic Sciences. (A) A glass microchip, which has two domains for dual sample analysis, uses magnetic silica-based beads to capture DNA from lysed buccal cells, and contains microfabricated heaters and resistance temperature detectors for STR PCR. (B) Full STR profiles (9-plex, 4-color) from two dried buccal swabs obtained after a fully integrated run in under 3 h. (From Liu, P. et al., *Lab Chip* 11: 1041–8.)

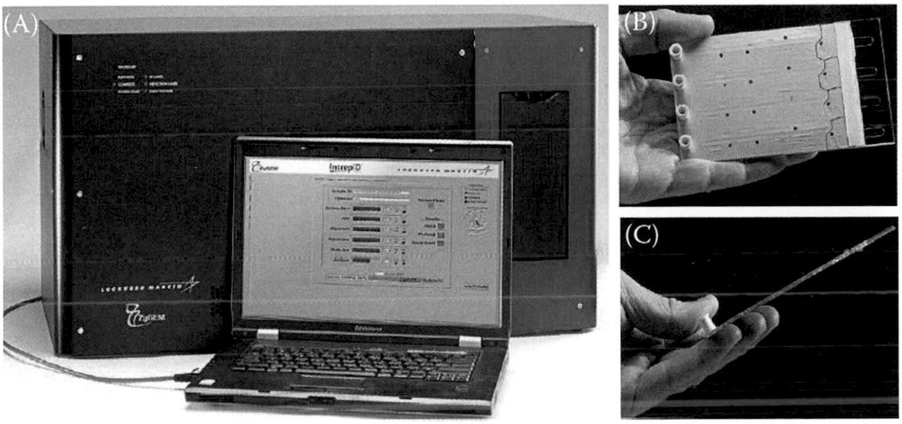

Figure 15.9 The IntrepID S2A90™ system under development by MicroLab Diagnostics, a subsidiary of ZyGEM, and Lockheed Martin. (A) Photograph of the second generation prototype showing the size of the system relative to the laptop needed for instrument operation. Additional photographs show the (B) top and (C) side views of the disposable, integrated cartridge used, simultaneously analyzing up to four buccal swab samples. (Root, B. E., C. R. Reedy, K. A. Hagan et al., 2011, *Presentation—15th Int Conf Miniaturized Systems for Chemistry and Life Sciences,* Seattle, WA.)

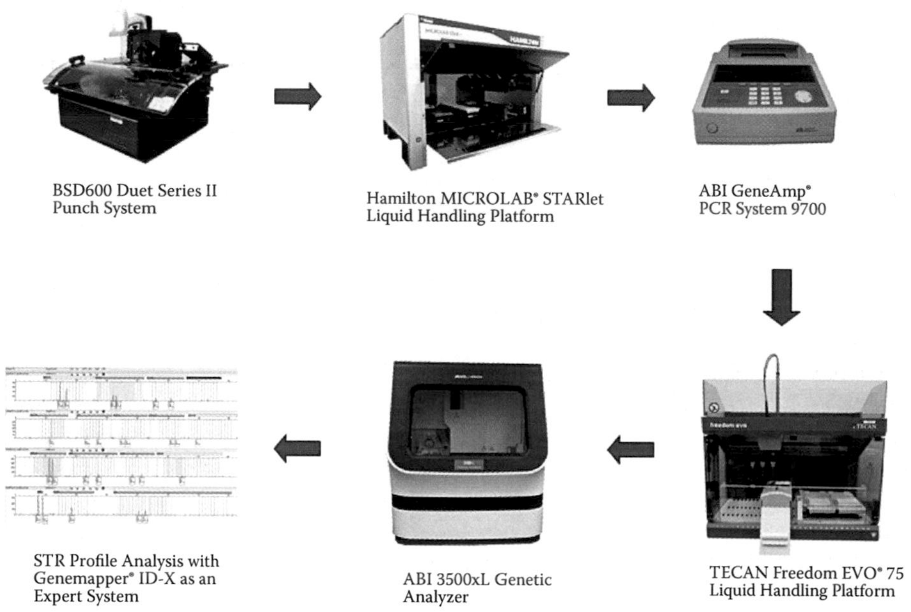

Figure 16.1 Optimized and validated workflow for high-throughput processing of reference samples for generating STR profile.

Presciuttini, S., F. Ciampini, M. Alu et al. 2003. Allele sharing in first-degree and unrelated pairs of individuals in the Ge.F.I. AmpFlSTR Profiler Plus database. *Forensic Sci Int* 131:85–9.

Pumpernik, D., B. Oblak, and B. Borstnik. 2008. Replication slippage versus point mutation rates in short tandem repeats of the human genome. *Mol Genet Genomics* 279:55–61.

Ruitberg, C., D. Reeder, and J. Butler. 2001. STRBase: Aa short tandem repeat DNA database for the human identity testing community. *Nucleic Acids Res* 29:320–2.

Schlötterer, C. and D. Tautz. 1992. Slippage synthesis of simple sequence DNA. *Nucleic Acids Res* 20:211–15.

Shinde, D., Y. Lai, F. Sun et al. 2003. Taq DNA polymerase slippage mutation rates measured by PCR and quasi-likelihood analysis: (CA/GT)n and (A/T)n microsatellites. *Nucleic Acids Res* 31:974–80.

Sinden, R., V. Potaman, E. Oussatcheva et al. 2002. Triplet repeat DNA structures and human genetic disease: Dynamic mutations from dynamic DNA. *J Biosci* 27:53–65.

Sun, G., S. McGarvey, R. Bayoumi et al. 2003. Global genetic variation at nine short tandem repeat loci and implications on forensic genetics. *Eur J Hum Genet* 11:39–49.

Sun, J., J. Mullikin, N. Patterson et al. 2009. Microsatellites are molecular clocks that support accurate inferences about history. *Mol Biol Evol* 26:1017–27.

Taylor, J., J. Durkin, and F. Breden. 1999. The death of a microsatellite: A phylogenetic perspective on microsatellite interruptions. *Mol Biol Evol* 16:567–72.

Tishkoff, S. and K. Kidd. 2004. Implications of biogeography of human populations for "race" and medicine. *Nat Genet* 36:521–7.

Tran, N., B. Bharaj, E. Diamandis et al. 2004. Short tandem repeat polymorphism and cancer risk: Influence of laboratory analysis on epidemiologic findings. *Cancer Epidemiol Biomarkers Prev* 13:2133–40.

Trujillo, R. and S. Amelon. 2009. Development of microsatellite markers in *Myotis sodalis* and cross-species amplification in *M. gricescens*, *M. leibii*, *M. lucifugus*, and *M. septentrionalis*. *Conserv Genet* 10:1965–8.

Vanderheyden, N., A. Mai, A. Gilissen et al. 2007. Identification and sequence analysis of discordant phenotypes between AmpFlSTR SGM Plus and PowerPlex 16. *Int J Legal Med* 121:297–310.

Venter, J., M. Adams, E. Myers et al. 2001. The sequence of the human genome. *Science* 291:1304–51.

Walsh, P., N. Fildes, and R. Reynolds. 1996. Sequence analysis and characterization of stutter products at the tetranucleotide repeat locus vWA. *Nucleic Acids Res* 24:2807–12.

Walsh, S., S. Robinson, G. Turbett et al. 2003. Characterisation of variant alleles at the HumD21S11 locus implies unique Australasian genotypes and re-classification of nomenclature guidelines. *Forensic Sci Int* 135:35–41.

Wang, J., L. Wang, C. Lin et al. 2005. Association study using combination analysis of SNP and STRP markers: CD14 promoter polymorphism and IgE level in Taiwanese asthma children. *J Hum Genet* 50:36–41.

Weber, J. and K. Broman. 2001. Genotyping for human whole-genome scans: Past, present, and future; Rao DC, Province MA (Eds) *Advances in Genetics* pp 77–96.

Weber, J. and C. Wong. 1993. Mutation of human short tandem repeats. *Hum Mol Genet* 2:1123–8.

Wooster, R., A. Cleton-Jansen, N. Collins et al. 1994. Instability of short tandem repeats (microsatellites) in human cancers. *Nat Genet* 6:152–6.

Xu, X., M. Peng, and Z. Fang. 2000. The direction of microsatellite mutations is dependent upon allele length. *Nat Genet* 24:396–9.

Zhivotovsky, L., D. Goldstein, and M. Feldman. 2001. Genetic sampling error of distance (delta(mu))2 and variation in mutation rate among microsatellite loci. *Mol Biol Evol* 18:2141–5.

Zhivotovsky, L., P. Underhill, C. Cinnioglu et al. 2004. The effective mutation rate at Y chromosome short tandem repeats, with application to human population-divergence time. *Am J Hum Genet* 74:50–61.

Additional Y-STRs in Forensics: Why, Which, and When

11

KAYE N. BALLANTYNE
MANFRED KAYSER

Contents

11.1 Y-STRs in Forensics and the Need for Additional Markers 222
11.1.1 The Y-Chromosome in Forensics 222
11.1.2 Currently Applied Y-STRs in Forensics 223
11.1.3 Need for Additional Y-STRs in Forensics 224
11.2 Forensic Value of Additional Y-STRs 226
11.2.1 Male-Relative Differentiation 227
11.2.2 Male-Lineage Differentiation 227
11.2.3 Paternity/Familial Testing 232
11.3 Availability of Additional Y-STRs for Forensic Applications 233
11.4 Practical Implications of Introducing Additional Y-STRs to Forensics 240
11.5 Next-Generation Commercial Y-STR Kits 241
Acknowledgments 242
References 242

Abstract: Male-specific DNA profiling using nonrecombining Y-chromosomal genetic markers is becoming ubiquitous in forensic genetics, with many laboratories and jurisdictions taking advantage of the benefits that Y-chromosome short tandem repeat (Y-STR) profiling can bring. The current suite of 9–17 core Y-STRs, available as commercial kits, perform adequately for identifying male lineages in many populations, a feature highly suitable for excluding a male suspect from involvement in rape cases where autosomal STR profiling often is troubled. However, there is a growing need to achieve higher resolution in paternal-lineage differentiation as adventitious matches between unrelated males are becoming increasingly common with the increasing size of Y-STR haplotype frequency databases. Furthermore, with the currently used Y-STRs, male relatives (both close and distant) usually cannot be separated, marking a strong limitation in forensic applications, as conclusions cannot be drawn on the individual level as desired. Performing Y-chromosome analysis in familial testing, which outperforms autosomal STR profiling in certain deficiency cases, with the current Y-STR sets can be troubled by observed mutations that complicate relationship-probability estimations. To overcome these limitations, considerable research has been performed over recent years to identify and characterize additional Y-STRs. This chapter summarizes the forensic performance of current sets of Y-STRs, points out their limitations in the three main areas of forensic Y-STR applications (male-lineage differentiation, male-relative differentiation, and paternity/familial testing), lays out the need for improvements by adding additional Y-STRs to the forensic portfolio, discusses which additional Y-STRs are suitable, and notes how industry has already picked up on the latest knowledge by developing next-generation commercial Y-STR kits that allow certain improvements in forensic Y-chromosome analysis in the near future.

11.1 Y-STRs in Forensics and the Need for Additional Markers

11.1.1 The Y-Chromosome in Forensics

By virtue of its nonrecombining status, haploid nature, and strong linkage to certain cultural practices, the genetic diversity of the major part of the human Y chromosome (nonrecombining part of the Y-chromosome, NRY) has been a useful tool to trace population history and movements, link paternally related males to each other, and differentiate between unrelated male lineages (Coble et al. 2009; Kayser 2003; Kayser et al. 2001, 2005, 2007; King and Jobling 2009; Roewer 2009; Shi et al. 2010). Furthermore, the Y-chromosome provides the ability for DNA-based sex testing, particularly when combined with X-chromosomal analysis examining X-Y homologues that display length or sequence differences between both chromosomes suitable for DNA-based sex determination (Santos et al. 1998; Sullivan et al. 1993). Forensic geneticists have utilized X-Y homologous genes, Y-chromosomal short tandem repeats (Y-STRs), and Y-chromosomal single nucleotide polymorphisms (Y-SNPs) for these purposes. The most common application of Y-STRs is in sexual assault cases, where the female component can greatly overshadow the male component, making autosomal STR typing difficult or impossible (Corach et al. 2001; Henke et al. 2001; Prinz et al. 2001; Prinz and Sansone 2001; Tsuji et al. 2001). Current autosomal STR kits have a routine mixture detection limit of 1:10, whereas Y-STRs (only detected in male DNA) have been shown to be able to detect the male component in a male-female mixture ratio of up to 1:2,000 (Prinz et al. 1997). This is particularly relevant in the investigation of sexual assault cases, where differential DNA extractions (attempting to extract sperm DNA separately from the DNA of other cells in vaginal swab material) cannot be used to separate the male-derived nonsperm cells from the more numerous female cells, especially in cases involving azoospermic males or nonintimate samples (Shewale et al. 2003; Sibille et al. 2002). Furthermore, mixtures with multiple male contributors can be extremely difficult, if not impossible, to resolve with autosomal STR profiling, while haploid Y-STRs are commonly able to resolve mixtures produced from two or three unrelated males (Cerri et al. 2003; Parson et al. 2001; Prinz et al. 1997). Additionally, in paternity or familial testing Y-STRs can be highly useful, particularly for deficiency cases involving male offspring where the alleged father is not available for genetic testing and none of the male relatives are potential fathers. In such cases, autosomal STRs may not be informative unless both parents of the diseased alleged father are available for testing. In contrast, when using Y-STRs any of the available paternal relatives can be used to replace the deceased father in the analysis as paternal relatives usually share the same Y-STR haplotype (Kayser et al. 2000b; Kayser and Sajantila 2001; Rolf et al. 2001). For missing persons or disaster victim identification, Y-STR profiling of male relatives can be useful when antemortem reference samples are unavailable (Alonso et al. 2005; Bradford et al. 2011).

Analytically, Y-STRs are genotyped in the same manner as autosomal STRs; the same DNA extract is used for both analyses. As with forensically used autosomal STRs, PCR amplification of Y-STRs is usually performed with commercially available multiplex assays covering several markers in a single reaction via employing fluorescently labeled primers, and the products are detected by capillary electrophoresis with DNA sequencers (genetic analyzers) under similar conditions used with autosomal STR kits. Thus, the implementation of Y-STR profiling within a forensic laboratory does not require any large changes in methodology, skills, or equipment.

Statistically, the significance of any Y-STR haplotype match between suspect and crime sample, or between a son and his alleged father, must be evaluated in a different way than autosomal STRs. As the NRY is inherited usually intact from father to son, the Y-STR haplotype represents a single locus, rather than a set of independent markers to which the product rule can be applied. Because Y-STR haplotypes are much more polymorphic than single STR loci, it is necessary to have much larger databases to estimate Y-STR haplotype frequencies, from which either the number of observations can be used directly as the weight of evidence (the counting method), or to estimate the frequency distribution of the haplotype (frequency surveying method) (Brenner 2010; Buckleton et al. 2011; Roewer et al. 2000; Willuweit et al. 2011). With any nonexclusion constellation, regardless of the rarity of the haplotype, it is vital that any conclusion includes the caveat that any paternal relatives of the suspect cannot be excluded from contributing the crime sample instead of the suspect (De Knijff 2003). Thus, the weight of evidence that Y-STRs can provide is generally lower than that provided by the autosomal STRs used in forensics. Furthermore, in paternity testing, when differences between the alleged father (or his male relatives) and the son are observed, the number of Y-STR loci displaying the differences, as well as the repeat-wise allelic differences, need to be considered to differentiate a true exclusion constellation from an inclusion with mutations, as is done with autosomal STRs. However, in contrast to autosomal STR analysis, mutations at Y-STRs can be clearly determined (with DNA sequencing confirmation) due to the strict paternal inheritance of the Y-chromosome. Thus, the uncertainty of the origin of an allelic difference observed between offspring and parents does not apply for Y-STR mutations. This usually leads to more accurate mutation rate estimates for Y-STRs relative to autosomal STRs as long as a large number of (autosomal) DNA-confirmed father-son pairs is investigated.

It is also possible to infer Y haplogroups from Y-STR haplotype data, although direct Y-haplogroup determination via dedicated Y-SNP analysis is more accurate, and should therefore be preferred over indirect inference with Y-STRs wherever possible. As the major Y-SNP haplogroups show striking differences in geographic distributions, it can be possible to infer an individual's paternal biogeographic ancestry from Y-chromosome analysis (Karafet et al. 2008; Y Chromosome Consortium 2002). In rare cases this is also possible solely from the Y-STR haplotype and without the use of Y haplogroups, but Y haplogroups in general are more informative for biogeographic ancestry inferences. In forensics, this can be useful in tracing unknown suspects via concentrated police investigation in cases where the evidentiary autosomal STR profile does not match that of any of the known suspects, or any in the forensic DNA database.

11.1.2 Currently Applied Y-STRs in Forensics

As with autosomal STRs, the discovery of Y-STRs and their introduction to forensic genetics has been a long and slow process that for Y-STRs started somewhat later than for autosomal STRs. One reason for the time discrepancy is that autosomal STRs were initially sought for linkage mapping to find (disease) genes. Due to its nonrecombining nature, this approach does not work for the Y-chromosome, so it was initially left out in the search for human STR markers. The first Y-STR was described in 1992 (Roewer and Epplen 1992a), and immediately applied in a forensic context (Roewer and Epplen 1992b). By 1997, the total number of Y-STRs known was only 15, when the 7 core loci for forensic use were first defined: DYS19, DYS389I, DYS389II, DYS390, DYS391, DYS392, and DYS393 (Kayser et al. 1997). This set of "minimal haplotype" (MH) Y-STR loci was able to differentiate 74–90% of unrelated males in European populations, which increased to 91–97% with the addition of the multicopy Y-STR DYS385a/b to generate an "extended haplotype" (Kayser et al. 1997; Roewer et al. 2001). The vast majority of Y-STR data produced to date has been with these 7–9 markers. The online Y-chromosome Haplotype Reference Database (YHRD) (www.yhrd.org) currently (as of October 2012) lists 104,174 7-locus MH haplotypes, and 102,377 9-locus extended haplotypes (release 40).

As adventitious matches (which we define here in the strict sense as any match occurring just by chance) between unrelated males can be common with only 7–9 Y-STRs, it was important to expand the number of markers routinely tested. Therefore, DYS438 and DYS439 (Ayub et al. 2000) were added to the core set in 2003, with the further addition of DYS437, DYS448, DYS456, DYS458, DYS635, and/or Y-GATA-H4 (Ayub et al. 2000; Redd et al. 2002; White et al. 1999) to commercial multiplexes such as PowerPlex Y (Promega; Krenke et al. 2005) in 2003 and Yfiler (Applied Biosystems 2011; Mulero et al. 2006) in 2004. The 17 most commonly utilized Y-STR loci available with the Yfiler system all share key molecular characteristics (Table 11.1). Most are tetranucleotide repeats, to minimize stutter formation, while still obtaining sufficient variation within each locus. All have low- to midrange mutation rates, between 10^{-3} (meaning a few mutations every 1,000 generations per locus) and 10^{-4} (meaning a few mutations every 10,000 generations per locus), ensuring that mutations between fathers and sons are infrequent (Goedbloed et al. 2009). Diversity estimates within European populations are low- to midrange, while in some populations particular markers show extremely low diversity, such as DYS19 in Finns (0.31; Hedman et al. 2004), DYS392 in Mozambicans (0.018; Alves et al. 2003), and DYS437 in Native Americans (0.2; Budowle et al. 2005). For the most common forensic application, namely identifying male lineages to exclude suspects in criminal cases, the current set of Y-STRs performs well in most outbred populations. Globally, haplotype resolution of 0.813 can be achieved with the 9 MH Y-STRs, and 0.905 with the 17 Yfiler Y-STRs. In particular, European populations show extremely high haplotype resolution, at 0.989 with 17 Yfiler Y-STRs (Vermeulen et al. 2009).

11.1.3 Need for Additional Y-STRs in Forensics

Not all populations show the same high degree of resolution achieved with the currently used Y-STR sets. Well-known examples are Finns (haplotype diversity of 0.835 with 16 Y-STRs; Hedman et al. 2011), Tunisians (0.75 with 12 Y-STRs; Onofri et al. 2008), Xhosa (0.627 with 7 Y-STRs; D'Amato et al. 2010) and Polynesians (0.9 with 7 Y-STRs; Kayser et al. 2000a), representing examples of inbred populations or those that went through a recent bottleneck, both resulting into low overall (including Y-STR) genetic diversity. Furthermore, even in outbred populations the discrimination capacity does not approach that usually reached with the autosomal STRs used in forensics with values of ~1 x 10^{-19} (Butler 2009). Hence, there is a clear need to further extend the currently used sets of Y-STR markers to increase the paternal lineage resolution in populations where the current sets provide limited help. Roewer (2009) provides an example of an Afro-Caribbean

Table 11.1 Y-STRs Commonly Used for Forensic Applications

Marker	Repeat Motif	Allele Range	Mutation Rate (95% CI)*	Gene Diversity (European)**	Initial Reference
DYS19	$(TAGA)_3(TAGG)_1(TAGA)_{6-16}$	9–19	4.37×10^{-3} $(1.98 \times 10^{-3} - 8.23 \times 10^{-3})$	0.541	(Roewer and Epplen 1992a)
DYS389I	$(TCTG)_3(TCTA)_{6-14}$	9–17	5.51×10^{-3} $(2.72 \times 10^{-3} - 9.74 \times 10^{-3})$	0.575	(Kayser et al. 1997)
DYS389II	$(TCTG)_{4-5}(TCTA)_{10-14}N_{28}(TCTG)_3(TCTA)_{6-14}$	24–36	3.83×10^{-3} $(1.61 \times 10^{-3} - 7.49 \times 10^{-3})$	0.703	(Kayser et al. 1997)
DYS390	$(TCTG)_8(TCTA)_{9-14}(TCTG)_1(TCTG)_4$	17–29	1.52×10^{-3} $(3.52 \times 10^{-4} - 4.09 \times 10^{-3})$	0.713	(Kayser et al. 1997)
DYS391	$(TCTG)_3(TCTA)_{6-15}$	5–16	3.23×10^{-3} $(1.26 \times 10^{-3} - 6.65 \times 10^{-3})$	0.54	(Kayser et al. 1997)
DYS392	$(TAT)_{4-20}$	4–20	9.70×10^{-4} $(1.43 \times 10^{-4} - 3.23 \times 10^{-3})$	0.615	(Kayser et al. 1997)
DYS393	$(AGAT)_{7-18}$	7–18	2.11×10^{-3} $(6.21 \times 10^{-4} - 5.00 \times 10^{-3})$	0.412	(Kayser et al. 1997)
DYS385a/b	$(AAGG)_4N_{14}(AAAG)_3N_{12}(AAAG)_3N_{29}(AAGG)_{6-7}(GAAA)_{7-23}$	6–28	2.08×10^{-3} $(6.24 \times 10^{-4} - 5.06 \times 10^{-3})$, 4.14×10^{-3} $(1.75 \times 10^{-3} - 8.09 \times 10^{-3})$	0.855	(Kayser et al. 1997)
DYS438	$(TTTTC)_{7-16}$	7–18	9.56×10^{-4} $(1.37 \times 10^{-4} - 3.18 \times 10^{-3})$	0.622	(Ayub et al. 2000)
DYS439	$(GATA)_3N_{32}(GATA)_{5-19}$	5–19	3.84×10^{-3} $(1.63 \times 10^{-3} - 7.54 \times 10^{-3})$	0.663	(Ayub et al. 2000)
DYS437	$(TCTA)_{4-12}(TCTG)_2(TCTA)_4$	10–18	1.53×10^{-3} $(3.54 \times 10^{-4} - 4.10 \times 10^{-3})$	0.624	(Ayub et al. 2000)
DYS448	$(AGAGAT)_{11-13}N_{42}(AGAGAT)_{8-9}$	14–24	3.94×10^{-4} $(1.41 \times 10^{-5} - 2.11 \times 10^{-3})$	0.651	(Redd et al. 2002)
DYS456	$(AGAT)_{11-23}$	5–23	4.94×10^{-3} $(2.35 \times 10^{-3} - 8.97 \times 10^{-3})$	0.703	(Redd et al. 2002)
DYS458	$(GAAA)_{11-24}$	11–24	8.36×10^{-3} $(4.80 \times 10^{-3} - 1.34 \times 10^{-2})$	0.808	(Redd et al. 2002)
DYS635	$(TCTA)_4(TGTA)_2(TCTA)_2(TGTA)_2(TCTA)_2(TATG)_{0-2}(TCTA)_{4-17}$	16–30	3.85×10^{-3} $(1.63 \times 10^{-3} - 7.55 \times 10^{-3})$	0.682	(White et al. 1999)
Y-GATA-H4	$(TAGA)_3N_{12}(TAGG)_3(TAGA)_{8-15}N_{22}(TAGA)$	8–15.1	3.22×10^{-3} $(1.28 \times 10^{-3} - 6.62 \times 10^{-3})$	0.604	(White et al. 1999)

* From Goedbloed, M. et al., 2009, Comprehensive mutation analysis of 17 Y-chromosomal short tandem repeat polymorphisms included in the AmpFlSTR Yfiler PCR amplification kit, *Int J Legal Med* 123 (6):471–82.

** From Applied Biosystems, 2011, *AmpFℓSTR Yfiler PCR Amplification Kit Users Manual.*

suspect matching a crime stain at 11 Y-STRs; only with additional testing of another 10 (noncore) Y-STRs was a single mismatch discovered that excluded the innocent man.

However, the most obvious limitation of currently used Y-STR sets in forensics is that male relatives of a suspect can not be excluded from having deposited the matching crime scene material. Relatives separated by up to 20 generations have been shown to have identical

17 Y-STR profiles (Ballantyne et al. 2012). It could be predicted that Y-STRs with much higher mutation rates than the loci currently used would reduce this problem, as they may allow differentiation of paternal relatives by means of more frequently observed mutations. In addition, as more males are profiled and added to Y-STR frequency databases such as YHRD, adventitious matches are increasing between unrelated males. Identical haplotypes are observed throughout continents and across biogeographic ancestry groups, even with 17 locus Y-STR haplotypes. The most striking example is a particular MH Y-STR haplotype that is shared among 3,562 out of 104,174 males (3.4%) currently stored in the YHRD (release 40, haplotype DYS19-14, DYS389I-13, DYS389II-29, DYS390-24, DYS391-11, DYS392-13, and DYS393-13). In the same way that increasing the number of autosomal STRs has improved individualization, increasing the number of Y-STRs tested in cases of nonexclusion improves accuracy and resolution capacity. In the example above, the most common Yfiler haplotype that includes the most common MH Y-STR haplotype is only shared among 0.1% of all males in YHRD (43 of 44,469, release 40). Clearly, this reflects a strong improvement in male lineage differentiation, at least for this group of unrelated males.

Furthermore, with the current sets of forensically applied Y-STRs, mutations are occasionally (albeit rarely) observed in familial-testing applications, because particular loci within the current set have medium-level mutation rates (Goedbloed et al. 2009), and their consideration complicates relationship-probability estimations (Rolf et al. 2001). Expectedly, Y-STRs with significantly lower mutation rates, if available, would limit this problem. Thus, there has been considerable research and many publications searching for additional Y-STR loci to increase the role and utility of Y-STRs in forensic biology.

11.2 Forensic Value of Additional Y-STRs

As outlined above, the main uses of the Y-chromosome in forensic biology can be separated into three areas: male-lineage differentiation, male-relative differentiation, and familial/paternity testing (Roewer 2009). For all these applications, Y-STR loci with high diversity within and across populations would be desirable to reduce the probability of adventitious matches. However, for nonrecombining DNA such as the Y chromosome, genetic diversity is largely governed by mutation rate, as loci with higher mutation rates are substantially more likely to have greater numbers of alleles. On the other hand, for paternity testing and missing persons identification, high mutation rates would increase the risk of false exclusions (Kayser and Sajantila 2001). Although there was no explicit intention to (because many currently used Y-STRs were ascertained from a very limited set and without having much knowledge on the locus-specific mutation rates), a serendipitous compromise between the two aspects has been achieved with the current sets of Y-STRs. These markers show reasonably high diversities, but the mutation rates are low enough to usually allow paternity testing without problems, albeit with occasional mutations between father and son. However, numerous publications on novel Y-STR loci and multiplexes in various combinations have demonstrated that it is possible to achieve improved results for many populations and in different applications (Ballantyne et al. 2012; Butler et al. 2002; D'Amato et al. 2010, 2011; Decker et al. 2007; Ehler et al. 2010; Hall and Ballantyne 2003; Hanson and Ballantyne 2004; Hanson et al. 2006; Hedman et al. 2011; Iida and Kishi 2005; Lessig et al. 2009; Lim et al. 2007; Maybruck et al. 2009; Palo et al. 2008; Park et al. 2009; Rodig et al. 2008; Vermeulen et al. 2009; Xu et al. 2010). Obviously, with the vastly increased numbers of Y-STRs available, including emerging knowledge on their mutation rate and diversity estimates, it would now be possible to select sets of markers specifically tailored to each application to achieve optimal results across many populations.

11.2.1 Male-Relative Differentiation

The most significant limitation of current Y-STR profiling is its inability to differentiate between males of the same paternal lineage. For the Yfiler set, there is a probability of only 0.047 in observing a mutation between a given father-son pair at any one of the 17 Y-STRs considered (Ballantyne et al. 2010). The relatively low mutation rates of Yfiler Y-STRs results in an extremely low rate of differentiation of close male relatives (7–8% of fathers and brothers, Figure 11.1) (Ballantyne et al. 2012). Even when considering distant paternal relationships of 10–20 meioses, differentiation by 1 or more mutation is relatively low at only 44% (range 0–57%) (Ballantyne et al. 2012; Kayser et al. 2007). Obviously, if male relative differentiation is required, a panel of Y-STRs with high mutation rates would be desirable to increase the probability of observing a mutation in at least one marker in any given meiosis event. To achieve this aim, a set of Rapidly Mutating (RM) Y-STRs has recently been described and multiplex genotyping tools were developed (Ballantyne et al. 2010, 2012). All 13 RM Y-STR markers within the set have mutation rates above 10^{-2} (meaning a few mutations every 100 generation per each marker), giving an expected combined (for the entire set) father-son differentiation probability of 19.5%—four times higher than for Yfiler (Ballantyne et al. 2010). Nearly 50% of father-son pairs could be differentiated with the RM Y-STRs, as could 60% of brothers and 75% of cousins, while any relationship with 9 or more meioses separation was always differentiated (Ballantyne et al. 2010) (Figure 11.1). Although the number of male relatives being investigated for the empirical effect of the RM Y-STRs on male-relative differentiation is still relatively small with less than 200 pairs, it can be increased in future studies to strengthen the reliability of the estimates, it can be seen already that this RM Y-STR set represents a significant advance on the most informative current Y-STR panel (Yfiler) in allowing distant and on many occasions close male relatives to be excluded from a match statement. Casework situations where RM Y-STRs are expected to provide more information than current Y-STRs include any situation where close (or distant) male paternal relatives of the suspect may be involved. This includes cases where male-relative involvement is indicated by prior information, but also instances where it cannot be excluded or cannot be investigated but nevertheless may exist. Hence, applying RM Y-STRs in all cases where conventional Y-STR analysis (such as Yfiler) has revealed a matching haplotype between a suspect and the crime scene sample is expected to at least narrow the potential number of relatives not excluded, if not providing complete individualization.

11.2.2 Male-Lineage Differentiation

While Y-STR profiles are commonly shared by descent, they can also be identical by state (IBS), describing a scenario where recurrent mutations cause unrelated males to share particular combinations of Y-STRs, and sometimes even the same Y-STR haplotype. The frequency of observing IBS is dependent on the number and genetic diversity of loci examined as well as the populations in question. As noted above, IBS occurs more frequently in some populations than others and, obviously, applying only a small number of Y-STRs

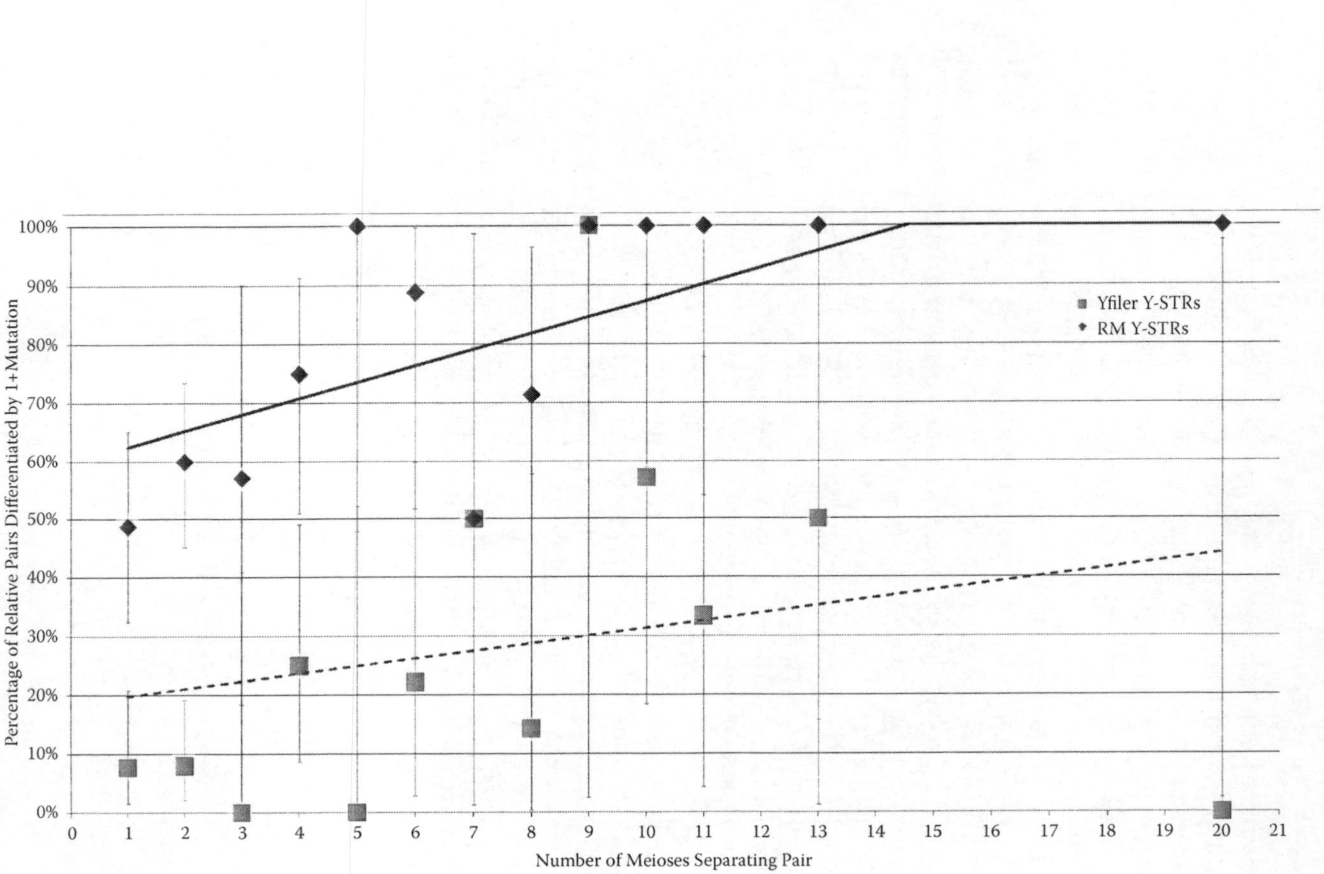

Figure 11.1 Male-relative differentiation with 13 RM Y-STRs and 17 Yfiler Y-STRs. Male relatives (156 pairs) separated by 1–20 generations were assessed for the presence of differentiating mutations in both the RM Y-STR set and the Yfiler Y-STR set. Lines represent the average differentiation between pairs across generations, with error bars representing 95% binomial confidence intervals. (Figure adapted from Ballantyne, K. N. et al., 2012, A new future of forensic Y-chromosome analysis: Rapidly mutating Y-STRs for differentiating male relatives and paternal lineages, *Forensic Sci Int Genet* 6:208–18.)

will lead to IBS being observed more often than if a large number of Y-STRs are used for haplotype construction. To reduce the chance of IBS, as is desired in male-lineage differentiation, it is necessary to either increase the number of Y-STRs tested by adding noncore Y-STRs to the current sets of core Y-STRs, or to increase the power of lineage differentiation by developing completely new Y-STR sets including more powerful Y-STRs than currently used. Several different strategies have been employed to find Y-STR loci able to increase haplotype resolution. The most commonly applied approach is to select markers with high genetic diversity in the population(s) of interest, while disregarding the diversity in other, nontested populations. This strategy can produce panels of Y-STRs that are highly effective in the population(s) under study, but may not be as informative in other nontested populations as this approach runs the risk to select for population-specific effects. Examples include a Finnish 7-locus panel increasing haplotype diversity from 0.992 with Yfiler to 0.999 with the additional loci (Hedman et al. 2011), a South African panel to increase diversity in Indian, European, and Xhosa populations from 0.934 with the minimal haplotype loci to 0.967 (D'Amato et al. 2011), or several multiplexes of 10–21 loci to increase resolution in U.S. European and African Americans (Decker et al. 2007; Hanson and Ballantyne 2007; Maybruck et al. 2009). Unsurprisingly, each set has different combinations of markers, reflecting differences in the Y-chromosome structure and phylogenetic history of the targeted population(s). Hence, such a strategy can result in highly useful sets for some populations—usually those used for set design—which, on the other hand, may not be as useful in other nontarget populations, which is a clear disadvantage when applying such sets to nontarget populations. This is problematic in the forensic situation, where usually it is not known from which population the sample donor originates, especially in countries where the number of individuals not belonging to the initial target population(s) is not low.

A more universal strategy is to select Y-STR markers based on global diversity. This may ensure that the selected markers work well in many different populations, rather than only one or a few target populations. However, only a few global sample panels have been used for Y-STR analysis, such as the YCC (Y Chromosome Consortium) set of 76 males and the HGDP-CEPH (Human Genome Diversity Panel—Centre d'Etude du Polymorphisme Humain) of ~600 males, as well as self-collected panels (Kayser et al. 2001). Although the precise diversity estimates are highly dependent on the size and composition of the panel, different analyses have produced remarkably similar results. The top four Y-STRs (DYS481, DYS570, DYS576, and DYS643) for global haplotype resolution were identical between the HGDP and YCC sample panels (Lim et al. 2007; Vermeulen et al. 2009) and these Y-STRs have shown consistently high diversity values across additional population studies (for examples, see Hanson and Ballantyne 2007; Lessig et al. 2009; Rodig et al. 2008). The results of the hill-climbing approach, a statistical method for selecting the most informative markers out of bulk data (Vermeulen et al. 2009), elegantly demonstrates the differences in Y-STR resolution capabilities between worldwide HGDP populations. Of the 49 simple, single-copy Y-STRs tested in this study, some markers showed extreme differences. For example, DYS533, the sixth-best marker globally, was not informative in Europeans, Native Americans, and Oceanians, but was in East Asians, the Middle East, North Africans, South Asians, and sub-Saharan Africans. Likewise, DYS549, the fifth-best global marker, was not informative in East Asians, Oceanians, Native Americans, and North Africans, but was in Europeans, the Middle East, South Asians, and sub-Saharan Africans. The four markers listed above, although displaying high diversity, were not among the most effective in some populations in separate studies; for example, DYS481 was not optimal in Finns (Hedman et al. 2011), nor were DYS449, DYS481, and DYS570 considered informative in the major U.S. population groups (Maybruck et al. 2009). This illustrates limitations also in the more global approaches, which usually are limited by the number and worldwide coverage of the populations investigated as well as by the number of individuals per population tested.

Therefore, selecting Y-STRs based on diversity values, either population-specific or global, can be problematic. While this approach has been useful in identifying a small number of Y-STRs that show consistently high diversity values across populations, proving their utility has required a large number of studies covering significant numbers of samples, which are not easy to perform for all ~500 known Y-STRs. Hence, it would be more efficient to identify the sequence characteristics of loci known to be highly informative, and use this knowledge to select additional loci for population studies as a prerequisite for future applications such as forensics. It is known that Y-STR diversity is strongly correlated with mutation rates, as high levels of mutations are required to generate large allele spectra, and to ensure that they are maintained within the population (Burgarella and Navascues 2011; Jarve et al. 2009; Zhivotovsky et al. 2004). Furthermore, the sequence features that cause high mutation rates at Y-STRs and autosomal STRs have been elucidated, such as the total number of repeats within a locus (including both variable and nonvariable arrays), the sequence complexity of a locus, and the length of the repeat unit, with smaller repeats generating higher rates (Ballantyne et al. 2010; Kayser et al. 2000b; Kelkar et al. 2008; Lai and Sun 2003; Xu et al. 2000). Thus, with accurate mutation rates known for 186 Y-STRs (Ballantyne et al. 2010), and the ability to predict mutation rates for the remaining Y-STRs currently without mutation information (Ballantyne et al. 2010; Burgarella and Navascues 2011), a set of Y-STRs with high mutation rates, and therefore predicted universally high diversity, can be selected for testing.

As described above, a set of Y-STRs with high mutation rates (RM Y-STRs) has already been selected and characterized (Ballantyne et al. 2010, 2012). In addition to providing the ability to differentiate male relatives, the RM Y-STR panel shows consistently high haplotype diversity and resolution across a global panel as demonstrated recently (Ballantyne et al. 2012). The RM Y-STRs provide global increases of 0.6% in diversity and 7.9% in resolution compared to the 17 Yfiler Y-STRs (Figure 11.2). The RM Y-STRs were also consistently selected as more informative than Yfiler Y-STRs using the hill-climbing approach (Ballantyne et al. 2012). Thus, the RM Y-STRs were able to provide higher haplotype resolution with fewer markers than the current "gold standard" Yfiler set.

Not unexpectedly, although the so-called ultra-high discrimination (UHD) set of Y-STRs (Hanson and Ballantyne 2007) was selected based on diversity values in major U.S. populations, the markers selected had a higher average mutation rate (6.5×10^{-3}, 95% confidence interval 3.88×10^{-3}–1.10×10^{-2}) than the Yfiler set (3.22×10^{-3}, 95% CI 1.37×10^{-3}–6.53×10^{-3}), and therefore were able to increase haplotype resolution by 1.3%. Likewise, the 10 Y-STRs selected by D'Amato et al. (2011) increased haplotype resolution by 4.6% in certain populations, displaying a much higher average mutation rate of 8.7×10^{-3} (95% CI 4.83×10^{-3}–1.33×10^{-2}) relative to Yfiler with an average mutation rate of 2.2×10^{-3} (95% CI 1.9×10^{-3}–2.6×10^{-3}) (Goedbloed et al. 2009). However, the 13 RM Y-STR set, with the highest average mutation rate of 2.16×10^{-2} (95% CI 1.43×10^{-2}–2.74×10^{-2}) that is about 10x higher than for Yfiler, showed the highest increase in male-lineage resolution of 7.9%, illustrating the trend for increasing lineage resolution with increasing set mutation rate. Within the

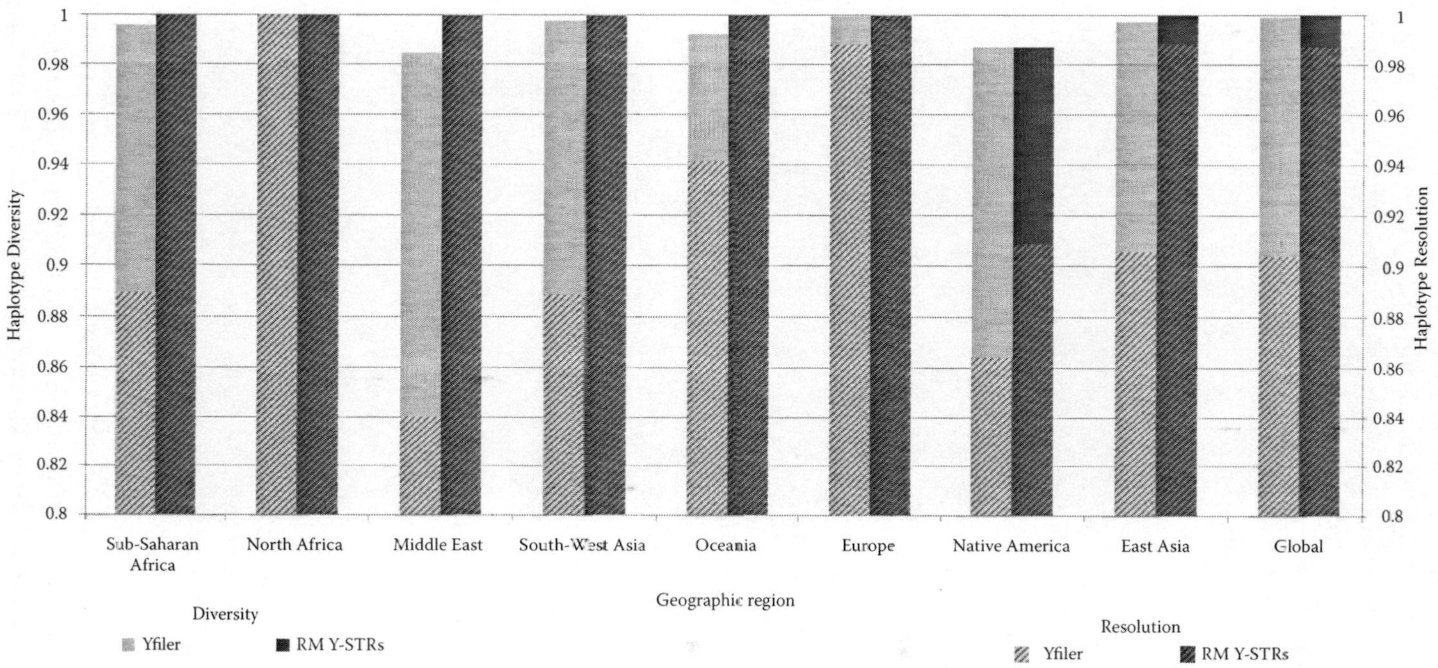

Figure 11.2 Comparison between haplotype resolution and genetic diversity for RM Y-STR and Yfiler Y-STR haplotypes. RM Y-STRs show exceptionally high diversity and resolution across all population groups and globally, exceeding that achieved with the Yfiler Y-STRs in seven of eight geographic regions, and an 8% increase in resolution and 0.6% increase in diversity on a global scale. (Figure adapted from Ballantyne, K. N. et al., 2012, A new future of forensic Y-chromosome analysis: Rapidly mutating Y-STRs for differentiating male relatives and paternal lineages, *Forensic Sci Int Genet* 6:208–18.)

worldwide HGDP-CEPH dataset of 604 males, 33 haplotypes were shared among 85 males from various populations when tested with the 17 Yfiler Y-STRs (Ballantyne et al. 2012). This may be partially explained by the limited power of Yfiler to discriminate between unrelated males, and partially by the inclusion of (some) small and isolated populations in the HGDP-CEPH panel where individuals can be expected to be distantly related (close relatives within HGDP-CEPH were already excluded from this analysis). However, further testing with the 13 RM Y-STRs revealed only 3 haplotypes shared among 8 of these 604 worldwide males (Ballantyne et al. 2012), illustrating the power of RM Y-STRs in improving male-lineage differentiation. In a case situation, the increased discriminatory power would prevent incorrect nonexclusions, avoiding investigation or prosecution of innocent suspects.

Although there are now several Y-STR multiplexes available for increasing lineage differentiation in specific populations and globally, extensive population testing needs to be conducted for all the novel Y-STRs proposed for forensic applications. It is only through large amounts of data that the community can fully evaluate the value of each Y-STR, and make informed decisions on the inclusion of new loci to the core set of Y-STRs. Furthermore, such population data are needed for reliable frequency estimates of the novel Y-STR haplotypes, as currently established for conventional Y-STR haplotypes in YHRD, because reliable frequency estimates are required for calculating the weight of the evidence in the event of matching Y-STR haplotypes. This is challenging for RM Y-STR haplotypes, as the frequencies within a population change constantly due to the high mutation rates. Therefore, frequency databases for RM Y-STR haplotypes must be even larger than for conventional Y-STRs with lower mutation rates. Furthermore, a nearest-neighbor analysis, as already implemented in YHRD, would be useful for RM Y-STRs due to the strong and significant excess of single-repeat versus multi-repeat changes that has been observed with Y-STR mutations (Ballantyne et al. 2010). An international multicenter study on RM Y-STRs is currently (October 2012) ongoing, aiming to increase the knowledge on worldwide haplotype diversity but also on the power of male-relative differentiation from large datasets.

11.2.3 Paternity/Familial Testing

The existing sets of Y-STRs have been used for both paternity testing—especially in deficiency cases with deceased alleged fathers and male offspring—and missing persons/disaster victim identification applications, because these Y-STRs, with their low- to midrange mutation rates and relatively high diversity, perform adequately for linking close relatives as described above. However, mutations at these Y-STRs are occasionally observed between biologically true father-son pairs, with single mutations occurring in ~7% of cases (Ballantyne et al. 2012). At present, paternity calculations can account for single-step mutations, or mutations at two loci, with relative ease. However, true biological father-son pairs (as established from autosomal DNA analysis) with Y-STR mutations at three loci have been found (Goedbloed et al. 2009), which would result in a nonpaternity conclusion under current recommendations (Gjertson et al. 2007). Therefore, if the current Y-STR loci are maintained for familial testing there is a need to reconsider the stringency of exclusion criteria. Instead of a strict number of genetic inconsistencies as exclusion criteria, it would be more appropriate to develop a probabilistic method, based on known mutation rates, to determine the likelihood of multiple mutation events or nonpaternity.

To date, there have been no novel Y-STRs suggested specifically for familial testing. However, as with male-relative and lineage differentiation, the increased knowledge regarding mutation rates in principle allows new panels to be selected that show fewer discrepancies between fathers and sons. Many Y-STRs with mutation rates of 10^{-4} (meaning a few mutations every 10,000 generation per each locus) and lower have recently been identified (Ballantyne et al. 2010). Such slowly mutating (SM) Y-STRs would be most suitable for familial testing, expectedly more so than currently used Y-STRs with higher mutation rates, and would avoid the need for statistical consideration of observed mutation in establishing paternity probabilities. However, given the close relationship between mutation rate and diversity, SM Y-STRs are less diverse, so to avoid adventitious matches it will be necessary to increase the current number of markers when SM Y-STRs are applied for familial testing including missing person/disaster victim identification. An example of a case where SM Y-STRs could be utilized is the father-son pair described above with three allelic differences from 17 Yfiler Y-STRs. Additional genotyping revealed allelic identity at a further 157 Y-STR loci, clearly confirming biological paternity. An analysis of 10–15 SM Y-STRs instead of 17 Yfiler Y-STRs would have reduced the uncertainty regarding nonpaternity in cases such as this. Obviously, RM Y-STRs are not suitable for paternity/familial testing as the higher chance of mutations observed will trouble statistical interpretations in relationship-probability estimation, even more than the currently used Y-STR sets do.

11.3 Availability of Additional Y-STRs for Forensic Applications

With around 500 Y-STRs now described (for examples see Hanson and Ballantyne 2006; Kayser et al. 2004), there are many candidate Y-STR markers available to improve forensic applications of Y-chromosome analysis. Most of these candidates were discovered in a single study (Kayser et al. 2004) that scanned the entire then available Y-chromosome sequence for short tandem repeats (excluding dinucleotides), allowing completely *in silico* identification followed up by laboratory testing. This study identified 166 practically useful new Y-STR markers, of which 139 were polymorphic in a small number of haplogroup-selected males. Detailed information including sequences, mutation rates, and diversity estimates currently exists for ~180 markers, allowing an informed assessment of each locus' suitability for forensic (and other) usage. Although an exhaustive list could not be displayed here, Table 11.2 summarizes the most commonly used markers included in published multiplexes. Of the novel Y-STRs discovered over the last decade, some are clearly not suitable for forensic applications. Loci such as DYS499, DYS581, and DYS615 show little or no variation across large numbers of males (Ballantyne et al. 2010; Kayser et al. 2004). Others such as DYS464 have been strongly linked with male infertility, and as such would not be appropriate for routine testing.

Of all noncore Y-STRs, DYS576 appears to have the strongest support for inclusion in future multiplexes and applications. First discovered by Kayser et al. (2004), it displays high diversity in various populations such as northern Chinese (Xu et al. 2010), Baltic (Lessig et al. 2009), Finnish (Hedman et al. 2011), European American and African American (Decker et al. 2007; Hanson et al. 2006; Maybruck et al. 2009) populations, as well as in diverse global panels (Lim et al. 2007; Vermeulen et al. 2009). In all studies it was ranked one of the most effective loci for increasing haplotype resolution among unrelated males. This may be explained by its high mutation rate of 1.43×10^{-2}, leading to its inclusion in

Table 11.2 Novel Y-STRs Proposed for Inclusion as Core Y-STR Loci

Locus [Initial Reference]	Repeat Type	Repeat	Mutation Rate (Credible Interval, Number of Meioses Examined)	Diversity Estimates (Population with Reference)
DYS444 (Kayser et al. 2004)	Single-copy, simple	$(TAGA)_{9-16}$	5.45×10^{-3} ($2.68 \times 10^{-3} - 9.65 \times 10^{-3}$, 1775)	0.61 (Eur/AfAm) (Hanson and Ballantyne 2007)
DYS446 (Redd et al. 2002)	Single-copy, simple	$(TCTCT)_{8-21}$	2.67×10^{-3} ($9.38 \times 10^{-4} - 5.87 \times 10^{-3}$, 1747)	0.86 (Afr) (Redd et al. 2002), 0.72–0.85 (Eur) (Redd et al. 2002; Rodig et al. 2008), 0.84 (YCC) (Redd et al. 2002), 0.76 (Asn) (Redd et al. 2002), 0.72 (Eur/AfAm) (Hanson and Ballantyne 2007),
DYS447 (Redd et al. 2002)	Single-copy, complex	$(TTATA)_{8-21}(TTATT)_1\,(TTATA)_{8-13}(TTATT)_1\,(TTATA)_{5-9}$	2.12×10^{-3} ($6.28 \times 10^{-4} - 5.11 \times 10^{-3}$, 1722)	0.77 (Eur/AfAm) (Hanson and Ballantyne 2007), 0.70–0.82 (Eur) (Rodig et al. 2008; D'Amato et al. 2009; Redd et al. 2002), 0.78 (YCC) (Redd et al. 2002), 0.68 (Afr) (Redd et al. 2002), 0.72 (Asn) (Redd et al. 2002), 0.87 (Ind) (D'Amato et al. 2009), 0.77 (Xho) (D'Amato et al. 2009)
DYS449 (Kayser et al. 2004; Redd et al. 2002)	Single-copy, complex	$(TTCT)_{13-19}N_{22}(TTCT)_3N_{12}\,(TTCT)_{13-19}$	1.22×10^{-2} ($7.54 \times 10^{-3} - 1.85 \times 10^{-2}$, 1617)	0.84 (Eur/AfAm) (Hanson and Ballantyne 2007), 0.78–0.88 (Eur) (Rodig et al. 2008; D'Amato et al. 2009; Redd et al. 2002; Ehler et al. 2010), 0.86 (Ind) (D'Amato et al. 2009), 0.81 (Xho) (D'Amato et al. 2009), 0.87 (YCC) (Redd et al. 2002), 0.89 (Afr) (Redd et al. 2002), 0.88 (Asn) (Redd et al. 2002), 0.88 (HGDP) MK/KB unpub
DYS459 (Redd et al. 2002)	Single-copy, simple	$(ATTT)_{6-11}$	2.67×10^{-3} ($9.36 \times 10^{-4} - 5.86 \times 10^{-3}$, 1741)	0.75 (Eur/AfAm) (Hanson and Ballantyne 2007), 0.73 (YCC), (Redd et al. 2002), 0.74 (Afr) (Redd et al. 2002), 0.61 (Eur) (Redd et al. 2002), 0.55 (Asn) (Redd et al. 2002)
DYS460 (Kayser et al. 2004; White et al. 1999)	Single-copy, simple	$(TAGA)_{8-13}$	6.22×10^{-3} ($3.19 \times 10^{-3} - 1.07 \times 10^{-2}$, 1717)	0.46 (Finnish) (Hedman et al. 2004)
DYS471 (DYS610) (Hanson and Ballantyne 2006)	Single-copy, complex	$(TTC)_{22}TT(TTC)_4$ (Genbank)	N/A	0.89 (Eur/AfAm) (Maybruck et al. 2009)

DYS481 (Kayser et al. 2004)	Single-copy, simple	$(CTT)_{22-32}$	4.97×10^{-3} ($2.36 \times 10^{-3} - 9.03 \times 10^{-3}$, 1744)	0.84 (Eur/AfAm) (Hanson and Ballantyne 2007), 0.71–0.84 (Eur) (Rodig et al. 2008; D'Amato et al. 2009; Leat et al. 2007), 0.7 (Ind) (D'Amato et al. 2009), 0.81 (Xho) (D'Amato et al. 2009), 0.9 (YCC) (Lim et al. 2007), 0.64–0.85 (Chi) (Xu et al. 2010)
DYS487 (Kayser et al. 2004)	Single-copy, simple	$(AAT)_{10-16}$	1.77×10^{-3} ($4.08 \times 10^{-4} - 4.78 \times 10^{-3}$, 1511)	0.62 (YCC) (Lim et al. 2007), 0.38 (Eur/AfAm) (Maybruck et al. 2009)
DYS488 (Kayser et al. 2004)	Single-copy, simple	$(ATA)_{10-16}$	4.40×10^{-4} ($1.60 \times 10^{-5} - 2.32 \times 10^{-3}$, 1576)	0.58 (YCC) (Lim et al. 2007), 0.20 (Eur/AfAm) (Maybruck et al. 2009), 0.08–0.28 (Chi) (Xu et al. 2010)
DYS504 (Kayser et al. 2004)	Single-copy, complex	$(CCTT)_{10-20}N_7(CCCT)_3$	3.24×10^{-3} ($1.26 \times 10^{-3} - 6.62 \times 10^{-3}$, 1746)	0.81 (Eur/AfAm) (Maybruck et al. 2009)
DYS505 (Kayser et al. 2004)	Single-copy, simple	$(TCCT)_{9-15}$	1.51×10^{-3} ($3.50 \times 10^{-4} - 4.07 \times 10^{-3}$, 1760)	0.68 (Eur/AfAm) (Hanson and Ballantyne 2007), 0.60 (Eur) (Rodig et al. 2008), 0.71 (YCC) (Lim et al. 2007)
DYS508 (Kayser et al. 2004)	Single-copy, simple	$(TATC)_{8-15}$	3.03×10^{-3} ($1.05 \times 10^{-3} - 6.63 \times 10^{-3}$, 1544)	0.73 (Eur/AfAm) (Hanson and Ballantyne 2007), 0.60 (Eur) (Rodig et al. 2008), 0.71 (YCC) (Lim et al. 2007)
DYS518 (Kayser et al. 2004)	Single-copy, complex	$(AAAG)_3(GAAG)_1(AAAG)_{14-22}(GGAG)_1(AAAG)_4N_6(AAAG)_{11-19}N_{27}(AAGG)_4$	1.84×10^{-2} ($1.25 \times 10^{-2} - 2.60 \times 10^{-2}$, 1556)	0.85 (Ind) (D'Amato et al. 2009), 0.79–0.81 (Eur) (D'Amato et al. 2009; Leat et al. 2007), 0.87 (Xho) (D'Amato et al. 2009), 0.87 (HGDP) (MK/KB unpub)
DYS522 (Kayser et al. 2004)	Single-copy, simple	$(ATAG)_{8-15}$	1.04×10^{-3} ($1.53 \times 10^{-4} - 3.44 \times 10^{-3}$, 1620)	0.64 (Eur/AfAm) (Hanson and Ballantyne 2007), 0.74 (YCC) (Lim et al. 2007)
DYS526 (Kayser et al. 2004)	Multi-copy (2), complex	$(CCTT)_{10-17}/(CCCT)_3N_{20}(CTTT)_{11-17}(CCTT)_{6-10}N_{113}(CCTT)_{10-17}$	2.72×10^{-3} ($9.52 \times 10^{-4} - 5.97 \times 10^{-3}$, 1716)/$1.25 \times 10^{-2}$ ($7.88 \times 10^{-3} - 1.87 \times 10^{-2}$, 1651)	0.78/0.88 (HGDP) (MK/KB unpub)

(continued)

Table 11.2 Novel Y-STRs Proposed for Inclusion as Core Y-STR Loci (Continued)

Locus [Initial Reference]	Repeat Type	Repeat	Mutation Rate (Credible Interval, Number of Meioses Examined)	Diversity Estimates (Population with Reference)
DYS527 (Hanson and Ballantyne 2004)	Single-copy, simple	$(GAAA)_4(AGAA)_1(GGAA)_3(ATGA)_1(AACA)_1(AGAA)_1(AGGA)_1(AAGA)_{14}(AGG)n(AAAG)$ (Hanson and Ballantyne 2007)	N/A	0.92 (Eur/AfAm) (Hanson and Ballantyne 2007)
DYS532 (Kayser et al. 2004)	Single-copy, complex	$(TCCC)_3N_5(TTCC)_5N_9(TTCT)_3(TTCC)_1(TTCT)_{6-17}N_{17}(TTCT)_3N_{13}(TTCC)_4N_{70}(TTCT)_3N_6(TTCT)_3$	3.24×10^{-3} ($1.13 \times 10^{-3} - 7.10 \times 10^{-3}$, 1441)	
DYS533 (Kayser et al. 2004)	Single-copy, simple	$(TATC)_{9-14}$	5.01×10^{-3} ($2.39 \times 10^{-3} - 9.11 \times 10^{-3}$, 1730)	0.62 (Eur/AfAm) (Hanson and Ballantyne 2007), 0.57 (Eur) (Rodig et al. 2008), 0.72 (YCC) (Lim et al. 2007)
DYS534 (Kayser et al. 2004)	Single-copy, complex	$(CTTT)_3N_8(CTTT)_{9-20}N_9(CTTT)_3$	6.51×10^{-3} ($3.44 \times 10^{-3} - 1.10 \times 10^{-2}$, 1794)	
DYS547 (Kayser et al. 2004)	Single-copy, complex	$(CCTT)_{9-13}T(CTTC)_{4-5}N_{56}(TTTC)_{10-22}N_{10}(CCTT)_4(TCTC)_1(TTTC)_{9-16}N_{14}(TTTC)_3$	2.36×10^{-2} ($1.70 \times 10^{-2} - 3.18 \times 10^{-2}$, 1679)	0.87 (HGDP) (MK/KB unpub)
DYS549 (Kayser et al. 2004)	Single-copy, simple	$GATA)_{9-15}$	4.55×10^{-3} ($2.05 \times 10^{-3} - 8.58 \times 10^{-3}$, 1684)	0.72 (Eur/AfAm) (Hanson and Ballantyne 2007), 0.61 (Eur) (Rodig et al. 2008), 0.72 (YCC) (Lim et al. 2007)
DYS552 (Kayser et al. 2004)	Single-copy, complex	$(TCTA)_3(TCTG)_1(TCTA)_{7-12}N_{40}(TCTA)_{11-16}$	2.69×10^{-3} ($9.21 \times 10^{-4} - 5.87 \times 10^{-3}$, 1742)	0.58 (Eur/AfAm) (Hanson and Ballantyne 2007), 0.64 (Eur) (Rodig et al. 2008)
DYS570 (Kayser et al. 2004)	Single-copy, simple	$(TTTC)_{14-24}$	1.24×10^{-2} ($7.52 \times 10^{-3} - 1.91 \times 10^{-2}$, 1426)	0.79 (Eur/AfAm) (Hanson and Ballantyne 2007), 0.78–0.80 (Eur) (Leat et al. 2007; Rodig et al. 2008), 0.86 (YCC) (Lim et al. 2007), 0.83 (HGDP) (MK/KB unpub)

DYS576 (Kayser et al. 2004)	Single-copy, simple	$(AAAG)_{13-22}$	1.43×10^{-2} ($9.41 \times 10^{-3} - 2.07 \times 10^{-2}$, 1727)	0.81–0.83 (Eur/AfAm) (Hanson and Ballantyne 2007; Maybruck et al. 2009), 0.77 (Eur) (Rodig et al. 2008), 0.82 (YCC) (Lim et al. 2007), 0.53–0.80 (Chi) (Xu et al. 2010), 0.81 (HGDP) (MK/KB unpub)
DYS607 (Hanson et al. 2006)	Single-copy, complex	$(GAAG)_{15}(GAAAGAAG)_2(GATG)_1(GAAG)_2$ (Leat et al. 2007)	N/A	0.77 (Eur/AfAm) (Hanson and Ballantyne 2007), 0.70 (Eur) (Leat et al. 2007)
DYS612 (Kayser et al. 2004)	Single-copy, complex	$(CCT)_5(CTT)_1(TCT)_4(CCT)_1(TCT)_{19-31}$	1.45×10^{-2} ($9.61 \times 10^{-3} - 2.09 \times 10^{-2}$, 1767)	0.81 (Ind) (D'Amato et al. 2009), 0.74–0.78 (Eur) (D'Amato, Benjeddou, and Davison 2009; Leat et al. 2007), 0.84 (Xho) (D'Amato et al. 2009), 0.84 (HGDP) (MK/KB unpub)
DYS626 (Kayser et al. 2004)	Single-copy, complex	$(GAAA)_{14-23}N_{24}(GAAA)_3N_6(GAAA)_5(AAA)_1(GAAA)_{2-3}(GAAG)_1(GAAA)_3$	1.22×10^{-2} ($7.70 \times 10^{-3} - 1.82 \times 10^{-2}$, 1689)	0.82 (Ind) (D'Amato et al. 2009), 0.84–0.85 (Eur) (D'Amato et al. 2009; Leat et al. 2007), 0.81 (Xho) (D'Amato et al. 2009), 0.85 (HGDP) (MK/KB unpub)
DYS627 (Kayser et al. 2004)	Single-copy, complex	$(AGAA)_3N_{16}(AGAG)_3(AAAG)_{12-24}N_{81}(AAGG)_3$	1.23×10^{-2} ($7.80 \times 10^{-3} - 1.81 \times 10^{-2}$, 1766)	0.86 (Eur/AfAm) (Hanson and Ballantyne 2007), 0.85 (HGDP) (MK/KB unpub)
DYS643 (Kayser et al. 2004)	Single-copy, simple	$(CTTTT)_{6-15}$	1.50×10^{-3} ($3.49 \times 10^{-4} - 4.05 \times 10^{-3}$, 1773)	0.78 (Eur/AfAm) (Hanson and Ballantyne 2007), 0.65 (Eur) (Rodig et al. 2008), 0.82 (YCC) (Lim et al. 2007)
DYS644 (Kayser et al. 2004)	Single-copy, complex	$(TTTTA)_{10-11}(TTTA)_{0-1}(TTTTA)_{0-13}$	3.22×10^{-3} ($1.25 \times 10^{-3} - 6.62 \times 10^{-3}$, 1761)	0.78 (Ind) (D'Amato et al. 2009), 0.70–0.72 (Eur) (D'Amato et al. 2009; Leat et al. 2007), 0.78 (Xho) (D'Amato et al. 2009)
DYS685 (Hanson and Ballantyne 2007)	Single-copy, complex	$(TTTC)_{18}TC(TTTC)_2N_7)TTTC)_2TC(TTTC)_{11}TT(TTTC)_5$ (Genbank)	N/A	0.84 (Eur/AfAm) (Maybruck et al. 2009)
DYS688 (Hanson and Ballantyne 2007)	Single-copy, complex	$(CTT)_5T(CTT)_{30}CT(CCT)_3TCTC(CTT)_3N_{30}(CTT)_{25}N_6(CTT)_3$ (Genbank)	N/A	0.91 (Eur/AfAm) (Maybruck et al. 2009)
DYS703 (Hanson and Ballantyne 2007)	Single-copy, simple	$(AAT)_{10}$ (Genbank)	N/A	0.60 (Eur/AfAm) (Maybruck et al. 2009)

(continued)

Table 11.2 Novel Y-STRs Proposed for Inclusion as Core Y-STR Loci (Continued)

Locus [Initial Reference]	Repeat Type	Repeat	Mutation Rate (Credible Interval, Number of Meioses Examined)	Diversity Estimates (Population with Reference)
DYS707 (Hanson and Ballantyne 2007)	Single-copy, complex	$(ATTCA)_{12}(ACTGC)_2N_{10}(ACTCC)_2(ATTCC)_{10}$ (Genbank)	N/A	0.68 (Eur/AfAm) (Maybruck et al. 2009)
DYS710 (Leat et al. 2007)	Single-copy, complex	$(AAAG)_{17}(AG)_{13}(AAAG)_{11}$ (Leat et al. 2007)	N/A	0.94 (Indian) (D'Amato et al. 2009), 0.90–0.93 (Eur) (D'Amato et al. 2009; Leat et al. 2007), 0.83 (Xho) (D'Amato et al. 2009)
DYF387S1 (Kayser et al. 2004)	Multi-copy (2), complex	$(AAAG)_3(GTAG)_1(GAAG)_4N_{16}(GAAG)_9(AAAG)_{13}$	1.59×10^{-2} ($1.08 \times 10^{-2} - 2.24 \times 10^{-2}$, 1804)	0.95 (HGDP) (MK/KB unpub)
DYF399S1 (Kayser et al. 2004)	Multi-copy (3), complex	$(GAAA)_3N_{7-8}(GAAA)_{10-23}$	7.73×10^{-2} ($6.51 \times 10^{-2} - 9.09 \times 10^{-2}$, 1794)	0.99 (HGDP) (MK/KB unpub)
DYF403S1 a/b (Kayser et al. 2004)	Multi-copy (3/1), complex	$(TTCT)_{10-17}N_{2-3}(TTCT)_{3-17}$/ $(TTCT)_{12}N_2(TTCT)_8(TTCC)_9(TTCT)_{14}N_2(TTCT)_3$	3.10×10^{-2} ($2.30 \times 10^{-2} - 4.07 \times 10^{-2}$, 1504)/$1.19 \times 10^{-2}$ ($7.05 \times 10^{-3} - 1.86 \times 10^{-2}$, 1402)	0.99/0.89 (HGDP) (MK/KB unpub)
DYF404S1	Multi-copy (2), complex	$(TTTC)_{10-20}N_{42}(TTTC)_3$	1.25×10^{-2} ($7.92 \times 10^{-3} - 1.84 \times 10^{-2}$, 1739)	0.92 (HGDP) (MK/KB unpub)
DYF406S1 (Kayser et al. 2004)	Multi-copy, simple	$(TATC)_{8-14}$	3.82×10^{-3} ($1.61 \times 10^{-3} - 7.48 \times 10^{-3}$, 1744)	0.74 (Eur) (Rodig et al. 2008), 0.75 (YCC) (Lim et al. 2007)

Note: Repeat designations and mutation rates were obtained from Ballantyne et al. (2010), unless otherwise specified. In several instances, sequences have not yet been published (DYS471, DYS685, DYS688, DYS703, and DYS707). In these cases, sequence information from Genbank (http://www.ncbi.nlm.nih.gov/genbank/) was used to define the repeat unit in accordance with the rules proposed by (Kayser et al. 2004). Population abbreviations for diversity estimates are as follows: Eur (European), Eur/AfAm (European American/African American combined), Afr (African), Asn (Asian), Ind (Indian), Xho (Xhosa African), Chi (Chinese), YCC (YCC Global Panel), HGDP (CEPH-HGDP Global Panel). Data from the HGDP panel is derived from data presented in Ballantyne et al. (2012), and is accessible as raw genotypes at http://www.cephb.fr/en/hgdp/.

the RM Y-STR set (Ballantyne et al. 2010, 2012). Likewise, DYS570 has been demonstrated to be effective in many of the populations listed above (Decker et al. 2007; Hanson et al. 2006; Kayser et al. 2004; Lessig et al. 2009; Lim et al. 2007; Rodig et al. 2008; Vermeulen et al. 2009), and also has a high mutation rate (1.24×10^{-2}) and therefore is included in the RM Y-STR set. Both Y-STRs have consistently high gene diversities across multiple populations, unlike many markers that show substantial variation between groups. This once again underlines the strong relationship between mutation rate and diversity of Y-STRs. However, there are also exceptions demonstrating that the full complement of factors driving a Y-STR to be highly polymorphic are not yet fully understood. One example is DYS481, which has a midrange mutation rate of 4.97×10^{-3} but nevertheless has been shown to have high diversities and haplotype resolution capacities in various populations such as Chinese (Xu et al. 2010), Europeans (Rodig et al. 2008), U.S. European and African Americans (Hanson and Ballantyne 2007), South Africans (D'Amato et al. 2010), Baltic populations (Lessig et al. 2009), and also in global panels such as YCC and HGDP (Lim et al. 2007; Vermeulen et al. 2009).

Simple, single-copy (ss) Y-STRs (for instance, DYS570, DYS576, and DYS481) satisfy the requirements for forensic Y-STRs in that interpretation of mixtures is simplified, and the presence of a single variable-repeat unit prevents internal sequence variation generating the appearance of the same allele from different combinations of repeats within the marker. Because of these advantages for forensic applications, there has been a substantial bias in past research toward ssY-STRs: Kayser et al. (2004), Lim et al. (2007), and Vermeulen et al. (2009) have all focused on ssY-STRs, while other studies such as Rodig et al. (2008) and Hanson et al. (2006) have examined significantly more ssY-STRs than complex or multilocus markers. However, complex or compound Y-STRs, by virtue of having multiple-repeat segments that may vary between individuals or populations, can often show higher diversity than simple repeats. The complexity of an STR has previously been shown to be a significant predictor of its mutation rate (Ballantyne et al. 2010), and correlates strongly with the allele length. Highly complex Y-STRs will usually be significantly longer, and have higher mutation rates, than simple Y-STRs. This higher mutation rate will often generate greater numbers of alleles, and thus higher diversity and haplotype-resolution power. Indeed, complex Y-STRs from the Yfiler and RM Y-STR panels were frequently more informative for both haplotype diversity and discrimination in a worldwide sample set than the simple Y-STRs examined (Ballantyne et al. 2012). Notably, there are 11 complex Y-STRs included in the Yfiler panel, which do not present any additional challenges for genotyping compared to the simple Y-STRs also in the panel.

The most commonly examined, and one of the potentially most useful, noncore complex Y-STRs is DYS449. There are in fact two separate variable loci within this marker (according to the definition of an STR locus by Kayser et al. [2004]), as the repeat structure is $(TTCT)_{13\text{-}19}N_{22}(TTCT)_3N_{12}(TTCT)_{13\text{–}19}$ amplified with one primer pair. It has previously been described as one of the most diverse Y-STRs, with estimates ranging from 0.812 to 0.876 in various populations (Butler et al. 2002; Ehler et al. 2010; Redd et al. 2002; Rodig et al. 2008). It is an RM Y-STR with a mutation rate of 1.22×10^{-2}, and is one of the largest Y-STRs under consideration, with an allele range of 24–37 repeats. It is likely that DYS449 actually carries considerably more variation than is currently being detected, as size homoplasy has been found (D'Amato et al. 2010; Ballantyne and Kayser, unpublished data). Numerous studies have successfully genotyped DYS449 within a range of multiplexes, and this marker can successfully increase the haplotype resolution of common 9

Y-STR locus profiles (Redd et al. 2002). Other complex Y-STRs that have been shown to be informative in a range of populations are DYS447, DYS471, DYS504, DYS518, DYS527, DYS532, DYS534, DYS536, DYS547, DYS552, DYS607, DYS612, DYS626, DYS627, DYS644, DYS685, DYS688, DYS703, DYS707, and DYS710.

To date, the forensic genetic community has aimed to avoid the use of multicopy Y-STRs, with the notable exceptions of DYS385a/b and DYS389I/DYS389II that are part of the commercial Y-STR kits. However, additional male-specific alleles at usually single-copy Y-STRs have been observed for many conventional Y-STRs (see YHRD), which occur occasionally due to the multicopy tendency of the Y-chromosome. The presence of multiple male-specific alleles, caused by the multiplication of the particular Y-chromosome sequence in which the Y-STR is embedded, can complicate mixture analysis, requiring the deconvolution of sets of alleles to their respective contributors based on peak height or the amplified alleles. However, multicopy markers often display the highest diversity values and mutation rates due to the increased probability of mutation across two or more loci within the marker. They also provide the advantage that more markers can be multiplexed together; with fewer primer pairs to be incorporated, restrictions caused by primer design are minimized. If multicopy markers with large allele ranges (such as DYF403S1 with 12–59 repeats) are used, the probability of IBS is reduced, simplifying the interpretation if reference samples are available. Five multicopy markers were included in the RM Y-STR panel, and were consistently the most informative markers to differentiate both unrelated paternal lineages and related males in a global panel of males (Ballantyne et al. 2010, 2012). Thus, if highly diverse markers with exceptional discrimination capacity are required, certain multicopy markers should be considered for future multiplexes. However, if applied, care needs to be provided to mixture interpretation, which could be practiced by combining single-copy Y-STRs (which can clearly highlight mixtures) and multicopy Y-STRs in the same set (such as done for the RM Y-STR set).

11.4 Practical Implications of Introducing Additional Y-STRs to Forensics

While there is clearly a need to increase the number of Y-STRs tested in forensic applications, as we have outlined above, there are a number of implications to their introduction to forensic case work. Most significant is the issue of generating appropriate reference databases required for frequency interpretation of the haplotypes generated from a crime scene sample with such novel markers. There are currently more than 127,000 7-locus MH haplotypes and 57,000 Yfiler haplotypes on publicly available Y-STR frequency databases (www.yhrd.org and http://www.usystrdatabase.com). Given such large numbers, replacing these profiles with completely new sets of Y-STRs would not be feasible due to time, cost, and unavailability of many of the samples used. In addition, and as outlined before, the currently applied Y-STR panels work well in many cases, so there is no need to completely replace the existing systems. However, wherever a nonexclusion is encountered, there should always be a question regarding the source of the matching haplotype—whether it has originated from a relative of the true perpetrator, or whether an adventitious match has occurred as a result of limited haplotype resolution, for instance because the sample donor comes from a population with reduced Y-chromosome diversity. Thus, advocating a two-stage process for Y-STR analysis seems more reasonable, where if initial testing with conventional Y-STR sets fail to exclude a suspect, further testing with an extended set of Y-STRs (e.g., the RM Y-STR set) should be performed to reduce the probability of the above possibilities, and to increase certainty in the accuracy of the conclusions drawn. Similarly, if initial paternity/familial testing with conventional Y-STR sets reveals a relatively small number of allelic differences so that the involvement of mutations is assumed, further testing with an extended set of Y-STRs (i.e., SM Y-STRs) should be performed so as not to complicate relationship estimations by mutations. While this will increase the number of tests needed to be performed, it will increase the accuracy of the obtained conclusion, hence providing improved value to forensic Y-chromosome analysis in the future.

11.5 Next-Generation Commercial Y-STR Kits

In recent months, two new Y-STR commercial multiplexes have been announced or released. The first, Promega's PowerPlex Y23, has extended the core 17 Y-STRs previously only available with Yfiler to 23 with the additional inclusion of DYS570, DYS576, DYS481, DYS533, DYS549, and DYS643. As discussed above, these six additional loci were the top ranked for male-lineage differentiation of the 49 novel simple single-copy Y-STR loci examined by Vermeulen et al. (2009) and two of them, DYS570 and DYS576, are considered RM Y-STRs (Ballantyne et al. 2010). All six also have extremely high diversity estimates of 0.7 or greater in the global YCC panel (Lim et al. 2007). From the limited data available so far for this new Promega Y23 kit, it can already be seen, as expected from the studies that led to the selection of the additional markers for the kit, that the discrimination power is increased such as in U.S. Americans (comprising African and European Americans, Hispanics, and Native Americans) to 0.9968, from 0.9959 with the 17 Yfiler Y-STRs (Davis et al. 2013). The second next-generation Y-STR kit from Applied Biosystems/Life Technologies, is currently (as of October 2012) under development. However, from initial details released by company representatives at conference presentations, it is expected that this multiplex will contain the 17 Yfiler Y-STRs plus several additional highly discriminating Y-STRs, including RM Y-STRs. The development of these extended commercial Y-STR kits is highly welcomed by the forensic Y-STR community. Although it may take some time to generate sufficient population data for frequency estimations of these extended Y-STR haplotypes, the rates of adventitious matches and false exclusions is expected to decrease with the application of these new Y-STR kits. However, it is likely that greater haplotyping capacity will be needed than is available with these two next-generation Y-STR kits, particularly in the specialized cases of relative differentiation and paternity testing as outlined above. Additional suitable Y-STR markers are already available as discussed in this chapter. Therefore, it will mostly depend on future technological advances allowing inclusion of even more Y-STRs in future commercial multiplex systems to allow forensic geneticists to use the full potential of the human Y-chromosome in their future casework applications.

Acknowledgments

We are grateful to the numerous colleagues who have published on Y-STRs and their application to forensic questions, and whose work we had the privilege to partly summarize here. MK was supported by funding from the Netherlands Forensic Institute (NFI), and by a grant from the Netherlands Genomics Initiative (NGI)/Netherlands Organization for Scientific Research (NWO) within the framework of the Forensic Genomics Consortium Netherlands (FGCN).

References

Alonso, A., P. Martin, C. Albarrán et al. 2005. Challenges of DNA profiling in mass disaster investigations. *Croat Med J* 46 (4):540–8.

Alves, C., L. Gusmao, J. Barbosa, and A. Amorim. 2003. Evaluating the informative power of Y-STRs: A comparative study using European and new African haplotype data. *Forensic Sci Int* 134 (2):126–33.

Applied Biosystems. 2011. *AmpFℓSTR Yfiler PCR Amplification Kit Users Manual.*

Ayub, Q., A. Mohyuddin, R. Qamar et al. 2000. Identification and characterisation of novel human Y-chromosomal microsatellites from sequence database information. *Nucleic Acids Res* 28 (2):e8.

Ballantyne, K. N., M. Goedbloed, R. Fang et al. 2010. Mutability of Y-chromosomal microsatellites: Rates, characteristics, molecular bases, and forensic implications. *Am J Hum Genet* 87 (3):341–53.

Ballantyne, K. N., V. Keerl, A. Wollstein et al. 2012. A new future of forensic Y-chromosome analysis: Rapidly mutating Y-STRs for differentiating male relatives and paternal lineages. *Forensic Sci Int Genet* 6 (2):208–18.

Bradford, L., J. Heal, J. Anderson, N. Faragher, K. Duval, and S. Lalonde. 2011. Disaster victim investigation recommendations from two simulated mass disaster scenarios utilized for user acceptance testing CODIS 6.0. *Forensic Sci Int Genet* 5 (4):291–6.

Brenner, C. H. 2010. Fundamental problem of forensic mathematics—The evidential value of a rare haplotype. *Forensic Sci Int Genet* 4 (5):281–91.

Buckleton, J. S., M. Krawczak, and B. S. Weir. 2011. The interpretation of lineage markers in forensic DNA testing. *Forensic Sci Int Genet* 5 (2):78–83.

Budowle, B., M. S. Adamowicz, X. G. Aranda et al. 2005. Twelve short tandem repeat loci Y chromosome haplotypes: Genetic analysis on populations residing in North America. *Forensic Sci Int* 150 (1):1–15.

Burgarella, C. and M. Navascues. 2011. Mutation rate estimates for 110 Y-chromosome STRs combining population and father-son pair data. *Eur J Hum Genet* 19 (1):70–5.

Butler, J. M. 2009. *Fundamentals of forensic DNA typing*. San Diego, CA: Elsevier Academic Press.

Butler, J. M., R. Schoske, P. M. Vallone, M. C. Kline, A. J. Redd, and M. F. Hammer. 2002. A novel multiplex for simultaneous amplification of 20 Y chromosome STR markers. *Forensic Sci Int* 129 (1):10–24.

Cerri, N., U. Ricci, I. Sani, A. Verzeletti, and F. De Ferrari. 2003. Mixed stains from sexual assault cases: Autosomal or Y-chromosome short tandem repeats? *Croat Med J* 44 (3):289–92.

Coble, M. D., O. M. Loreille, M. J. Wadhams et al. 2009. Mystery solved: The identification of the two missing Romanov children using DNA analysis. *PLoS One* 4 (3):e4838.

Corach, D., L. Filgueira Risso, M. Marino, G. Penacino, and A. Sala. 2001. Routine Y-STR typing in forensic casework. *Forensic Sci Int* 118 (2–3):131–5.

D'Amato, M. E., V. B. Bajic, and S. Davison. 2011. Design and validation of a highly discriminatory 10-locus Y-chromosome STR multiplex system. *Forensic Sci Int Genet* 5 (2):122–5.

D'Amato, M. E., M. Benjeddou, and S. Davison. 2009. Evaluation of 21 Y-STRs for population and forensic studies. *Forensic Sci Int Genet Supp Ser* 2 (1):446–7.

D'Amato, M. E., L. Ehrenreich, K. Cloete, M. Benjeddou, and S. Davison. 2010. Characterization of the highly discriminatory loci DYS449, DYS481, DYS518, DYS612, DYS626, DYS644 and DYS710. *Forensic Sci Int Genet* 4 (2):104–10.

Davis, C., J. Ge, C. Sprecher et al. 2013. Prototype PowerPlex Y23 System: A concordance study. *Forensic Sci Int Genet*, 7 (1):204–8.

De Knijff, P. 2003. Son, give up your gun: Presenting Y-STR results in court. *Profiles in DNA* 7 (1):3–5.

Decker, A. E., M. C. Kline, P. M. Vallone, and J. M. Butler. 2007. The impact of additional Y-STR loci on resolving common haplotypes and closely related individuals. *Forensic Sci Int Genet* 1 (2):215–7.

Ehler, E., R. Marvan, and D. Vanek. 2010. Evaluation of 14 Y-chromosomal short tandem repeat haplotype with focus on DYS449, DYS456, and DYS458: Czech population sample. *Croat Med J* 51 (1):54–60.

Gjertson, D. W., C. H. Brenner, M. P. Baur et al. 2007. ISFG: Recommendations on biostatistics in paternity testing. *Forensic Sci Int Genet* 1 (3–4):223–31.

Goedbloed, M., M. Vermeulen, R. N. Fang et al. 2009. Comprehensive mutation analysis of 17 Y-chromosomal short tandem repeat polymorphisms included in the AmpFlSTR Yfiler PCR amplification kit. *Int J Legal Med* 123 (6):471–82.

Hall, A. and J. Ballantyne. 2003. The development of an 18-locus Y-STR system for forensic casework. *Anal Bioanal Chem* 376 (8):1234–46.

Hanson, E. K. and J. Ballantyne. 2004. A highly discriminating 21 locus Y-STR "megaplex" system designed to augment the minimal haplotype loci for forensic casework. *J Forensic Sci* 49 (1):40–51.

Hanson, E. K. and J. Ballantyne. 2006. Comprehensive annotated STR physical map of the human Y chromosome: Forensic implications. *Leg Med (Tokyo)* 8 (2):110–20.

Hanson, E. K. and J. Ballantyne. 2007. An ultra-high discrimination Y chromosome short tandem repeat multiplex DNA typing system. *PLoS One* 2 (8):e688.

Hanson, E. K., P. N. Berdos, and J. Ballantyne. 2006. Testing and evaluation of 43 "noncore" Y chromosome markers for forensic casework applications. *J Forensic Sci* 51 (6):1298–314.

Hedman, M., A. M. Neuvonen, A. Sajantila, and J. U. Palo. 2011. Dissecting the Finnish male uniformity: The value of additional Y-STR loci. *Forensic Sci Int Genet* 5 (3):199–201.

Hedman, M., V. Pimenoff, M. Lukka, P. Sistonen, and A. Sajantila. 2004. Analysis of 16 Y STR loci in the Finnish population reveals a local reduction in the diversity of male lineages. *Forensic Sci Int* 142 (1):37–43.

Henke, J., L. Henke, P. Chatthopadhyay et al. 2001. Application of Y-chromosomal STR haplotypes to forensic genetics. *Croat Med J* 42 (3):292–7.

Iida, R. and K. Kishi. 2005. Identification, characterization and forensic application of novel Y-STRs. *Leg Med (Tokyo)* 7 (4):255–8.

Jarve, M., L. A. Zhivotovsky, S. Rootsi et al. 2009. Decreased rate of evolution in Y chromosome STR loci of increased size of the repeat unit. *PLoS One* 4 (9):e7276.

Karafet, T. M., F. L. Mendez, M. B. Meilerman, P. A. Underhill, S. L. Zegura, and M. F. Hammer. 2008. New binary polymorphisms reshape and increase resolution of the human Y chromosomal haplogroup tree. *Genome Res* 18 (5):830–8.

Kayser, M. 2003. The human Y-chromosome—Introduction into genetics and applications. *Forensic Sci Rev* 15 (77):78–90.

Kayser, M., S. Brauer, G. Weiss et al. 2000a. Melanesian origin of Polynesian Y chromosomes. *Curr Biol* 10 (20):1237–1246.

Kayser, M., A. Caglia, D. Corach et al. 1997. Evaluation of Y-chromosomal STRs: A multicenter study. *Int J Legal Med* 110 (3):125–33, 141–9.

Kayser, M., R. Kittler, A. Erler et al. 2004. A comprehensive survey of human Y-chromosomal microsatellites. *Am J Hum Genet* 74 (6):1183–97.

Kayser, M., M. Krawczak, L. Excoffier et al. 2001. An extensive analysis of Y-chromosomal microsatellite haplotypes in globally dispersed human populations. *Am J Hum Genet* 68:990–1018.

Kayser, M., O. Lao, K. Anslinger et al. 2005. Significant genetic differentiation between Poland and Germany follows present-day political borders, as revealed by Y-chromosome analysis. *Hum Genet* 117 (5):428–43.

Kayser, M., L. Roewer, M. Hedman et al. 2000b. Characteristics and frequency of germline mutations at microsatellite loci from the human Y chromosome, as revealed by direct observation in father/son pairs. *Am J Hum Genet* 66 (5):1580–8.

Kayser, M. and A. Sajantila. 2001. Mutations at Y-STR loci: Implications for paternity testing and forensic analysis. *Forensic Sci Int* 118 (2–3):116–21.

Kayser, M., M. Vermeulen, H. Knoblauch, H. Schuster, M. Krawczak, and L. Roewer. 2007. Relating two deep-rooted pedigrees from Central Germany by high-resolution Y-STR haplotyping. *Forensic Sci Int Genet* 1 (2):125–8.

Kelkar, Y.D., S. Tyekucheva, F. Chiaromonte, and K. D. Makova. 2008. The genome-wide determinants of human and chimpanzee microsatellite evolution. *Genome Res* 18:30–8.

King, T. E. and M. Jobling. 2009. What's in a name? Y chromosomes, surnames and the genetic genealogy revolution. *Trends in Genetics* 25 (8):351–60.

Krenke, B. E., L. Viculis, M. L. Richard et al. 2005. Validation of male-specific, 12-locus fluorescent short tandem repeat (STR) multiplex. *Forensic Sci Int* 151 (1):111–24.

Lai, Y. and F. Sun. 2003. The relationship between microsatellite slippage mutation rate and the number of repeat units. *Mol Biol Evol* 20 (12):2123–31.

Leat, N., L. Ehrenreich, M. Benjeddou, K. Cloete, and S. Davison. 2007. Properties of novel and widely studied Y-STR loci in three South African populations. *Forensic Sci Int* 168 (2–3):154–61.

Lessig, R., J. Edelmann, J. Dressler, and M. Krawczak. 2009. Haplotyping of Y-chromosomal short tandem repeats DYS481, DYS570, DYS576 and DYS643 in three Baltic populations. *Forensic Sci Int Genet* Supp Ser 2 (1):429–30.

Lim, S. K., Y. Xue, E. J. Parkin, and C. Tyler-Smith. 2007. Variation of 52 new Y-STR loci in the Y Chromosome Consortium worldwide panel of 76 diverse individuals. *Int J Legal Med* 121 (2):124–7.

Maybruck, J. L., E. Hanson, J. Ballantyne, B. Budowle, and P. A. Fuerst. 2009. A comparative analysis of two different sets of Y-chromosome short tandem repeats (Y-STRs) on a common population panel. *Forensic Sci Int Genet* 4 (1):11–20.

Mulero, J. J., C. W. Chang, L. M. Calandro et al. 2006. Development and validation of the AmpFlSTR Yfiler PCR amplification kit: A male specific, single amplification 17 Y-STR multiplex system. *J Forensic Sci* 51 (1):64–75.

Onofri, V., F. Alessandrini, C. Turchi, M. Pesaresi, and A. Tagliabracci. 2008. Y-chromosome markers distribution in Northern Africa: High-resolution SNP and STR analysis in Tunisia and Morocco populations. *Forensic Sci Int Genet* Supp Ser 1 (1):235–6.

Palo, J. U., M. Pirttimaa, A. Bengs et al. 2008. The effect of number of loci on geographical structuring and forensic applicability of Y-STR data in Finland. *Int J Legal Med* 122 (6):449–56.

Park, S. W., C. H. Hwang, E. M. Cho, J. H. Park, B. O. Choi, and K. W. Chung. 2009. Development of a Y-STR 12-plex PCR system and haplotype analysis in a Korean population. *J Genet* 88 (3):353–8.

Parson, W., H. Niederstatter, S. Kochl, M. Steinlechner, and B. Berger. 2001. When autosomal short tandem repeats fail: Optimized primer and reaction design for Y-chromosome short tandem repeat analysis in forensic casework. *Croat Med J* 42 (3):285–7.

Prinz, M., K. Boll, H. J. Baum, and B. Shaler. 1997. Multiplexing of Y chromosome specific STRs and performance for mixed samples. *Forensic Sci Int* 85 (3):209–18.

Prinz, M., A. Ishii, A. Coleman, H. J. Baum, and R. C. Shaler. 2001. Validation and casework application of a Y chromosome specific STR multiplex. *Forensic Sci Int* 120 (3):177–88.

Prinz, M., and M. Sansone. 2001. Y chromosome-specific short tandem repeats in forensic casework. *Croat Med J* 42 (3):288–91.

Redd, A. J., A. B. Agellon, V. A. Kearney et al. 2002. Forensic value of 14 novel STRs on the human Y chromosome. *Forensic Sci Int* 130 (2–3):97–111.

Rodig, H., L. Roewer, A. M. Gross et al. 2008. Evaluation of haplotype discrimination capacity of 35 Y-chromosomal short tandem repeat loci. *Forensic Sci Int* 174 (2–3):182–8.

Roewer, L. 2009. Y chromosome STR typing in crime casework. *Forensic Sci Med Pathol* 5 (2):77–84.

Roewer, L. and J. T. Epplen. 1992a. Simple repeat sequences on the human Y chromsome are equally polymorphic as their autosomal counterparts. *Hum Genet* 89 (4):389–394.

Roewer, L. and J. T. Epplen. 1992b. Rapid and sensitive typing of forensic stains by PCR amplification of polymorphic simple repeat sequences in casework. *Forensic Sci Int* 53 (2):163–71.

Roewer, L., M. Kayser, P. de Knijff et al. 2000. A new method for the evaluation of matches in non-recombining genomes: Application to Y-chromosomal short tandem repeat (STR) haplotypes in European males. *Forensic Sci Int* 114 (1):31–43.

Roewer, L., M. Krawczak, S. Willuweit et al. 2001. Online reference database of European Y-chromosomal short tandem repeat (STR) haplotypes. *Forensic Sci Int* 118 (2–3):106–13.

Rolf, B., W. Keil, B. Brinkmann, L. Roewer, and R. Fimmers. 2001. Paternity testing using Y-STR haplotypes: Assigning a probability for paternity in cases of mutations. *Int J Legal Med* 115 (1):12–5.

Santos, F. R., A. Pandya, and C. Tyler-Smith. 1998. Reliability of DNA-based sex tests. *Nat Genet* 18 (2):103.

Shewale, J. G., S. C. Sikka, E. Schneida, and S. K. Sinha. 2003. DNA profiling of azospermic semen samples from vasectomized males by using Y-PLEX 6 amplification kit. *J Forensic Sci* 48 (1):127–9.

Shi, W., Q. Ayub, M. Vermeulen, et al. 2010. A worldwide survey of human male demographic history based on Y-SNP and Y-STR data from the HGDP-CEPH populations. *Mol Biol Evol* 27 (2):385–93.

Sibille, I., C. Duverneuil, G. Lorin de Grandmaison et al. 2002. Y-STR DNA amplification as biological evidence in sexually assaulted female victims with no cytological detection of spermatozoa. *Forensic Sci Int* 125 (2–3):212–16.

Sullivan, K. M., A. Mannucci, C. P. Kimpton, and P. Gill. 1993. A rapid and quantitative DNA sex test: Fluorescence-based PCR analysis of X-Y homologous gene amelogenin. *Biotechniques* 15 (4):636–638, 640–1.

Tsuji, A., A. Ishiko, N. Ikeda, and H. Yamaguchi. 2001. Personal identification using Y-chromosomal short tandem repeats from bodily fluids mixed with semen. *Am J Forensic Med Pathol* 22 (3):288–91.

Vermeulen, M., A. Wollstein, K. van der Gaag et al. 2009. Improving global and regional resolution of male lineage differentiation by simple single-copy Y-chromosomal short tandem repeat polymorphisms. *Forensic Sci Int Genet* 3 (4):205–13.

White, P. S., O. L. Tatum, L. L. Deaven, and J. L. Longmire. 1999. New, male-specific microsatellite markers from the human Y chromosome. *Genomics* 57 (3):433–7.

Willuweit, S., A. Caliebe, M. M. Andersen, and L. Roewer. 2011. Y-STR frequency surveying method: A critical reappraisal. *Forensic Sci Int Genet* 5 (2):84–90.

Xu, X., M. Peng, and Z. Fang. 2000. The direction of microsatellite mutations is dependent upon allele length. *Nat Genet* 24 (4):396–9.

Xu, Z., H. Sun, Y. Yu, et al. 2010. Diversity of five novel Y-STR loci and their application in studies of north Chinese populations. *J Genet* 89 (1):29–36.

Y Chromosome Consortium. 2002. A nomenclature system for the tree of human Y-chromosomal binary haplogroups. *Genome Res* 12:339–348.

Zhivotovsky, L. A., P. A. Underhill, C. Cinnioğlu et al. 2004. The effective mutation rate at Y chromosome short tandem repeats, with application to human population-divergence time. *Am J Hum Genet* 74 (1):50–61.

Expanding the Genotyping Capabilities IV

Forensic Mitochondrial DNA Analysis

Current Practice and Future Potential

12

MITCHELL HOLLAND
TERRY MELTON
CHARITY HOLLAND

Contents

12.1 Introduction 250
12.2 Current Practice 251
12.2.1 Demand, Customers, and Sample Type 251
12.2.2 Evolution of Protocols 252
12.2.3 Interpretation Guidelines 255
12.2.4 Mixture Interpretation 255
12.2.5 Statistics and Databases 261
12.2.6 Courtroom Experiences 262
12.2.7 Regulation and Accreditation 262
12.3 Alternative Methods and Future Potential 263
12.3.1 Screening Methods 263
12.3.2 Expanded Sequence Analysis 265
12.3.3 Mixtures and Heteroplasmy Investigation: DGGE and dHPLC 267
12.3.4 Mass Spectrometry 268
12.3.5 Pyrosequencing and Deep Sequencing 269
12.4 Conclusions 272
Acknowledgments 273
References 273

Abstract: Current practices for performing forensic mitochondrial DNA (mtDNA) sequence analysis, as employed in public and private laboratories across the United States, have changed remarkably little over the past 20 years. Alternative approaches have been developed and proposed, and new technologies have emerged, but the core methods have remained relatively unchanged. Once DNA has been recovered from biological material (for example, from older skeletal remains and hair shafts), segments of the mtDNA control region are amplified using a variety of approaches, dictated by the quality of the sample being tested. The amplified mtDNA products are subjected to Sanger-based sequencing and data interpretation is performed using one of many available software packages. These relatively simple methods, at least in retrospect, have remained robust, and have stood the test of time. However, alternative methods for mtDNA analysis remain viable options (for example, linear array assays and dHPLC), and should be revisited as the desire to streamline the testing process, interpret heteroplasmy, and deconvolute mixed mtDNA profiles intensifies. Therefore, it is important to periodically reassess the alternative methods available to the mtDNA practitioner, and to evaluate newer technologies being put forth by the scientific community, for example, next-generation sequencing. Although the basic mitochondrial DNA protocols and practices of public and private laboratories are similar, an overview of the current practices of forensic mtDNA analysis is provided, helping to frame the path forward.

12.1 Introduction

Refinement of methods for forensic mitochondrial DNA (mtDNA) analysis that were introduced in the early 1990s has led to the present "golden age" of mtDNA testing in public or government laboratories such as the Federal Bureau of Investigation (FBI) and the Armed Forces DNA Identification Laboratory (AFDIL), as well as in private-sector laboratories such as Mitotyping Technologies, the Bode Technology Group, and Orchid Cellmark. The success of mtDNA in forensic DNA analysis can be gauged from the fact that a laboratory such as Mitotyping has completed over 1,000 mtDNA cases since its inception in 1998; hundreds have resulted in resolution of criminal cases, contributory and relevant trial testimonies, and postconviction exonerations. Although it is not possible to retroactively review the history or present the workings of the entire forensic mtDNA community of test providers, a retrospective analysis of the operation at Mitotyping provides a comprehensive overview of the testing process. In this review we will describe how the "state of the art" has evolved since 1996 when mtDNA testing was introduced to the criminal justice system in the case of *Tennessee v. Ware* (Davis 1998), review alternative methods for mtDNA analysis, and describe forthcoming new methods with the potential to change the ways in which casework is carried out.

Certain portions of the control region of mtDNA are highly variable among individuals. Forensic analysis typically involves examination of the sequence variation within two hypervariable (HV) regions, HV1 and HV2. While laboratories may work with slightly different ranges, HV1 spans at least from position ~16024 to ~16365 and HV2 from position ~73 to ~340. Mitochondrial DNA analysis is employed when degraded skeletal remains or hairs without roots are encountered in forensic casework or human identification cases. Mitochondrial DNA offers two primary advantages over nuclear DNA analysis: (1) thousands of copies of mtDNA are present in a cell compared to two copies of nuclear DNA, leading to higher sensitivity and (2) mtDNA is maternally inherited, enabling distant maternal relatives to be compared to the analyzed samples for relationship hypothesis testing or when the original depositor of the sample is not available. However, the discrimination power of mtDNA analysis is limited compared to that of short tandem repeat (STR) analysis. Readers are advised to refer to earlier reviews for a detailed description of molecular biology, genetics, sequence determination procedures, interpretation practices, and utility of mtDNA sequence analysis in forensic casework/human identification (Holland and Parsons 1999; Melton 2004).

12.2 Current Practice

12.2.1 Demand, Customers, and Sample Type

Mitochondrial DNA forensic laboratories, regardless of whether they are private or public, share many attributes, particularly the application of common scientific approaches to typical samples requiring mtDNA testing. Federal and state laboratories such as the FBI or AFDIL, and academic providers such as the University of North Texas, may have specific or unique mandates for testing, such as current criminal cases, military identifications, or missing persons cases. In general, the clientele served by private service providers include state and defense agencies, more or less evenly divided. Prosecutors and law enforcement agencies with time-critical cases proceeding quickly to investigative conclusions or trial dates are in need of a faster alternative to public labs such as the FBI that carry backlogs, even though those services are free of charge. Coroners and medical examiners, especially those who need supporting documentation for body identifications and family notifications, may also need a quick turnaround time. Defense and postconviction testing clients are often legally unable to access the public sector labs, and this population is always at risk of being underserved by the forensic crime laboratory system. Some countries do not have mtDNA laboratories and hence depend solely on private service providers. Custom mtDNA analysis remains costly and time-consuming because of hands-on analysis of individual samples, some of which are degraded due to age and environmental challenges. Cost is a significant consideration, and can be an obstacle for the agencies or individuals submitting samples to the private laboratory.

A large number of cases analyzed by our laboratory involve mtDNA analysis of samples from old or cold crimes, small crime scene hairs less than 10 mm, nonhuman samples, and canine mtDNA analysis (Melton 2006; Melton and Nelson 2005; Nelson and Melton 2007). Analysis of hairs less than 1 cm is routine (Melton et al. 2012); some public laboratories have minimal hair size limits that prohibit acceptance of affected cases. Regardless of the case-specific approach, the methods for extraction, amplification, and sequencing are largely shared among all mitochondrial DNA testing laboratories, with minor differences. These minor differences permit ready comparison between results from different laboratories when necessary.

Over a period of years, and largely due to educational efforts by the laboratory and the public sector, clients have adapted to the realities that (a) mtDNA analysis cannot usually be used effectively on samples that are impossible to clean prior to testing, such as stains, swabs, and swatches, due to the likelihood of complex mixtures occurring, and (b) there is limited statistical power to a failure to exclude with mtDNA compared to that of STRs. Even now, frequent requests or inquiries are made about testing of stains that were unsuccessfully tested for STRs. Conversely, laboratories have learned that there is a continuum of cuttings, swabs, and swatches that may successfully, and rarely, undergo mtDNA testing if the original surface was pristine prior to deposit of a sample. For example, a bloodstain on a UV-exposed surface such as the hood of a vehicle may be too degraded for STRs but suitable for mtDNA analysis, and a single type may be easily obtained. Stains on clothing almost always result in an uninterpretable complex mtDNA mixture, with some, such as stains on shoes, being the most complex given their extensive environmental exposure.

The limited statistical power of mtDNA analysis compared to STR analysis results in the submission of many fewer samples than in the early days of testing. Interested parties now inquire about testing for only the most highly probative samples, as opposed to early submissions of many crime scene hairs. Directing the choice of samples to be submitted for testing is a frequent service. In our longstanding experience, the number of samples within a case averages about four, but we have encountered several cases with 30–60 samples and one case with approximately 200 samples. Microscopic preliminary evaluation of hairs is recommended in the interest of collecting the most information about a sample, such as size, color, diameter, and hair structure, prior to destructive testing. Since the late 1990s it is no longer customary for probative hair evidence to go to trial without confirmatory mtDNA analysis, based on one published study of the relative value of hair microscopy and its potential to give erroneous results (Houck and Budowle 2002).

The cost of testing in the private sector limits the number of samples submitted but also forces the client to carefully consider the relative value of any single piece of evidence. Costs of doing mtDNA casework have increased for public and private sector laboratories. For example, the accredited fee-for-service laboratory is required to adhere to all standards and guidelines promulgated by overseeing accrediting bodies, and the number of these rules and regulations is increasing. Fee structures take into account the nearly 30% of resources that are expended each year to accommodate all these requirements in areas such as training, quality control, quality assurance, accreditation, and proficiency testing in addition to the actual costs of doing casework. Public laboratories also contend with funding challenges, and often receive external funding from federal agencies such as the National Institute of Justice for projects such as cold-case investigation, missing persons identification, and post-conviction testing, all areas in which mtDNA analysis is frequently required.

12.2.2 Evolution of Protocols

Laboratory methods for all mitochondrial DNA test providers have functionally changed very little since the original Sanger sequencing protocols were applied in mtDNA testing by both the FBI and AFDIL in the early to mid-1990s (Holland and Parsons 1999; Irwin et al. 2009). Preextraction sample prep is a critical part of the analysis. Hairs are washed multiple times in an ultrasonic water bath and rinsed with sterile water and ethanol. Bones are prepped with sanding of the exterior surface before cutting or powdering, and 10% bleach washes can be applied to teeth and bone fragments. An organic in-house or kit extraction protocol is applied to samples such as hair, bone, buccal, or blood samples, followed by polymerase chain reaction (PCR) amplification using strategies designed to capture extracted control-region mtDNA template from within the two hypervariable regions (Melton and Nelson 2001). As an example, a preliminary amplification of region 16160–16400 can be carried out and a yield gel run to determine if any product has been obtained, and if so, how robust a product it is. The amount of DNA extraction product then drives the required expenditure of remaining template for the additional three amplifications on a questioned sample. Other laboratories carry out a microarray-based quantification step at this point, such as that provided by Agilent Technologies (Mueller et al. 2000). Regardless of quantitation method, if a product is obtained, the input of PCR product to a cycle sequencing reaction is titered to obtain the best quality sequence via electrophoresis in a genetic analyzer such as an Applied Biosystems 310, 3130, or 3130xl (Life Technologies, Foster City, California) after cycle-sequencing with

Big Dyes v. 1.1 (Life Technologies). In general, the profile obtained from multiple PCR reactions that capture overlapping segments of the hypervariable regions are reconstructed during sequence editing into the "mitochondrial DNA profile" that characterizes that particular sample. Sequence editing is typically carried out by two qualified forensic examiners using DNA editing software such as Lasergene (DNAStar, Madison, Wisconsin), Sequencher (Genecodes, Ann Arbor, Michigan), or Mutation Surveyor (SoftGenetics, State College, Pennsylvania).

Methods to quantitate mtDNA after DNA extraction and prior to PCR amplification via use of probes and real-time PCR are available (Alonso et al. 2004). In the case of skeletal remains, nuclear DNA quantitation protocols are applied (Quantifiler, Life Technologies, Foster City, California) because these samples are often eligible for STR analysis. Postextraction mtDNA quantification methods are found most often in laboratories where both mtDNA and STR profiles are desired for skeletal remains that will be entered into missing persons databases, and can save time and money in determining the best triage approach when the amount of nuclear DNA in a sample is unknown and mtDNA analysis might be required as well.

A significant protocol change for the handling of skeletal remains was the introduction in 2008 of an EDTA demineralization protocol described by AFDIL (Loreille et al. 2007). This approach has been adopted by a number of mtDNA test providers. Skeletal material is cleaned with rotary tools, cut, and powdered via blender and then incubated at 56°C overnight in 0.5 M EDTA with rotation in a hybridizing oven. Full dissolution of the bone powder occurs, releasing DNA from the bony matrix. An organic extraction and silica cleanup yields significantly larger quantities of DNA template than previous methods where the bone powder was not completely dissolved. Early work with the PrepFiler BTA kit (Life Technologies, Foster City, California) seems to indicate that mtDNA results can be obtained from as little as 50 mg of bone, even those that are hundreds of years old (unpublished data).

For almost all laboratories, a standard examination on questioned samples captures HV1 and HV2 in four overlapping amplification products, whereas a standard examination on known samples captures these regions in two longer amplification products. The four amplification products are approximately 300 base pairs long, and all questioned samples are presumed to have somewhat degraded DNA, defining this approach. While not used in many public or private laboratories, "miniprimers" were developed to capture template from degraded samples (Gabriel et al. 2001). With this approach, a set of eight primer pairs, four for each hypervariable region, target amplification products less than 200 base pairs in size; the eight amplification products together cover nearly all of hypervariable regions 1 and 2. The miniprimer approach was pioneered at Penn State University in the early 1990s and an early adopter of this method was AFDIL. Additional primers were developed to cover difficult sequencing regions, capture degraded DNA, and add more control region data to the regions normally tested for further discrimination of common types. Further, the additional primers provide replicate overlapping sequence coverage in cases where there are deletions (e.g., at positions 249, 290, 291) or homopolymeric C-stretches (16189, 309.1, etc.), because these phenomena typically result in either a failure to amplify any product with standard primer pairs or generation of only single-stranded data unless secondary coverage is obtained via the use of internal primers.

Two additional primer pairs cover regions designated as "VR1" and "VR2" (Variable Regions, nucleotide positions 16471–16562 and 424–548, sometimes called "HV3," respectively) (Lutz et al. 2000). In a rare case, these regions can aid in discriminating between two samples with the same common profile, particularly the H1 haplogroup "263G, 315.1C" haplotype that is observed in about 7% of individuals with European/Caucasian maternal origins. There are other "common" types observed in other ethnic groups. Although minisequencing and single nucleotide polymorphism (SNP) assays have been developed for discriminating between common types by examining mtDNA coding regions (Coble et al. 2006; Mosquera-Miguel et al. 2009; Nilsson et al. 2008), there appears to be relatively little cost benefit to using these assays for the low-throughput laboratory that may have only a rare case needing the application. The method would require costly multiplex amplification protocols, additional instrumentation, validation when no kit is available, and annual proficiency testing. In addition, for a single hair, there is often insufficient DNA template to set up the required multiplex reactions. A simple linear array assay for SNPs in the coding region has been developed and may fulfill this need in future (Kline et al. 2005). A more comprehensive review of these methods is provided in a subsequent section of this chapter.

Amplification and sequencing of a ~150 base pair (bp) fragment of mtDNA that codes for 12S ribosomal RNA was developed to identify the species origin of nonhuman casework samples, particularly mammalian hair (Melton and Holland 2007). The ~100-bp sequence product is searched at http://www.ncbi.nlm.nih.gov/BLAST and the species match is reported. The use of this assay has halved the number of samples for which no mtDNA results are obtained and is useful on all mammalian hairs, especially because preliminary hair microscopy is applied less frequently by submitting clients with each passing year. The size of the 12S amplification product is in line with those of the mtDNA miniprimer sets used for degraded samples, meaning that the assay is successful even on highly degraded samples. Species determination aids forensic investigators in opening or closing off lines of inquiry where a highly probative hair is submitted. The assay is frequently required in casework, and crime-scene hairs have yielded a range of species including sheep, pig, mouse, and raccoon. While, to our knowledge, only one mtDNA testing laboratory provides this assay, it is utilized in other biological disciplines.

A frequent crime scene sample is hair from domestic dogs and cats. Pet hairs have the potential to connect victims, suspects, and crime scenes. However, while STR analyses for canine and feline blood or saliva samples provide near-individualization of a domesticated pet, much like in human STR typing (Eichmann et al. 2005), mtDNA analysis must be used on rootless or naturally shed fur just as for human hair. Mitochondrial DNA diversity in both these species is very restricted compared to that in humans due to the short history of domestication (Gundry et al. 2007; Himmelberger et al. 2008; Tarditi et al. 2011). However, using precisely the same extraction, amplification, and sequencing methods as those used for human-specific samples, three canine hypervariable regions can be analyzed (HV1: nucleotide position [np] 15431–15782; HV2: np 15739–16092; HV3: np 16451–00014). A reference dog sequence is available (Kim et al. 1998), and matches are searched from within *C. familiaris* mtDNA control region data compiled by us from GenBank for a frequency statistic much as is computed in human mtDNA testing. There are two providers of canine mtDNA analysis in the United States: Mitotyping Technologies and the University of California Davis Veterinary Genetics Laboratory.

12.2.3 Interpretation Guidelines

Interpretation guidelines for the mtDNA forensic arena were published by the FBI and the European agency EDNAP in 2003 and 2001, respectively, but no formal guidelines have been promulgated since then (SWGDAM 2003; Tully et al. 2001). The FBI guidelines primarily addressed the interpretation of sequence data, including basic nomenclature of base-calling for polymorphisms, homopolymeric C-stretches, and insertions/deletions (indels). Ensuing discussions in the forensic literature have revolved around more complex treatments of length variation in homopolymeric stretches (Bandelt and Parson 2008; Wilson et al. 2002a,b) as well as consistent nomenclature for difficult-to-assign base-number calls when indels occur (Budowle et al. 2010; Rock et al. 2011).

In the early years of forensic mtDNA analysis, mixtures and heteroplasmy were not discussed, likely due to the poorer quality of sequence data obtained from early reagents and instruments. With early genetic analyzers such as the ABI 373, or early dye chemistries used for the ABI 310, noisy sequence baselines did not always permit easy recognition of either phenomenon. As chemistry and sensitivity improved, heteroplasmy captured a great deal of attention beginning around 1994, with a glut of publications debating the actual and relative effects on forensic application of mtDNA analysis (see for example, Comas et al. 1995; Grzybowski 2000; Grzybowski et al. 2003; Melton 2004; Stoneking 1996). Since that time, with research indicating that low-level heteroplasmy is widespread in all tissues of the body yet is manageably interpreted in forensic applications due to the common major variant that most individuals display (Roberts and Calloway 2011), forensic laboratories have validated their own interpretational protocols that cover reporting of sequence mixtures and heteroplasmy. For example, based on internal validation studies, heteroplasmy can be defined as the presence of a single mixed-nucleotide position within the region reported and a mixture from two or more individuals as the presence of two or more mixed positions for this region (Melton 2004). There is now substantial literature on both length and site heteroplasmy in human mtDNA (Andrew et al. 2011; de Camargo et al. 2011; Forster et al. 2010; Li et al. 2010; Roberts and Calloway 2011; Seo et al. 2010; Wang et al. 2011).

12.2.4 Mixture Interpretation

Interpreting mtDNA mixtures continues to present itself as one of the major challenges in forensic DNA analysis. A typical mixture in forensic casework samples may be defined as DNA originating from more than one individual. Possible explanations for a DNA mixture result can range from contamination of a sample during collection or DNA testing to a "naturally occurring" mixture such as an intimate sample or a biological sample taken from an article of clothing worn or stained by more than one person. In STR analysis it is often possible and appropriate to deconvolute mixtures in order to determine the number of contributors and sometimes also determine the major and minor components of the mixture. With mtDNA mixtures, such deconvolution can prove to be difficult if not impossible using current methods. In addition, because mtDNA analysis is inherently more susceptible to contamination, many laboratories choose not to interpret mixtures at all, and simply categorize the result as inconclusive.

To date, there has been very little published in the literature about interpreting mtDNA mixtures. In fact, the only published information specific to forensic casework involves two cases where STR analysis failed to detect a minor component of the mixed sample and therefore mtDNA analysis was performed using cloning techniques (Hatsch et al. 2007) or analysis of mismatch primer-induced restriction sites (Szibor et al. 2003). Another study on resolving mtDNA mixtures using cloning methods identified potential pitfalls with this approach, including the possibilities of overestimating the number of contributors due to naturally occurring heteroplasmies or underestimating the number due to individuals with identical haplotypes (Walker et al. 2004). Cloning methods are also generally labor-intensive and low-throughput, and therefore not practical for forensic casework. To the best of our knowledge, no published research exists on interpreting mtDNA mixtures derived from Sanger sequencing data, which is the current method utilized by most forensic mtDNA laboratories. There is also no mention of interpretation of mtDNA mixtures in the SWGDAM guidelines (SWGDAM 2003).

Using decades of experience and caution to ensure that any conclusions drawn from a mixture are conservative and not overstated, it is possible to interpret mtDNA mixtures. Mixtures in hair and bone analyses can be cautiously interpreted for the purposes of exclusions, using a validated concept that if the mixture profiles that can be generated from all possible combinations of the mixed sites exclude an individual, especially within individual amplicons, this is fair evidence of exclusion. However, each polymorphism must also be considered individually. In general, Sanger sequencing and the current instruments and chemistry associated with that analytical approach allow for detection of a secondary minor nucleotide where the minor variant is at least 5–10% of the total component of the mixture, whether due to heteroplasmy or a mixture of mtDNAs from two or more individuals. Below that level, there is a possibility of mixed-base dropout (akin to allelic dropout in STR analysis) for the minor templates in one or more amplification products that constitute a full analysis. Therefore, interpretations of hair and bone mixtures must be applied judiciously.

Disregarding an mtDNA mixture profile may result in discarding useful information. The following is an excerpt from an SOP regarding interpretation of mtDNA mixtures: "The clear presence of mixed nucleotide bases at two or more positions will be assumed to represent a mixture of two or more mitochondrial DNA types, which probably originate from two or more individuals. If the known sample cannot be excluded as one of the many possible contributors to the mixture, the report will clearly reflect that multiple contributors to the mixture are possible. The report will state that when a mixture profile is obtained, the number of potential mtDNA types that may be derived from that mixture is equal to 2^n, where n is equal to the number of nucleotide positions at which two different nucleotides have been observed, and that all of these types are not equally probable. The report will state how many possible types there are for the mixture observed. Mixtures may be used with care to exclude an individual as contributor of a sample. Because of the phenomenon of base dropout, mixtures containing any of the substitutions characterizing the known comparison sample must be interpreted with extreme caution in order to conclude that there is an exclusion" (Mitotyping Technologies, SOP, 2013, p. 98). Mixed sites can be included in a case report, and in the event that a known sample is included as a contributor to the mixture, no database search or statistical analysis would be provided. Therefore, this result is more for investigative purposes, since court testimony cannot be provided regarding an inclusion in a mixture because statistics are not provided to put the inclusion into context or add appropriate weight to the evidence.

A more progressive approach to mtDNA mixture interpretation has recently been proposed, including a possible method for providing mixture statistics (Egeland and Salas 2011). This statistical approach uses phylogenetic knowledge to deconvolute mtDNA mixtures, noting that not all combinations of variants are equally possible because many of them do not fit into the accepted human evolutionary phylogenetic tree. Two different statistical methods are recommended, one using categorized or qualitative data and the other using quantitative data. The method using categorized qualitative data only takes into account the nucleotide positions where the mixed sites occur, assumes a given number of contributors, and uses a likelihood ratio approach with the following two hypotheses: (1) the mixture comes from the haplotypes of two known individuals (e.g., suspect and victim), or (2) the mixture comes from two random individuals (or victim and an unknown donor). Because the qualitative approach has limitations in cases where there may be several different haplotype combinations or an unknown number of contributors, the quantitative method uses peak heights or areas and regression models to estimate the contributor fractions to the mixture. More specifically, the challenge of mixture deconvolution is determining which haplotypes may have contributed to forming the mixed mtDNA profile. If several haplotype combinations are possible, further quantitative data is needed. By quantitatively measuring the contribution of different nucleotide variants to the same position using signal strength and peak height estimates, it may be possible to deconvolute the mixture into its individual haplotypes. This can be achieved by using standard linear regression analysis to distinguish between two competing hypotheses (as in the example above) based on how well the data fits each hypothesis. As the authors note, however, Sanger sequencing is not a pure quantitative method. Therefore, it is suggested that by replicating the PCR and subsequent sequencing reactions, the information pertaining to the relative proportion of contributors may improve. Also, the use of next-generation sequencing or other methods that allow the sequencing of single copies of DNA would facilitate a more precise determination of the donor contributions. In recent years, researchers have been pursuing efforts to quantify mtDNA mixtures using pyrosequencing (Andreasson et al. 2006) and to resolve mtDNA mixtures by denaturing HPLC (dHPLC) (Danielson et al. 2007). Also, newer methods such as next-generation sequencing and SNPs may allow for better methods of mtDNA mixture deconvolution in the future (Holland et al. 2011; Homer et al. 2008).

A significant component of the experience necessary to interpret mtDNA mixtures is the ability to recognize and contend with postmortem DNA damage. DNA decays rapidly after death in biological samples and chemical damage begins to accumulate in the DNA (Lindahl 1993). This damage can take many different forms, including strand breakage or fragmentation, oxidative damage that may inhibit PCR, and the generation of miscoding lesions (Hoss et al. 1996; Lindahl 1993; Paabo 1989). These miscoding lesions can manifest as base modifications, which can in turn lead to erroneous substitutions (and/or mixed sites in Sanger sequencing). The mechanism causing these base modifications is deamination, which is one of the most common forms of DNA damage. Deamination is particularly rapid for cytosine (Hofreiter et al. 2001), which results in the conversion of cytosine to uracil, an analog of thymine. Deamination of adenine to hypoxanthine (HX), an analog of guanine, has also been documented as a common form of DNA damage (Hansen et al. 2001). These deamination conversions result in two complementary groups of transitions, termed "type 1" (A-G/T-C) and "type 2" (C-T/G-A) (Hansen et al. 2001). It is generally reported that the "type 2" transitions resulting from the deamination of cytosine occur more frequently than the "type 1" transitions (Gilbert et al. 2003b; Hansen et al. 2001; Lindahl 1993). Most importantly, these deamination events cannot be properly repaired postmortem, and are therefore something that must be considered during mtDNA sequence analysis.

DNA extraction of an old or degraded sample may yield a low number of template molecules, some of which may have base modifications due to postmortem damage. Depending on the distribution of molecules that are incorporated into the PCR reaction, the sequencing of this amplified product may result in three possible scenarios: (1) the number of original templates in the reaction is significantly outnumbered by the modified templates, yielding the incorrect sequence; (2) the number of original templates in the reaction significantly outnumbers the modified templates, yielding the correct sequence; and (3) a mixture of the two is observed (Gilbert 2006). Most, if not all, published data on postmortem DNA damage relates to "ancient" DNA from paleontological and archeological remains. It is hypothesized that if one starts with a DNA sample that is not "ancient" but still "old" (~20–50 years), as is commonly seen in forensic casework samples, there may be fewer damaged DNA molecules that would be starting templates for PCR amplification as compared to ancient DNA samples. This would hypothetically result in mtDNA sequence data that is most often the "correct" undamaged sequence or a mixture sequence of damaged and undamaged DNA. Therefore, mtDNA damage resulting in sequence misidentification should in theory be a rarity in forensic casework. Indeed, it is generally considered that when the initial template number is >1,000 copies, postmortem damage rates are unlikely to bias results (Handt et al. 1996; Krings et al. 1997). However, when few DNA templates initiate a PCR, the resulting sequences are likely to contain base modifications. Several instances of damaged mtDNA have been observed in older hair samples and skeletal remains that were manifested as a mixed mtDNA sequence (unpublished data). Postmortem DNA damage manifesting itself as a mixture has not been previously discussed as an mtDNA interpretation issue in ancient DNA studies, most likely because cloning methods (resulting in single-molecule profiles) are typically utilized in ancient DNA research whereas Sanger cycle sequencing (resulting in a pooled molecule profile), is most often used in forensic mtDNA laboratories.

There are obviously many forensic implications associated with interpreting mtDNA mixtures, especially with regard to postmortem DNA damage. Because there are currently no guidelines set forth on this subject, up until recently mtDNA damage mixed sites have been treated no differently than typical mixed sites. All mixed sites have been included in case reports and regular mixture guidelines were typically followed according to protocol. Over time, however, experience has allowed for the recognition of key differences between DNA damage and typical mixtures. Three core observations have emerged that have become the foundation for our enhanced mixture guidelines relating to damaged DNA: (1) noting at which base positions the mixed sites are occurring; (2) noting whether or not the mixed sites are reproducible through repeat extractions or amplifications; and (3) noting other substitutions that show no signs of a mixture.

Evidentiary samples received by laboratories for mtDNA testing typically consist of single hair shafts and skeletal remains. Skeletal samples usually provide ample opportunity for replicate testing, both in reamplification(s) and reextraction(s). Conversely, with hair shaft evidence, there is generally very limited opportunity for duplication of results. Because of this, mixed sites from potential damage in hair-shaft samples are conservatively reported as a regular mixture. However, due to the ability to duplicate results from bones and teeth, mixture interpretation policies can be changed with respect to skeletal remains samples that may have damaged DNA.

For example (taken from a Mitotyping Technologies protocol):

- The full profile will be developed using regular primers and/or miniprimers.
- The full profile will be edited in a first pass, noting the locations of unmixed polymorphisms and mixed sites, if any.
- If there are no mixed sites, the profile will be edited by both examiners and reported.
- If more than one mixed site is noted, there will be an attempt to determine if there are any unmixed polymorphic sites present.
- If all polymorphic sites or most polymorphic sites are mixed, the mixture will be assumed to be a true mixture of DNA from two or more individuals and reported as a mixture.
- If there are unmixed polymorphic sites, the region(s) containing the mixture will be reamplified.
- The products of these reamplifications will be sequenced and then edited in the original project along with the previously amplified and sequenced amplification products.
- If the mixed sites disappear in the second-round amplification and/or if new mixed sites appear in the second-round amplification, the mixed sites will be assumed to have resulted from damaged mtDNA template being captured in early rounds of amplification.
- The layout (printout of analyzed sequence data) containing all data will note the unmixed polymorphisms or persistent mixed polymorphisms as highlighted and labeled sites.
- If more than one mixed site persists as mixed, along with the unmixed polymorphic sites, this sample will be reported as a mixture.
- The layout containing all data will have an asterisk below mixed sites that are not reproducible in any subsequent amplification. This asterisk will reference an accompanying case note.
- The accompanying case note will show a table of the nonreproducible mixed sites and which amplifications they were observed in. This table will be created by one examiner and co-signed by a second examiner.
- The final report will not need to show these nonreproducible mixed sites.
- Three conditions are required to report this kind of sample as an unmixed profile:
  - The mixed sites are not reproducible in any subsequent amplification.
  - The unmixed polymorphisms remain unmixed in all amplifications.
  - A single mixed site is permitted due to the possibility of heteroplasmy.
- The appropriate call as to whether a sample is composed of a mixture or not will be left up to the discretion of both examiners, with both examiners making and agreeing on the determination.
- In general, the inability to reproduce mixed sites will lead to a conclusion that these sites are due to damage and not to additional DNA templates from a second individual.

Nine skeletal remains cases with sample mtDNA sequence mixtures were interpreted in the past three years following these guidelines (unpublished data). Six of these cases involved DNA damage, one was a true mixture, and two were likely DNA damage but there was not enough template DNA to replicate the data. The six cases with DNA-damaged samples each had unmixed, duplicated substitution sites that were reported as the true mtDNA profile. Five of these cases resulted in inclusions with a known sample (known samples were tested after the skeletal remains in each case), and the sixth case was a historical case consisting of Late Prehistoric Native American skeletal remains that had no known reference samples for comparison. Presumed mtDNA damage has also occasionally been observed in hair samples. Although these hairs have been reported as regular mixtures, the following specific mtDNA-casework examples involving aged hair samples clearly exhibited DNA damage.

Case #1

A questioned hair (~25 years old) had the following mtDNA profile: 16188 Y(C/T), 16218 Y(C/T), 152 C, 214 R(A/G), 263 G, and 315.1 C. The three mixed sites were in overlapping regions where one PCR product resulted in a mixture and the other PCR product resulted in the rCRS base. Suspecting that DNA damage was causing at least two of the three mixed sites (with potential heteroplasmy at 214), another piece of this same hair was reextracted. The two mixed sites in HV1 disappeared, while the three clean substitutions along with the 214 potential heteroplasmic site remained (in both PCR products). The known sample from the suspect's buccal swab gave the same three substitutions as the questioned hair and showed no heteroplasmy at position 214. Knowing that heteroplasmy is more prolific in hairs than in body fluids (Melton 2004; Melton and Nelson 2005), a known hair from the same suspect was tested in order to validate that position 214 was a true heteroplasmic site in the questioned hair. When the known suspect hair was typed, heteroplasmy was observed at position 214 and interestingly, another heteroplasmy was seen at position 16222.

Case #2

Eight questioned hairs (~15 years old) were tested. Three hairs collected from the same item of evidence gave the same profile (unmixed substitutions at positions 16126, 152, and 263). One of the three hairs also showed five mixed damaged sites (at positions 16234, 100, 140, 269, and 307). The suspect and victim were both excluded as contributors of these three hairs.

There have been several trends observed with regard to DNA damage. First, in agreement with other studies (Gilbert et al. 2003b; Hansen et al. 2001; Lindahl 1993), most, if not all, of the damaged sites in casework samples have been observed at either cytosine or guanine rCRS positions, resulting in C/T or G/A mixed sites (Type 2 damage). Another noteworthy observation is in regard to identifying postmortem damage "hotspots." Previous studies on this topic have reported conflicting results. For example, Hofreiter et al. (2001) reported that "there is no evidence for 'hotspots' for mis-incorporation in the resulting sequences." However, Gilbert et al. (2003a, 2005) suggested that there are postmortem damage hotspots and that they correspond with sites of elevated *in vivo* mutation rates. Other data suggest that most damaged sites occur at random sites that are not associated with *in vivo* mutational hotspots (unpublished data). A possible explanation for these conflicting results could be the differing definitions of which sites constitute mutational hotspots (Excoffier and Yang 1999; Meyer et al. 1999; Soares et al. 2009).

In our laboratory, two instances have been observed where DNA damage resulted in complete substitutions, rather than just mixed sites. Both cases involved old skeletal remains, one of which was approximately 25 years old and showed two C to T substitutions at sites 16107 and 16112. As there was not enough template DNA to replicate this result, these two substitutions were included in the case report with an asterisk, indicating that they were not duplicated and therefore might not be true substitutions. The submitted bone sample from the second case was approximately 15 years old and gave several G to A substitutions at random (nonmutational) sites. Replicated amplification products gave the same profile, but without the G to A substitutions observed in the first PCR, further indicating that they were a result of damaged DNA. This highlights the importance of duplicating results whenever possible—a recommendation also noted in ancient DNA research. For example, Hofreiter et al. (2001) recommended that when extracts of ancient specimens contain only a few template molecules for PCR that DNA sequences are determined from at least two independent amplification products. It is interesting to note that an increase in DNA damage has been observed from skeletal remains soon after the incorporation of a new bone demineralization extraction procedure in 2008. Two possible reasons for this may be: (1) a longer incubation period (overnight) subjects the bone samples to additional heat, and (2) the increased efficiency of the protocol results in more DNA recovered overall, including more damaged DNA. A more detailed study examining damaged DNA is forthcoming (manuscript in preparation).

12.2.5 Statistics and Databases

In the United States, the criminal justice community including both Combined DNA Index System (CODIS) and private laboratories have relied on the FBI's Scientific Working Group on DNA Analysis Methods (SWGDAM) database of human mtDNA sequences to derive the statistical weight of a failure to exclude with mtDNA results (Monson et al. 2002). Using common statistical equations (Holland and Parsons 1999) to estimate the upper-bound proportion of a population that cannot be excluded as having the casework profile in question with 95% or 99% confidence has been the method of choice in courtroom presentations since the mid- to late 1990s. In most cases, this approach permits the exclusion of well over 99% of individuals in a population as donors of a sample, due to the large number of mtDNA types present in the world population, but it is also conservative enough to counteract normally small effects of mtDNA population substructure. With current technologies emerging to rapidly type whole mtDNA genomes, a good understanding of the phylogenetic structure and saturation levels of human mtDNA haplotypes has been gained, indicating that databases can be more accurately planned to establish frequency estimates of haplotypes within different populations (Egeland and Salas 2008). Sequencing assays, at least on pristine samples, are faster and less expensive than ever, and an estimated 6,700 whole human genomes were reported to be present in GenBank and the scientific literature as of 2010 (Salas and Amigo 2010).

The FBI and others are currently evaluating whether a slightly different calculation may be applied for haploid lineage markers like mtDNA and Y-STRs (Butler 2012). The Clopper and Pearson method provides a two-tailed upper 95% confidence limit, and can be equally applied to cases where the profile observed is one not previously observed, as well as to cases where the profile has been observed before (Clopper and Pearson 1934). The resulting number is slightly more beneficial to a defendant, especially when database sizes are small. Other methods of presenting the weight of mtDNA "matches" have been proposed, including match probabilities and likelihood ratios (Brenner 2010).

As of early 2012, the SWGDAM CODIS database that had been available online at www.FBI.gov for use in the public sector was no longer available, although there are plans to make it available via the National Institute of Standards and Technology (NIST) after review of the data therein at the FBI (Eric Pokorak, FBI DNAU2, personal communication). CODIS laboratories continue to have use of this database, called CODIS 7.0, for criminal casework as well as for missing persons cases. Alternatives exist for database searching, most notably the online searchable database EMPOP (empop.org), which contains at present over 15,000 human profiles from highly vetted forensic datasets as well as separate datasets from the published anthropology literature (Parson and Dür 2007). Searching via DNA text strings or lists of polymorphisms in a sample is possible. As of January 2013, the database contained over 29,000 mtDNA sequences from all over the world, with approximately 9,000 forensic sample profiles from North America. Outputs classifying sequence-match results with respect to geographic origins, such as "European," "West Asian," "Sub-Saharan," and so forth, as well as for metapopulations such as African, Asian, European, and so forth, provide a very helpful approach for forensic applications. For example, EMPOP may be searched for African American samples collected within North America or within Africa. Recent funding of Lakehead University to produce "MitoNorth," a forensic mtDNA database for Canada, will provide a resource for North American cases outside the United States. In addition, a Korean Web site (mtmanager.yonsei.ac.kr) is available for searching 9,294 human sequences, although there is overlap with data found in other databases.

12.2.6 Courtroom Experiences

After a flood of admissibility hearings under the Kelly-Frye and Daubert scientific evidence rules between 1996 and about 2005, mtDNA testing appears to be well accepted in the criminal justice system, although many jurisdictions have not yet tried an mtDNA court case. Many of the written decisions on notable and early cases may be found at www.denverda.org. The only federal appellate decision on mtDNA to date is *United States v. Beverly* in the 6th Circuit Court of Appeals in 2004. This case was tried in Columbus, Ohio, in 2000. Beverly was convicted in part based on mtDNA comparison of a hair recovered from a hat in a getaway vehicle after bank surveillance cameras captured Beverly wearing the hat during the commission of a bank robbery. The court's primary finding was that the maternal inheritance and nonunique characteristic of mtDNA can be suitably explained by both cross-examination and well-accepted statistical analysis, guaranteeing that the result is not more prejudicial than probative.

Presentation to juries remains a critical feature of forensic mtDNA usage. Emphasis on the nonunique status of the marker is important so that juries are not confused about the differences between the powerful statistics used for nuclear DNA and the more modest statistics possible with mtDNA results (Kaye et al. 2007). To date, no court decision has been overturned due to any misrepresentation during testimony about the strength of the statistical conclusion or a failure to represent the nonunique haploid mode of mtDNA inheritance.

12.2.7 Regulation and Accreditation

Although accreditation remains optional for non-CODIS laboratories in the United States, most if not all laboratories performing mtDNA testing have chosen to become accredited under the ISO 17025:2005 Standards titled "General Requirements for the Competence of Testing and Calibration Labs." Various accrediting bodies exist such as the American Society of Crime Laboratory Directors/Laboratory Accreditation Board (ASCLD/LAB) and Forensic Quality Services (FQS), and laboratories have a choice of which agency will provide their accreditation. During the accrediting body inspections under ISO, external auditors also audit laboratories under the FBI's Quality Assurance Standards for Forensic DNA Testing Laboratories that are required operating standards to be a CODIS provider. In particular, most private laboratories are undergoing annual or semiannual visits that cover a range of standards and guidelines, internal audits, management reviews, proficiency testing, and numerous other required activities. The single most striking change in this area over the last 14 years has been the rapid increase in standards and guidelines, and the increased rigor of these programs.

Private laboratories that serve multiple states are also required to conform to those states' individual forensic testing standards. For example, the New York State Department of Health has separate Forensic Identity Standards that a laboratory must follow to perform casework shipped from New York, and laboratories must undergo audits biannually by this agency as well as participate in their mandated proficiency tests. In contrast, other states such as Texas simply require proof of current accreditation via ASCLD/LAB or another entity for cases to be sent to the private sector. Private laboratories are ineligible to use CODIS 7.0, although there are steps that could be taken to allow a CODIS laboratory to take ownership of a private contractor laboratory's data for a search in a current criminal case or missing persons case.

12.3 Alternative Methods and Future Potential

It is clear that conventional mtDNA sequence analysis is robust and reliable for routine forensic investigations (Holland and Parsons 1999). However, there are a number of alternative methods that can be employed to enhance or advance current practices. Some of these methods are currently available to the practitioner, while others are still in development and have the potential to significantly impact the testing process in the future. The alternative methods presented here include screening techniques to identify potential mtDNA matches prior to full-scale sequence analysis, expanded analysis of the mtDNA genome within and outside of the control region, technologies that provide for a deeper assessment of mtDNA mixtures and heteroplasmy, and a second-generation sequencing approach that provides a more sensitive means for detecting and quantifying heteroplasmic variants and mixture components, and thus may allow for the deconvolution of mixtures.

12.3.1 Screening Methods

Conventional mtDNA sequence analysis is often considered relatively time-consuming and expensive, but remains the most comprehensive approach to developing a forensic mtDNA profile. A quick assessment of work performed by population geneticists to classify population structure through the clustering of mtDNA sequences into haplogroups illustrates the value of obtaining complete sequence information (see, for example, Soares et al. [2009] for a recent worldwide mtDNA phylogeny). While haplogroup designations can clarify the relationship between and within population groups, private polymorphisms found within haplotypes that typically do not contribute to haplogroup assignment can significantly increase discrimination potential and make conventional mtDNA sequence analysis a highly informative typing system.

Although potential obstacles of time and expense for conventional sequence analysis can be mitigated by using a variety of screening approaches (Butler 2012), only the Roche mtDNA Linear Array has been adopted in a limited way within the forensic community, especially within the community of laboratories doing extensive missing persons projects. The value and usefulness of linear arrays has been the focus of investigation (Divne et al. 2005; Kline et al. 2005). However, the proposed extent of their use has also spurred debate (Melton et al. 2006). The Roche Applied Science LINEAR ARRAY Mitochondrial DNA HVI/HVII Region-Sequence Typing Kit examines 18 polymorphic sequence positions along the length of HV1 and HV2. Ten short sequences are targeted, encompassing 18 polymorphic sites. Each of the 10 targets has multiple sequence combinations, or alleles, for a total of 30 possible alleles. The individual alleles are interrogated using 33 sequence-specific oligonucleotide probes immobilized as 31 lines or strips on a nylon membrane. The HV1 and HV2 segments are coamplified from input DNA using PCR primers with biotin moieties attached to the 5'-end of each of the four primers. Products of PCR amplification bind to the probe-bound strips on the nylon membrane through allele-specific hybridization and are detected using an enzyme-conjugate-based development process similar to the HLA DQA1 and Polymarker systems (Saiki et al. 1989). The subsequent interpretation of linear array results is relatively simple, and the result takes the form of a barcode-like profile.

The specific application of the linear array assay has been debated. While some believe that the assay is useful as a routine screening tool to eliminate samples from needing full sequence analysis, others contend that the potential uses should be limited. For example, the assay was used in 16 adjudicated forensic cases containing 57 evidence samples and 33 references to exclude 56% of the samples as potential matches; thus, less than half of the samples required further sequencing (Divne et al. 2005). Of the samples that were originally excluded through sequence analysis, 79% could be omitted using the array system alone. These results were the impetus for suggesting the use of linear arrays to decrease sequencing efforts and turnaround time, and thereby reduce the cost of analysis in routine casework. Therefore, laboratories with limited capabilities or instrumentation may consider using linear arrays to conduct mtDNA analysis. However, practitioners working in the field have questioned the wisdom of using the array system for routine forensic casework, or whether the system is substantially more cost-effective or efficient (Melton et al. 2006).

The time and expense required for extraction of DNA from hair shafts (or any biological specimen) will be similar for each of the two typing methods (conventional sequencing versus linear arrays). The amplification process, including post-PCR product gel analysis, will also be the same, and the reagent costs for DNA sequencing are basically offset by the kit costs for linear array analysis. Therefore, one must look further to pinpoint the efficiencies and limitations with each method. For example, the current linear array system is more susceptible to cross-hybridization anomalies or null results, potentially complicating the interpretation process. To minimize cross-hybridization artifacts, a reasonably precise post-PCR quantification system is required to estimate the amount of product for hybridization. At least 2 cm of hair shaft are ideally required for linear array analysis to obtain a full profile. In contrast, conventional mtDNA sequencing is relatively insensitive to the amount of input DNA added to the sequencing reaction, and typing results can routinely be obtained from less than 1 cm of hair shaft (Melton et al. 2012). In a study of more than 2,500 freshly collected head hairs, the success rate of developing a linear array profile never exceeded 75% (Roberts and Calloway 2007), whereas when using conventional DNA sequencing the success rate routinely exceeds 92%, including 80% of casework hairs less than 1 cm in length (Melton and Nelson 2005). Much of this difference in rates can be attributed to the size of the target amplicons, which varies between the two systems: approximately 400 base pairs for the linear array assay and 100–300 base pairs for sequencing. Of course, a limitation of conventional sequence analysis is the significantly longer time to complete the analysis (instrument and interpretation time), especially when considering high-volume scenarios. Therefore, the linear array method may be a valuable tool in missing persons or human rights investigations that require rapid reassociation of thousands of commingled remains where there is the potential for multiple reextractions (Gabriel et al. 2003). However, for routine mtDNA testing of crime scene hairs, where DNA template is limited and complete profiles are desirable for courtroom presentation, conventional sequence analysis may be the appropriate method of choice.

As an alternative to the linear array approach, fractional mtDNA sequencing can also be used as a screening tool, given judicious selection of appropriate samples. Assuming the full mtDNA profiles of reference samples are known and the screening region displays a low-frequency profile, informative polymorphic sites can be targeted to identify which evidence samples are potential matches or exclusions. For example, given the HV1 profile of the following victim reference sample as 16093C, 16189C, 16278T, and 16311C, the practitioner can amplify and sequence nucleotide positions 16160–16400 in evidence samples for comparison to the reference profile. Evidence samples that share the 16189C, 16278T, and 16311C polymorphisms can be assumed to be from the victim and discounted if not probative (for example, victim hairs on victim clothing). Although use of this system relies heavily on the relative rarity of a particular profile in the screening region utilized (so as not to falsely assume the profile belongs to the known individual), an approach like this will significantly reduce the workload of a full-length sequence analysis, while maintaining the discrimination potential of the system. In addition, when this partial profile is not exclusionary and is probative, continuing quickly on to develop the remaining portion of the full-sequence profile is possible.

12.3.2 Expanded Sequence Analysis

In the early 1990s, it became clear that a handful of common mtDNA sequence profiles within HV1 and HV2 were being encountered when performing mtDNA analysis on population groups of European Caucasian descent that were collected in the United States (unpublished observations). When these common profiles are encountered today, whether revealed by the linear array assay (Kline et al. 2005) or through conventional DNA sequencing (see Coble et al. [2004] for a list of common mtDNA sequence types), they can impede the practitioner's ability to differentiate between two individuals, or two samples, if sequence analysis is limited to HV1 and HV2. Therefore, an expanded investigation of mtDNA sequence information is required to resolve these identical profiles. There are two principal approaches that can be employed, neither of which must be performed to the exclusion of others: expand the range of sequence being analyzed in the control region (Lutz et al. 1998, 2000) and query SNP sites in the coding region of the mitochondrial genome (Coble et al. 2004; Parsons and Coble 2001).

As an example, an early population study of 200 unrelated individuals from Germany revealed 88 variable nucleotide positions in HV1 (26% of the total sites) and 65 variable sites in HV2 (24% of the total) (Lutz et al. 1998). A third segment of the control region, HV3 (encompassing positions 438–574, and sometimes referred to as variable region two, VR2), exhibited lower variability, with 25 polymorphic sites (18% of the total), but in contrast to other segments of the control region was quite informative; only 7% of polymorphic sites occurred between positions 16,366–16,569 and 1–72 (sometimes referred to as variable region one, VR1), and 3% of polymorphic sites occurred between positions 341–437 (included in VR2). Approximately 20% of the identical HV1/HV2 sequences could be resolved through HV3 analysis. In addition, while VR1 has lower overall variability (7%), 19 polymorphic sites were identified in the dataset, revealing a high degree of discrimination potential primarily due to a highly polymorphic site at position 16519. This site is one of two in the entire control region that has a frequency approaching 50% in the population, and thus is highly informative.

If DNA sequence in the control region is not sufficient to resolve identical HV1/HV2 profiles, practitioners can look to the coding region for help. For example, mtDNA genomes from 241 individuals who matched common European HV1/HV2 profiles have been sequenced to identify polymorphisms that enhance forensic discrimination (Coble et al. 2004). Individuals with the same HV1/HV2 profile rarely matched across the entire genome. The 13 protein-coding genes in the mitochondrial DNA genome are composed of more than 11,000 nucleotides. When datasets on the mtDB Web site are queried (www.genpat.uu.se/mtDB), approximately 40% of the codon wobble positions show variable sequence (Coble et al. 2006). Therefore, when attention is placed on these neutral positions, eight panels with 7–11 multiplexed SNPs per panel can be designed to provide additional levels of discrimination. The appropriate panel can then be chosen because of its direct association with one or more of the common HV1/HV2 profiles, which helps to conserve sample extracts while providing for maximum discrimination. This added level of separation reduced the frequency of the most common European profiles from ~7% down to ~2%, and the 18 common profiles were resolved into 105 different haplotypes, 55 of which were seen only once (Coble et al. 2004). When including key nonsynonymous SNPs, the total number of haplotypes increased to 127 (Coble et al. 2006). Therefore, it is clear that expanding the range of mtDNA sequence to the coding region has a dramatic impact on resolving identical HV1/HV2 profiles. Of course, the ability to perform coding-region assays is dependent on the quantity of DNA extract available for analysis; said quantity is quite limited in the case of small fragments of hair shaft. In addition, because the majority of the mtDNA genome contains coding information vital to the survival of the cell and may therefore be medically relevant, this tactic for increasing discrimination may need to be assessed for use in forensic analysis. One of the challenges of performing new analytical procedures on forensic casework samples is the availability of commercial kits and technologies. Historically, forensic mtDNA analysis has not relied on commercialization of coupled amplification and sequencing kits, but instead has involved the development of in-house amplification reagent systems (Fisher et al. 1993; Holland et al. 1993; Holland and Parsons 1999). The same holds true for the development of assays that query DNA sequence within the mtDNA coding region. Although effective strategies have been developed by practitioners for the analysis of coding region SNPs using primer extension or SNaPshot approaches (Huang et al. 2010; Vallone et al. 2004), a major concern remains, as each in-house assay contains different target SNPs, and may not include the most informative loci. In addition, a forensic context requires a database upon which the statistical weight of the profiles is based, and while there are large forensic databases for the control region, no comparable searchable database has been prepared for SNP datasets. While many whole genomes are available for databasing, there has not been an effort to collect this information for forensic purposes. Until a kit-based system is commercially available that includes the most informative target loci and a frequency database of profiles is developed from those loci, it is doubtful that coding-region SNP assays will have widespread appeal in the forensic community.

12.3.3 Mixtures and Heteroplasmy Investigation: DGGE and dHPLC

When using the Sanger method of DNA sequence analysis, it is difficult to identify low-level heteroplasmic variants, and it is nearly impossible to resolve mixtures. Each of these phenomena has been the subject of assay development and has resulted in a number of potential solutions. However, no one technique has emerged with widespread acceptance. Other than the laborious yet effective technique of cloning (Gill et al. 1994), the verification and identification of heteroplasmic variants can be accomplished in a variety of ways, but not always by addressing both interests—that is, verification and identification. In the late 1990s, a denaturing gradient gel electrophoresis (DGGE) approach was developed to verify the presence of heteroplasmic variants in HV1, and to provide a means for isolating each variant for subsequent sequence analysis (Steighner et al. 1999). The DGGE technique allows for verification of heteroplasmy through the resolution of mismatch-containing heteroduplices from fully base-paired homoduplices in an increasingly denaturing environment. Detection of heteroplasmy was accomplished down to a minor component proportion of 1%. This level of detection is approximately onefold better than Sanger sequence analysis provides, where variants as low as 5–10% of the total can sometimes be reported. As an illustration of the effectiveness and reliability of the DGGE technique, mixtures of 49 pairs of HV1 sequences, each pair differing by a single polymorphism, were successfully verified as exhibiting heteroplasmy-like characteristics; that is, heteroduplex and homoduplex bands on a DGGE gel. In addition, heteroplasmy was successfully verified in 13 samples known to have heteroplasmic sites. Variant nucleotide positions were identified (confirmed) by reamplifying physically excised homoduplex bands of DNA from DGGE gels and performing sequence analysis on the products.

The DGGE assay has also been used effectively to determine the rate of control-region heteroplasmy in the population (Tully et al. 2000). Heteroplasmy in HV1 was observed in 35 of 253 randomly chosen individuals (sample source was whole blood), or 13.8% of those tested. Given the greater detection level of DGGE, it is not surprising that this is a higher rate of heteroplasmy than the reported rate of 11.4% for the hypervariable regions when using Sanger sequencing on casework samples involving hairs (Melton and Nelson 2005). The identified heteroplasmic sequences revealed single-nucleotide differences in 33 of the 35 individuals tested, whereas two individuals exhibited heteroplasmic sites at two different positions (triplasmy). Heteroplasmy occurred at 16 different nucleotide positions throughout HV1, with the most frequent observations at positions 16093 and 16129, consistent with prior and recently published studies (Holland et al. 2011; Isenberg and Moore 1999). In addition, the majority of heteroplasmic variants occurred at low frequency and could not be detected by conventional sequencing. The study from Tully et al. (2000) was the first to indicate that low-level heteroplasmy in HV1 was more common than was previously believed, and that it occurred across the entire control region, a finding that continues to have importance in evolutionary studies and forensic applications.

It became apparent that the gel-based DGGE system was too arduous for routine mtDNA analysis, so an advanced column-based system was developed using a dHPLC approach (Kristinsson et al. 2009). A total of 920 pairwise combinations of HV1/HV2 amplicons from 95 individuals were assessed for sequence concordance. For combinations of amplicons from individuals who shared identical HV1/HV2 sequences, dHPLC verified sequence concordance. However, for 849 combinations with different sequences, dHPLC was able to detect the presence of sequence nonconcordance in all but 13 samples (98.5%), including the detection of transitions, transversions, insertions, and deletions. This study clearly illustrated the utility of the dHPLC assay as an indicator of mtDNA sequence heteroplasmy, and by extension, the presence of a mixture. In addition, the dHPLC system provided a means for relatively simple fractionation of the individual components of a simple mixture from two individuals by enriching the homoduplices for one variant or another and allowing for subsequent sequence analysis of isolated DNA fragments representing the separate contributions of the two individuals. Given these capabilities, the dHPLC system could in theory be used for screening purposes to determine if two samples are concordant prior to sequence analysis. A dHPLC system, the Transgenomic Wave System 3500 or 4500, is commercially available, making it accessible to practitioners. However, some challenges have been encountered with the interpretation of dHPLC results. Length heteroplasmy, a common mtDNA phenomenon, broadens heteroduplex peaks and is often observed with shoulders representing the multiple variants. This can make the detection of neighboring-point heteroplasmy more difficult to interpret. In addition, the detection level for heteroplasmy using the dHPLC system is only marginally better than Sanger sequencing. Nonetheless, the ability of the dHPLC system to fractionate individual variants is significant because the only other option, historically, has been cloning of heteroplasmic variants or excision of variants from DGGE gels for sequencing.

12.3.4 Mass Spectrometry

The use of mass spectrometry to identify variants of mtDNA sequence has been investigated by forensic laboratories to analyze simple mixtures and heteroplasmy, as well as to develop single-source profiles (Oberacher et al. 2006). More recently, an automated system for high-resolution analysis has been developed (Hall et al. 2009). The nucleotide base composition of DNA fragments is determined after multiplex PCR amplification by electrospray ionization mass spectrometry (ESI-MS), and is commercially available using Abbott's PLEX-ID™ System. The ESI-MS method targets 1,051 nucleotides of DNA sequence within the control region, including HV1 and HV2. Twenty-four overlapping segments of DNA are amplified in eight triplex reactions with a sensitivity of less than 25 pg of genomic DNA per reaction. Automated PCR product purification occurs prior to injection onto the ESI-MS. Mass calculations of individual DNA fragments are converted into base composition values for each amplicon; the full profile is assembled with computer algorithms that recognize and link overlapping end-point homologies for the 24 fragments. The profile can be compared to population databases of composition profiles derived from sequence information, and therefore, can be subjected to the same statistical approach used for assessing the significance of matching mtDNA sequences. Although only 94% of the information obtained by direct sequencing of HV1 and HV2 is detected with the ESI-MS assay, ESI-MS is more informative overall because it covers more than 400 additional base pairs of the control region. The reduced discrimination potential within HV1/HV2 is due to reciprocal nucleotide changes that cause fragment masses to appear unchanged (e.g., C150T, T152C). More importantly, while the ESI-MS system can quantitatively deconvolute heteroplasmic sites, the precise nucleotide differences between samples cannot always be elucidated, as changes in sequence can happen across the length of the DNA fragment being analyzed. On the contrary, the ESI-MS system can effectively resolve length variants in homopolymeric stretches—an attractive feature, as a large percentage of mtDNA profiles include length variants. Therefore, while the assay is not hindered by length heteroplasmy, identifying the location of point heteroplasmy can be a challenge because mass weights are determined rather than the precise order of bases.

The robustness of the ESI-MS method has been tested on more challenging sample types (Howard et al. 2012). In 2009, a project was launched by the Commonwealth War Graves Commission to identify the remains of 250 World War I soldiers recovered from a mass grave in Fromelles, France. A comparative assessment of the performance of Sanger sequencing and the ESI-MS method was conducted on 225 of those skeletal remains. Assessment included the ability to amplify extracted DNA, to develop an mtDNA profile (sequence or base composition), and the ease-of-use associated with each method. The ESI-MS approach fared well during this comparative analysis. The smaller amplicon lengths when using the ESI-MS method are an advantage with degraded DNA (40–100 bp). More than 99% of the 225 skeletal samples produced at least partial results using the ESI-MS method, generating data for at least 75% of the target amplicons. Almost 60% of the samples produced full base-composition profiles. This was as good as or better than the Sanger sequencing results, and is even more compelling given that a miniprimer-set approach was used for the Sanger method (amplicon sizes of 150–225 bp) (Gabriel et al. 2001). It is quite possible that the ESI-MS method would have been superior to the routine primer-set approach using amplicon sizes of approximately 250 bp.

Given the amplification strategies of each method (sequencing or base-composition profiling), a more applicable comparison is the respective coverage rates of generated sequence information. The Sanger and ESI-MS methods produced equivalent levels of DNA sequence: coverage of approximately 98% of the respective ranges of sequence. The only exception was when a small stretch of sequence, which is not covered with the ESI-MS method, was considered (nucleotide positions 16251–16253). Overall, the ESI-MS method was easy to use, and was highly automated. However, the instrument is relatively complex from an engineering perspective, and is quite expensive to procure, so it is unclear how these factors will impact laboratory operations. In addition, while the ESI-MS method has a higher overall discrimination potential than Sanger sequencing of HV1 and HV2 alone, databases of base composition profiles do not currently exist that can be used for comparison purposes. Fortunately, work is progressing forward to address this deficiency. In the short term it has been recommended that the ESI-MS method be used as a rapid high-throughput screening tool prior to conventional sequence analysis (Hall et al. 2009).

12.3.5 Pyrosequencing and Deep Sequencing

Pyrosequencing techniques have been used to quantify SNP profiles in the mtDNA coding region and to detect mixture or heteroplasmic variants in both the coding and control regions (Andréasson et al. 2006, 2007). The pyrosequencing method involves sequential introduction of the four nucleotides (dATP, dCTP, dTTP, and dGTP), followed by a cascading series of events that will lead to the emission of light when a nucleotide is incorporated into the newly synthesized strand of DNA. With the incorporation of a nucleotide, through the action of DNA polymerase, pyrophosphate is released as a byproduct that is fed into a coupled enzymatic pathway. The route was initially a three-enzyme system, but in some applications moved to a two-enzyme approach to increase read lengths (Mashayekhi and Ronaghi 2007). The three original enzymes included *sulfurylase* to convert the pyrophosphate into ATP in the presence of ASP (adenosine-5'-phosphosulfate), then *luciferase* to convert luciferin to oxyluciferin in the presence of ATP with the release of photons of light, and finally, *apyrase* to digest unincorporated nucleotides between sequential steps in the sequencing process. However, it turns out that apyrase is an inefficient enzyme, so read lengths of the early pyrosequencing method were limited to approximately 100 bp. Replacing the use of apyrase with a wash step between nucleotide additions greatly increased read lengths, although the wash step reduced overall yield due to loss of template. As discussed below, the advent of next-generation sequencing instruments has helped to resolve this problem.

A pyrosequencing method for coding region analysis, comprising 17 sequencing reactions performed on 15 PCR fragments, was used to increase the potential for separating similar HV1/HV2 profiles (Andréasson et al. 2007). The assay was performed on 135 samples, 60 of which had zero to one difference from the HV1/HV2 reference sequence (Anderson et al. 1981; Andrews et al. 1999), while the other 75 samples had two differences from the reference. An average read length of 81 of 165 nucleotides was obtained from each sample, with a range of 20–120 bases. A total of 52 SNP sites were identified, of which 18 had a single SNP variant. This is a significant increase in discrimination potential when compared to methods that employ primer-extension assays, but it remains to be seen whether the target sites are well suited for a wide range of populations. Most importantly, for the 60 samples with zero or one sequence differences in the control region, only 12 samples (20%) could not be resolved through the addition of at least one coding-region difference. Therefore, the use of this pyrosequencing-based coding region approach may effectively enable the differentiation of samples with similar HV1/HV2 profiles.

An easy-to-use and rapid pyrosequencing method has also been developed to assess the linear relationship between incorporated nucleotides and released light, allowing for quantification of variants in mtDNA mixtures or samples with heteroplasmy (Andréasson et al. 2006). The assay was designed for five PCR amplified targets, ranging in size from 200–310 bp, to query seven variable positions in the control and coding regions. For all detected SNPs, the measured mixture ratios were consistent with the expected, providing reliable quantification data. However, a significant drawback of this method is the relatively low level of mixture detection, similar to that of Sanger sequencing. Minor mixture components less than 10% of the total DNA content are not well resolved, reducing the value of the pyrosequencing system.

When investigating mtDNA mixtures or heteroplasmy, the practitioner would like access to methods that are sensitive enough to detect low-level variants, and are precise enough to identify the variants. It is quite possible that deep-DNA-sequencing approaches will address both of these interests in the future. However, before the forensic community can rely on the next-generation sequencing platforms for routine mtDNA analysis, forensic standards must be applied, and forensic concerns addressed. For example, in a *Nature* paper from 2010 it was reported that intra-individual heteroplasmic variation was frequently observed at levels of around 1–2% when employing the Genome Analyzer from Illumina (San Diego, California) to sequence the mtDNA genome of CEPH families and soft-tissue samples from the same person (He et al. 2010). While this claim was consistent with previous studies using DGGE analysis (unpublished data from M. Holland), later phylogenetic assessment of the reported data revealed that, on average, at least five polymorphic sites were missed in the reported sequences (Bandelt and Salas 2012). To address the higher error rates commonly observed with deep-DNA-sequencing approaches, a second study reported an assessment of the accuracy of mtDNA-sequencing results generated on the Genome Analyzer II from Illumina (Li et al. 2010). It was determined that minor components down to 1–2% could be reliably reported. Therefore, it is possible that the samples from the previous study were contaminated (He et al. 2010), or that errors in interpretation of the data had occurred. Regardless of the reasons for the discrepancies, neither of these studies suitably addressed forensic standards or concerns.

In a more recent study, the HV1 segment of the mtDNA control region was sequenced using the pyrosequencing-based 454 GS Junior instrument from Roche Applied Science (Indianapolis, Indiana) (Holland et al. 2011). Mock mixtures were employed to evaluate the ability of the 454 method to deconvolute variant components and to reliably detect heteroplasmy. Amplicon sequencing was performed on full-length HV1 amplicons (approximately 400 bp). The amplification primers included multiplex identifier (MID) sequences to allow for multiplexing, and adaptor sequences for the three-enzyme pyrosequencing process. The 454 method uses a wash step to enhance the removal of residual dNTPs and ATP, to assure longer read lengths. Sensitivity levels of the 454 instrument are maintained by reducing loss of template during the washing step between each addition of a new nucleotide, a process that is accomplished by confining the DNA template to a fixed microbead. A lens array is used to focus the generated light or luminescence from each well of the picotiter plate onto the chip of a CCD (charge-coupled device) camera. The CCD camera captures light and records it in a raw data output file, resulting in a series of pyrograms. Intensity of peaks in the pyrograms is proportional to the number of nucleotides incorporated during the sequencing events. Using this method on mock mixtures, the different contributors were detected down to a minor component ratio of 1:250, or 0.4% (40 minor variant copies with a coverage rate of 10,000 sequences), and could be identified down to a 1:1,000 ratio (0.1%) with expanded coverage.

The 454 method was also used to analyze 30 individuals from 25 different maternal lineages (Holland et al. 2011). Low-level heteroplasmy was detected for 11 of the 25 lineages, a 44% rate of observed heteroplasmy. Minor component variants ranged from 0.33% to 20% of the total sequencing reads. When using the Sanger method on these same samples, only one sample exhibited heteroplasmy at a detectable level (the one sample with 20% heteroplasmy), equating to a 4% rate of observed heteroplasmy. The nucleotide positions where heteroplasmy was observed were consistent with mutational hotspots or sites where forensic polymorphisms and heteroplasmy have been observed in past studies (Parsons et al. 1997; Tully et al. 2000). Concerns regarding the reliability of the sequence data were at least partially addressed through reproducibility studies. Multiple samples were run in either duplicate or triplicate, with results confirming the positions of heteroplasmy, and at very similar minor component percentages. Therefore, polymerase-driven artifacts were ruled out as the source of the relatively high frequency of observed heteroplasmy. In addition, the coverage rates and total number of reads for all reported positions of heteroplasmy were high. In part, this was due to the fact that all reported instances of low-level heteroplasmy resulted from at least 40 reads of sequence (most with more than 100 reads), and with a balanced ratio of forward and reverse reads. These data and observations allowed for the development of initial standards for reporting low-level heteroplasmic variants in a forensic setting. Recommended reporting criteria include a requirement that low-level variants be reported only when at least 40 reads are generated, and when the ratio of forward to reverse reads is consistent with the total read ratio. For example, mixtures of 1:100, 1:250, 1:500, and 1:1,000 would require total coverage of at least 4,000, 10,000, 20,000, and 40,000 reads, respectively. This level of coverage is well within the capability of the 454 instruments.

The concept of resolving mtDNA mixtures is quite different than identifying low-level heteroplasmic variants. Depending on the mixture ratio of the two or more components, it may be difficult to identify which component goes with which contributor, especially when the ratios are closer to 1:1. Therefore, a phylogenetic approach may be necessary to identify which sequences from which amplicons are associated with the same individual. Algorithms can be developed to perform this type of assessment, although it will become increasingly more difficult to tease apart the components when the individuals involved originate from the same related population group. As a precautionary step, it is currently prudent to restrict the reporting of mixture data using the 454 method to minor component ratios above 1:100, and when the ratio stays below 1:5. In addition, more work is necessary to address sequencing error rates for the 454 system, and how the components of mixtures can be effectively resolved. For example, the pyrosequencing approach results in poor resolution of homopolymeric sequences, and both PCR and sequencing artifacts require a filtering mechanism similar to that for STR stutter and spectral bleed-through. In addition, chimeric sequences from jumping PCR are quite commonly observed and must be addressed before making the 454 method an operational system in forensic laboratories.

12.4 Conclusions

The most significant challenge for mtDNA analysis remains the high cost and low throughput for evidentiary samples. Many samples require some special handling of one form or another, whether it is accommodation for degradation via miniprimer sets, additional internal sequence replication to cover indels or length heteroplasmy on duplicate strands, extra scrutiny of site heteroplasmy, or management of minimal contamination that is detected in reagent blanks. The system is robust when a sample contains good-quality abundant DNA, but also works well for difficult samples when all possible extra steps for recovering a profile are applied. Cleaning and DNA extraction of samples are the most critical steps, and it is difficult to automate these steps without risking cross-contamination due to the high number of manipulations required for hairs and skeletal samples. Postextraction improvements in PCR amplification and sequencing, or the introduction of new methods that do not require abundant template for detection of variation at the level of fine discrimination, are desirable. Several methods that confront these challenges have been proposed, such as PCR multiplexing of additional hypervariable-region fragments either for mtDNA alone or in conjunction with STRs (Berger and Parson 2009; von Wurmb-Schwark et al. 2009). The advantage of these methods is that they recognize the level of challenge presented by the worst samples (the least common denominator) and could in theory be applied to all samples. Overall, the limitations are more profound for single shed hairs than for skeletal remains, because skeletal material is often virtually unlimited in any single case.

Significant growth of forensic mtDNA databases to serve as foundations for population statistics continues to be a priority. These datasets, such as the one exemplified by EMPOP, should be freely available online for all users and highly vetted for quality data, and they will continue to be newly developed around the world for the growing forensics community. There is need for a statistical method that appropriately weights a failure to exclude when a forensic case contains both questioned and known samples with the same site heteroplasmy. A very good understanding of the mutation rates of most polymorphic human hypervariable sites has been gained over the last 20 years since the original *Forensic Science Review* article on the validation of forensic mtDNA (Holland and Parsons 1999); simply rank-ordering polymorphic sites by their rate of change could be easily accomplished from published data (Soares et al. 2009) to determine an effective, realistic, and conservative multiplier or additive for current statistical applications.

Newly developed and forensically validated methods such as ESI-MS have great potential for high-throughput applications where samples are abundant, such as skeletal remains from mass graves, missing persons projects, mass disasters, and other situations where rapid reassociation of thousands of skeletal remains is desired. Of course, all such applications should be firmly grounded in an understanding of the mtDNA variation that is present in the populations to which they are applied; for example, a description and frequency estimates of common mtDNA types and any population substructuring. Next-generation sequencing is promising for the elucidation of mixture characteristics and heteroplasmy, as the high throughput with this system allows scrutiny of DNA in, effectively, thousands of single strands. Using appropriate computer algorithms, overlapping amplicons, and phylogenetic contexts, mixture deconvolution may for the first time become routine, expanding the pool of samples with degraded and insufficient nuclear DNA for mitochondrial DNA analysis.

Acknowledgments

The authors would like to thank their many colleagues in the forensic mtDNA community who have contributed greatly to the research and improvements of the mtDNA analysis process over the last two decades.

References

Alonso, A., P. Martin, C. Albarrán et al. 2004. Real-time PCR designs to estimate nuclear and mitochondrial DNA copy number in forensic and ancient DNA studies. *Forensic Sci Int* 139:141–9.

Anderson, S., A. T. Bankier, B. G. Barrell et al. 1981. Sequence and organization of the human mitochondrial genome. *Nature* 290:457–65.

Andréasson, H., M. Nilsson, B. Budowle, S. Frisk, and M. Allen. 2006. Quantification of mtDNA mixtures in forensic evidence material using pyrosequencing. *Int J Legal Med* 120:383–90.

Andréasson, H., M. Nilsson, H. Styrman, U. Pettersson, and M. Allen. 2007. Forensic mitochondrial coding region analysis for increased discrimination using pyrosequencing technology. *Forensic Sci Int Genet* 1:35–43.

Andrew, T., C. D. Calloway, S. Stuart et al. 2011. A twin study of mitochondrial DNA polymorphisms shows that heteroplasmy at multiple sites is associated with mtDNA variant 16093 but not with zygosity. *PLoS One* 6:e22332.

Andrews, R. M., I. Kubacka, P. F. Chinnery et al. 1999. Reanalysis and revision of the Cambridge reference sequence for human mitochondrial DNA (Correspondence). *Nat Genet* 23:147.

Bandelt, H. J. and W. Parson. 2008. Consistent treatment of length variants in the human mtDNA control region: A reappraisal. *Int J Leg Med* 122:11–21.

Bandelt, H. J. and A. Salas. 2012. Current next generation sequencing technology may not meet forensic standards. *Forensic Sci Int Genet* 6:143–5.

Berger, C. and W. Parson. 2009. Mini-midi-mito: Adapting the amplification and sequencing strategy of mtDNA to the degradation state of crime scene samples. *Forensic Sci Int Genet* 3:149–53.

Brenner, C. H. 2010. Fundamental problem of forensic mathematics—The evidential value of a rare haplotype. *Forensic Sci Int Genet* 4:281–91.

Budowle, B., D. Polanskey, C. L. Fisher, B. K. Den Hartog, R. B. Kepler, and J. W. Elling. 2010. Automated alignment and nomenclature for consistent treatment of polymorphisms in the human mitochondrial DNA control region. *J Forensic Sci* 55:1190–5.

Butler, J. 2012. Mitochondrial DNA analysis. In Butler JM: *Advanced Topics in Forensic DNA Typing: Methodology.* Elsevier Academic Press: San Diego, CA, pp. 405–456.

Butler, J. M., M. R. Wilson, and D. J. Reeder. 1998. Rapid mitochondrial DNA typing using restriction enzyme digestion of polymerase chain reaction amplicons followed by capillary electrophoresis separation with laser-induced fluorescence detection. *Electrophoresis* 19:119–24.

Clopper, C. and E. Pearson. 1934. The use of confidential or fiducial limits illustrated in the case of the binomial. *Biometrika* 26:404–13.

Coble, M. D., R. S. Just, J. E. O'Callaghan et al. 2004. Single nucleotide polymorphisms over the entire mtDNA genome that increase the power of forensic testing in Caucasians. *Int J Leg Med* 118:137–46.

Coble, M. D., P. M. Vallone, R. S. Just, T. M. Diegoli, B. C. Smith, and T. J. Parsons. 2006. Effective strategies for forensic analysis in the mitochondrial DNA coding region. *Int J Legal Med* 120:27–32.

Comas, D., S. Pääbo, and J. Betranpetit. 1995. Heteroplasmy in the control region of human mitochondrial DNA. *Genome Res* 5:89–90.

Danielson, P. B., H.-Y. Sun, T. Melton, and R. Kristinsson. 2007. Resolving mtDNA mixtures by denaturing high-performance liquid chromatography and linkage phase determination. *Forensic Sci Int Genet* 1:148–53.

Davis, C. L. 1998. Mitochondrial DNA: State of Tennessee v. Paul Ware. *Profiles in DNA* 1:6–7.

de Camargo, M. A., G. G. Paneto, A. C. de Mello, J. A. Martins, W. Barcellos, and R. M. Cicarelli. 2011. No relationship found between point heteroplasmy in mitochondrial DNA control region and age range, sex and haplogroup in human hairs. *Mol Biol Rep* 38:1219–23.

Divne, A.-M., M. Nilsson, C. Calloway, R. Reynolds, H. Erlich, and M. Allen. 2005. Forensic casework analysis using the HVI/HVII mtDNA linear array assay. *J Forensic Sci* 50:548–54.

Egeland, T. and A. Salas. 2008. Estimating haplotype frequency and coverage of databases. *PLoS One* 3:e3988.

Egeland, T. and A. Salas. 2011. A statistical framework for the interpretation of mtDNA mixtures: Forensic and medical applications. *PLoS One* 6:e26723.

Eichmann, C., B. Berger, M. Steinlechner, and W. Parson. 2005. Estimating the probability of identity in a random dog population using 15 highly polymorphic canine STR markers. *Forensic Sci Int* 151:37–44.

Excoffier, L. and Z. Yang. 1999. Substitution rate variation among sites in mitochondrial hypervariable region I of humans and chimpanzees. *Mol Biol Evol* 16:1357–68.

Fisher, D. L., M. M. Holland, L. Mitchell et al. 1993. Extraction, evaluation, and amplification of DNA from decalcified and undecalcified United States Civil War bone. *J Forensic Sci* 38:60–8.

Forster, L., P. Forster, S. M. Gurney et al. 2010. Evaluating length heteroplasmy in the human mitochondrial DNA control region. *Int J Legal Med* 124:133–42.

Gabriel, M. N., C. D. Calloway, R. L. Reynolds, and D. Primorac. 2003. Identification of human remains by immobilized sequence-specific oligonucleotide probe analysis of mtDNA hypervariable regions I and II. *Croat Med J* 44:293–8.

Gabriel, M. N., E. F. Huffine, J. H. Ryan, M. M. Holland, and T. J. Parsons. 2001. Improved mtDNA sequence analysis of forensic remains using a "mini-primer set" amplification strategy. *J Forensic Sci* 46:247–53.

Gilbert, M., R. C. Janaway, D. J. Tobin, A. Cooper, and A. C. Wilson. 2006. Histological correlates of post mortem mitochondrial DNA damage in degraded hair. *Forensic Sci Int* 156:201–7.

Gilbert, M. T., E. Willerslev, A. J. Hansen et al. 2003a. Distribution patterns of postmortem damage in human mitochondrial DNA. *Am J Hum Genet* 72:32–47.

Gilbert, M. T. P., A. J. Hansen, E. Willerslev et al. 2003b. Characterization of genetic miscoding lesions caused by postmortem damage. *Am J Hum Genet* 72:48–61.

Gilbert, M. T. P., B. Shapiro, A. Drummond, and A. Cooper. 2005. Post-mortem DNA damage hotspots in Bison (*Bison bison*) provide evidence for both damage and mutational hotspots in human mitochondrial DNA. *J Archaeol Sci* 32:1053–60.

Gill, P., P. L. Ivanov, C. Kimpton et al. 1994. Identification of the remains of the Romanov family by DNA analysis. *Nat Genet* 6:130–5.

Grzybowski, T. 2000. Extremely high levels of human mitochondrial DNA heteroplasmy in single hair roots. *Electrophoresis* 21:548–53.

Grzybowski, T., B. Malyarchuk, J. Czarny, D. Miscicka-Sliwka, and R. Kotzbach. 2003. High levels of mitochondrial DNA heteroplasmy in single hair roots: Reanalysis and revision. *Electrophoresis* 24:1159–65.

Gundry, R. L., M. W. Allard, T. R. Moretti et al. 2007. Mitochondrial DNA analysis of the domestic dog: Control region variation within and among breeds. *J Forensic Sci* 52:562–72.

Hall, T. A., K. A. Sannes-Lowery, L. D. McCurdy et al. 2009. Base composition profiling of human mitochondrial DNA using polymerase chain reaction and direct automated electrospray ionization mass spectrometry. *Anal Chem* 81:7515–26.

Handt, O., M. Krings, R. H. Ward, and S. Paabo. 1996. The retrieval of ancient human DNA sequences. *Am J Hum Genet* 59:368–76.

Hansen, A., E. Willerslev, C. Wiuf, T. Mourier, and P. Arctander. 2001. Statistical evidence for miscoding lesions in ancient DNA templates. *Mol Biol Evol* 18:262–5.

Hatsch, D., S. Amory, C. Keyser, R. Hienne, and L. Bertrand. 2007. A rape case solved by mitochondrial DNA mixture analysis. *J Forensic Sci* 52:891–4.

He, Y., J. Wu, D. C. Dressman et al. 2010. Heteroplasmic mitochondrial DNA mutations in normal and tumour cells. *Nature* 464:610–4.

Himmelberger, A. L., T. F. Spear, J. A. Satkoski et al. 2008. Forensic utility of the mitochondrial hypervariable region 1 of domestic dogs, in conjunction with breed and geographic information. *J Forensic Sci* 53:81–9.

Hofreiter, M., V. Jaenicke, D. Serre, A. von Haeseler, and S. Paabo. 2001. DNA sequences from multiple amplifications reveal artifacts induced by cytosine deamination in ancient DNA. *Nucleic Acids Res* 29:4793–9.

Holland, M. M., D. L. Fisher, L. G. Mitchell et al. 1993. Mitochondrial DNA sequence analysis of human skeletal remains: Identification of remains from the Vietnam War. *J Forensic Sci* 38:542–53.

Holland, M. M., M. R. McQuillan, and K. A. O'Hanlon. 2011. Second generation sequencing allows for mtDNA mixture deconvolution and high resolution detection of heteroplasmy. *Croat Med J* 52:299–313.

Holland, M. M. and T. J. Parsons. 1999. Mitochondrial DNA sequence analysis—Validation and use for forensic casework. *Forensic Sci Rev* 11:21–50.

Homer, N., S. Szelinger, M. Redman et al. 2008. Resolving individuals contributing trace amounts of DNA to highly complex mixtures using high-density SNP genotyping microarrays. *PLoS Genet* 4:e1000167.

Hoss, M., P. Jaruga, T. H. Zastawny, M. Dizdaroglu, and S. Paabo. 1996. DNA damage and DNA sequence retrieval from ancient tissues. *Nucleic Acids Res* 24:1304–7.

Houck, M. A. and B. Budowle. 2002. Correlation of microscopic and mitochondrial DNA hair comparisons. *J Forensic Sci* 47:964–7.

Howard, R., V. Encheva, J. Thomson et al. 2012. Comparative analysis of human mitochondrial DNA from World War I bone samples by DNA sequencing and ESI-TOF mass spectrometry. *Forensic Sci Int Genet;* In press (doi: 10.1016/j.fsigen.2011.05.009).

Huang, D., C. Gui, S. Yi, Q. Yang, R. Yang, and K. Mei. 2010. Typing of 24 mtDNA SNPs in a Chinese population using SNaPshot minisequencing. *J Huazhong Univ Sci Technolog Med Sci* 30:291–8.

Irwin, J. A., J. L. Saunier, H. Niederstatter et al. 2009. Investigation of heteroplasmy in the human mitochondrial DNA control region: A synthesis of observations from more than 5000 global population samples. *J Mol Evol* 68:516–27.

Isenberg, A. R. and J. M. Moore. 1999. Mitochondrial DNA analysis at the FBI laboratory. *Forensic Science Communication* 1 (http://www2.fbi.gov/hq/lab/fsc/backissu/july1999/dnalist.htm).

Kaye, D., V. Hans, B. Dann, E. Farley, and S. Albertson. 2007. Statistics in the jury box: How jurors respond to mitochondrial DNA match probabilities. *J Empir Legal Stud* 4:797–834.

Kim, K. S., S. E. Lee, H. W. Jeong, and J. H. Ha. 1998. The complete nucleotide sequence of the domestic dog (*Canis familiaris*) mitochondrial genome. *Mol Phylogenet Evol* 10: 210–20.

Kline, M. C., P. M. Vallone, J. W. Redman, D. L. Duewer, C. D. Calloway, and J. M. Butler. 2005. Mitochondrial DNA typing screens with control region and coding region SNPs. *J Forensic Sci* 50:377–85.

Krings, M., A. Stone, R. W. Schmitz, H. Krainitzki, M. Stoneking, and S. Pääbo. 1997. Neandertal DNA Sequences and the Origin of Modern Humans. *Cell* 90:19–30.

Kristinsson, R., S. E. Lewis, and P. B. Danielson. 2009. Comparative analysis of the HV1 and HV2 regions of human mitochondrial DNA by denaturing high-performance liquid chromatography. *J Forensic Sci* 54:28–36.

Li, M., A. Schonberg, M. Schaefer, R. Schroeder, I. Nasidze, and M. Stoneking. 2010. Detecting heteroplasmy from high-throughput sequencing of complete human mitochondrial DNA genomes. *Am J Hum Genet* 87:237–49.

Lindahl, T. 1993. Instability and decay of the primary structure of DNA. *Nature* 362:709–15.

Loreille, O. M., T. M. Diegoli, J. A. Irwin, M. D. Coble, and T. J. Parsons. 2007. High efficiency DNA extraction from bone by total demineralization. *Forensic Sci Int* 1:191–5.

Lutz, S., H.-J. Weisser, J. Heizmann, and S. Pollak. 1998. Location and frequency of polymorphic positions in the mtDNA control region of individuals from Germany. *Int J Leg Med* 111:67–77.

Lutz, S., H. Wittig, H.-J. Weisser et al. 2000. Is it possible to differentiate mtDNA by means of HVIII in samples that cannot be distinguished by sequencing the HVI and HVII regions? *Forensic Sci Int* 113:97–101.

Mashayekhi, F. and M. Ronaghi. 2007. Analysis of read length limiting factors in Pyrosequencing chemistry. *Anal Biochem* 363:275–87.

Melton, T. 2004. Mitochondrial DNA heteroplasmy. *Forensic Sci Rev* 16:1–20.

Melton, T. 2006. Mitochondrial DNA examination of cold case crime scene hairs. In Walton R (ed.): *Cold Case Homicides: Practical Investigative Techniques*. CRC Press: Boca Raton, FL.

Melton, T., G. Dimick, B. Higgins, M. Yon, and C. Holland. 2012. Mitochondrial DNA analysis of 114 hairs measuring less than 1 cm from a 19-year-old homicide. *Investigative Genetics* 3:12 (http://www.investigativegenetics.com/content/3/1/12).

Melton, T. and C. Holland. 2007. Routine forensic use of the mitochondrial 12S ribosomal RNA gene for species identification. *J Forensic Sci* 52:1305–7.

Melton, T., C. Holland, and K. Nelson. 2006. Commentary on: Divne, A.-M., M. Nilsson, C. Calloway, R. Reynolds, H. Erlich, and M. Allen. Forensic casework analysis using the HVI/HVII mtDNA linear array assay. *J Forensic Sci* 51:935–6.

Melton, T. and K. Nelson. 2001. Forensic mitochondrial DNA analysis: Two years of commercial casework experience in the United States. *Croat Med J* 42:298–303.

Melton, T. and K. Nelson. 2005. Forensic mitochondrial DNA analysis of 691 casework hairs. *J Forensic Sci* 50:73–80.

Meyer, S., G. Weiss, and A. von Haeseler. 1999. Pattern of nucleotide substitution and rate heterogeneity in the hypervariable regions I and II of human mtDNA. *Genetics* 152:1103–10.

Monson, K. L., K. W. P. Miller, M. R. Wilson, J. A. DiZinno, and B. Budowle. 2002. The mtDNA population database: An integrated software and database resource for forensic comparison. *Forensic Sci Commun* 4 (http://www.fbi.gov/about-us/lab/forensic-science-communications/fsc/april2002/miller1.htm).

Mosquera-Miguel, A., V. Alvarez-Iglesias, M. Cerezo, M. Laureu, A. Carracedo, and A. Salas. 2009. Testing the performance of mtSNP minisequencing in forensic samples. *Forensic Sci Int Genet* 3:261–4.

Mueller, O., K. Hahnenberger, M. Dittmann et al. 2000. A microfluidic system for high-speed reproducible DNA sizing and quantitation. *Electrophoresis* 21:128–34.

Nelson, K. and T. Melton. 2007. Forensic mitochondrial DNA analysis of 116 casework skeletal samples. *J Forensic Sci* 52:557–61.

Nilsson, M., H. Andréasson-Jansson, M. Ingman, and M. Allen. 2008. Evaluation of mitochondrial DNA coding region assays for increased discrimination in forensic analysis. *Forensic Sci Int Genet* 2:1–8.

Oberacher, H., H. Niederstatter, C. G. Huber, and W. Parson. 2006. Accurate determination of allelic frequencies in mitochondrial DNA mixtures by electrospray ionization time-of-flight mass spectrometry. *Anal Bioanal Chem* 384:1155–63.

Paabo, S. 1989. Ancient DNA: Extraction, characterization, molecular cloning, and enzymatic amplification. *Proc Natl Acad Sci USA* 86:1939–43.

Parson, W. and A. Dür. 2007. EMPOP—A forensic mtDNA database. *Forensic Sci Int Genet* 1:88–92.

Parsons, T. J. and M. D. Coble. 2001. Increasing the forensic discrimination of mitochondrial DNA testing through analysis of the entire mitochondrial DNA genome. *Croat Med J* 42:304–9.

Parsons, T. J., D. S. Muniec, K. Sullivan et al. 1997. A high observed substitution rate in the human mitochondrial DNA control region. *Nat Genet* 15:363–8.

Roberts, K. and C. Calloway. 2011. Characterization of mitochondrial DNA sequence heteroplasmy in blood tissue and as a function of hair morphology in four population groups. *J Forensic Sci* 56: 46–60.

Roberts, K. A. and C. Calloway. 2007. Mitochondrial DNA amplification success rate as a function of hair morphology. *J Forensic Sci* 52:40–7.

Röck, A., J. Irwin, A. Dür, T. Parsons, and W. Parson. 2011. SAM: String-based sequence search algorithm for mitochondrial DNA database queries. *Forensic Sci Int Genet* 5:126–32.

Saiki, R. K., P. S. Walsh, C. H. Levenson, and H. A. Erlich. 1989. Genetic analysis of amplified DNA with immobilized sequence-specific oligonucleotide probes. *Proc Natl Acad Sci USA* 86:6230–4.

Salas, A. and J. Amigo. 2010. A reduced number of mtSNPs saturates mitochondrial DNA haplotype diversity of worldwide population groups. *PLoS One* 5:e10218.

Seo, S. B., B. S. Jang, A. Zhang et al. 2010. Alterations of length heteroplasmy in mitochondrial DNA under various amplification conditions. *J Forensic Sci* 55:719–22.

Soares, P., L. Ermini, N. Thomson et al. 2009. Correcting for purifying selection: An improved human mitochondrial molecular clock. *Am J Hum Genet* 84:740–59.

Steighner, R. J., L. A. Tully, J. D. Karjala, M. D. Coble, and M. M. Holland. 1999. Comparative identity and homogeneity testing of the mtDNA HV1 region using denaturing gradient gel electrophoresis. *J Forensic Sci* 44:1186–98.

Stoneking, M. 1996. Mitochondrial DNA heteroplasmy: Out of the closet. *Biol Chem* 377:603–4.

SWGDAM: Guidelines for mitochondrial DNA (mtDNA) nucleotide sequence interpretation. 2003. *Forensic Sci Commun* 5 (http://www.fbi.gov/about-us/lab/forensic-science-communications/fsc/april2003/swgdammitodna.htm).

Szibor, R., M. Michael, I. Plate, H. Wittig, and D. Krause. 2003. Identification of the minor component of a mixed stain by using mismatch primer-induced restriction sites in amplified mtDNA. *Int J Leg Med* 117:160–4.

Tarditi, C. R., R. A. Grahn, J. J. Evans, J. D. Kurushima, and L. A. Lyons. 2011. Mitochondrial DNA sequencing of cat hair: An informative forensic tool. *J Forensic Sci* 56(Suppl 1):S36–46.

Tully, G., W. Bär, B. Brinkmann et al. 2001. Considerations by the European DNA profiling (EDNAP) group on the working practices, nomenclature and interpretation of mitochondrial DNA profiles. *Forensic Sci Int* 124:83–91.

Tully, L. A., T. J. Parsons, R. J. Steighner, M. M. Holland, M. A. Marino, and V. L. Prenger. 2000. A sensitive denaturing gradient-gel electrophoresis assay reveals a high frequency of heteroplasmy in hypervariable region 1 of the human mtDNA control region. *Am J Hum Genet* 67:432–43.

Vallone, P. M., R. S. Just, M. D. Coble, J. M. Butler, and T. J. Parsons. 2004. A multiplex allele-specific primer extension assay for forensically informative SNPs distributed throughout the mitochondrial genome. *Int J Leg Med* 118:147–57.

von Wurmb-Schwark, N., A. Preusse-Prange, A. Heinrich, E. Simeoni, T. Bosch, and T. Schwark. 2009. A new multiplex-PCR comprising autosomal and y-specific STRs and mitochondrial DNA to analyze highly degraded material. *Forensic Sci Int Genet* 3:96–103.

Walker, J. A., R. K. Garber, D. J. Hedges, G. E. Kilroy, J. Xing, and M. A. Batzer. 2004. Resolution of mixed human DNA samples using mitochondrial DNA sequence variants. *Anal Biochem* 325:171–3.

Wang, J., V. Venegas, F. Li, and L. J. Wong. 2011. Analysis of mitochondrial DNA point mutation heteroplasmy by ARMS quantitative PCR. *Curr Protoc Hum Genet* Jan; Chapter 19:Unit 19.6.

Wilson, M. R., M. W. Allard, K. Monson, K. W. P. Miller, and B. Budowle. 2002a. Recommendations for consistent treatment of length variants in the human mitochondrial DNA control region. *Forensic Sci Int* 129:35–42.

Wilson, M. R., M. W. Allard, K. L. Monson, K. W. P. Miller, and B. Budowle. 2002b. Further discussion of the consistent treatment of length variants in the human mitochondrial DNA control region. *Forensic Sci Commun* 4 (http://www.fbi.gov/about-us/lab/forensic-science-communications/fsc/oct2002/wilson.htm).

Applications of Autosomal SNPs and Indels in Forensic Analysis

13

CHRISTOPHER PHILLIPS

Contents

13.1 Introduction 280
13.1.1 A Brief Overview of the Definition of SNPs 280
13.1.2 Viable Forensic SNP Genotyping Systems in 2012 280
13.1.3 The Original Rationale of Forensic SNP Analysis: Typing Highly Degraded DNA 282
13.2 Indels: Combining the Advantages of Direct PCR-to-CE Typing and Short Binary Loci 286
13.3 Enhancing the Discrimination Power of STRs Using SNPs 287
13.3.1 SNPs In or Around STRs 287
13.3.2 Adding SNPs to Extend the Polymorphisms Available for Complex Relationship Testing 289
13.4 Utility of SNP Databases 291
13.4.1 Online SNP Data and How to Access It for Forensic Purposes 291
13.4.2 Fine-Scale Genetic Maps for Forensic Markers Using HapMap High-Density SNP Data 295
13.5 Ancestry-Informative Marker SNPs 297
13.6 Tri-Allelic SNPs 303
13.7 SNP Typing Applied to Predictive Forensic Tests for Common Pigmentation Variation 304
References 307

Abstract: The potential applications of short binary markers to forensic analysis are reviewed. Short binary markers are the most common human genomic variation and include single nucleotide polymorphisms (SNPs) and insertion/deletion polymorphisms (indels). This review outlines their use and performance in typing highly degraded DNA—the original rationale for developing SNPs for forensic analysis—as well as their ability to infer the ancestry or likely pigmentation characteristics of an individual not present on a national DNA database, thus potentially providing investigative leads. Throughout the review, reference is made to short indels as a new and potentially powerful alternative to SNPs for enhancing short tandem repeat (STR) results by using a simple amplification to capillary electrophoresis (PCR-to-CE) technique that retains the direct relationship between input DNA and signal strength, offering much-improved mixture-detection capabilities while retaining the favorable characteristics of short-amplicon polymerase chain reaction (PCR).

13.1 Introduction

13.1.1 A Brief Overview of the Definition of SNPs

This review covers markers that can be described as short binary variants. Such a broad definition allows the review to encompass both single nucleotide polymorphisms (SNPs) and small-scale binary insertion/deletion polymorphisms (indels). Indels are of increasing forensic interest due to their very simple direct genotyping approach, their good performance in typing degraded DNA, and their markedly improved sensitivity and precision detecting mixtures compared to SNPs. The term SNP sometimes also refers to *simple* nucleotide polymorphisms, notably in the Santa Cruz genome variant catalog (URL1). However, this definition is designed to be inclusive of all short markers and can thus be confusing, therefore, Figure 13.1 outlines the difference between the definition of *simple* and *single* nucleotide polymorphisms in this context. Short binary indels include loci with a mean allele size range of 1 to 5 bases, with rarer longer alleles up to ~25 bases, while indels sometimes show three or more alleles. Herein, the term indel refers to loci with short alleles, commonly 2 to 5 base differences, but occasionally extending beyond this range. For example, three component loci of the first commercial forensic indel kit are insertion alleles of 15, 18, and 22 nucleotides in length.

13.1.2 Viable Forensic SNP Genotyping Systems in 2012

Despite the ever-widening choice of SNP genotyping systems available in the vast field of genetic analysis, the strategy adopted by the majority of association studies applied to medical genetics research has centered on a consensus approach that has relevance for many aspects of forensic SNP typing. First, the ancestry of study subjects can be assessed, generally using panels comprising sets of small multiplexes of SNPs highly informative for ancestry (AIM-SNPs) to ensure that case and control groups are not stratified for ancestry—such stratification potentially creates false associations between any population-differentiated loci and the disease or trait investigated (Campbell et al. 2005; Marchini et al. 2004). Rebalancing the ancestry compositions of case and control groups by rearranging or excluding individuals that are incorrectly assigned to a population can save time

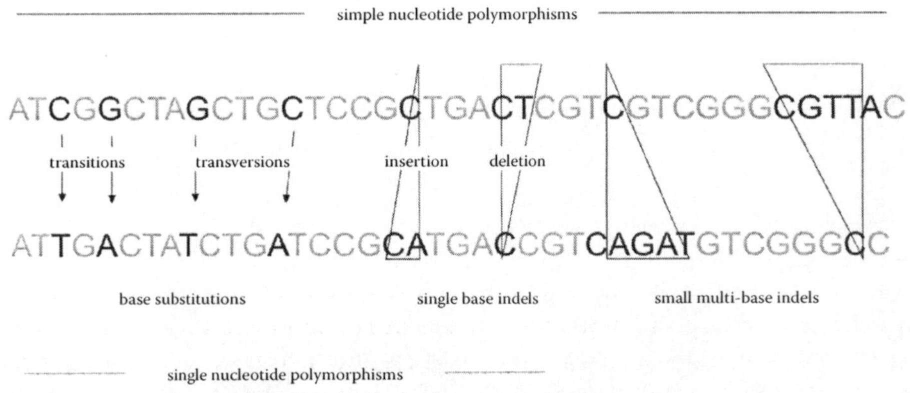

Figure 13.1 The structure of simple nucleotide polymorphisms showing the relationship between simple and single-nucleotide sequence variation.

and money by focusing the expensive genome-wide high-density SNP typing on the most useful study donors (Sladek et al. 2007). The important point for forensic SNP analysis is the potential application of such AIM-SNPs for ancestry inference. If the AIM-SNPs are well differentiated across populations, then relatively small-scale typing approaches can be used and forensic analysis can benefit from the publication of marker lists primarily designed for association study purposes. Following ancestry checks, whole-genome scans provide genome-wide screens for association (Phase I studies) and this provides another key source of data for forensic analysis. In the last 10 years, a comprehensive range of association studies has pinpointed genomic regions strongly associated with particular traits (Donnelly 2008; Hirschhorn and Gajdos 2011). This can potentially accelerate the building of SNP sets able to predict common physical characteristics, with coding or promoter SNPs strongly associated with their expression. Identification of the key SNPs underlying the studied traits represents the last stage (Phase II) of association studies where very small numbers of SNPs are typed in much larger cohorts of study subjects to confirm/reject the associations or find new low-frequency SNPs. Here the choice of small-scale typing platforms can feed new technologies into forensic analysis, although many high-throughput systems used for Phase II studies still depend on prohibitively large amounts of input DNA.

Currently, the choice of typing platforms includes Life Technologies Taqman and Sequenom iPlex MALDI-TOF that together make up the core Phase II approaches but which respectively have the limitations of singleplex scale and input DNA requirements above those of most forensic casework (Hughes-Stamm et al. 2011). This leaves three alternative systems often used for Phase II SNP typing: SNaPshot dye-linked primer extension, SNPstream chip-based capture-probe detection, and GenPlex, an adaptation of SNPlex that reverses this system's chemistry steps with PCR first and oligoligation second (Phillips et al. 2007b). The SNPstream and GenPlex systems are no longer commercially available. Furthermore, they are dependent on the original manufacturer's kits, consumables, and custom oligonucleotide supply and therefore cannot currently be applied to any forensic casework scenarios. It is important to note that one previous supplier of SNP-based ancestry inference services to police investigators, DNAprint (Frudakis 2007), used the now defunct *SNPstream* system, therefore previous SNP-based casework analyses they performed cannot be reexamined due to this SNP typing technology being closed down. Therefore, any forensic laboratory wishing to instigate SNP analysis only has two realistic choices: Taqman, typing one SNP per amplification, and SNaPshot, applying primer extension based on published multiplexes or those developed by the laboratory themselves. We will make reference to several SNaPshot typed SNP sets: for forensic identification, 52-plex and 44-plex sets; for forensic ancestry inference, a 34-plex and two 12-plex sets, and for eye-color prediction, a 6-plex recently extended to include 18 additional hair-color–predictive SNPs in a 24-plex assay.

Last, it is interesting to note that a sample-tracking test offered since early 2012 by Sequenom for the iPlex platform (URL2) makes use of 45 forensic SNPs previously developed by SNP*for*ID using SNaPshot (Sanchez et al. 2006). The PCR primers are the same for both assays showing that, regardless of the detection platform used, well-optimized amplification chemistry is the main factor in achieving robust and sensitive forensic SNP typing assays.

13.1.3 The Original Rationale of Forensic SNP Analysis: Typing Highly Degraded DNA

Because SNPs are short binary loci, they allow allele detection of comparatively small amplicon size ranges down to a theoretical limit of 41 bp and therefore should provide better performance with highly degraded DNA than most STRs, which often require the accommodation of extensive allele size ranges (Budowle and van Daal 2008). Studies of the degradation process suggest target DNA in casework material that is not stored properly or is exposed to harsh environmental conditions is rapidly reduced to 120–150 bp (Dixon et al. 2006). However, it is notable that, so far, SNPs have not secured a niche as the markers of choice for highly degraded DNA, largely because considerable efforts have been made in the last 6 years to shorten STR amplicon lengths and reformulate kit chemistries; the combined effect of these developments has led to enhanced profiling performance for mainstream forensic STRs (Budowle and van Daal 2009). The benefits of these efforts include better DNA database searches by allowing comparisons with more complete profiles and enhanced abilities to detect and deconvolute mixed-source DNA. This latter aspect of forensic analysis represents a major drawback of SNP typing with SNaPshot because primer extension chemistry involves a double reaction, so that the direct relationship between the proportions of input DNA and signal (peak height ratios) is largely lost. The same shortcoming does not apply to indels because these are detected using end-labeled PCR primers, so that products go directly to CE. Therefore, the most suitable application of SNPs is the identification of missing persons, where mixtures are much less likely and relationship-testing comparisons are usually made to surviving relatives rather than database entries.

Two SNP multiplexes have been developed for forensic identification: a 52-SNP assay developed by the SNP*for*ID Consortium, comprising a 52-plex PCR followed by tandem 21- and 29-plex primer extension reactions (Musgrave-Brown et al. 2007; Sanchez et al. 2006), and a 44-plex PCR followed by tandem 18- and 26-plex extensions (Lou et al. 2011) based on the Kiddlab forensic identification marker panels, consisting of a list with almost twice as many ID-SNPs than the 44 collated in this assay (Pakstis et al. 2007, 2010). The 52-plex multiplex has been demonstrated to work well with degraded DNA (Fondevila et al. 2007; Sanchez et al. 2008), and this was subsequently extended to successful analysis of DNA that underwent extreme double degradation processes as well as extracts from skeletal material more than 35 years old or recovered from the sea after many months' immersion (Fondevila et al. 2008; Phillips et al. 2012a). Furthermore, the SNP*for*ID 52-plex set is now in established use in multiple laboratories and is applied in some under the ISO 17025 accreditation standard (Børsting et al. 2009). Although SNP multiplexes tend to fail more readily when the multiplex is large and the DNA input to PCR is very limited, application of SNaPshot-based tests to casework where the DNA was at very low levels have successfully produced full profiles (Phillips et al. 2009). Therefore, the application of such SNP typing systems to missing persons identification appears highly suitable. Two further qualifications to the use of SNPs to identify missing persons should be discussed here: the relative balance of variability among different population groups and the number of SNPs required to obtain high enough relationship-testing likelihoods.

First, the SNP components of the 44-plex show better balance between populations than the 52-plex SNPs. This is partly a result of the original SNP*for*ID selection process that searched SNP databases lacking complete population data, and partly because

SNP*for*ID placed emphasis on the best possible context sequence and good multiplex performance in the component SNPs, in addition to population variation. Therefore, many "ideal" SNP candidates in terms of polymorphism were rejected due to suboptimal amplification. Nevertheless, both of the above multiplexes provide random match probabilities (RMP) from their profiles far exceeding global uniqueness in all worldwide population groups (i.e., RMPs higher than ~1 in 9 billion). As well as offering ready-to-use SNaPshot protocols, both multiplexes have extensive population data accessible online from which to construct profile frequencies or paternity statistics in the ALlele FREquency Database (ALFRED) (Rajeevan et al. 2003) and the SNP*for*ID allele frequency browser (Amigo et al. 2008). Figure 13.2 shows the comparative RMPs of a series of identification multiplexes; the nucleosome-based SNP and the ID-indel marker sets are discussed later. The plots show that each marker set is largely comparable in their ability to provide RMP values well above a uniqueness baseline and only African population variability in the SNP*for*ID SNPs shows a detectably reduced level of individual differentiation compared to other population groups in both the 52-plex and 44-plex multiplexes.

Second, it is important to consider that the number of SNPs required to match the informativeness of STRs is higher in relationship testing than in identification applications. This is because relationship testing compares the alleles shared by two individuals whereas identification uses the whole genotype, so at any one locus only half the genetic information is used to establish a statistical likelihood of relatedness. Charles Brenner has published an elegant "thought experiment" that provides a simple framework for comparing the information content of SNPs and STRs in both applications (URL3). If the SNPs are perfect (0.5:0.5 allele frequency distributions), then one STR is equivalent to about 2.6 SNPs for identification and 4 SNPs for relationship testing. Therefore, approximately 40 SNPs are sufficient to match the efficacy of STRs for identification purposes but over 60 are necessary for relationship testing, not allowing for the reduced power of the majority of SNPs having less informative allele frequencies than a perfect 0.5:0.5.

To finish the review of identification applications of SNPs suited to the analysis of extremely degraded DNA, we review an interesting recent study that sought to identify SNPs that could benefit from the protective effect of nucleosome binding during the DNA degradation process. A relatively small-scale SNaPshot multiplex of 18 SNPs has been developed comprising loci selected to have a high probability to be sited in the center of nucleosome binding sequence complexes (Freire-Aradas et al. 2012). These complexes have recently been characterized and mapped in the human genome by making use of diagnostic sequence motifs found in all of them (Levitsky 2004). The forensic nucleosome SNP multiplex was constructed from SNPs known to be in an optimum position with reference to the nucleosome motif sequences. There is growing evidence that the DNA-histone complexes of nucleosomes infer considerable protection to the surrounding DNA during degradation, both from environmental insults and apoptotic cellular processes (the latter triggered by extreme heat and necrosis during putrefaction). The performance of the nucleosome SNP 18-plex was tested using artificially degraded DNA and 24 casework samples where the likely state of degradation of DNA was established by comparison to profile completeness in three other forensic identification assays: Identifiler 15-plex STR kit, MiniFiler 8-plex miniaturized STR kit, and the SNP*for*ID 52-plex SNP assay. The nucleosome SNP assay gave genotyping success rates 6% higher than the best existing forensic SNP assay (the 29-plex extension component of the 52-plex) and significantly higher than the MiniFiler STR assay. Therefore, the nucleosome SNPs that were located and combined provide a new

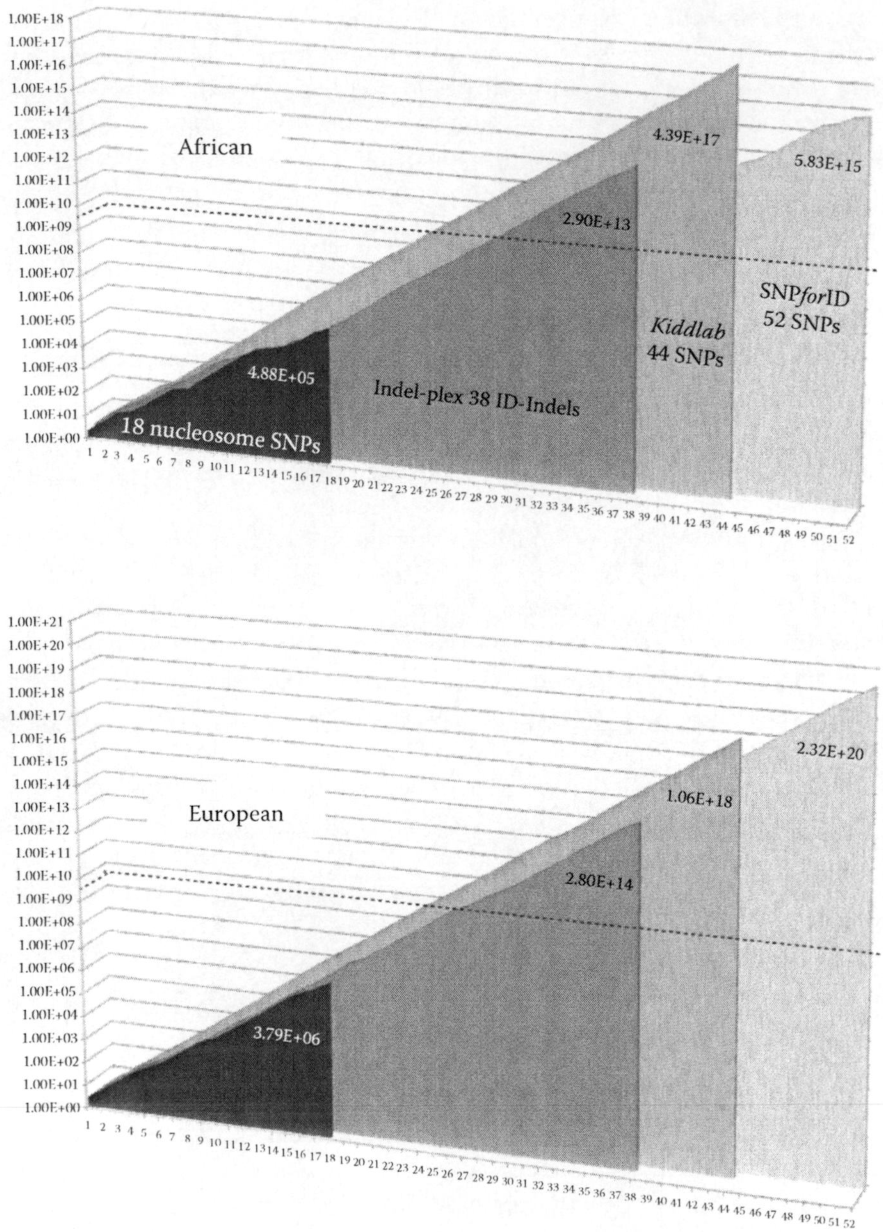

(A)

Figure 13.2 Simple comparisons of cumulative random match probabilities (RMP: the probability of an identical profile from another individual from the same population, the standard measure of forensic discrimination) obtained with different short-binary marker sets in four population groups. The plots indicate accumulating RMP values as components are added (in random order). In nearly all cases the 52 SNP*for*ID SNPs, 40 Kiddlab SNPs, and 38 indels from the 38-plex of Pereira et al. (2009) provide comparable values given as the final values at the head of each plot. SNP*for*ID SNPs show slightly less consistent discrimination for Africans compared with other groups, but all three large-scale forensic identification multiplexes provide discrimination well in excess of the value representing uniqueness (assumes global population of 7×10^{-9}), shown as the dotted line on each plot. Relative RMP values indicate the nucleosome SNP 18-plex would be an additional multiplex of choice for the 52 SNP*for*ID markers when analyzing highly degraded DNA. (Continued)

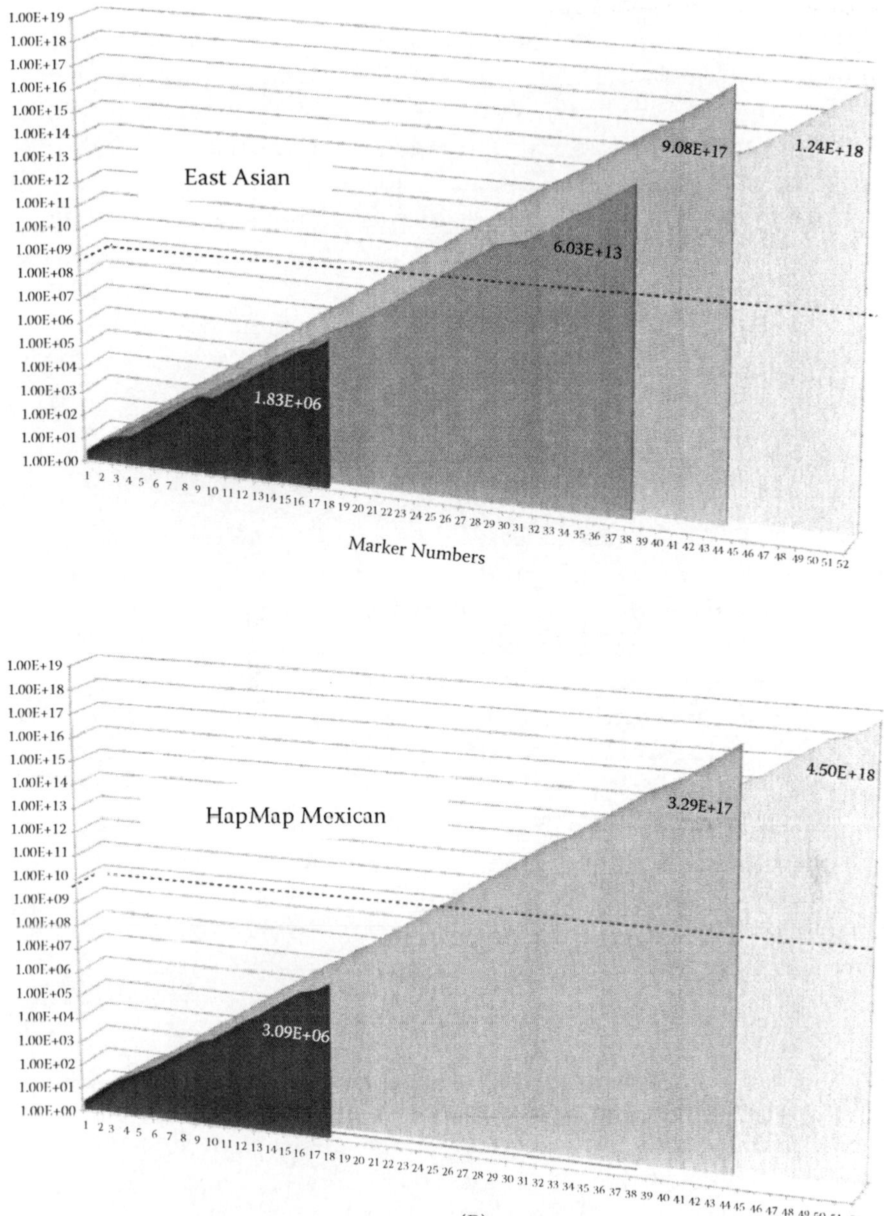

(B)

Figure 13.2 (Continued) Simple comparisons of cumulative random match probabilities (RMP: the probability of an identical profile from another individual from the same population, the standard measure of forensic discrimination) obtained with different short-binary marker sets in four population groups. The plots indicate accumulating RMP values as components are added (in random order). In nearly all cases the 52 SNP*for*ID SNPs, 40 Kiddlab SNPs, and 38 indels from the 38-plex of Pereira et al. (2009) provide comparable values given as the final values at the head of each plot. SNP*for*ID SNPs show slightly less consistent discrimination for Africans compared with other groups, but all three large-scale forensic identification multiplexes provide discrimination well in excess of the value representing uniqueness (assumes global population of 7×10^{-9}), shown as the dotted line on each plot. Relative RMP values indicate the nucleosome SNP 18-plex would be an additional multiplex of choice for the 52 SNP*for*ID markers when analyzing highly degraded DNA.

type of forensic ID-SNP that can be used to supplement existing approaches when the analyzed DNA is likely to be extremely degraded and may fail to give sufficient information from STR or SNP genotypes for reliable identification. With only 18 components and lower levels of polymorphism than the other ID-SNPs in some populations, this multiplex has reduced discrimination (see Figure 13.2), but recommends itself as a supplement to incomplete genotyping results from first-strike marker sets. Therefore, with the emphasis very much on genotyping performance, the 52-plex plus 18-plex nucleosome SNPs represents an ideal combination of discrimination power and predictable amplification success from the most degraded DNA.

13.2 Indels: Combining the Advantages of Direct PCR-to-CE Typing and Short Binary Loci

Up to 5% of known polymorphisms in the human genome consist of short indel loci (Mullaney et al. 2010). In the last 2 years indels have demonstrated their considerable potential applied to forensic identification, since they combine the best characteristics of both SNPs and STRs. First, they can be easily typed from very short amplicon sizes, comparable to those of SNPs, because indel alleles consist of sequence segments that are rarely longer than 4–5 nucleotides. As they generate amplicons at the lowest end of the size range, indels offer levels of typing success from highly degraded target DNA comparable to most of the SNPs discussed above. Second, the simplicity of indel typing, by the detection of dye-labeled PCR products sent directly from the PCR amplification reaction to capillary electrophoresis (PCR-to-CE), exactly mirrors the proven genotyping system of forensic microsatellites. The major advantage of PCR-to-CE typing approaches is the restoration of a direct relationship between the electropherogram peak height ratios and the input DNA. This characteristic can be lost with SNP typing using SNaPshot single-base extension as it is a dual-reaction system of PCR followed by extension—allowing two separate stochastic events to interfere with the balance of signal ratios. So SNaPshot profiles are largely unable to distinguish imbalanced heterozygote peak pairs from combined homozygous donors of a mixed-source DNA sample, a considerable drawback for much of forensic DNA typing. However, it should be noted that some laboratories report a suitable level of peak signal balance using SNaPshot, and can readily differentiate heterozygote peak height ratio variation from mixed-source homozygote pairs within certain ratios (Børsting et al. 2009). SNaPshot has other disadvantages circumvented by PCR-to-CE typing, including a dependence on complete inactivation of PCR primers and unlabeled bases to avoid interference with the extension-reaction dynamics; differences in base-extension efficiencies leading to sharp contrasts in yields of A and G extension products (blue and green signal strengths) compared to C and T (yellow and red); and multiple tube transfers (four times more than PCR-to-CE). Two forensic identification indel (ID-indel) multiplexes have been developed. The Investigator DIPplex kit (Qiagen, Hilden, Germany) of 30 indels plus the amelogenin gender marker (URL4) and a 38-indel multiplex developed by Pereira et al. (2009) concurrently with Qiagen's kit, termed indel-plex. In routine forensic applications, both indel sets show noticeable improvement in genotyping success when typing highly degraded DNA extracted from skeletal materials or paraffin-embedded tissues, and furthermore make ideal complementary marker sets for relationship testing because the allelic mutational stability of indels matches that of SNPs. There are certain differences in characteristics that have emerged between each multiplex, including overlapping size ranges between certain neighboring alleles in the DIPplex (i.e., in some cases one component's slow allele falls behind another's fast allele, making it impossible to assign mobility-predictive size bins in GeneMapper); two nonstandard alleles observed in the DIPplex that have been sequenced, revealing stable mobility variants at low population frequencies (Fondevila et al. 2012a); and an overall 2–3 orders of magnitude lower cumulative RMP for Qiagen's 30 loci compared to the 38 of Pereira's indel-plex—as would be expected from comparing two different-sized multiplexes (Fondevila et al. 2012a). Despite these minor differences, both sets appear to be viable approaches to indel typing for laboratories considering expanding their identification marker repertoire in a straightforward way without the need to invest time and effort to optimize SNP typing dependent on the less robust SNaPshot system.

Figure 13.3B shows a typical electropherogram of the 38-plex ID-indel multiplex using end-labeled primers and it is evident that there is an improved level of heterozygote-peak and dye balance when compared to the SNaPshot electropherogram (a 34-plex AIM-SNP mutliplex) shown alongside. Recent studies of the Qiagen DIPlex performance detecting and deconvoluting simple two-donor mixtures also indicate the direct PCR-to-CE approach of indel typing is efficient at allowing this aspect of profile interpretation commonly encountered in forensic casework (Fondevila et al. 2012a).

Most proposed forensic ancestry inference approaches are based on SNP typing but indels also show some population differentiation, though in general to a lesser extent than the most differentiated SNPs. Two forensic Ancestry Informative Marker Indel (AIM-Indel) panels have been developed: a single-reaction 46-plex by Pereira et al. (2012) and three 16-plex reactions by Santos et al. (2010). In addition to the obvious benefit of a single reaction compared to three, the AIM-Indels chosen by Pereira were designed to differentiate four population groups with a high overall level of population differentiation across each group comparison, while those of Santos show weak East Asian differentiation. Last, an AIM set published by Yang et al. (2005) included many indels likely to be useful for building forensically viable multiplexes.

13.3 Enhancing the Discrimination Power of STRs Using SNPs

13.3.1 SNPs In or Around STRs

The characterization of SNPs at STR sites is desirable for a number of reasons but is difficult to achieve without recourse to direct sequencing. There are three types of SNPs in or around STR loci. The first kind comprises low-frequency SNP variants in PCR primer-binding sequences. Such SNPs block or impair the amplification of the targeted allele on the same strand so it is passed on as a null allele and can explain much of the allele dropout that is infrequently but consistently observed in paternity analyses. When a SNP occurs in the primer site of one STR multiplex but not in others, so-called kit discordancy is observed. Since the SNP is often rare or population-specific, even the most thorough testing will fail to capture a proportion of these SNPs and the forensic community will increasingly need to consult full genome-sequence SNP catalogs such as 1000 Genomes (The 1000 Genomes Project Consortium 2010) to check for low-frequency primer SNPs possibly only present in certain populations or more widely distributed but at extremely low frequencies.

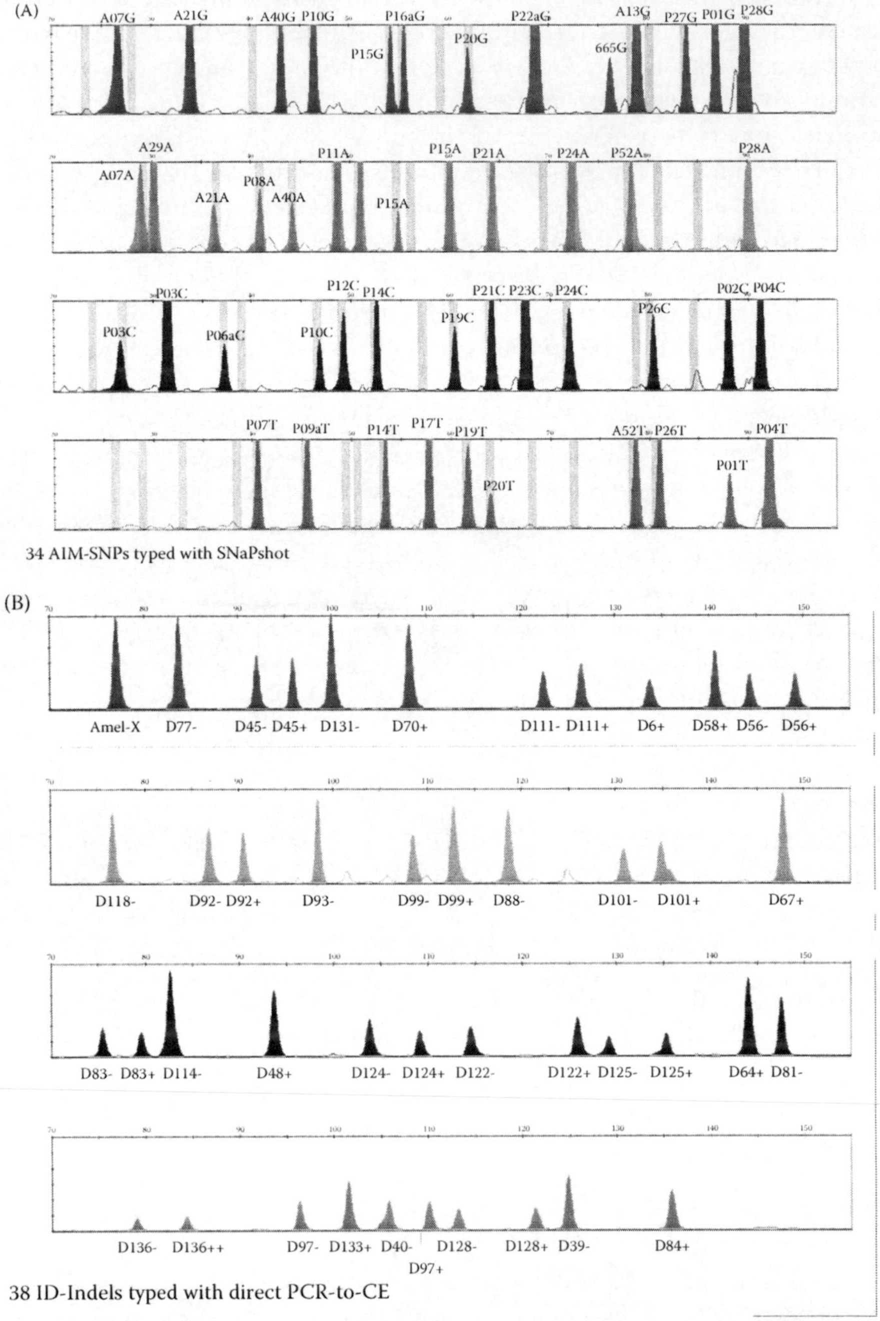

Figure 13.3 (See color insert.) Two electropherograms representing typical forensic short-binary marker analysis: SNaPshot SNP genotyping patterns (upper pane) and PCR-to-CE indel genotyping patterns. The uppermost electropherogram (A) shows the SNP*for*ID 34-plex genotypes where heterozygotes have peaks labeled with different dyes depending on varying base-terminator combinations. The lower electropherogram (B) shows the 38-plex ID-Indel multiplex where heterozygotes have two successive peaks labeled with the same dye, separated by indel base motif size. The 34-plex component SNP codes used to identify each allele peak are outlined at: http://mathgene.usc.es/snipper/default3_34_new.html; the codes used for the 38-plex ID-Indels are outlined in Periera et al. (2009).

The second type of SNP associated with STRs consist of nucleotide variation embedded in the repeat units of STRs. Detection of repeat-unit SNPs is not possible with capillary electrophoresis because the size and thus the mobility of the allele is not changed beyond the fractionation range of the allele by nucleotide substitutions, only the mass. Therefore, ICE-MS mass spectrometry sensitive to the minor percentage differences in molecular weight of a base change is the only system presently able to easily detect them (Oberacher et al. 2008; Planz et al. 2009). ICE-MS can still fail to detect SNPs in STRs with long repeat alleles harboring multiple SNPs with common alleles, since symmetrical substitutions can be more likely (e.g., simultaneous C to T and T to C changes on the same strand).

The third type of SNPs associated with STRs also occur in repeat regions and result from an STR that consists of consecutive arrays of repeat units differing by a single base, for example, $[GATA]_n$ followed by $[GACA]_m$. If the sum of n and m repeat numbers is the same but each strand has different values for n and m there will be SNPs created at each site where the repeat motifs are different (in the above example, a C/T SNP at each repeat-unit third base where GATA and GACA overlap). This might seem an uncommon occurrence but it is dependent on the complexity of the overall repeat structure of any one STR. Our sequencing studies of the newly established European Standard Set STR D12S391 (Phillips et al. 2011a) revealed a large number of repeat-motif overlap SNPs among numerous apparent homozygotes created by two tandem arrays of AGAT and AGAC repeat-unit runs with a wide range of repeat numbers each. Therefore, the observed 8–21 AGAT repeats and 5–11 AGAC repeats potentially create 98 combinations each, making a unique allele in terms of composition, if not length. While the AGAT-AGAC length ratios are not detectable by capillary electrophoresis means, again, ICE-MS is able to characterize the total nucleotide variability within the STR repeats. So, the range of alleles detectable in D12S391 alone could potentially be increased fivefold by using typing systems sensitive to base-composition variation.

The 1000 Genomes Project currently displays somewhat patchy coverage of SNPs associated with STRs—particularly the latter two types embedded in STR repeat region sequences. This is because the short-read chemistry of high-throughput sequencing systems is ineffective at detecting base substitutions positioned within tandemly repeated DNA sequences. Therefore, classical Sanger sequencing will be required for the immediate future to properly map and characterize this potentially very informative extra SNP variation embedded in STRs.

13.3.2 Adding SNPs to Extend the Polymorphisms Available for Complex Relationship Testing

When testing deficient pedigrees—those where only a few members of the kinship are available for testing—it is sometimes the case that a single STR marker gives an indication of an exclusion of the claimed relationship that is contrary to the results obtained from all the other markers typed. Because deficient pedigrees limit the likelihood values obtained, such single-exclusion cases represent ambiguous findings; they are either consistent with a true exclusion of a person, likely related to the true father such that the number of excluding markers is severely reduced, or the results are consistent with the tested man being the true father and the exclusion is a genotype inconsistency arising from an STR mutation—nearly always a single repeat-unit addition or diminution occurring at meiosis. Since STRs have several orders of magnitude higher mutation rates than SNPs, while additional

STRs, for example, SE33, often introduced to raise likelihood values in a complex relationship case, can show higher mutability than first-strike STRs, ambiguous single-exclusion cases can be difficult to resolve. In such cases it is tempting to calculate the residual inclusion probability from the other STR genotypes and, where the exclusion is a single-step or two-step difference between tested man and offspring, to infer a mutation event has occurred and ignore the exclusion completely. However, adding a panel of SNPs—markers with much higher mutational stability—can provide greatly improved security to the interpretation made by providing more excluding genotypes or by raising the inclusion probability well above likelihoods that give virtual proof of the relationship (Børsting and Morling 2011; Phillips et al. 2008, 2012b).

The logical extension of the use of SNP sets as supplementary markers to STRs in complex pedigree analyses is the complete replacement of STRs by the much denser marker panels of whole-genome scan arrays such as those of Affymetrix and Illumina. Applying such dense SNP data is really only justified where the relationship is very distant, because STRs generally can provide sufficiently strong likelihoods to discount alternative hypotheses to the claimed or challenged relationship under analysis (Phillips et al. 2012b). However, when only pairwise comparisons are possible—for example when, as is commonly the case, most other close relatives are dead, or not available to test—STRs will begin to fail to give strong statistical support for a relationship if it is also distant. Distant relationships can be defined as ones bridging several generations, where the root, that is, the relatives shared by the tested individuals, are separated by several segregations on both branches of the pedigree. Examples are given in Figure 13.4 and these all show at least four segregation events across two branches. The nomenclature coding for the number of steps between the tested individuals helps to outline the degree of relatedness by reference to the distance between each branch and its root. So cousins, written as S-2-2, share a grandparental root and each have two segregations between their positions in the pedigree and their shared root. Second cousins sharing a great grandparent are an S-3-3 pair, while your relationship with your cousin's offspring can be coded as S-3-2. When S-2-2 through S-3-3 pairs are simulated using allele frequency data for core ID-STRs and, even when this data is supplemented with an equal number of additional STR loci or with small-scale SNP sets, the

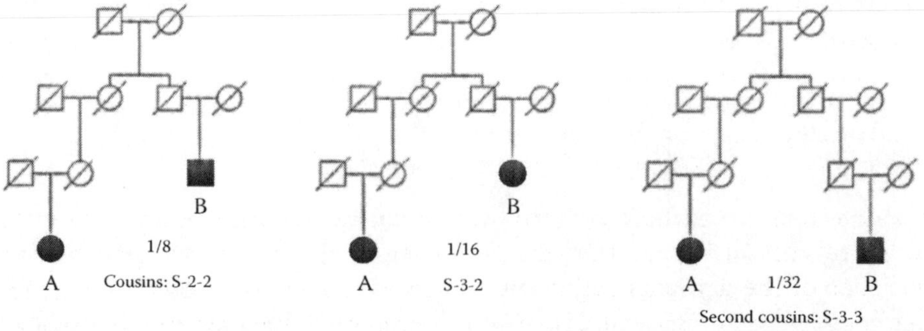

Figure 13.4 Three pedigrees showing some of the more distant pairwise relationships in extended pedigrees: S-2-2 (cousins); S-3-2 (e.g., the relationship between you and offspring of your cousin), and S-3-3 (second cousins). These related pairs represent the statistical limit of relationship testing when other pedigree members can be tested, but such pedigrees will not be properly resolved when deficient (i.e., only the paired individuals can be tested). These cases require marker densities approaching those of whole-genome-scan SNP typing systems.

relationship likelihoods are largely overlapping with those of random pairwise relatedness likelihoods within the population as a whole (Phillips et al. 2012b). In other words, distant relatives and random pairs in the same population have near-identical relationship likelihoods using limited numbers of STRs, so both hypotheses are equally supported by the overlapping distributions of their likelihoods. Simulations help avoid applying limited STR or SNP marker sets to distant pairwise relationships that will not be resolvable, but also help explore the ability of much larger SNP arrays to separate likelihood distributions into clearly differentiated ranges. We have made such evaluations for S-2-2 through to S-3-3 for the Affymetrix 6.0 SNP array, comprising a thinned dataset of 30,564 SNPs at minimum 0.1 minor allele frequencies and 0.1 cM inter-marker genetic distances (Lareu et al. 2012). Results show a second-cousin relationship can be reliably separated from random pairwise relationships by measuring allele-sharing values, with almost no overlap between the highest random-pair allele sharing and the lowest second-cousin allele-sharing levels. An actual second-cousin claim was also analyzed in the same study and this separated slightly more from three random pairs from controls of the same population, with the recorded value of the tested pair in the upper 5% of the allele-sharing distribution generated by S-3-3 pair simulations. This latter observation could be explained by some consanguinity within the original pedigree but is most likely to be a failure to adequately model linkage among the Affymetrix 6.0 components in the simulations that mapped the expected allele-sharing distributions. To summarize, if we consider a second-cousin pair to be close to the limit of identifiable distant relationships and the analyst is confronted with a deficient pedigree lacking all other members, then dense SNP arrays—readily available as off-the-shelf genotyping systems—can be expected to resolve the claim in nearly all cases (Lareu et al. 2012; Phillips et al. 2012b).

13.4 Utility of SNP Databases

13.4.1 Online SNP Data and How to Access It for Forensic Purposes

A particular advantage of SNP analysis is the wealth of online genetic data relating to each marker that can be freely accessed from a researcher's computer. In forensic DNA typing most of the characterization of the established markers used for identification, that is, autosomal/Y-chromosome STRs and mitochondrial variation, was accomplished by the forensic genetics community, whereas SNPs already come with an extensive body of data relating to their genetics, population variability, context sequence, and the role of nucleotide changes (for those SNPs in coding regions). Even primer details are listed marker by marker, providing a readymade system for building initial tests of SNPs of interest. In fact, up until relatively recently much more was known about the genomic characteristics of the approximately 5 million SNPs with common variation than the STRs that had been in routine forensic use 10 years previously. Comprehensive and easily accessible databases that catalog SNP variation provide a very straightforward way to collect SNPs with suitable characteristics for the intended forensic purpose (e.g., identification, ancestry inference). Online data also provide an efficient way to design SNP multiplexes, as it is possible to know enough about the surrounding sequence to make the component's performance in PCR multiplex more predictable. This allows much of the design of forensic SNP assays to be achieved in silico before committing to the expense of primer manufacture. This has benefits in reducing the number of components that may fail in multiplex because of context-sequence problems and underlies much of the SNP selection of the SNP*for*ID 52-plex ID-SNP set. Approximately 60% of candidates were rejected in the 52-plex selection process due to sequence features considered likely to diminish the efficiency of amplification of the SNP in large-scale multiplexes. Therefore, data mining of the largest genomic database, dbSNP (URL5), is an important first step in SNP assay development, either as an initial screening approach for SNPs sharing particular forensically desirable characteristics using Entrez, or simply as a tool for checking loci one by one in their individual SNP report pages (Phillips 2005, 2007, 2009).

While checking SNP databases one locus at a time gives the necessary depth of detail, once the markers have been collated this is not an efficient way to compare levels of polymorphism in different populations or to compare large sets of SNPs in one step. Since SNPs are increasingly important population-genetics markers, the forensic researcher interested in SNPs should consider accessing the growing catalogs of HapMap and 1000 Genomes as a means to review patterns of allele-frequency distributions among a global range of populations. Again, the single biggest obstacle to efficient in silico analysis of online SNP data is the need to access this population data one marker at a time. In order to overcome this problem, we adapted the original SNP*for*ID allele frequency browser developed for the 52 ID-SNP and 34 AIM-SNP sets (Amigo et al. 2008) to extend to all current SNP catalogs with multiple-population frequency data. The original browser framework of user-defined population groupings has now evolved into the SPSmart (SNPs for Population Studies data mart) SNP browser (Amigo et al. 2008, 2009) that allows the review of the population-variability details of any manageable number of SNPs in a single query. The SPSmart browser currently accesses four databases, in order of SNP numbers: 1000 Genomes (15 million), HapMap (3 million), Perlegen (1.2 million), and the HGDP-CEPH Illumina 650K studies of Stanford (Li et al. 2008) and Michigan universities (650,000). As well as the advantage of allowing data queries for large multiple-SNP sets, SPSmart also allows the downloading of the raw data in the form of genotype tables directly to the end-user's computer. Therefore, reference population genotypes from the above extensive SNP catalogs can be placed straight into standard population-analysis systems such as Structure and Principal Component Analysis (PCA) alongside the user's own study-population data. This is a particularly useful way to upload SNP population data for the CEPH human genome diversity panel (HGDP-CEPH), which, as the most extensive global population survey currently available, is a de facto standard reference panel for population genetics studies. SPSmart accesses 650,000 SNPs with variation data for 52 populations of the HGDP-CEPH panel downloadable in a single procedure for up to ~1000 SNPs at a time. In the same single query, 1000 Genomes and HapMap population coverage for the same loci can be compared side-by-side and also downloaded as a concurrent dataset of similar populations. The recent inclusion of the 1000 Genomes Project SNP data into the SPSmart framework as ENGINES *(*ENtire Genome INterface for Exploring SNPs) has brought the advantages of upscaling the coverage to nearly every human SNP with allele frequencies at minimum polymorphism-defining levels of ≥1% (Amigo et al. 2011).

The 1000 Genomes Project aims to fully sequence, at high coverage, whole genomes from a wide range of global populations, including established sample panels previously characterized by HapMap. The sequence data will provide a deep catalog of human genetic variation, encompassing variants with minor allele frequencies as low as ~1%. Data are currently being released for the initial study phases: Pilots 1–3, based on different sequencing strategies used in each phase. Pilot 1 comprises low-coverage whole-genome sequencing (2–6x) of 270 individuals from three populations, now expanded to 629 from 12 populations; Pilot 2 comprises high-coverage (>60x) sequencing of an African trio and a European trio, across multiple platforms and centers. There is a browser for the six Pilot 2 genomes, but no browser exists for Pilot 1 to allow users to explore SNP variation in different populations. Therefore, ENGINES allows users to find established and new SNP sites in defined segments or genes, or by submitting lists of RefSNP (rs) numbers—allowing the browsing of positions, allele frequencies, or other genome details of any human SNP set. Therefore, 1000 Genomes represents the most comprehensive survey of human variation currently available and should be of interest to all forensic geneticists. We outline four examples illustrating how ready access to this extended catalog might permit forensically orientated analysis of sequence variability around STRs, forensic SNPs not covered by HapMap, primer redesign, and low-frequency variation in genes of relevance to physical-trait prediction.

Figure 13.5A–F demonstrates four aspects of typical ENGINES use: Figure 13.5A, comparison of different database marker coverages; Figure 13.5B, exploration of low-frequency SNPs in or around primer sequence; Figure 13.5C–D, finding new SNP variation in the example STR of vWA; and Figure 13.5E, exploration of new coding-region SNPs in the example gene of MC1R. Prior to 1000 Genomes, the HapMap catalog showed some patchy coverage of SNP variation. In the first example query, we select and list Mexican and Puerto Rican population data for the two established ID-SNP sets of SNP*for*ID and Kiddlab plus the established SNP*for*ID 34-plex AIM-SNP set, showing that 3/40, 2/52, and 6/34 SNPs, respectively, were not previously typed by HapMap. Interestingly, two 52-plex ID-SNPs: rs1029047 and rs938283, give a "not genotyped" notice from 1000 Genomes (though both are typed by Perlegen and rs1029047 also by HapMap). The second example describes the search made to explain a rare primer-dropout problem in a 34-plex AIM-SNPs component marker. SNP rs2304925 (17:75551667) has shown allelic imbalance in certain American populations. *ENGINES* enabled a search of the region 100 bp up/downstream and detected SNP rs77154082 within the reverse primer-binding site and in linkage disequilibrium with rs2304925 in Americans, prompting a redesign. The third example queries the immediate region around vWA, the segment 12:6093096-6093282. This chromosome segment query reveals several new low-frequency SNPs on or within the Promega PowerPlex-16 primer sequences. In the sequence segment shown in Figure 13.5D, purple SNPs are novel 1000 Genomes discoveries, while red SNPs represent established dbSNP loci. Most of these sites actually consist of TAGA-TGGA or TAGA-CAGA repeat-unit overlaps. The fourth example relates to scrutiny of SNPs in coding regions, an important preamble to developing SNP-based predictive tests for common physical traits such as hair color. In a query of the gene MC1R in the European populations of Finland, Great Britain, and Tuscany, it can be seen that the number of characterized SNPs has expanded 2.5-fold from the 18 of HapMap to 45 with some detected minor allele variation in 1000 Genomes. Since much hair-color variation may be explained by the interaction of two or more low-frequency variants at different positions in MC1R that may change the equilibrium in the conversion pathway of pheomelanin to eumelanin, it is informative to know the distribution of many of these minor variants in northern and southern European populations, where the distribution of red and fair hair is different.

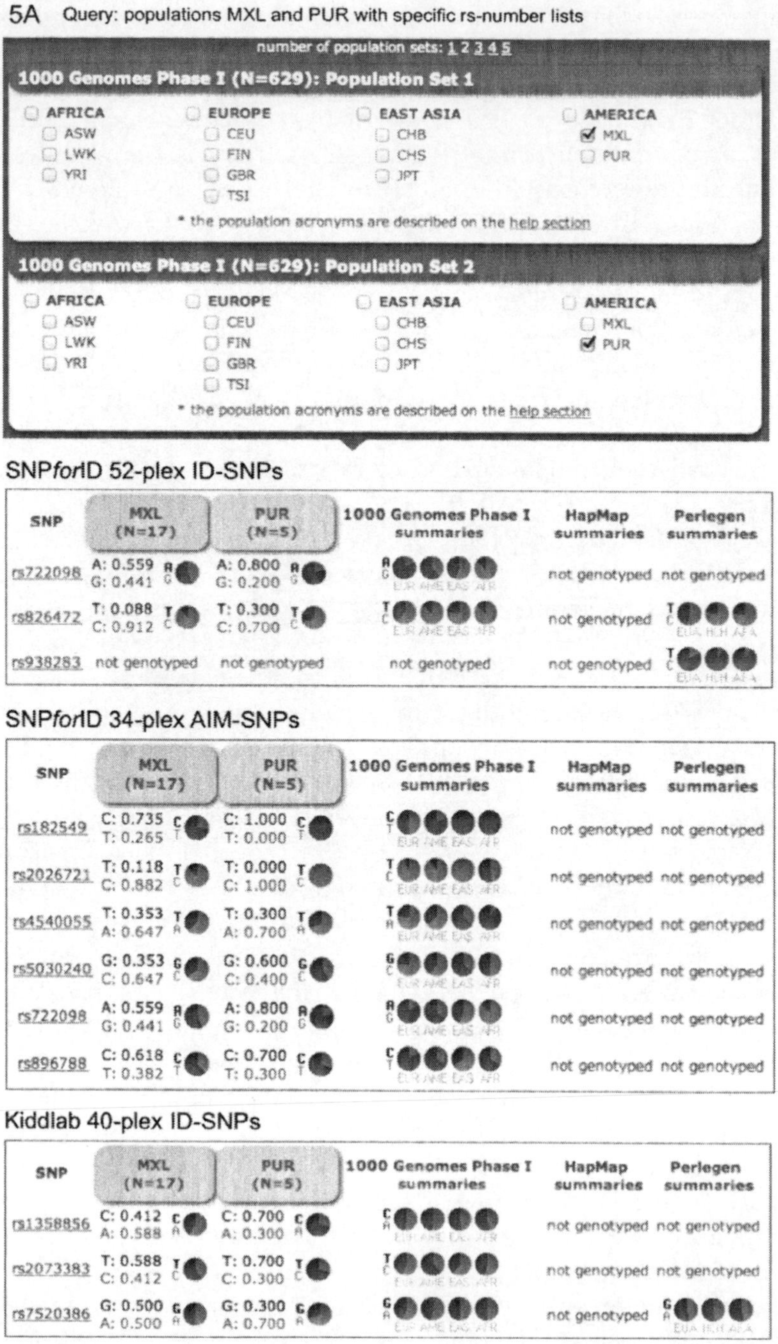

Figure 13.5 (See color insert.) Five examples of studies using the SPSmart *ENGINES* SNP browser to explore 1000 Genomes data relevant to forensic analyses. (A) comparison of different database coverage of components of SNP*for*ID ID 52-plex, AIM 34-plex, and the Kiddlab 40-plex SNPs. (B) exploration of low-frequency SNPs in or around the PCR primer sequences of AIM-SNP rs2304925 (17:75551667). (C and D) Finding new SNP variation around the STR vWA. (E) Exploration of low-frequency coding region SNPs in the MC1R gene. Finally, (F) shows examples of tri-allelic SNPs now curated and listed in 1000 Genomes; data shows Japanese population frequencies found using the *ENGINES* SNP browser. (Continued)

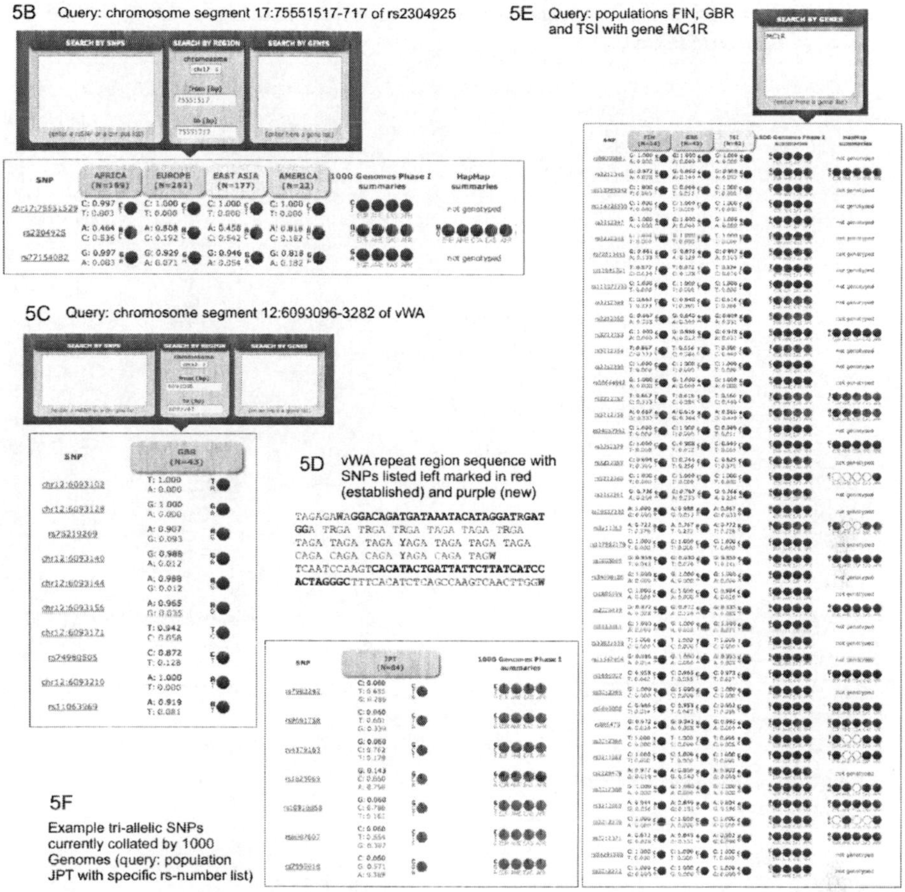

Figure 13.5 (Continued) (See color insert.) Five examples of studies using the SPSmart *ENGINES* SNP browser to explore 1000 Genomes data relevant to forensic analyses. (A) comparison of different database coverage of components of SNP*for*ID ID 52-plex, AIM 34-plex, and the Kiddlab 40-plex SNPs. (B) exploration of low-frequency SNPs in or around the PCR primer sequences of AIM-SNP rs2304925 (17:75551667). (C and D) Finding new SNP variation around the STR vWA. (E) Exploration of low-frequency coding region SNPs in the MC1R gene. Finally, (F) shows examples of tri-allelic SNPs now curated and listed in 1000 Genomes; data shows Japanese population frequencies found using the *ENGINES* SNP browser.

Last, it is interesting to highlight the recent inclusion of microsatellites in the dbSNP database where the variant cataloging program has now begun to assign rs numbers to many, if not most, forensic STRs. Table 13.1 lists the rs numbers currently assigned to common forensic STRs. Using rs reference numbers when accessing dbSNP brings up a cluster report page for the STR and therefore provides a very convenient way for forensic geneticists to scrutinize the genomic landscape around core forensic microsatellite markers.

13.4.2 Fine-Scale Genetic Maps for Forensic Markers Using HapMap High-Density SNP Data

In addition to cataloging SNPs, mapping haplotype block structure and validating much of the SNP data previously listed in dbSNP, the HapMap project has constructed a fine-scale

Table 13.1 dbSNP RS-Number Identifiers for Core and Supplementary Autosomal STRs Used in Forensic Identification[a]

Chrom.	Core STRs	dbSNP RS-Number Identifier for STR	Chrom.	Supplementary STRs	Kit	dbSNP RS-Number Identifier for STR
C1	D1S1656	rs113633160	C1	F13B	CS7[b]	*rs10643350*
C2	TPOX	rs113475620	C2	D2S1360	HDplex[c]	rs113680434
C2	D2S1338	rs112111672	C3	D3S1744	HDplex[c]	rs113865588
C2	D2S441	*rs10203882*	C4	D4S2366	HDplex[c]	rs113820309
C3	D3S1358	rs111694514	C5	D5S2500	HDplex[c]	rs111362704
C4	FGA	*rs67296980*	C4	D4S2366	HDplex[c]	rs113820309
C5	D5S818	rs112497490	C6	D6S1043	SinoFiler[d]	rs111544865
C5	CSF1PO	rs113729910	C6	F13A01	CS7	*rs71817584*
C6	SE33	*rs71021371*	C6	D6S474	HDplex	rs113991233
C7	D7S820	rs112714641	C7	D7S1517	HDplex	rs112397288
C8	D8S1179	*rs67563232*	C8	D8S1132	HDplex	*rs71307053*
C10	D10S1248	rs113518246	C8	LPL	CS7	*rs10558335*
C11	TH01	*rs71029110*	C9	D9S1120	in-house	rs112358545
C12	D12S391	rs113002069	C9	Penta C	CS7	*rs72398274*
C12	vWA	*rs10579907*	C10	D10S2325	HDplex	no SNPs found
C13	D13S317	rs111980288				
C15	Penta E	*rs8036258*	C15	FES-FPS	CS7	*rs6229*
C16	D16S539	rs112689398				
C18	D18S51	*rs10560567*				
C19	D19S433	rs113951851				
C21	D21S11	rs113145752	C21	D21S2055	HDplex	rs113225349
C21	Penta D	*rs7279663*				
C22	D22S1045	rs112790319				

[a] Loci are arranged by chromosome; italic rs-numbers indicate proxy SNPs (close to or within repeat region), where dbSNP has not assigned an rs-number to the STR locus. Data taken from Phillips, C., D. Ballard, P. Gill, D. S. Court, Á. Carracedo, and M. V. Lareu, 2011a, The recombination landscape around forensic STRs: Accurate measurement of genetic distances between syntenic STR pairs using HapMap high density SNP data, *Forensic Sci Int Genet* 6: 354–65.

[b] Powerplex CS7 kit from Promega (Madison, WI).

[c] Investigator HDplex kit from Qiagen (Hilden, Germany).

[d] SinoFiler kit from Life Technologies/Applied Biosystems (Foster City, CA).

recombination map of the human genome. The map was made using a subset of 1.6 million SNPs with data for three population panels from Europe, Africa, and East Asia, and genome-wide genetic distances were estimated from sex-averaged and population-averaged recombination rates. The HapMap recombination rate estimates indicated highly variable rates at the local kilobase scale as a result of extremely high recombination activity concentrated in short chromosome segments that form so-called "hotspots" occurring on average every 200 kb. There were also much larger regions of relatively low activity; the average length of regions showing the lowest 10% of recombination activity was over 91 kb. Therefore, this comprehensive measurement of recombination rates at both fine and coarse scale provides the most detailed human genetic map and can be used to assess the degree of linkage between syntenic forensic STRs, that is, pairs on the same autosome or together on the X chromosome. This becomes an increasingly important check to make as more STRs are introduced for extended forensic analyses in deficient-pedigree relationship testing and a larger range of miniSTRs are likely to become available in the near future. By providing much finer scale measurements at far greater accuracy, HapMap data allows straightforward checking of new STRs for linkage without the need for extensive family studies. Pedigree-based linkage analysis by typing extended families can be time-consuming, expensive, and often uninformative if one or both parental alleles are identical or recombination cannot be detected due to ambiguous phase. Work to generate such pedigree data can therefore represent wasted efforts that cannot be foreseen, so the freely available HapMap data is a much more efficient and easier way to measure recombination at greater accuracy between syntenic STRs.

To assemble a genetic map of 29 syntenic forensic STRs identified among 39 STRs in validated kits, we collated the local SNP position: combined rate—that is, the localized recombination rate in centiMorgans per megabase (cM/Mb) measured between that SNP and the SNP that succeeds it; and genetic map—that is, the cM map distance accumulated from the first SNP (closest to the p-arm telomere) and that position (Phillips et al. 2011a). We identified the closest HapMap SNP site upstream or downstream of the STR position and used these markers as the STR proxies, recording the genetic map distance in cumulative centiMorgans between each SNP proxy. The individual SNP rs-number identifiers are not listed in the HapMap recombination map data but can be easily obtained using the "chromosome location" query function of SPSmart's HapMap browser (Amigo et al. 2008). The forensic STR genetic map does not tell us anything new about recombination distribution on a chromosome-wide scale; marker pairs close to the telomere have higher recombination than those close to the centromere and the 1 cM ≈ 1 Mb ≈ 1% recombination rule-of-thumb is a remarkably good approximation. More importantly, this process has allowed better analysis of certain aspects of close linkage among forensic marker sets: a rigorous scrutiny of the particularities of recombination on the X chromosome; further evidence that vWA-D12S391, as the closest core STR pair, are separated by enough genetic distance for linkage not to interfere with their treatment as independent loci even in relationship testing; and the idea that high-density SNP data provides a highly efficient means to map recombination around any future novel forensic markers before they are introduced.

13.5 Ancestry-Informative Marker SNPs

Autosomal SNPs have been the principal focus for building forensic multiplexes for the inference of ancestry—making use of population-differentiated loci widely referred to as ancestry informative markers (AIMs). Finding the most population-differentiated autosomal SNPs will identify the best AIMs of any type of common genomic variation: SNPs, STRs, or indels. Furthermore, SNPs have the benefit of much more extensive catalogs than these other markers and each SNP database allows scrutiny of the major population groups of Africa, Europe, and East Asia. Currently, there are two forensic multiplexes of AIM-SNPs available for laboratories to apply to casework, both using SNaPshot: one typing 34 SNPs in a single test published by Phillips et al. (2007b) and the other typing 24 SNPs in two tests published by Lao et al. (2010). The 34-plex forensic ancestry test of Phillips et al. (2007b) has recently been enhanced by swapping out the uninformative and underperforming AIM-SNP: rs727811, with the much more powerful marker rs3827760 (Fondevila et al. 2012b). A third, widely publicized, forensic AIM panel that comprised 176 SNPs published by Halder et al. (2008) was developed prior to the above two smaller-scale AIM-SNP sets. A set of 176 SNPs is clearly able to provide higher overall ancestry informativeness, but for at least 4 years the composite marker identifiers and primer details were proprietary to a U.S. forensic services company known as DNAprint™ Genomics Inc. (Sarsota, Florida) and so remained outside the public domain. However, despite eventual publication, the 176-SNP set described by Halder et al. (2008) was entirely based on discontinued genotyping chemistry: the Beckman-Coulter SNPstream system. Although SNPstream showed good forensic sensitivity and could type up to 48 SNPs per reaction, it was hampered by requiring proprietary chip-based consumables/instruments as well as the need to group identical SNP substitutions and strands together into discrete PCRs (e.g., G/A, C/T, G/T, C/A, etc.). Thus, the minimum number of reactions for the 176 loci (still not publicly reported) is likely to be between 5 and 8, as the published set comprises 16 G/T-A/C SNPs, 16 G/C, and 144 C/T-A/G. Last, mention should be made of published AIM-SNP lists that lack multiplex designs or are based on singleplex Taqman genotyping. Two such SNP sets have been proposed for forensic use: 128 AIM-SNPs collated by Kosoy et al. (2009) and further analyzed by Kidd et al. (2011) and 47 AIM-SNPs mainly overlapping the 24 of Lao above, originally described by Kersbergen et al. (2009).

Table 13.2 lists the most informative component AIM-SNPs of the five forensic sets described above, ranked by a standard population genetics ancestry information metric: *In* (Rosenberg et al. 2003), sometimes referred to as Divergence. The two values listed, *In3* and *In4*, refer to differentiation of three and four population groups: *In3* comparing African-European-East Asian and **In4** these three plus Native Americans, estimated using 1000 Genomes and HGDP-CEPH panel SNP data, respectively (the Lao/Kersbergen *In4* data is omitted due to missing HGDP-CEPH panel–characterized SNPs in these lists). It is notable that two interesting characteristics are revealed by these SNP set comparisons: first, there is a very low degree of marker commonality among sets, that is, the top AIMs chosen by multiple studies comprise just two SNPs: rs2814778 and rs16891982; second, nearly all the top 24–34 AIM-SNP components have similar ***In*** values. The lack of many common core AIM-SNPs may partly reflect the different selection strategies chosen by the studies. The 34-plex and Halder AIM-SNPs were accumulated over several years searching for highly differentiated allele frequencies in online databases, whereas the Lao/Kersbergen and Kosoy studies examined whole genome scan (WGS) data of 10,000 SNPs and 300,000 SNPs, respectively. The comparability of the ancestry informativeness levels among the best markers in each set indicates that different AIM selection approaches can each provide informative sets; in other words, there are sufficient numbers of ancestry-informative SNPs available to select markers that can multiplex well for forensic use. The noticeably small number of extremely differentiated AIM-SNPs, for example, $In3 \geq 0.75$, indicates it is difficult to find such SNPs but also reflects their absence from WGS SNP arrays due to their lack of association power, since these loci tend to lack variability within any one population group. Last, there are noticeable differences in the cumulative ***In4*** values between 34-plex and the two large-scale SNP sets, indicating different approaches in marker selection. The 34-plex AIMs were originally selected solely for three group differentiations, but have now been supplemented with another 28-plex AIM-SNP multiplex dedicated to differentiating Native American populations. This used the CEPH panel genotypes for over 650,000 SNPs accessible via SPSmart and enabled the identification of many more SNPs with allele

Table 13.2 Component SNP Details of Five Ancestry-Informative Marker Sets Applicable to Forensic Analysis[a] (Cumulative [cum.] *In3* and *In4* Values Are Calculated for 34 SNPs and from the Total SNP Data from Each Set or Multiple)

	Phillips				Kersbergen[b]			Halder				Kosoy				Lao[b]	
	34-plex	*In3*	*In4*		47 SNPs	*In3*		178 SNPs	*In3*	*In4*		128 SNPs	*In3*	*In4*		24-plex	*In3*
1	rs2814778	0.865	0.783	1	rs1369290	0.704	1	rs2814778	0.865	0.783	1	rs4891825	0.680	0.540	1	rs16891982	0.781
2	rs1426654	0.833	0.562	2	rs1478785	0.521	2	rs590086	0.493	0.476	2	rs11652805	0.554	0.477	2	rs1369290	0.704
3	rs16891982	0.781	0.550	3	rs2052760	0.429	3	rs236336	0.475	0.499	3	rs3784230	0.547	0.416	3	rs1448484	0.654
4	rs3827760	0.759	0.468	4	rs722869	0.426	4	rs984654	0.443	0.361	4	rs10007810	0.542	0.424	4	rs1478785	0.521
5	rs881929	0.525	0.303	5	rs153264	0.407	5	rs212498	0.422	0.349	5	rs9522149	0.511	0.428	5	rs2052760	0.429
6	rs12913832	0.520	0.194	6	rs3843776	0.403	6	rs361055	0.416	0.365	6	rs260690	0.494	0.441	6	rs722869	0.426
7	rs182549	0.515	0.157	7	rs1405467	0.388	7	rs361065	0.415	0.375	7	rs2416791	0.489	0.369	7	rs3843776	0.403
8	rs239031	0.438	0.291	8	rs725667	0.385	8	rs6003	0.406	0.300	8	rs9845457	0.473	0.459	8	rs1405467	0.388
9	rs2303798	0.375	0.294	9	rs1371048	0.369	9	rs1337038	0.403	0.405	9	rs7554936	0.468	0.432	9	rs1876482	0.386
10	rs773658	0.367	0.394	10	rs721352	0.311	10	rs662117	0.402	0.322	10	rs9530435	0.460	0.433	10	rs1371048	0.369
11	rs2572307	0.366	0.206	11	rs1465648	0.296	11	rs9032	0.399	0.310	11	rs4908343	0.423	0.426	11	rs1907702	0.363
12	rs722098	0.344	0.182	12	rs1461227	0.284	12	rs593226	0.394	0.254	12	rs1040045	0.420	0.426	12	rs952718	0.357
13	rs4540055	0.333	0.303	13	rs1048610	0.277	13	rs523200	0.385	0.301	13	rs6548616	0.409	0.427	13	rs714857	0.312
14	rs2065982	0.319	0.479	14	rs1391681	0.268	14	rs2242480	0.382	0.438	14	rs3745099	0.395	0.313	14	rs721352	0.311
15	rs5997008	0.310	0.181	15	rs1000313	0.221	15	rs1888952	0.377	0.259	15	rs772262	0.390	0.334	15	rs1465648	0.296
16	rs730570	0.307	0.217	16	rs951378	0.220	16	rs595961	0.371	0.279	16	rs798443	0.388	0.339	16	rs1858465	0.292
17	rs2026721	0.305	0.275	17	rs2179967	0.195	17	rs3176921	0.371	0.352	17	rs7657799	0.379	0.389	17	rs1461227	0.284
18	rs1978806	0.304	0.218	18	rs1823718	0.177	18	rs1800410	0.364	0.288	18	rs316598	0.377	0.335	18	rs1048610	0.277
19	rs1335873	0.297	0.289	19	rs950257	0.153	19	rs1415680	0.363	0.282	19	rs7803075	0.361	0.284	19	rs1391681	0.268
20	rs3785181	0.260	0.212	20	rs1363933	0.150	20	rs869337	0.349	0.405	20	rs4821004	0.356	0.275	20	rs2179967	0.195
21	rs2065160	0.255	0.088	21	rs340199	0.144	21	rs830599	0.344	0.474	21	rs6422347	0.348	0.388	21	rs1667751	0.095
22	rs2040411	0.206	0.176	22	rs721568	0.142	22	rs1426208	0.342	0.328	22	rs10496971	0.348	0.295	22	rs1344870	0.082
23	rs1321333	0.193	0.165	23	rs721361	0.141	23	rs1399272	0.335	0.277	23	rs7421394	0.348	0.261	23	rs926774	0.073
24	rs1886510	0.183	0.106	24	rs2014131	0.115	24	rs883055	0.329	0.254	24	rs2330442	0.336	0.288	24	rs1808089	0.011
25	rs10141763	0.162	0.132	25	rs240358	0.097	25	rs1034290	0.318	0.256	25	rs9809104	0.331	0.259		***cum. value of 24***	**8.276**
26	rs1573020	0.161	0.310	26	rs3857144	0.096	26	rs913258	0.318	0.309	26	rs1760921	0.330	0.288			
27	rs1498444	0.155	0.064	27	rs1667751	0.095	27	rs733563	0.312	0.226	27	rs10108270	0.328	0.409			

(*continued*)

Table 13.2 Component SNP Details of Five Ancestry-Informative Marker Sets Applicable to Forensic Analysis[a] (Cumulative [cum.] *In3* and *In4* Values Are Calculated for 34 SNPs and from the Total SNP Data from Each Set or Multiple) (Continued)

Phillips				**Kersbergen[b]**			**Halder**				**Kosoy**				**Lao[b]**	
	34-plex	*In*3	*In*4		47 SNPs	*In*3		178 SNPs	*In*3	*In*4		128 SNPs	*In*3	*In*4	24-plex	*In*3
28	rs917118	0.147	0.126	28	rs721723	0.092	28	rs730570	0.307	0.346	28	rs6451722	0.324	0.389		
29	rs10843344	0.111	0.284	29	rs2060319	0.085	29	rs434504	0.300	0.321	29	rs948028	0.323	0.254		
30	rs7897550	0.110	0.109	30	rs1328732	0.078	30	rs67302	0.296	0.313	30	rs385194	0.312	0.266		
31	rs896788	0.106	0.091	31	rs950303	0.077	31	rs735050	0.295	0.245	31	rs7238445	0.309	0.250		
32	rs1024116	0.097	0.083	32	rs926774	0.073	32	rs1395579	0.290	0.237	32	rs1513181	0.308	0.441		
33	rs5030240	0.093	0.195	33	rs1384037	0.064	33	rs553950	0.284	0.216	33	rs2125345	0.301	0.361		
34	rs2304925	0.075	0.078	34	rs824198	0.052	34	rs959858	0.279	0.225	34	rs2504853	0.298	0.224		
cumulative value of 34		**11.176**	**8.562**	35	rsl1427108	0.050	*cum value of 34*		**12.840**	**11.432**	*cum value of 34*		**13.658**	**12.338**		
				36	rs1411380	0.048	35/36rs 285/rs3317		0.275	0.263	36	rs12544346	0.292	0.223		
				37	rs2376366	0.045	37	rs960709	0.272	0.176	37	rs2986742	0.291	0.309		
				38	rs967895	0.043	38	rs721825	0.264	0.262	38	rs7844723	0.291	0.219		
				39	rs1386212	0.036	39	rs1375229	0.264	0.196	39	rs8035124	0.289	0.275		
				40	rs896386	0.035	40	rs1528037	0.264	0.151	40	rs6464211	0.287	0.217		
				41	rs723801	0.033	41	rs667508	0.263	0.152	41	rs8113143	0.286	0.222		
				42	rs724149	0.027	42	rs1467044	0.258	0.248	42	rs4670767	0.286	0.238		
				43	rs727518	0.025	43	rs2065160	0.255	0.418	43	rs6556352	0.283	0.212		
				44	rs2903752	0.021	44	rs173537	0.255	0.227	44	rs11227699	0.281	0.246		
				45	rs16091	0.020	45	rs1800498	0.248	0.281	45	rs874299	0.280	0.251		
				46	rs2868198	0.016	46	rs10852218	0.242	0.190	46	rs3907047	0.273	0.325		
				47	rs1808089	0.011	47	rs3287	0.242	0.218	47	rs13400937	0.270	0.282		
				cum. value of 47		**8.343**	48	rs522287	0.241	0.224	48	rs9319336	0.262	0.302		
							49	rs1476597	0.241	0.189	49	rs1040404	0.258	0.238		
							50	rs915056	0.241	0.164	50	rs1503767	0.256	0.195		
							51	rs783064	0.240	0.231	51	rs2397060	0.255	0.237		
							52	rs1076160	0.233	0.151	52	rs1296819	0.251	0.203		
							53	rs697212	0.229	0.213	53	rs7745461	0.246	0.191		

54	rs1650999	0.227	0.231	54	rs4781011	0.241	0.229
55	rs1937025	0.223	0.129	55	rs4984913	0.238	0.189
56	rs251741	0.216	0.155	56	rs2946788	0.234	0.257
57	rs1004571	0.210	0.152	57	rs7997709	0.232	0.377
58	rs1368872	0.208	0.202	58	rs1408801	0.230	0.243
59	rs1385851	0.204	0.255	59	rs10839880	0.230	0.218
60	rs1221172	0.203	0.295	60	rs1871428	0.230	0.178
178	rs920915	0.011	0.080	128	rs10954737	0.007	0.246
cum. Value of 178		**30.967**	**31.487**	***cum. Value of 128***		**38.759**	**41.362**

[a] The *In*3 metric measures population divergence comparing HGDP-CEPH Africans-Europeans-East Asians, *In*4 compares these three population groups plus Native Americans. All data obtained from *ENGINES* SPSmart SNP browser accessing 1000 Genomes population groups. (Amigo, J. et al. 2011, ENGINES: Exploring single nucleotide variation in entire human genomes, *BMC Bioinformatics* 12: 105.)

[b] *In4* data is omitted from Kersbergen and Lao SNP sets due to a large proportion of SNPs not being typed for HGDP-CEPH panel samples.

frequency distributions close to fixation (i.e., alleles at or near frequencies of 1 exclusively in one or more groups) in both Americans and Oceanians (Phillips et al. 2011b).

Once AIM-SNPs have been chosen, combined into a multiplex, and typed in a forensic case, there is the question of applying a simple statistical analysis regime to the profile(s) to arrive at a predicted ancestry based on comparisons with the allele frequencies recorded in reference populations. A naïve Bayesian system that directly equates probability of membership of a class to the recorded allele frequency distributions in each class (and is termed "naïve" because this approach assumes each component marker is independent) provides the simplest framework for assigning a likely ancestry to a profile. Allele frequencies previously recorded for reference populations or obtained by online searches via SPSmart of global population panels (e.g., HGDP-CEPH) provide training sets for the likelihood calculators. A number of factors influence the extent to which a Bayesian classifier can provide reliable estimations of the likely ancestry of a sample; principal among these are the problems of incomplete geographic coverage in training sets used and deficiencies in the differentiations shown by the AIMs—for example, the Kosoy AIMs mentioned above show less efficient East Asian differentiations than the other groups (Kosoy et al. 2009). The best approach is to ensure the most complete possible coverage of population variability, and the HGDP-CEPH panel is a good start here, as well as properly assessing the performance of the ancestry markers typed. The Snipper open-access Bayesian classification portal offers the possibility to make an ancestry inference of a SNP (or indeed an Indel) profile in a straightforward, profile-by-profile manner. Furthermore, the training sets can be simply organized by the end user to reflect the population comparisons required. Downloading and adapting the allele frequency data from HGDP-CEPH, HapMap, or 1000 Genomes SNP databases just requires a reformatting step in Excel; then the data can be uploaded to the portal and a comparison made with single or sets of casework profiles (Phillips et al. 2012b). The power of the markers used to make particular population differentiations is equally easily achieved ahead of the profile analysis by cross-validation of the training set data to gauge the success/error of reclassifying the reference populations—that is, how many reference Europeans classify correctly as European (success) and what proportion classify as East Asian or African (collectively, the error). Clearly, the range of AIMs typed and the individual levels of differentiation shown by the component loci influences the power of the AIM-SNP set to make reliable classifications. Most importantly, this approach does not handle population admixture well, so an individual with parentage from more than one ancestry will tend to show very low likelihoods to one or both of the contributing ancestries, though this also depends on the admixture ratio and any differentiation bias in the AIM set used (for example, the SNP*for*ID 34-plex does not differentiate American and East Asian ancestry as efficiently as European-African-East Asian comparisons). Another beneficial feature of Snipper is the ability to make the same kind of Bayesian analysis with training sets generated from allele-frequency data alone. This extends the classification system beyond binary (or triallelic) loci to multiple-allele systems such as STRs and the "counted" frequencies of haplotype combinations typical of uniparental Y loci or mitochondrial variation. This opens up the possibility of combining SNPs, Indels, uniparental loci, and STRs—potentially a flexible and informative system for ancestry inference only limited by the quantity of DNA available for the chosen tests.

In summary, the prospects are good for further development of AIM-SNP multiplexes that provide forensic viability and scale; different types of forensic markers can now be readily combined into a single statistical inference of ancestry—making use of core identification loci such as STRs, routinely applied to nearly all forensic cases in the first place; and the power of forensic AIM panels to make reliable inferences of ancestry looks set to continue to rise, ensuring maximum use of any DNA available to the investigation after identification tests have been completed.

13.6 Tri-Allelic SNPs

Tri-allelic SNPs, loci with three substitution alleles recorded at the variant nucleotide position, offer several potentially useful characteristics for forensic analysis: (1) having six genotypes (three more than binary SNPs) provides better discrimination per locus so identification multiplexes can be smaller and more informative; (2) they allow the analysis and interpretation of simple mixtures by the detection of third alleles in a profile; and (3) their allele frequencies regularly show high population stratification, making them particularly well-differentiated ancestry markers. One disadvantage is that SNaPshot typing must accommodate one tri-allelic SNP per size position, losing the benefit of combining two overlapping SNPs with complimentary allele pairs (e.g., placing an AC with a GT SNP in the same mobility position). Most SNP genotyping technologies used by HapMap and in whole genome scans fail to detect more than two alleles per SNP whether present or not (Hüebner et al. 2007), while recent studies of resequenced genes suggest tri-allelic SNPs occur at twice the frequency that base-substitution rates predict (Hodgkinson and Eyre-Walker 2010). Therefore, the deep catalogue of variation that 1000 Genomes seeks to compile could be expected to offer a comprehensive list of validated tri-allelic SNPs for selecting the best forensic markers, since whole genome sequencing should identify tri-allelic SNPs that previously escaped detection due to rare third alleles. Unfortunately, the 1000 Genomes project has appeared to have difficulties successfully identifying known tri-allelic sites in the human genome, so this represents a lost opportunity to gain insights and better knowledge of the population variability and biology of human tri-allelic SNPs. Two such examples of failed detection of known tri-allelics are SNPs rs4540055 and rs5030240, both composite markers in the SNP*for*ID 34-plex AIM-SNP set and listed as normal binary AC allele SNPs by 1000 Genomes. Interestingly, 1000 Genomes has listed, then withdrawn, and is now reintroducing a small number of tri-allelic SNPs as their sequence-variant checks progress and examples are shown in Figure 13.5F. This suggests that eventually many more sites will be fully characterized from the comparison of whole-genome sequence data from multiple centers and platforms when clear indications of third alleles can emerge.

Two studies have collected and characterized tri-allelic SNPs for forensic purposes and given some indications of their potential to detect mixed-source DNA when applying SNaPshot typing to casework analyses (Phillips et al. 2004; Westen et al. 2009). Each study identified three loci in common: rs5030240, rs3091244, and rs2069945. The study of Phillips et al. (2004) identified nine tri-allelic SNPs, eight of which were tri-allelic in both African and European populations (one in Africans only). Simple mixture-detection experiments indicated ratios up to 3:1 could be detected and in combination the nine loci gave a very high probability of detecting a mixture via a third allele: 99.999% and 99.99999% in Europeans and Africans, respectively. The study of Westen et al. (2009) identified 15 tri-allelic SNPs and showed they can reveal the presence of a second DNA donor in mixed-source samples up to a ratio of 1:8 by indicating a third allele as well as disrupted peak-height ratios and uncalled peaks above the detection threshold (Børsting et al. 2009). Finally, both studies suggested that tri-allelic SNPs show highly population-differentiated allele frequency distributions—making them very useful ancestry-informative markers.

13.7 SNP Typing Applied to Predictive Forensic Tests for Common Pigmentation Variation

In the last 4 years, considerable progress has been made toward establishing predictive tests for common variation in the pigmentation traits of eye color and hair color. Two reviews of new developments in forensic genetics by Kayser and Schneider (2009) and by Kayser and de Knijff (2011) provide good surveys. Almost all eye- and hair-color variation occurs among European populations (Sulem et al. 2007, 2008). Arguably, skin color variation can also be included in common pigmentation variation, but this is predominantly an adaptive trait rather than a polymorphic one, that is, the bulk of skin-color variation is confined to differences between population groups rather than within any one group, although Europeans and South Asians are part of the Eurasian metapopulation and show some of the widest ranges in skin tone, making them ideal subjects for association studies (Stokowski et al. 2007). Furthermore, the between-group differences in skin tone are shared by multiple population groups: Africans and Oceanians are both dark-skinned, Europeans and East Asians light-skinned, yet the genetic bases of dark and light skin are different for each population group and these have not been elucidated in sufficient depth to locate the sets of key SNPs in each case, though the study of Eurasians by Stokowski et al. (2007) firmly points to three coding SNPs important in this population group: rs1426654 (substitution A111T in gene SLC24A5), rs16891982 (L374F in SLC45A2), and rs1042602 (S192Y in TYR). It is also the case that AIM sets can provide reliable ancestry assignments in unadmixed individuals and it is reasonable from this to make an inference of likely skin color, regardless of its genetic basis. In contrast, the prediction of the intermediate skin-color tones of admixed individuals is certain to remain an inexact analysis in the same way that gauging degrees of admixture ratios is error-prone, based as it is on typing limited-scale AIM multiplexes dictated by scant casework DNA.

Forensic eye-color analysis has become the most developed SNP-based trait prediction system, following a series of association studies that all successfully located a single promoter SNP in HERC2: rs12913832 (Eiberg et al. 2008; Kayser et al. 2008; Sturm et al. 2008). This SNP promotes the expression of the closely sited OCA2 gene and is responsible for the greatest proportion of blue and dark-brown eye-color variation. At this point, it is important to highlight an interesting feature of eye-color variation (and to a lesser extent hair color, too) that makes the analysis of both eye and hair pigmentation traits more complicated than it may at first appear. This is the presence of a direct and strong association of one SNP with blue-brown eye color phenotypes embedded within a much more complex phenotypic range and series of associations: that of the intermediate eye colors of hazel, green, and light brown. This complexity overlaying a simpler SNP-trait relationship is likely to involve the interaction of several genes harboring both coding and promoter SNPs as well as the possibility of complex gene–gene interactions, as yet undiscovered variants plus their associations and even the environmental effect of age-related darkening of iris color (Liu et al. 2010). It is also important to remember that OCA2 is among the largest of human genes and contains 24 exons and more than 2,300 SNPs in these coding regions (Duffy et al. 2007; Frudakis et al. 2007; Sturm 2009). One other layer of complexity is contributed by the different phenotyping systems employed by various studies (e.g., light and dark used in place of brown and blue (Mengel-From et al. 2010). A similar pattern of complexity is also recognizable in hair-color variation with a clear and direct association found many years previously between SNP variation in the MC1R gene and red hair. This discovery led to the first forensic trait-predictive test, a SNaPshot assay comprising 12 MC1R SNPs (Grimes et al. 2001). Overlaying this simple MC1R–red hair association is a more complex group of gene associations and interactions. Similarly, a continuum of hair color in the blond–light brown–mid-brown range is likely to lead to the same problems of phenotype definition as that seen for intermediate eye color. Overall, the development of forensic pigmentation-predictive tests has been hindered by the need to describe the traits subjectively in the same way an eyewitness is asked to do. Most medical genetic-association tests that preceded forensic association studies have benefited from more objective systems to assign a phenotype. Alternatively, the differentiation of phenotypes is deliberately inflated by excluding the intermediate range of measurements and an example of this approach is seen in the skin-color variation studies of Stokowski et al. (2007) and the definition of complex intermediate eye colors employed in the studies of Ruiz et al. (2012).

Despite the complications of trait complexity and phenotyping ambiguity, progress has been made in optimizing and validating a single forensic eye-color test using SNaPshot and anchored around the key HERC2 SNP of rs12913832. This test is termed Irisplex (Walsh et al. 2011a,b, 2012b) and is based on a large association study of SNPs in pigmentation genes (Liu et al. 2009). The 6-SNP Irisplex assay has also now been supplemented by the 18 SNPs most strongly associated with hair color (11 of these in MC1R) identified previously by the same study group (Branicki et al. 2011) to create the HIrisplex assay (Walsh et al. 2012a). In principle, a single assay covering all the key coding and promoter SNPs most strongly associated with pigmentation variation in Eurasians should be readily achievable, as there is a degree of commonality between each of the three traits of skin-, eye-, and hair-color variation, as shown in Figure 13.6. One problem, however, is the number of rare MC1R variants that can be realistically captured by the limitations of small-scale SNaPshot tests. Aiming to include all strong-effect/high-penetrance alleles (MC1R R) and those most common weak-effect/low-penetrance (MC1R r) alleles seems the most prudent strategy—particularly as both strong- and weak-effect MC1R variants as compound heterozygotes (i.e., different positions on opposite DNA strands) will affect red-hair phenotypes and are associated with other light hair colors (Grimes et al. 2001).

Irisplex has been comprehensively covered in three papers by Walsh et al. (2011a,b, 2102b) outlining results of several studies: constructing the multiplex validating its forensic use, and analyzing its applicability across a range of European populations, where the ratio of blue to brown eye color varies considerably. The HIrisplex assay will doubtless go through the same validation steps and the developers recommend that both the assay and the logistic regression-based predictive model applied to the HIrisplex SNP genotypes can replace that made for Irisplex with the contingency to apply the Irisplex SNaPshot assay separately if the DNA quality leads to locus dropouts in HIrisplex. This possibility in routine forensic analyses does, however, highlight the problem of the Excel-based prediction calculator developed for HIrisplex, and previously Irisplex, of currently being unable to adjust for missing genotypes. An independent study of additional eye-color SNPs in HERC2 and OCA2, plus extra markers among other associated genes (Ruiz et

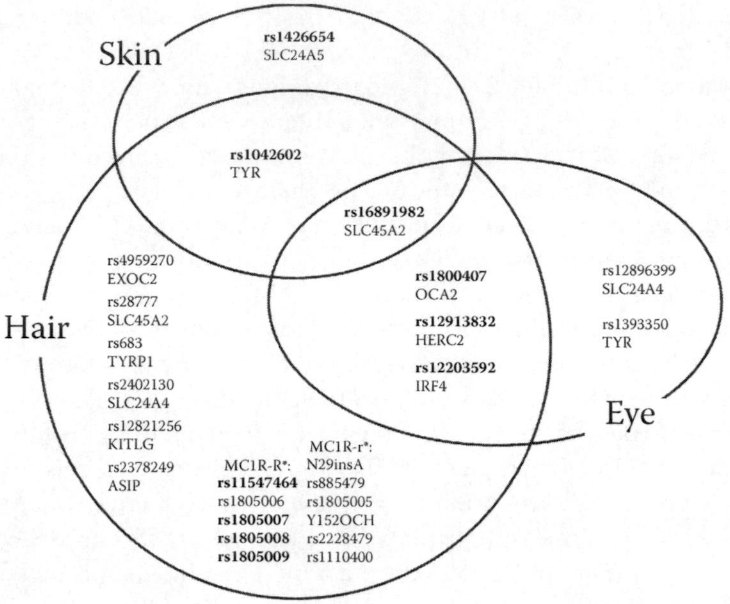

Figure 13.6 The relationship between the major coding and promoter SNPs currently known to be most strongly associated with common pigmentation variation in Eurasian populations. SNP identifiers in bold show markers most strongly associated with each trait. MC1R R and r descriptions denote high- and low-penetrance characteristics but these labels have not been used in the HIrisplex marker descriptions and this test's system for hair-color prediction. (From Walsh, S. et al. 2012a, The HIrisPlex system for simultaneous prediction of hair and eye colour from DNA, *Forensic Sci Int Genet*, DOI: http://dx.doi.org/10.1016/j.fsigen.2012.07.005.)

al. 2012), suggested that typing other SNPs not in Irisplex could marginally improve the classification success of blue and brown eye colors, with the highest increase seen in intermediate eye-color prediction rates. However, these rises in predictive power are small. Furthermore, there is the complication of dealing with closely linked SNPs in the HERC2/OCA2 complex where the effect of rs12913832 is difficult to separate from the much lesser effects of its close neighbors—notably, rs1129038 and rs11636232, both showing detectable effects but evident from very low recombination rates between rs12913832. However, an advantage of the studies of Ruiz et al. (2012) was the adaptation of the Snipper likelihood classifier for eye-color prediction by applying this classifier's frequency-based option to assess the closely linked rs12913832–rs1129038 HERC2 SNP pair as frequency input from counted haplotypes. Likelihood classifiers like Snipper also allow the assessment of incomplete profiles, providing some flexibility as well as a complimentary statistical approach to the logistic regression Excel calculators that accompany the Irisplex/HIrisplex SNP tests. Since this is a rapidly developing field, it is likely that other associated SNPs will be discovered, interactive effects will begin to emerge among many of the component loci, and the statistical frameworks for prediction of common-trait variation will improve. HIrisplex has also begun the move toward treating common pigmentation variation as continuously variable rather than attempting to characterize the variation as categorical (e.g., black, brown, blond hair, etc.) in the same way eyewitness descriptions are invariably termed.

References

Amigo, J., C. Phillips, M. Lareu, and Á. Carracedo. 2008. The SNP*for*ID browser: An online tool for query and display of frequency data from the SNP*for*ID project. *Int J Legal Med* 122: 435–40.

Amigo, J., C. Phillips, A. Salas, and Á. Carracedo. 2009. Viability of in-house datamarting approaches for population genetics analysis of SNP genotypes. *BMC Bioinformatics* 10 Suppl 3: S5.

Amigo, J., A. Salas, and C. Phillips. 2011. ENGINES: Exploring single nucleotide variation in entire human genomes. *BMC Bioinformatics* 12: 105.

Amigo, J., A. Salas, C. Phillips, and Á. Carracedo. 2008. SPSmart: Adapting population based SNP genotype databases for fast and comprehensive web access. *BMC Bioinformatics* 9: 428.

Børsting, C. and N. Morling. 2011. Mutations and/or close relatives? Six case work examples where 49 autosomal SNPs were used as supplementary markers. *Forensic Sci Int Genet* 5: 236–41.

Børsting, C., E. Rockenbauer, and N. Morling. 2009. Validation of a single nucleotide polymorphism (SNP) typing assay with 49 SNPs for forensic genetic testing in a laboratory accredited according to the ISO 17025 standard. *Forensic Sci Int Genet* 4: 34–42.

Branicki, W., F. Liu, K. van Duijn et al. 2011. Model-based prediction of human hair color using DNA variants. *Hum Genet* 129: 443–54.

Budowle, B. and A. van Daal. 2008. Forensically relevant SNP classes. *Biotechniques* 44: 603–8.

Budowle, B. and A. van Daal. 2009. Extracting evidence from forensic DNA analyses: Future molecular biology directions. *Biotechniques* 46: 339–40.

Campbell, C. D., E. L. Ogburn, K. L. Lunetta et al. 2005. Demonstrating stratification in a European American population. *Nat Genet* 37: 868–72.

Cheung, K. H., M. V. Osier, J. R. Kidd, A. J. Pakstis, P. L. Miller, and K. K. Kidd. 2000. ALFRED: An allele frequency database for diverse populations and DNA polymorphisms. *Nucleic Acids Res* 28: 361–3.

Dixon, L. A., A. E. Dobbins, H. K. Pulker et al. 2006. Analysis of artificially degraded DNA using STRs and SNPs—Results of a collaborative European (EDNAP) exercise. *Forensic Sci Int* 164: 33–44.

Donnelly, P. 2008. Progress and challenges in Genomewide Association studies in humans. *Nature* 456: 728–31.

Duffy, D. L., G. W. Montgomery, W. Chen et al. 2007. A three-single-nucleotide polymorphism haplotype in intron 1 of OCA2 explains most human eye-color variation. *Am J Hum Genet* 80: 241–52.

Eiberg, H., J. Troelsen, M. Nielsen et al. 2008. Blue eye color in humans may be caused by a perfectly associated founder mutation in a regulatory element located within the HERC2 gene inhibiting OCA2 expression. *Hum Genet* 123: 177–87.

Fondevila, M., C. Phillips, N. Naveran et al. 2007. Challenging DNA: Assessment of a range of genotyping approaches for highly degraded forensic samples. *Forensic Sci Int Genet* Supp Ser 1: 83–5.

Fondevila, M., C. Phillips, N. Naveran et al. 2008. Case report: Identification of skeletal remains using short-amplicon marker analysis of severely degraded DNA extracted from a decomposed and charred femur. *Forensic Sci Int Genet* 2: 212–8.

Fondevila, M., C. Phillips, C. Santos et al. 2012a. Forensic performance of two insertion-deletion marker assays. *Int J Legal Med* 126: 725–37.

Fondevila, M., C. Phillips, C. Santos et al. 2012b. Revision of the SNP*for*ID 34-plex forensic ancestry test: Assay enhancements, standard reference sample genotypes and extended population studies. *Forensic Sci Int Genet* epub, http://dx.doi.org/10.1016/j.fsigen. 2012.06.007.

Freire-Aradas, A., M. Fondevila, A. K. Kriegel et al. 2012. A new SNP assay for identification of highly degraded human DNA. *Forensic Sci Int Genet* 6: 341–9.

Frudakis, T. 2007. *Molecular Photofitting: Predicting Ancestry and Phenotype Using DNA*: Academic Press/Elsevier: Burlington, MA.

Frudakis, T., T. Terravainen, and M. Thomas. 2007. Multilocus OCA2 genotypes specify human iris colors. *Hum Genet* 122: 311–26.

Grimes, E. A., P. J. Noake, L. Dixon, and A. Urquhart. 2001. Sequence polymorphism in the human melanocortin 1 receptor gene as an indicator of the red hair phenotype. *Forensic Sci Int* 122: 124–9.

Halder, I., M. Shriver, M. Thomas, J. R. Fernandez, and T. Frudakis. 2008. A panel of ancestry informative markers for estimating individual biogeographical ancestry and admixture from four continents: Utility and applications. *Hum Mutat* 29: 648–58.

Hirschhorn, J. N., and Z. K. Gajdos. 2011. Genomewide Association studies: Results from the first few years and potential implications for clinical medicine. *Annu Rev Med* 62: 11–24.

Hodgkinson, A. and A. Eyre-Walker. 2010. Human triallelic sites: Evidence for a new mutational mechanism? *Genetics* 184: 233–41.

Hüebner, C., I. Petermann, B. L. Browning, A. N. Shelling, and L. R. Ferguson. 2007. Triallelic single nucleotide polymorphisms and genotyping error in genetic epidemiology studies: MDR1 (ABCB1) G2677/T/A as an example. *Cancer Epidem Biomar Prev* 16: 1185–92.

Hughes-Stamm, S. R., K. J. Ashton, and A. van Daal. 2011. Assessment of DNA degradation and the genotyping success of highly degraded samples. *Int J Legal Med* 125: 341–8.

Kayser, M. and P. de Knijff. 2011. Improving human forensics through advances in genetics, genomics and molecular biology. *Nat Rev Genet* 12: 179–92.

Kayser, M. and P. M. Schneider. 2009. DNA-based prediction of human externally visible characteristics in forensics: Motivations, scientific challenges, and ethical considerations. *Forensic Sci Int Genet* 3: 154–61.

Kayser, M., F. Liu, A. C. Janssens et al. 2008. Three Genomewide Association studies and a linkage analysis identify HERC2 as a human iris color gene. *Am J Hum Genet* 82: 411–23.

Kersbergen, P., K. van Duijn, A. D. Kloosterman, J.T. den Dunnen, M. Kayser, and P. de Knijff. 2009. Developing a set of ancestry-sensitive DNA markers reflecting continental origins of humans. *BMC Genet* 10: 69.

Kidd, J. R., F. R. Friedlaender, W. C. Speed, A. J. Pakstis, F. M. De La Vega, and K. K. Kidd. 2011. Analyses of a set of 128 ancestry informative single-nucleotide polymorphisms in a global set of 119 population samples. *Invest Genet* 2: 1.

Kidd, K. K., A. J. Pakstis, W. C. Speed et al. 2006. Developing a SNP panel for forensic identification of individuals. *Forensic Sci Int* 164: 20–32.

Kosoy, R., R. Nassir, C. Tian et al. 2009. Ancestry informative marker sets for determining continental origin and admixture proportions in common populations in America. *Hum Mutat* 30: 69–78.

Lao, O., P. M. Vallone, M. D. Coble et al. 2010. Evaluating self-declared ancestry of U.S. Americans with autosomal, Y-chromosomal and mitochondrial DNA. *Hum Mutat* 31: E1875–1893.

Lareu, M. V., M. Garcia-Magarinos, C. Phillips, I. Quintela, Á. Carracedo, and A. Salas. 2012. Analysis of a claimed distant relationship in a deficient pedigree using high-density SNP data. *Forensic Sci Int Genet* 6: 350–3.

Levitsky, V. G. 2004. RECON: A program for prediction of nucleosome formation potential. *Nucleic Acids Res* 32: W346–9.

Li, J. Z., D. M. Absher, H. Tang et al. 2008. Worldwide human relationships inferred from genome-wide patterns of variation. *Science* 319: 1100–4.

Liu, F., K. van Duijn, J. R. Vingerlinget et al. 2009. Eye color and the prediction of complex phenotypes from genotypes. *Curr Biol* 19: R192–3.

Liu, F., A. Wollstein, P. G. Hysi et al. 2010. Digital quantification of human eye color highlights genetic association of three new loci. *PLoS Genet* 6: e1000934.

Lou, C., B. Cong, S. Li et al. 2011. A SNaPshot assay for genotyping 44 individual identification single nucleotide polymorphisms. *Electrophoresis* 32: 368–78.

Marchini, J., L. R. Cardon, M. S. Phillips, and P. Donnelly. 2004. The effects of human population structure on large genetic association studies. *Nat Genet* 36: 512–7.

Mengel-From, J., C. Børsting, J. J. Sanchez, H. Eiberg, and N. Morling. 2010. Human eye colour and HERC2, OCA2 and MATP. *Forensic Sci Int Genet* 4: 323–8.

Mullaney, J. M., R. E. Mills, W. S. Pittard, and S. E. Devine. 2010. Small insertions and deletions (INDELs) in human genomes. *Hum Mol Genet* 19: R131.

Musgrave-Brown, E., D. Ballard, K. Balogh et al. 2007. Forensic validation of the SNPforID 52-plex assay. *Forensic Sci Int Genet* 1: 186–90.

Oberacher, H., F. Pitterl, G. Huber, H. Niederstatter, M. Steinlechner, and W. Parson. 2008. Increased forensic efficiency of DNA fingerprints through simultaneous resolution of length and nucleotide variability by high-performance mass spectrometry. *Hum Mutat* 29: 427–32.

Pakstis, A. J., W. C. Speed, R. Fang et al. 2010. SNPs for a universal individual identification panel. *Hum Genet* 127: 315–24.

Pakstis, A. J., W. C. Speed, J. R. Kidd, and K. K. Kidd. 2007. Candidate SNPs for a universal individual identification panel. *Hum Genet* 121: 305–17.

Pereira, R., C. Phillips, C. Alves, A. Amorim, Á. Carracedo, and L. Gusmão. 2009. A new multiplex for human identification using insertion/deletion polymorphisms. *Electrophoresis* 30: 3682–90.

Pereira, R., C. Phillips, N. Pinto et al. 2012. Straightforward inference of ancestry and admixture proportions through ancestry-informative insertion deletion multiplexing. *PLoS One* 7: e29684.

Phillips, C. 2005. Using online databases for developing SNP markers of forensic interest. *Methods Mol Biol* 297: 83–106.

Phillips, C. 2007. Online resources for SNP analysis: A review and route map. *Mol Biotechnol* 35: 65–97.

Phillips, C. 2009. SNP Databases. *Methods Mol Biol* 578: 43–71.

Phillips, C., D. Ballard, P. Gill, D. S. Court, Á. Carracedo, and M. V. Lareu. 2011a. The recombination landscape around forensic STRs: Accurate measurement of genetic distances between syntenic STR pairs using HapMap high density SNP data. *Forensic Sci Int Genet* 6: 354–65.

Phillips, C., R. Fang, D. Ballard et al. 2007a. Evaluation of the GenPlex SNP typing system and a 49plex forensic marker panel. *Forensic Sci Int Genet* 1: 180–5.

Phillips, C., L. Fernandez-Formoso, M. Garcia-Magarinos et al. 2011b. Analysis of global variability in 15 established and 5 new European Standard Set (ESS) STRs using the CEPH human genome diversity panel. *Forensic Sci Int Genet* 5: 155–69.

Phillips, C., M. Fondevila, M. Garcia-Magarinos et al. 2008. Resolving relationship tests that show ambiguous STR results using autosomal SNPs as supplementary markers. *Forensic Sci Int Genet* 2: 198–204.

Phillips C., M. Fondevila, and M. V. Lareu. 2012a. A 34-plex autosomal SNP single base extension assay for ancestry investigations. *Methods Mol Biol* 830: 109–26.

Phillips, C., M. Garcia-Magarinos, A. Salas, Á. Carracedo, and M. V. Lareu. 2012b. SNPs as supplements in simple kinship analysis or as core markers in distant pairwise relationship tests: When do SNPs add value or replace well-established and powerful STR tests? *Transfus Med Hemother* 39: 202–10.

Phillips, C., M. V. Lareu, A. Salas, and Á. Carracedo. 2004. Nonbinary single-nucleotide polymorphism markers. *Int Congr Ser* 1261: 27–9.

Phillips, C., L. Prieto, M. Fondevila et al. 2009. Ancestry analysis in the 11-M Madrid bomb attack investigation. *PLoS One* 4: e6583.

Phillips, C., A. Salas, J. J. Sanchez et al. 2007b. Inferring ancestral origin using a single multiplex assay of ancestry-informative marker SNPs. *Forensic Sci Int Genet* 1: 273–80.

Planz, J. V., B. Budowle, T. Hall, A. J. Eisenberg, K. A. Sannes-Lowery, and S. A. Hofstadler. 2009. Enhancing resolution and statistical power by utilizing mass spectrometry for detection of SNPs within the short tandem repeats. *Forensic Sci Int Genet* Supp Ser 2: 529–31.

Rajeevan, H., M. V. Osier, K. H. Cheung et al. 2003. ALFRED: The ALelle FREquency Database. Update. *Nucleic Acids Res* 31: 270–1.

Rosenberg, N. A., L. M. Li, R. Ward, and J. K. Pritchard. 2003. Informativeness of genetic markers for inference of ancestry. *Am J Hum Genet* 73: 1402–22.

Ruiz, Y., C. Phillips, A. Gomez-Tato et al. 2012. Further development of forensic eye color predictive tests. *Forensic Sci Int Genet.* DOI: http://dx.doi.org/10.1016/j.fsigen. 2012.05.009

Sanchez, J. J., C. Børsting, K. Balogh et al. 2008. Forensic typing of autosomal SNPs with a 29 SNP-multiplex—Results of a collaborative EDNAP exercise. *Forensic Sci Int Genet* 2: 176–83.

Sanchez, J. J., C. Phillips, C. Børsting et al. 2006. A multiplex assay with 52 single nucleotide polymorphisms for human identification. *Electrophoresis* 27: 1713–24.

Santos, N. P., E. M. Ribeiro-Rodrigues, A. K. Ribeiro-Dos-Santos et al. 2010. Assessing individual interethnic admixture and population substructure using a 48-insertion-deletion (INSEL) ancestry-informative marker (AIM) panel. *Hum Mutat* 31: 184–90.

Sladek, R., G. Rocheleau, J. Rung et al. 2007. A Genomewide Association study identifies novel risk loci for type 2 diabetes. *Nature* 445: 881–5.

Stokowski, R. P., P. V. Pant, T. Dadd et al. 2007. A Genomewide Association study of skin pigmentation in a South Asian population. *Am J Hum Genet* 81: 1119–32.

Sturm, R. A. 2009. Molecular genetics of human pigmentation diversity. *Hum Mol Genet* 18:R9.

Sturm, R. A., D. L. Duffy, Z. Z. Zhao et al. 2008. A single SNP in an evolutionary conserved region within intron 86 of the HERC2 gene determines human blue-brown eye color. *Am J Hum Genet* 82: 424–31.

Sulem, P., D. F. Gudbjartsson, S. N. Stacey et al. 2007. Genetic determinants of hair, eye and skin pigmentation in Europeans. *Nat Genet* 39: 1443–52.

Sulem, P., D. F. Gudbjartsson, S. N. Stacey et al. 2008. Two newly identified genetic determinants of pigmentation in Europeans. *Nat Genet* 40: 835–7.

The 1000 Genomes Project Consortium. 2010. A map of human genome variation from population-scale sequencing. *Nature* 467: 1061–73.

URL1. Santa Cruz genome browser: http://genome.ucsc.edu/

URL2. http://www.mysequenom.com/Home/Webinars/iPLEX-Sample-ID-Panel-Webinar

URL3. Charles Brenner: The Power of SNP's - Even Without Population Data: http://dna-view.com/SNPpost.htm

URL4. Qiagen Investigator DIP-plex (indel multiplex) product guide: http://www.qiagen.com/products/investigatordipplexkit.aspx

URL5. dbSNP: http://www.ncbi.nlm.nih.gov/snp/

Walsh, S., A. Lindenbergh, S. B. Zuniga et al. 2011a. Developmental validation of the IrisPlex system: Determination of blue and brown iris colour for forensic intelligence. *Forensic Sci Int Genet* 5: 464–71.

Walsh, S., F. Liu, K. N. Ballantyne, M. van Oven, O. Lao, and M. Kayser. 2011b. IrisPlex: A sensitive DNA tool for accurate prediction of blue and brown eye colour in the absence of ancestry information. *Forensic Sci Int Genet* 5: 170–80.

Walsh, S., F. Liu, A. Wollstein et al. 2012a. The HIrisPlex system for simultaneous prediction of hair and eye colour from DNA. *Forensic Sci Int Genet.* DOI: http://dx.doi.org/10.1016/j.fsigen.2012.07.005

Walsh, S., A. Wollstein, F. Liu et al. 2012b. DNA-based eye colour prediction across Europe with the IrisPlex system. *Forensic Sci Int Genet* 6: 330–40.

Westen, A. A., A. S. Matai, J. F. Laros et al. 2009. Tri-allelic SNP markers enable analysis of mixed and degraded DNA samples. *Forensic Sci Int Genet* 3: 233–41.

Yang, N., H. Li, L. A. Criswell et al. 2005. Examination of ancestry and ethnic affiliation using highly informative diallelic DNA markers: Application to diverse and admixed populations and implications for clinical epidemiology and forensic medicine. *Hum Genet* 118: 382–92.

Deep-Sequencing Technologies and Potential Applications in Forensic DNA Testing

14

ROXANNE R. ZASCAVAGE
SHANTANU J. SHEWALE
JOHN V. PLANZ

Contents

14.1 Introduction 312
14.2 Second-Generation Sequencing Technologies 314
14.3 Applications of Second-Generation Sequencing Technologies 319
14.4 Third-Generation Sequencing Technologies 322
14.5 Performance Comparison 324
14.6 Next-Generation Sequencing Technologies 329
14.7 Implications of Deep Sequencing in Forensic Investigations 332
14.7.1 Short Tandem Repeats 332
14.7.2 Mitochondrial DNA Analysis 333
14.7.3 Single Nucleotide Polymorphisms 335
14.7.4 RNA Body Fluid Identification 337
14.8 Conclusions 337
References 338

Abstract: Development of second- and third-generation DNA sequencing technologies have enabled an increasing number of applications in different areas such as molecular diagnostics, gene therapy, monitoring food and pharmaceutical products, biosecurity, and forensics. These technologies are based on different biochemical principles such as monitoring released pyrophosphate upon incorporation of a base (pyrosequencing), fluorescence detection subsequent to reversible incorporation of a fluorescently labeled terminator base, ligation-based approach wherein fluorescence of a cleaved nucleotide after ligation is measured, measuring the proton released after incorporation of a base (semiconductor-based sequencing), monitoring incorporation of a nucleotide by measuring the fluorescence of the fluorophore attached to the phosphate chain of the nucleotide, and by detecting the altered charge in a protein nanopore due to released nucleotide by exonuclease cleavage of a DNA strand. Analysis of multiple DNA fragments in parallel increases the depth of coverage while decreasing labor, cost, and time, highlighting some major advantages of deep-sequencing technologies. DNA sequencing has been routinely used in the forensic laboratories for mitochondrial DNA analysis. Fragment analysis, however, is the preferred method for short tandem repeat genotyping due to the cumbersome and costly nature of first-generation DNA sequencing methodologies. Deep-sequencing technologies have brought a new perspective to forensic DNA analysis. Studies include short tandem repeat (STR) analysis to reveal hidden variation in the repeat regions, mtDNA sequencing, single nucleotide polymorphism (SNP) analysis, mixture resolution, and body fluid identification. Recent publications reveal that attempts are being made to expand the capability.

14.1 Introduction

Forensic DNA testing most frequently includes analysis of short tandem repeat (STR) markers. STRs are routinely utilized for comparison of autosomal and Y-chromosomal DNA for paternity testing (Barrot et al. 2011a,b; Carboni et al. 2011; Chakraborty and Opitz 2005; H. Li et al. 2011; Pena and Chakraborty 1994), criminal investigation (Drobnic 2001; X. Li et al. 2011; Poetsch et al. 2011; Schneider and Martin 2001; Zech et al. 2012), human remains identification (Alvarez-Cubero et al. 2012; Gill et al. 1994; Hartman et al. 2011; Leclair et al. 2004; Rucinski et al. 2011; Schwark et al. 2011), and population structure analysis (Babiker et al. 2011; Bosch et al. 2000; Dulik et al. 2012; Liu et al. 2011; Trovoada et al. 2001). The current method for STR analysis involves polymerase chain reaction (PCR) amplification of the target STR loci using a commercialized kit followed by fragment analysis of fluorescently labeled amplicons via capillary electrophoresis (CE) (Butler 2007; Butler and Hill 2012; Liu et al. 2007; Yeung et al. 2006). The data resulting from STR analysis using the CE method consists of the number of repeat motifs present for alleles at each locus relative to an allelic ladder. However, analysis of the fluorescently labeled amplicons using CE cannot identify differences in nucleotide composition of the amplified region other than those that alter fragment size (e.g., insertions/deletions).

When resolution at the nucleotide level is required for comparative analyses, DNA sequencing is performed. At present, DNA sequence analysis is limited to the investigation of portions of the mitochondrial DNA (mtDNA) Control Region. Approximately 600–1,000 nucleotides of the 16.5-kb genome are typically compared between evidentiary and known samples for two hypervariable regions. Although these regions are of the highest diversity in the mitochondrial genome, the investigation of variants in the remainder of the genome is often necessary to differentiate between common haplotypes. Deep-sequencing technologies will have the capability to increase discrimination power of traditional forensic loci by providing sequence-based information that can increase allelic diversity and will permit the entire mitochondrial genome of an individual to be sequenced with resolved nucleotide determination. By directly sequencing forensic markers, inferences to allelic identity can be eliminated and allow for databasing and comparison of DNA data on its most discrete level.

Traditional DNA-sequencing methods originally evolved from three approaches. The original technique developed in 1975 by Sanger and Coulson was referred to as Plus-Minus Sequencing. To prepare for sequencing, the DNA region of interest was cloned to produce a single-stranded template. The template was then hybridized to a primer and incubated with DNA polymerase and a cocktail of deoxyribonucleotide triphosphates (dNTPs). The reaction created varying-length complimentary copies of the target region. Two subsequent sets of reactions were performed on the newly synthesized DNA copies, a "plus" reaction and a "minus" reaction. In the "plus" reactions the mixture of fragments and template DNA were incubated with DNA polymerase and a mixture of only three dNTPs. Four separate mixed reactions were performed, each omitting a different dNTP. In this manner, DNA synthesis will terminate each time the position of the missing nucleotide is reached. Alone, this system was insufficient for accurate sequencing; therefore, a series of four additional reactions, the "minus" reactions, was performed for sequence confirmation. In the minus reactions, the DNA samples are incubated with T4 polymerase and a single dNTP. The T4 polymerase degraded the DNA until it reached a position of the included dNTP (Sanger and Coulson 1975).

The first sequencing method to be heavily implemented was the Maxam-Gilbert method, which allowed for analysis of purified DNA without cloning and with the ability to analyze double-stranded and single-stranded DNA. The Maxam-Gilbert method introduced direct, preferential DNA cleavage through nucleotide modification. The chemical reaction results in targeted DNA damage and base cleavage. Four separate reactions were performed, and each reaction targeted a different dNTP (Maxam and Gilbert 1977).

The gold standard for DNA sequencing for approximately 30 years has been the Sanger dideoxynucleotide terminator method. First published in 1977, the "traditional" Sanger sequencing method utilizes a reaction containing dNTPs as well as a low concentration of dideoxynucleotide triphosphates (ddNTPs) (Sanger et al. 1977). Restriction enzymes were often used to fragment the template DNA and the individual fragments were separated on an acrylamide gel and cloned into a vector or used directly as a sequencing template. The DNA template was incubated with buffer, oligonucleotide primers, DNA polymerase, all four dNTPs, and a single ddNTP. Four separate incubations were performed, each containing a different ddNTP. The ddNTPs are randomly incorporated in lieu of a dNTP during elongation. A dideoxynucleotide contains a hydrogen on the 3' carbon of the nucleotide's pentose moiety instead of the hydroxyl group found in a dNTP. As a phosphodiester bond cannot form between two nucleotides without the hydroxyl group present on the 3' carbon, the incorporation of a ddNTP into an elongating DNA molecule will terminate extension of the individual strand. The end result of the sequencing reaction is a pool of DNA fragments of varying lengths. The final base of each DNA fragment in the pool is the ddNTP that terminated extension in the reaction (Sanger et al. 1977). In the original first-generation sequencing technologies, four reactions were performed, and a single dNTP in each reaction was labeled with ^{32}P for detection. The four reaction products were applied side by side onto a denaturing polyacrylamide-urea gel for fragment separation and analysis. An autoradiograph of the gel allowed for visualization of the bands on the gel. The varying-length fragment bands in a single lane all end in the same nucleotide. The bands of a single lane were used to determine all the genomic positions containing the nucleotide specific to that lane. The lanes were aligned and the sequence visually "read" from the smallest fragment to the largest, from the bottom of the gel to the top (Maxam and Gilbert 1977; Sanger and Coulson 1975; Sanger et al. 1977). The nucleotide sequence was determined by combining the information from all four gel lanes into a single read.

Dideoxynucleotide terminator sequencing has been the most implemented sequencing method for 30 years, and significant improvements have been made in streamlining the efficiency of the reactions and processes (Gharizadeh et al. 2006). Presently, the ddNTPs are labeled with fluorophor dye moieties that permit all four extension reactions to be performed simultaneously in a single sequencing reaction (Smith et al. 1986). The sequencing product from the single reaction is typically analyzed via capillary electrophoresis (Drossman et al. 1990). Sequence determination is automated using a base-calling software, reducing manual analysis of the data to quality checks of the automated reads. The data obtained from dideoxynucleotide terminator sequencing is generally of high quality; however, poor template quality often causes the first 15–20 bases following the primer to be uninterpretable, resulting in loss of data in these areas (Gharizadeh et al. 2006). A recurring issue with chain-terminator sequencing is the inability to resolve multiple-source samples efficiently. The difficulty with multiple-source analysis creates an inability to sequence through homopolymeric stretches of mtDNA containing length heteroplasmy. The length heteroplasmy that is observed in several regions of the mitochondrial genome results from the existence of varying-length mtDNA molecules being present within an individual. In a mixed sample with mitochondrial DNA or a diploid region of the nuclear genome, the linkage of multiple variations sequenced from the same DNA fragment cannot be verified due to the pooled nature of the sequencing reaction. The individual haplotypes cannot be directly determined (Browning and Browning 2011; Clark 1990; Excoffier and Slatkin 1995). On a large scale, cost and time also limit throughput capabilities when using Sanger sequencing. Future technologies for sequencing are aimed at resolving interpretational issues and generating highly reliable quantitative sequence data. Multiple approaches, termed "next generation," have been under development since the introduction of pyrosequencing in 1988 (Hyman 1988). This review summarizes the recent advancements in sequencing technologies and discusses the possible implications and applications of these approaches in forensic DNA analysis.

14.2 Second-Generation Sequencing Technologies

The advent of the next-generation sequencing (NGS) era began with Ronaghi et al.'s introduction of a sequencing method based on real-time pyrophosphate detection (Ronaghi et al. 1998). During DNA synthesis, when a dNTP is incorporated into the extending DNA strand, pyrophosphate is released. Ronaghi et al. (1998) developed a detection assay to measure and record the natural release of pyrophosphate through a coupled enzymatic reaction utilizing adenosine triphosphate (ATP) sulfurylase, luciferase, and the nucleotide-degrading enzyme apyrase. ATP sulfurylase generates ATP by incorporating the released pyrophosphate with 2 ADP. The ATP is necessary to drive the conversion of luciferin to oxyluciferin emitting visible light. The amount of light released is proportional to the amount of ATP present, and therefore proportional to the number of nucleotides incorporated into the DNA strand. The light is detected and recorded as a peak, with peak height being proportional to the number of incorporated nucleotides. Apyrase degrades any unincorporated dNTPs and ATP after incorporation and detection, returning the signal to baseline (Qiagen 2012; Ronaghi et al. 1998). The pyrosequencing technology was optimized and automated (Margulies et al. 2005) and in 2005, 454 Life Sciences Corporation (Branford, Connecticut), commercialized the first large-scale highly parallel pyrosequencing-based genome analysis platform (454 Life Sciences Corporation 2005). In 2003, 454 Life Sciences Corporation was granted a sole license to use pyrosequencing for its NGS technology (454 Life Sciences Corporation 2003) and the company was acquired by Roche Diagnostics (Indianapolis, Indiana) in 2007 (454 Life Sciences Corporation 2007b). The 454 Sequencing™ System utilizes a bead-capture approach for parallel creation and analysis of DNA polonies (polymerase colonies) (Mitra et al. 2003). The sequencing method requires the generation of a single-stranded DNA library that is created by fusing each template fragment to a 454 Sequencing adaptor sequence that will then hybridize to DNA capture beads. Optimal library development results in a single DNA template being bound to each bead. The DNA template-containing capture beads are emulsified with amplification reagents in a water-in-oil mixture. Each droplet of the emulsion undergoes

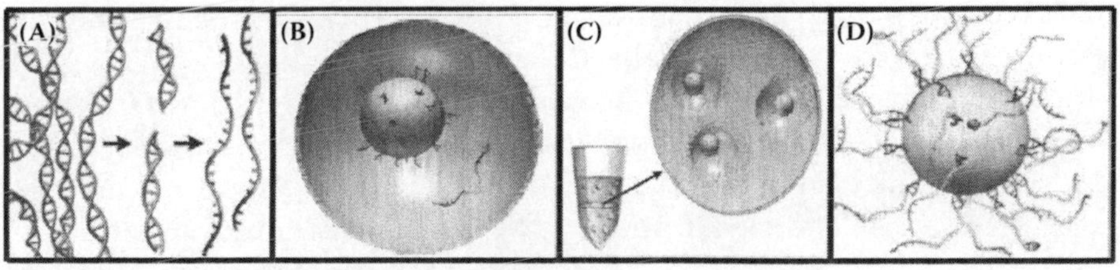

Figure 14.1 (See color insert.) DNA polony formation via bead amplification. (A) The DNA is fragmented and denatured. Adaptors are ligated to the ends of the single-stranded fragments. (B) The adaptors hybridize to probes on the DNA capture beads. A single fragment attaching to a single bead. (C) Emulsion PCR. The DNA template–containing beads are emulsified with amplification reagents in a water-in-oil mixture. Each water droplet contains a bead. The droplets are amplified simultaneously. (D) A DNA polony. Millions of copies of the original DNA fragment attached to a bead. (Image reproduced with permission from 454 Sequencing, Roche Diagnostics, Indianapolis, Indiana.)

amplification simultaneously (emulsion PCR), generating millions of copies of the original DNA fragment each attached to an individual bead (Figure 14.1). Once the emulsion is broken, the amplicons attached to the beads are sequenced on the proprietary PicoTiterPlate. The design allows each well to contain only a single bead, representing a single-fragment library. Sequencing progresses by flooding the plate sequentially with each dNTP, recording light emission after the introduction of each nucleotide (Figure 14.2) (Margulies et al. 2005).

The most commonly used second-generation sequencing technology is a sequence-by-synthesis method introduced in 2006 by Solexa® (Hayward, California), which was acquired by Illumina (San Diego, California) in 2007 (Illumina 2006). Illumina currently offers four sequencing platforms utilizing its TruSeq™ technology. As with pyrosequencing, the DNA template must be fragmented for library creation. Forward and reverse adapter sequences are ligated to both ends of the DNA fragments complimentary to the primer sequences that are attached to the flow cell. The template is immobilized on the flow cell by hybridization of the adapter sequences to the primers and amplicons are generated while attached to the flow cell via solid-phase bridge amplification to create the DNA library. During bridge amplification, elongation occurs in a single direction. The extended DNA molecule bends and attaches to the second PCR primer, forming a bridge. Elongation is initiated on the bridged strand from the reverse primer. A double-stranded bridge is formed from a single elongation step. The two strands of DNA are denatured, releasing nonbridged single-stranded templates. In the subsequent elongation steps, the strands may reattach to themselves, or reattach to other primers. This process is repeated for the desired number of cycles. Following amplification, the reverse strands are washed away. The result is a cluster of identical forward DNA molecules with an identical free adapter sequence on one end. A single flow-cell slide can contain hundreds of millions of clusters. Each cluster represents a different DNA fragment template with complementary strands of the fragment represented in different cells. The sequencing primer hybridizes to the adapter sequence remaining on the unattached end of the DNA molecule. The sequence-by-synthesis reaction involves the reversible incorporation of a fluorescently labeled terminator base. The fluorescence is detected and recorded, after which the fluorescent label is removed, allowing extension to continue (Figure 14.3) (Bentley 2006).

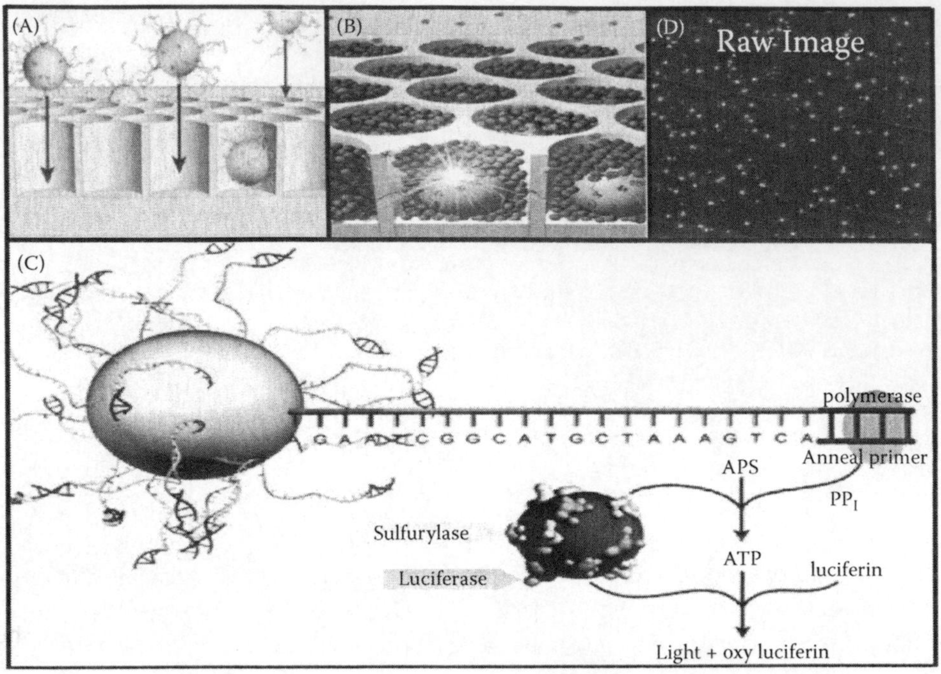

Figure 14.2 (See color insert.) Principles of 454 sequencing. (A) A single bead enters each well of the PicoTiterPlate. (B) Simultaneous, independent sequencing of each bead by flooding the plate sequentially with each dNTP. (C) Pyrosequencing chemistry. Pyrophosphate (PPi) is released with the incorporation of a dNTP into the extending DNA strand. ATP sulfurylase forms ATP by joining the PPi with ADP. Luciferace utilizes the ATP to convert luciferin to oxyluciferin, which emits visible light. (D) The visible light is detected and recorded. (Image reproduced with permission from 454 Sequencing, Roche Diagnostics, Indianapolis, Indiana.)

Applied Biosystems (Foster City, California), a subsidiary of Life Technologies (Carlsbad, California) released the SOLiD® sequencing system in 2007 (Applied Biosystems 2007). Similar to the Roche approach, the SOLiD System generates a DNA library via DNA capture beads and an emulsion PCR (Applied Biosystems 2010; Valouev et al. 2008). DNA template is fragmented and each fragment is attached to a DNA capture bead with an adapter sequence (P1 adapter) to allow for primer binding. The beads are attached to a slide or FlowChip. Instead of the sequence-by-synthesis method utilized by 454 Life Sciences and Illumina, SOLiD technology employs a sequence-by-ligation method. In the sequence-by-ligation method, a universal primer hybridizes to the P1 adapter sequence and the FlowChip is flooded with probes eight nucleotides in length. More than 1,000 different probes are utilized on a single chip. The first five nucleotides of the probe are sequence-specific, and the last three are universal. A two-nucleotide dye system is used with the fluorescence of the probe based solely on the combination of the first two nucleotides of the probe. After flooding the slide or FlowChip, complimentary probes are ligated to the DNA fragments beyond the primer. Following ligation, the fluorescent label and three universal nucleotides are cleaved, and the color of the fluorescence is detected and recorded by a camera and software. The first two nucleotides in the sequence are thus identified, followed by three unknown bases. The ligation process is repeated multiple times until the length of the entire fragment has been sequenced. Following the first series of ligation reactions, the first two nucleotides have been identified in every five-base string. Once the first round of

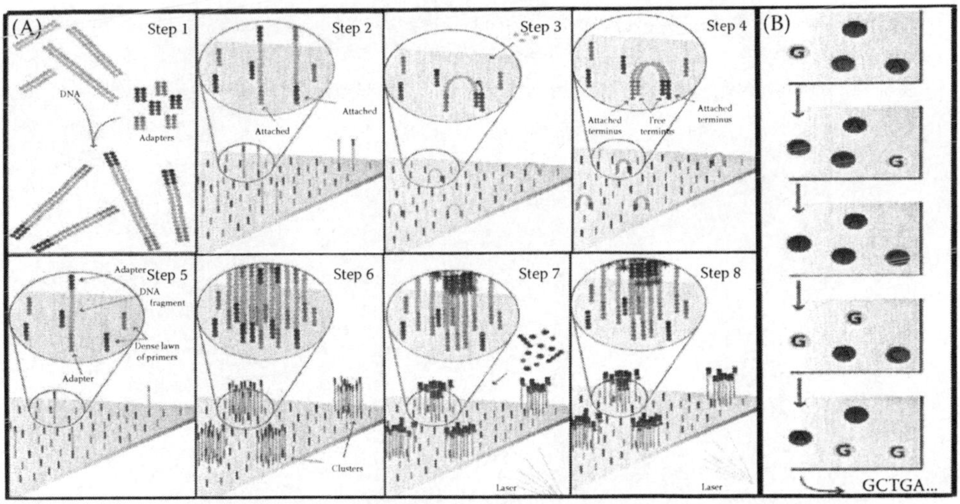

Figure 14.3 (See color insert.) Principles of Illumina sequencing. (A) TruSeq chemistry: Step 1: DNA is fragmented and adapters are ligated to both ends of the fragments. Step 2: The adapters hybridize to the primers on the flow cell, immobilizing the template fragments. Steps 3–4: Bridge amplification. Step 3: Elongation occurs unidirectionally from the primer on the flow cell. The extended DNA molecule bends and attaches to the reverse PCR primer on the flow cell. Step 4: Elongation occurs from the reverse primer. Step 5: The bridged strands are denatured, forming single-stranded templates. Step 6: Steps 3–5 are repeated to form a cluster containing millions of copies of the template fragment. Step 7: The flow cell is flooded with fluorescently labeled terminator dNTPs; the incorporated dNTPs are recorded. Step 8: The fluorescent label and terminator are removed, and elongation/sequencing continues. (B) The incorporated dNTP is determined based on the color of the fluorescence detected in each position following each round of flow-cell nucleotide flooding. (Image reproduced with permission from Illumina, Inc., San Diego, California.)

ligation steps is completed, the newly synthesized strand of DNA along with the primer is denatured from the template and the template is reset with a new primer complimentary to the n-1 position of the P1 adapter. This process of ligation and template reset is repeated for five separate primers. The template resetting in the n-1 position results in each base of the DNA template being analyzed two times (referred to as 2x or twofold coverage). The software merges the colors from the individual runs and converts them into nucleotide calls (Applied Biosystems 2010; Valouev et al. 2008). The new 5500 Series Genetic Analyzers released by Applied Biosystems introduced an upgrade from the previous SOLiD 4 System platform. The system upgrade includes a fifth round of primer resetting and ligation (Exact Call Chemistry, ECC). In the ECC ligation and detection round, a new set of three nucleotide color-coded probes is used to increase sequencing redundancy (Figure 14.4) (Applied Biosystems 2011b).

In early 2011, Life Technologies (Carlsbad, California) released the Ion Personal Genome Machine™ (PGM) benchtop sequencer under the subsidiary company Ion Torrent systems. On April 24, 2012, Life Technologies issued a press release announcing the installation of the first Ion Proton Sequencers, the second generation of Ion Torrent sequencing instrumentation (Life Technologies 2012a). The PGM and Proton™ still require the formation of a DNA library, identifying Ion Torrent systems as the latest second-generation-type technology. The Ion Torrent platforms represent the first completely semiconductor-based

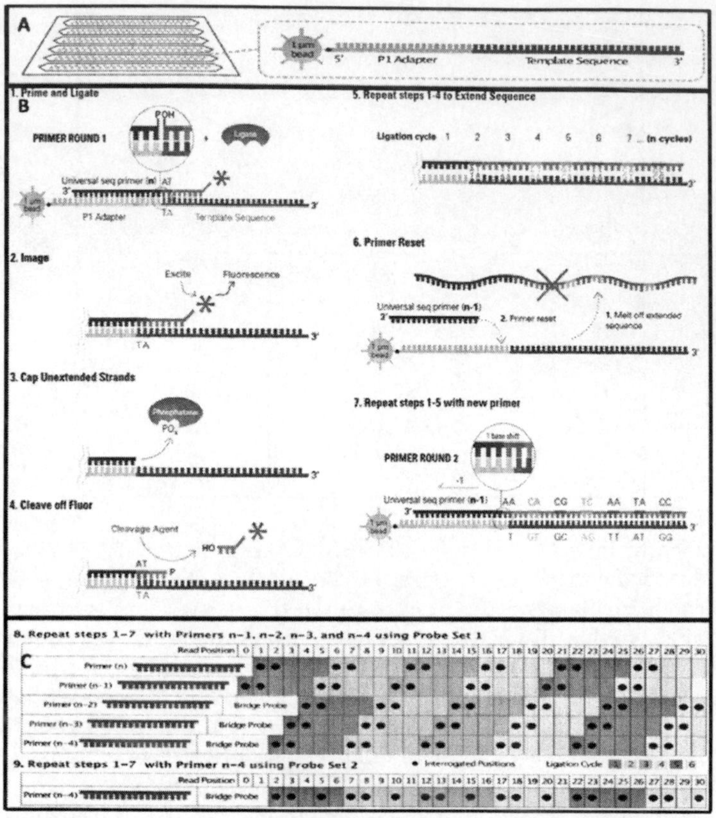

Figure 14.4 (See color insert.) Principles of SOLiD sequencing. (A) A FlowChip with millions of DNA capture beads attached. A single bead contains a polony of a single DNA fragment, each with a 5′ P1 adapter sequence. (B) Sequence-by-ligation chemistry. Step 1: A universal primer binds to the P1 adapter. The chip is flooded with 8-mer probes. The probe corresponding to the first five nucleotides of the sequence ligates to the template. Step 2: A two-nucleotide dye system is used. The color of the fluorescence is specific to the first two nucleotides of the probe. The first two nucleotides in the sequence are identified. Steps 3–4: The fluorescent tag, along with the last three nucleotides of the probe are removed. Step 5: Steps 1–4 are repeated. The nucleotides of the sequence complimentary to the first two in each probe are identified along the template. Step 6: The primer sequence is melted off. A new universal primer binds to the P1 adapter at the n-1 position. Steps 7–8: Steps 1–5 are repeated for six sets of primers. Step 8: A schematic of the nucleotides identified with each primer set. The identified nucleotides are marked with a dot. Step 9: An additional primer set, the ECC sequencing round, is utilized for additional coverage and increased quality assurance. (Image reproduced with permission from Life Technologies Corporation, Foster City, California).

sequencing technology, placing it in a category between current second-generation systems and third-generation platforms under development. The Ion Torrent system has the capability for fully automated template fragmentation, amplification, and library formation. The automation eliminates the labor and time dedicated to this process with other technologies and streamlines DNA sequencing. The Ion Sequencing array chips contain millions of micromachined wells, each with an ion-sensitive layer and detector in the base of the well. The structure and size of the wells allows only a single DNA template-containing bead to enter into each well. Sequencing is performed by flooding the chip sequentially with each of the four nucleotides. The incorporation of a dNTP into the growing DNA strand is

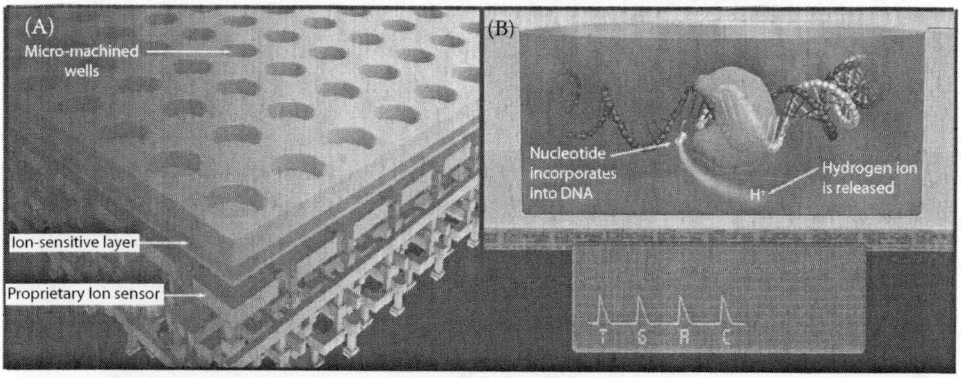

Figure 14.5 (See color insert.) Principles of Ion Torrent sequencing. (A) Ion sequencing array chip. The top layer contains millions of micromachined wells. Underneath the wells is an ion-sensitive layer. The base of the chip contains a pH detector. (B) Semiconductor sequencing chemistry. Sequencing is performed by sequentially flooding the chip with each nucleotide. The incorporation of a dNTP into the growing DNA strand is accompanied by the release of a hydrogen ion. The pH of the solution within the ion-sensitive layer of the chip changes with the release of the hydrogen ion. The change in pH is sensed by the detector and recorded as nucleotide incorporation. (Image reproduced with permission from Ion Torrent Systems, Inc., South San Francisco, California.)

accompanied by the release of a hydrogen ion that changes the pH of the solution within the ion-sensitive layer of the chip. A highly sensitive pH meter, Ion Sensor, detects the pH change. The specific nucleotide incorporations are thereby determined and recorded for each cycle of the sequencing process in each well of the array chip (Figure 14.5) without the use of fluorescence detection. If two identical bases are incorporated sequentially, the change in pH and the electrical signal detected increases proportionately (Rothberg et al. 2011).

14.3 Applications of Second-Generation Sequencing Technologies

The Human Genome Project (HGP) was started in 1990. The goal of the HGP was to sequence the entire human genome and identify all its genes. Sequencing was done primarily using the standard Sanger methods. The HGP took 13 years and approximately $3 billion to complete (Human Genome Project 2012). Second-generation sequencing technologies introduced a new era of whole-genome sequencing at a fraction of the cost of dideoxynucleotide terminator sequencing. In 2007, DNA structure co-discoverer James Watson's genome was sequenced using the 454 Genome Sequencer FLX™ System. The total time of the project was less than 2 months, and the cost under $1 million (454 Life Sciences Corporation 2007a).

Since the introduction of second-generation sequencing platforms, the technologies have been utilized for an array of sequencing studies, including *de novo* and resequencing whole genome analysis (Ghosh et al. 2011; Jiang et al. 2011; Lee et al. 2011; Qiu et al. 2012; Riccombeni et al. 2012), targeted enrichment (Besaratinia et al. 2012; Boers et al. 2012; Bras et al. 2012; Koudijs et al. 2011; Meyer and Salzburger 2012), as well as RNA and transcriptome evaluation of splicing (Twine et al. 2011; Wang et al. 2009) and gene expression

(Bainbridge et al. 2006; Chin et al. 2011; Jones et al. 2008; Zhou et al. 2012). The advancements of new sequencing systems motivated the initiation of the 1000 Genomes Project, a large-scale population study aimed at identifying human genetic variation in different populations. The launch of the 1000 Genomes Project utilized nine sequencing centers, each employing the Illumina/Solexa, 454, and/or the SOLiD platforms. The initial goal of the first set of studies was to collect 2x coverage of genomes from 1,000 individuals. Due to the relative ease of data collection using the second-generation platforms, upon completion of the pilot studies, 4x coverage of the whole genome and 20x coverage of the whole exome for 2,500 individuals, and 40x coverage of the whole genome for 500 individuals was accomplished (Clarke et al. 2012).

The introduction of second-generation sequencing technologies has allowed for the *de novo* sequencing of whole genomes for virtually any organism, not just humans. Rapid expansions in the available knowledge on the genetic makeup of hundreds of organisms have been made due to advancements in sequencing. Other whole nuclear genomes that have recently been characterized include: the channel catfish *Icatalurus punctatus* (Jiang et al. 2011) and the yeast *Candida orthopsilosis* (Riccombeni et al. 2012) using the 454 systems, a female yak, *Bos grunniens*, with the Illumina HiSeq® (Qiu et al. 2012), the woodland strawberry, *Fragaria vesca* (Shulaev et al. 2011) and the laryngotracheitis virus (Lee et al. 2011) using the SOLiD platform, and the endangered Tasmanian devil, *Sarcophilus harrisii* (Miller et al. 2011) using both the 454 and the Illumina technologies. Mitochondrial and chloroplast genomes have also been sequenced and characterized using second-generation technologies (He et al. 2010; Uthaipaisanwong et al. 2012; Van de Sande 2012; Zhang et al. 2011). The benefits of nuclear, mitochondrial, and chloroplast genome studies include improved insight into species evolution and phylogeny (Ghosh et al. 2011) and better characterization of a population's genetic diversity. Conservation genetics applications allow for causal evaluation of species endangerment, as with the Tasmanian devil (Miller et al. 2011) and Tasmanian tiger (Menzies et al. 2012), and assessment of environmental adaptation as observed in the yak (Qiu et al. 2012).

Next-generation approaches simplify identification of bacterial communities and assist in the characterization of potential causes for virulence in bacteria and viruses. A major focus for researchers has been methicillin-resistant *Staphylococcus aureus* (MRSA) due to the dangers associated with it. Any infection caused by the bacterium *S. aureus* can be dangerous if not treated. A skin infection can spread into the bones and the lungs and is potentially fatal. The ease of transmission via touch makes *Staphylococcus* infections even more dangerous because they can easily turn into an outbreak. MRSA is a strain of *S. aureus* that does not respond to the typical first line of defense against an infection, the antibiotic methicillin. The immunity against an initial treatment of the infection makes MRSA even more dangerous than a typical staph infection. It is important to rapidly detect MRSA in order to properly treat it before it spreads within the body or to other individuals (Centers for Disease Control 2012). Sequencing studies using all the second-generation platforms have focused on the evolution and detection of MRSA. Howden et al. (2010) sequenced the full genome of several methicillin-resistant strains of *S. aureus* using the SOLiD and the 454. The same group later utilized the Ion PGM to sequence regions of the genome for other drug-resistant strains of *S. aureus* and map the evolution of the resistance (Howden et al. 2011). Another study using the Illumina MiSeq® evaluated the ability to rapidly detect isolates belonging to varying clusters of *S. aureus*. Isolates were identified within 5 days of culturing and the clusters containing isolates of MRSA were able to be identified for early outbreak detection (Eyre et al. 2012). The MiSeq detection system was utilized to identify a specific MRSA cluster from an outbreak in a neonatal facility, demonstrating the capability of the MiSeq platform to rapidly provide outbreak-relevant information. The timely identification of the causative cluster allows for targeted treatment of the outbreak (Eyre et al. 2012). Other outbreaks that were rapidly identified via whole-genome sequencing were the virulent outbreak of Shiga toxin (Stx)-producing *Escherichia coli* O104:H4 within a month (Mellmann et al. 2011) as well as the *Klebsiella* Oxa-48 outbreak in a Dutch hospital (University of Munster 2011), both utilizing the Ion PGM. Improved analysis of cancerous tissues identifying underlying causes and predicted treatment responses in patients has also benefited from whole-genome sequencing, as seen in Ellis et al. (2012) comparing the genomes of 46 breast cancer tumors and normal tissue utilizing the Illumina system. The group was able to identify somatic alterations believed to cause the abnormal growth, and to chart their response to aromatase inhibitor, the common treatment for such cancers (Ellis et al. 2012). Other cancer studies have focused on epithelial cancer utilizing the SOLiD platforms (Hillmer et al. 2011) and mesothelioma sequencing on the 454 (Bueno et al. 2010). In 2008, Ley et al. (2008) utilized the Illumina Genome Analyzer® to compare normal tissue and tumorous tissue from an individual diagnosed with acute myeloid leukemia (AML). The whole-genome comparison revealed 10 tumor-specific mutations, two identified in previous AML patients and eight novel mutations present exclusively in the tumor, and in almost all of the cells analyzed from the tumor (Ley et al. 2008). Such studies are significant for mapping the origination and progression of cancerous tissues in order to better understand, diagnose, and treat patients.

Whole-genome sequencing data has provided a detailed characterization of most regions of the genome, giving way to sufficient information for well-established targeted DNA studies using the second-generation technologies. In a targeted DNA study, specific genes or regions of the genome are selected for sequencing. The advantages to a targeted study as opposed to a whole-genome analysis are decreased cost, increased coverage and turnaround time, and the capability to analyze and compare larger datasets due to the decrease in required throughput (Lin et al. 2012). This approach is often used for disease analysis when suspected disease causing genes or variants have been identified. In 2012, Bras et al. (2012) sequenced ATP13A2, a gene implicated in a class of metabolic storage disorders, in a large pedigree afflicted by the disorder. The study was able to identify a homozygous mutation that segregated with the disease, and has identified it as a recessive cause of the metabolic storage disorder (Bras et al. 2012). Targeted DNA enrichment has also been implemented in population studies for phylogenetic (Meyer and Salzburger 2012) and genetic diversity analyses. Meyer and Salzburger (2012) developed a primer set for classification of East African cichlids, a key model animal in evolutionary adaptation studies. Marklund and Carlborg (2010) identified SNPs specific to individual lines of chickens, and were able to use them to assess the level of diversity between and among lineages. Other examples of studies utilizing targeted DNA sequencing include bacterial strain identification (Boers et al. 2012) and cancer studies (Besaratinia et al. 2012; Kalender Atak et al. 2012; Koudijs et al. 2011; Robbins et al. 2011).

Another aspect of sequencing being utilized more widely since the implementation of second-generation technologies is RNA sequencing, most often in the form of transcriptome studies. While genome sequencing provides key insights into mutations and variations within individuals, genomic analysis does not directly infer gene expression. Transcriptome studies focus on the transcripts of the DNA, or the RNA sequences utilized for protein synthesis. The portion of the genome included in the transcriptome is only a fraction of the whole genome. Such studies are valuable for comparing epigenetic variations, such as alternative splicing (Twine et al. 2011; Wang et al. 2009), gene expression responses or adaptations to different stresses (Chen et al. 2012; Zhou et al. 2012) or to evaluate expression variations in diseased tissues (Bainbridge et al. 2006; Bueno et al. 2010; Jones et al. 2008). Twine et al. (2011) utilized postmortem brain samples for transcriptome analysis to compare gene expression in normal human brain to that of Alzheimer-affected brains. Differences in gene expression and the number of splicing variants were identified (Twine et al. 2011). In 2003, Juusola and Ballantyne (2003) demonstrated the ability to utilize mRNA for body fluid identification. Gene expression varies not just from person to person, but also from tissue to tissue and fluid to fluid in an individual. The study successfully isolated messenger RNA (mRNA) from various body fluids, and was able to identify the fluid from which the samples were obtained (Juusola and Ballantyne 2003). Combining the ability to measure gene expression in postmortem samples of tissues demonstrated by Twine et al. (2011) with the ability to identify specific fluids and tissues, and ultimately the gene expression and splicing variations within them, illustrates the potential for utilization of deep-sequencing transcriptome analysis for molecular autopsy.

14.4 Third-Generation Sequencing Technologies

The second generation of sequencing demonstrated improvements over the traditional Sanger method. Time and cost associated with DNA sequencing have been reduced, and automation has reduced labor intensity (Farias-Hesson et al. 2010). Data analysis has been simplified for new systems, and capabilities have been expanded to include mixture (Holland et al. 2011) and heteroplasmy resolution (He et al. 2010). DNA library preparation is necessary for accurate second-generation sequencing. The signal being recognized (light, fluorescence, or pH change) needs to be of sufficient intensity to be detected. The level of signal required for detection in second-generation systems cannot typically be obtained from a single DNA molecule. The cumulative release of signal from thousands of template copies enables recognition by the camera or detector. The requirements of DNA library creation may limit the quality and throughput efficiency in second-generation sequencing. DNA library formation typically requires a significant amount of starting DNA. Depending on the type of sequencing being performed, the required amount of template DNA for second-generation platforms ranges from 50 ng to 20 μg (Berglund et al. 2011; Metzker 2010; Quail et al. 2012). In research settings, large amounts of high-quality DNA are usually available. However, in a casework laboratory, the amount and quality of the DNA available is often limited. DNA library preparation can be tedious and time-consuming, requiring an experienced technician and up to 12–24 h. Advancements in the process allowing semiautomation or full automation can substantially streamline the process (Farias-Hesson et al. 2010). However, even an automated system, such as the Ion Torrent OneTouch™ system, requires 3.5 h for preparation, and 3 h for library construction (Life Technologies 2011). PCR amplification is often utilized to generate a targeted library pool for sequencing; however, this approach can introduce the potential for PCR bias (preferential amplification) and polymerase error (Schadt et al. 2010). The large data pool generated by the second-generation sequencing systems requires significantly greater computer power than traditional methods to process the data effectively (Treangen and

Salzberg 2011). Recent third-generation sequencing technologies have introduced single-molecule sequencing that does not require the preparation of a DNA library. Through several advancements, third-generation technologies boast faster turnaround times, longer read lengths, reduced template requirements (less than 1 μg for most sequencing reactions), improved data quality, and effective analysis pipelines (Metzker 2010; Pacific Biosciences 2012; Quail et al. 2012; Schadt et al. 2010).

Pacific Biosciences™ (Menlo Park, California) has developed the most promising third-generation sequencing system to date, the PacBio *RS*. The system was made commercially available in April 2011, to a select number of laboratories (Pacific Biosciences 2011). The PacBio *RS* utilizes the Pacific Biosciences Single Molecule, Real Time (SMRT®) Technology. As the name suggests, SMRT Technology is able to analyze the elongation of a single DNA molecule as it is being synthesized. The technology utilizes a sequencing chip called a SMRT Cell that is made up of ~75,000 zero-mode waveguides (ZMW). A ZMW is a hole tens of nanometers in diameter placed in a metal film on a glass substrate (Foquet et al. 2008). The diameter of the ZMW hole limits the passage of visible laser light completely through the ZMW. The visible light exponentially decays as it travels through the ZMW, allowing illumination only at the very bottom of the ZMW. Within each ZMW, a DNA polymerase attached to the SMRT Cell enables the sequencing of a different single DNA molecule in parallel. Sequencing occurs by flooding the array of ZMWs with a cocktail of four phospho-linked dNTPs containing a nucleotide-specific, laser-excited fluorophore attached to the phosphate chain. The incorporation of a nucleotide will result in the fluorescence specific to that nucleotide. A balance of rapid diffusion (occurring in microseconds) coupled with slower incorporation (occurring in milliseconds) of the dNTPs allows for a high signal-to-noise ratio for the incorporated vs. unincorporated dNTP. As fluorescence only occurs at the bottom of the ZMW, the diffusion of unincorporated dNTPs out of the well prevents fluorescence from unincorporated dNTPs, diminishing background fluorescence. Nucleotide incorporation is detected in real time. As the incorporation of a dNTP onto the growing DNA strand is accompanied by the release of the phosphate group, linking the fluorescent marker to the phosphate group results in the removal of the label after incorporation. The release of the fluorescent tag permits the continuation of elongation (Figure 14.6) (Eid et al. 2009; Korlach et al. 2010).

The PacBio *RS* has enhanced sequencing capabilities over the second-generation technologies. The long read lengths allowed by the technology make the PacBio *RS* valuable for whole-genome sequencing. Due to the relatively small size of their genomes, bacteria have been a primary focus of whole-genome sequencing with the Pacific Biosciences system. In 2010, Chin et al. (2011) compared strains of cholera from different geographic regions to determine the origination of a virulent outbreak in Haiti. The isolate was traced back to an outbreak-causing strain from Bangladesh in 2002 (Chin et al. 2011). Similarly, Rasko et al. (2011) were able to trace the logical formation of a hemolytic-uremic syndrome-causing strain of *E. coli* in Germany by comparing whole-genome sequences between *E. coli* sources from various locations. In 2012, Carneiro et al. (2012) demonstrated the usability of this single-molecule sequencing system for targeted DNA analysis as well. The group utilized the platform to analyze medically significant DNA regions, and evaluated the utility of the PacBio *RS* in terms of sensitivity and specificity for SNP and genetic-variation detection (Carneiro et al. 2012).

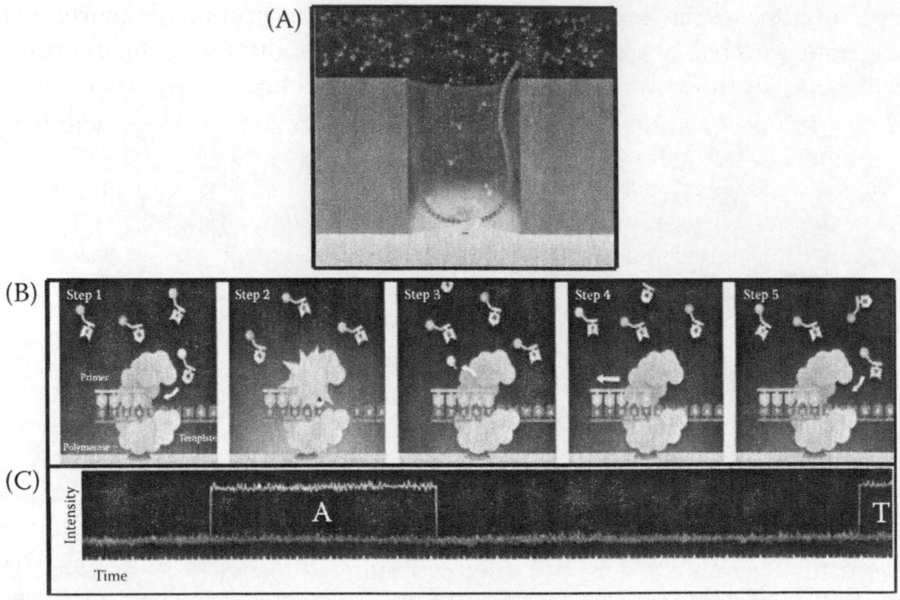

Figure 14.6 (See color insert.) Principles of Pacific Biosciences sequencing. (A) A ZMW with a DNA polymerase attached to the slide at the bottom of the well. Sequencing of a single DNA molecule occurs by flooding the array with all four fluorescently labeled dNTPs. Only the bottom of the well illuminates due to the rapid diffusion and the slower incorporation of the dNTPs. (B) SMRT sequencing chemistry. Step 1: All four fluorescently labeled dNTPs surround the polymerase active site. Step 2: The complimentary base is incorporated into the growing DNA strand, and the fluorescence emitted is specific to that base. Step 3: The fluorescent tag is removed to allow for elongation to continue. Step 4: The DNA strand moves as the DNA polymerase translocates to the next position on the template strand. Step 5: Steps 1–4 are repeated. (C) Real-time detection of the fluorescence from the fluorophore tag of the incorporated dNTP. (Image reproduced with permission from Pacific Biosciences of California, Inc., Menlo Park, California.)

14.5 Performance Comparison

Prior to the introduction of deep-sequencing technologies, the Sanger method was essentially implemented in all laboratories performing DNA sequencing. The new platforms allow laboratories to select the platform with the capabilities and parameters most suitable for the requirements of their experiments and goals. In order to accommodate research demands, most companies have commercialized multiple systems focusing on different applications. For high-throughput sequencing hubs requiring multiple megabases of sequence output, there are options such as the 454 GX FLX Titanium Systems (Titanium XLR70 and Titanium XL+), the three SOLiD platforms (5500 Genetic Analyzer, 5500xl-Genetic Analyzer, and SOLiD 4), the Illumina HiSeq and Genome Analyzer IIx (GA IIx), and the PacBio *RS*. These systems can provide relatively long read lengths and allow for the processing of extensive genome libraries, at the cost of long run times and expensive instrumentation. Many research laboratories would not need the capacity associated with these high-volume systems, or cannot justify the instrumentation expenses and cost associated when handling a lower volume of samples. Illumina, 454 Life Sciences, and Ion Torrent have developed smaller benchtop sequencers for personal laboratory sequencing. The 454

GS Junior, Illumina MiSeq, Ion PGM, and Ion Proton generally provide shorter maximum read lengths and decreased throughput per run; however, the size, cost, and run times of these benchtop sequencers are dramatically reduced from their workhorse counterparts.

One of the most important features of a sequencing system is the maximum read length obtained. Maximum read length refers to the longest string of consecutive nucleotides able to be sequenced accurately in a single pass. Many targeted DNA studies focus on short segments of DNA, and strategies for the resequencing of previously sequenced genomes can often be accomplished using short read lengths. Longer read lengths are valuable for *de novo* genome sequencing and where the assembly of sequence motifs is under investigation, for example, microsatellite markers. In order to accurately assemble a fragment of DNA containing a repeating motif, such as an STR or a homopolymeric stretch, the fragment must be long enough to fully span the entire repeating region. Assembling the sequence of an unknown genome is generally accomplished using overlapping sequence fragments. Larger sequence reads make it easier to quickly and correctly assemble a genome or gene rather than aligning many smaller pieces. The SOLiD systems have the shortest read lengths of the current deep-sequencing methods. When sequencing a fragmented genomic DNA library, a maximum read length of approximately 75 bp is reported (Rusk 2011). The current maximum read length for the Illumina platforms is reported to be approximately 2 x 150 bp. The library construction for Illumina prepares the DNA fragments for sequencing from both ends, with a maximum read length of 150 bp from each end. Combining the two reads provides a potential read of approximately 250 bp, allowing for a 50-bp overlap for accurate alignment of the two ends. Predictions for late 2012 included an upgrade to increase read lengths to 2 x 250 bp (Illumina 2010, 2012a,b). A recent study by Quail et al. (2012) confirmed the average 2 x 150 bp reads for the MiSeq, but averaged approximately 2 x 75 bp read lengths on the GA IIx and the Illumina HiSeq. The Ion PGM and Ion Proton claim an average read length of approximately 200 bp. This is a twofold increase in maximum read length from the original release of the PGM in 2011 (Life Technologies 2012). Quail et al. (2012), however, obtained average read lengths of 112–124 bp on the PGM. Lengths of up to 400 bp are consistently being attained in Ion Torrent's research and development laboratories, and these lengths are expected to be regularly attained by the end of 2012 (Life Technologies 2012b). The greatest benefit of the 454 technology over other second-generation technologies is its long read lengths. The original publication on the 454 technology in 2005 (Margulies et al. 2005) claimed the technology was optimized for a maximum of 100-bp reads, the best at the time for an NGS technology. The most current instrumentation commercialized by 454 Life Sciences Corporation boasts read lengths ranging from 400 bp on the benchtop GS Junior to up to 1,000 bp on the GS FLX Titanium XL+ (454 Life Sciences Corporation 2010). That said, none of the second-generation technologies can match the potential read lengths possible in the newest single-molecule systems. The average read length on the PacBio *RS* is approximately 2,200 bases, with a maximum read length of greater than 10 kb (Pacific Biosciences 2012).

It is necessary to find a balance between long read lengths and quality of data. Accuracy is a measure of the rate at which a correct base is assigned in the DNA sequence, typically measured as a percentage representing the number of bases correctly identified out of 100. Studies have demonstrated a higher error frequency associated with the long read lengths generated by the PacBio *RS* (Metzker 2010; Schadt et al. 2010). Early reports on the performance of the PacBio *RS* demonstrated an accuracy of 83% for a single sequencing pass on a 150-bp DNA fragment. The detected errors were stochastic, and resequencing the same fragment 15 times improved the accuracy to 99.3% (7 errors in 1,000 base assignments) (Eid et al. 2009). In late 2009, Pacific Biosciences reported improvements to their system, but no recent reports have been published on improved accuracy (Metzker 2010). Software systems for second-generation sequencing technologies utilize the Phred scoring system to assign a measurable quality to the assignment of each base. The consensus score for the entire sequence read is reported as a Q score. A Q10 score means the error rate is 1/10, Q20 error rate is 1/100, Q30 is 1/1,000, and a Q40 score designates an error rate of 1/10,000. By comparison, upon its release in July 2011, the average score for data from the Ion PGM on a 100-bp DNA fragment was Q23, meaning the accuracy of the system was 99.3%, equivalent to the PacBio *RS*. In 2012, the quality score of the same 100-bp fragment on the Ion PGM and Ion Proton was Q28, indicating an increase in accuracy to 99.8% (2 errors in 1,000 base assignments) (Life Technologies 2011). The Illumina systems do not report an average accuracy score for sequencing data obtained from their instrumentation. Illumina reported approximately 80% of the results for a 2 x 100 bp fragment having a Q30 or better accuracy score for the GA IIx and the HiSeq Systems (Illumina 2010, 2012a). The accuracy of the MiSeq is even higher, with 85% of the results for a 2 x 100 bp fragment having a score above Q30 (Illumina 2012b). For its pyrosequencing chemistry, 454 Life Sciences indicated a low misincorporation rate of natural nucleotides, resulting in a low substitution-error rate. The accuracy of the pyrosequencing system is as high as 99.997% (3 errors in 100,000 base assignments) for the recently released GS FLX Titanium XL+ platform. The GC FLX Titanium XLR70 produces similar quality data, with an accuracy of 99.995% (5 errors in 100,000 base assignments). The 454 benchtop sequencer, GS Junior, has lower accuracy compared to the larger instruments, with an average quality score of Q20 (99% accuracy) (454 Life Sciences Corporation 2010; Margulies et al. 2005). Although the read lengths on the SOLiD systems are short, the twofold coverage of each base, unique probe design, and two-nucleotide color-coding system, along with ECC, provides a quality check in the sequencing process that essentially eliminates base-substitution errors (Applied Biosystems 2011b). An average sequencing result obtained from any 5500 Genetic Analyzer series system is Q60 or above, meaning there is an incorrect base assignment less than once in every 1 million assignments designated by the system (Applied Biosystems 2011a).

One way to improve the accuracy of a system is to increase the coverage. Coverage is the average number of times a nucleotide position is interrogated during a sequencing run. For example, 2x coverage indicates that each base is interrogated twice. Eid et al. (2009) demonstrated the value of increased coverage in improving accuracy on the Pacific Biosciences platform; increasing coverage from 1x to 15x improved the accurate base-assignment percentage by 16.3%. The amount of coverage attained varies depending on the platform utilized, the number of overlapping fragments analyzed, and the number of sequencing cycles completed. Increasing the number of sequencing cycles will increase the coverage, and therefore the accuracy. However, increased number of cycles may introduce the possibility for more polymerase errors. The amount of required coverage for accurate sequence analysis varies depending on the type of project performed. When compiling a whole genome, short read lengths require an increased number of overlapping fragments for proper genome assembly. The more areas covered in the overlapping fragment regions, the more coverage required to sequence and assemble an entire genome. The longer read lengths allowed by the 454 platforms and PacBio *RS* provide the benefit of requiring fewer overlapping segments for genomic assembly (Koren et al. 2012; Schatz et al. 2012). An advantage with high coverage for forensic applications would be the ability to detect and measure low-level heteroplasmy in mitochondrial DNA or the ratio of contributors in DNA samples containing multiple sources (He et al. 2010).

Combining coverage and read length provides an estimate of the amount of throughput or output of the sequencing system. Throughput refers to the number of bases sequenced by the platform in a single run and is measured by multiplying the fragment read length by the number of reads, or coverage. High throughput is a benefit because it implies long read lengths, high coverage, or a combination of both. High output also requires a degree of efficiency in the sequencing system allowing for more data to be processed per run. The Illumina systems offer the capability of high coverage with short reads, giving them the capability for maximum output. The "lower throughput" benchtop MiSeq system produces approximately 1.5–2.0 Gb per run, with expected increases to as much a 7.0 Gb. The GA IIx falls in the moderate output category for an Illumina system with the potential to generate approximately 85–95 Gb per run. The high-output Illumina HiSeq produces approximately 600 Gb of sequencing data in a single run (Illumina 2010, 2012a,b). The Life Technologies 5500 Genetic Analyzer can produce approximately 7–9 Gb in a single day, and the most recent 5500xl Genetic Analyzer is capable of generating greater than 20 Gb per day (Applied Biosystems 2011a). However, the high amounts of data produced require adequate computer power to facilitate their analysis. Given the increased data output of deep-sequencing systems, lack of appropriate computational resources in terms of storage and processor capacity often impacts the difficulty of acquiring sufficient bioinformatics for efficiency of analysis. The longer read-length systems and the benchtop sequencers have a lower maximum output when comparing NGS platforms. The Ion Proton is currently able to process up to 10 Gb of data per run, a 100-fold increase from the original release of the Ion PGM in 2011 (Life Technologies 2012b). The 454 GS Junior, designed for long read lengths and rapid processing, averages approximately 35 Mb of sequencing data per run. The GS FLX Titanium XLR70 is capable of producing 450 Mb of data per run, and the newly released GS FLX Titanium XL+ can produce approximately 700 Mb in a single run. Pacific Biosciences designed its system for accurate results and genome evaluation with moderate throughput for easier data analysis. The maximum output per SMRT Cell is approximately 140 Mb (Pacific Biosciences 2012).

Run time is typically estimated as the time it takes to obtain sequencing results from a prepared DNA sample/library. Long read lengths, high coverage, and high throughput generally increase run time. For instance, the Illumina platforms have the longest run times, a byproduct of the deep coverage and throughput of the systems. The GA IIx requires approximately 2 weeks to complete a full 95-Gb run with read lengths of 2 x 150 bp (Illumina 2010). The same run on the HiSeq platform takes 11 days (Illumina 2012a). The MiSeq system can complete a 2 x 150 bp run in 27 h due to the reduction in output (Illumina 2012b). The run time on the 5500 Genetic Analyzer Series varies depending on the type of sequencing being performed. A single exome can be sequenced in 2 days, while a human genome with 4–5x coverage requires a week to complete (Applied Biosystems 2011a). The 454 platforms can produce data in less than a day; a 700-Mb run on the GS FLX Titanium XL+ takes 23 h, while in as little as 10 h, the GS Junior can complete a 35-Mb run or on the GS FLX Titanium XLR70 a 450-Mb run (454 Life Sciences Corporation 2010). The natural chemistry utilized by the Ion Torrent systems expedites the sequencing process. A run interrogating a 200-bp DNA fragment can be accomplished in 4.5 h on the Ion Proton (Life Technologies 2012b). The PacBio *RS* is not as rapid as the Ion Torrent systems, as it requires approximately 10 h to complete a run with fragment lengths above 1 Kb. The real-time sequencing, however, allows data to be analyzed as it is generated, results becoming available approximately 15 min into the run (Eid et al. 2009; Pacific Biosciences 2012).

Although deep coverage is beneficial for increasing sequencing accuracy, a point of diminishing returns is reached in which the accuracy of the data no longer improves with additional sequencing cycles. In order to maximize the efficiency of the NGS technologies and eliminate unnecessary processing, multiple samples can be analyzed simultaneously on the same chip or slide. Multiple samples can be evaluated on a single chip if there is a way to identify the different individuals and the coverage is adequate for accurate sequencing for all samples under interrogation. A barcode indexing system has been developed for all platforms that facilitates the labeling of multiple samples with unique identifier tags permitting simultaneous analysis. Sequencing multiple samples concurrently not only decreases sample turnaround time but also markedly reduces sequencing cost (Smith et al. 2010). Fordyce et al. (2011) evaluated 10 individuals together on the 454 GS Titanium FLX System. Each sample was interrogated an average of 160 times, providing sufficient coverage for accurate base assignments. Dudley et al. (2012) interrogated 48 individuals in a single run on the smaller 454 GS Junior and averaged greater than 800 and 1,000 reads per sample on two separate runs. Another group analyzed 96 individuals simultaneously on an older model SOLiD instrument, the SOLiD V3, averaging 262 reads per barcoded sample (Smith et al. 2010). The barcoding kits available on most NGS platforms allow for 96 identifying tags, allowing up to 96 samples to be processed simultaneously on a single chip; 454 Life Sciences recently achieved the capability to process 132 samples per chip (454 Life Sciences Corporation 2010).

Homopolymeric regions of DNA are often difficult to sequence accurately with the NG technologies. During dNTP incorporation with the chemistry utilized for the 454 and Ion Torrent platforms, the entire stretch of repeated nucleotides will be incorporated sequentially and detected simultaneously. The amount of signal emitted from the incorporation will increase, but the signal may be nonlinear. Studies utilizing pyrosequencing report decreased accuracy after more than 2 bp (Gharizadeh et al. 2006; Otto et al. 2010). Quail et al. (2012) reported a loss of accuracy with the Ion PGM after 8 identical bases, and a complete loss of coverage after 14 bp. In the comparison, nucleotide insertions were reported throughout the sequence data with the PacBio *RS*, including multiple insertions throughout homopolymeric stretches of sequence. Illumina has an increased capability to accurately sequence homopolymeric stretches of DNA due to their termination chemistry. The Illumina MiSeq is reported to be able to accurately sequence greater than 20-base homopolymers before producing any errors.

Oftentimes, a laboratory does not require increased performance. In a recent interview, Tim Stinear compared the Illumina MiSeq to the Ion PGM and suggested that the MiSeq outperformed the PGM in all areas, including read length, coverage, and output (Illumina 2012c). However, it was also expressed that the PGM did not underperform, as the PGM performed according to its expected specifications and performed satisfactorily in the project. Although read length, accuracy, coverage, and throughput are standard performance measures for comparing deep-sequencing platforms, many companies incorporated unique amenities into their products (Illumina 2012c). A highly competitive feature unique to the Applied Biosystems 5500 Series is the division of the glass slides into lanes, and the ability to utilize a single lane at a time. Single-lane analysis drastically decreases slide and reagent use, eliminating resource waste and decreasing cost while increasing the speed of a single run (Applied Biosystems 2011a). The Illumina instrument series provide the greatest adaptability in utilization. Four different platforms are offered, with each instrument allowing optimization for varying read lengths, faster run times, or increased output (Illumina 2010, 2012a,b). Due to the short run times, rapid improvements, and cost effectiveness, over 1,000 Ion PGM sequencers were sold in the first year of production. The Ion Torrent platforms are the most rapidly advancing, and the Ion Torrent community is quickly expanding (Life Technologies 2011, 2012b).

14.6 Next-Generation Sequencing Technologies

The latest advancement in DNA sequencing applications is by Oxford Nanopore Technologies® (Oxford, United Kingdom). Many of the sequencing methods discussed thus far utilize enzymes, fluorescently labeled nucleotides, and imaging equipment; most also require a first-round PCR or other library enhancement. Elimination of these steps and their required reagents will lead to a significant reduction in cost of sequencing reactions. At this writing, Oxford Nanopore Technologies was scheduled to release two systems called GridION and MinilON, with enhanced features and further reduction of the cost required for sequencing (Oxford Nanopore Technologies 2012k). The concept of using a nanopore as a biosensor was originally proposed in the mid-1990s (Oxford Nanopore Technologies 2012g). Oxford Nanopore Technologies was founded in 2005 based on the work of Hagan Baley of the University of Oxford (Oxford Nanopore Technologies 2012o).

Nanopore sequencing revolves around a hole that is approximately 1 nm in diameter, formed within a membrane. There are three classes of nanopores. A biological nanopore is created using bespoke, a proprietary pore; α-hemolysin heptameric protein creates pores in a lipid bilayer membrane (Oxford Nanopore Technologies 2012a). Protein pores are naturally found in cell membranes where they serve as ion channels used for the transport of molecules in and out of the cell. Bespoke can be produced and isolated on a large scale, allowing cost-effective commercialization of pore-containing membranes (Oxford Nanopore Technologies 2012a). Protein engineering techniques allow for the use of nanopores for multiple applications including strand sequencing, exonuclease sequencing, and protein analysis. Oxford Nanopore Technologies facilitated the sequencing of DNA by attaching a molecular motor to the pore to direct a strand of DNA through the pore, incorporating cyclodextrin within the nanopore to allow for exonuclease sequencing, and attaching ligands to the nanopore to allow targeted proteins to bind outside the pore (Oxford Nanopore Technologies 2012a).

Two other types of nanopores have been designed, solid state and hybrid, that have greater mechanical and chemical stability, improving on current biological nanopores (Oxford Nanopore Technologies 2012n). Solid-state nanopores may have variable-sized holes formed in synthetic membranes, such as graphene or silicon nitride, increasing the versatility for additional applications (Oxford Nanopore Technologies 2012g). Graphene is a single-atom-thick lattice of carbon that offers high electrical conductivity while being chemically inert (Oxford Nanopore Technologies 2012n). The pore is formed within this membrane by an ion or electron beam (Oxford Nanopore Technologies 2012n). Hybrid nanopores can be formed using the biological proteins to insert pores in a synthetic membrane.

Oxford Nanopore Technologies is developing two methods for DNA sequencing: strand sequencing and exonuclease sequencing (Figure 14.7). DNA sequencing is facilitated with a nanopore protein inserted across an electrically resistant, synthetically assembled lipid

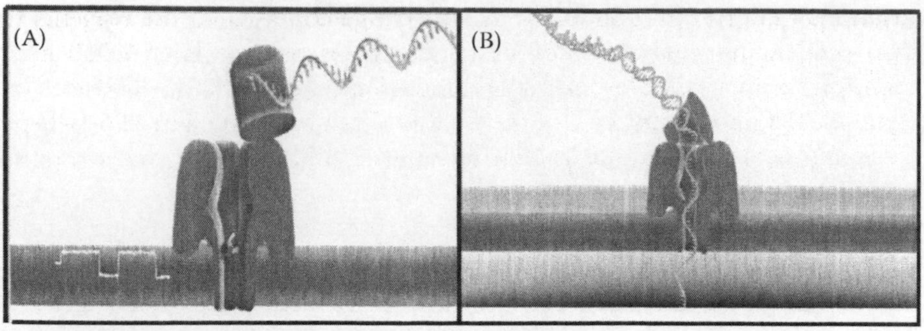

Figure 14.7 **(See color insert.)** Principles of nanopore sequencing. (A) Exonuclease sequencing. The nanopore embedded in the membrane bilayer is shown in blue. The processive enzyme, shown in green, cleaves the nucleotide bases in a sequential order. The cleaved nucleotide goes into the pore and undergoes a binding event with cyclodextrin, shown in red. This causes a change in current that is recorded and used to determine the order of nucleotides passing through the pore. (B) Strand sequencing. An enzyme complex, shown in green, binds to a double-stranded DNA fragment. The enzyme unzips the duplex strand and feeds the single-stranded DNA through the pore one base at a time. As the different nucleotides are fed through the pore, the change in current is measured to determine the order of nucleotides that has passed through the pore. (Image reproduced with permission from Oxford Nanopore Technologies, Boston, Massachusetts.)

bilayer membrane. An electrical potential is applied across the membrane where the current is allowed to flow through the nanopore protein (Oxford Nanopore Technologies 2012d). The passage of molecules through the pore causes a conformational change in the pore protein, altering the electrical current. The change in current is unique to the molecules that pass through the pore. In strand sequencing, a DNA-enzyme complex binds to the nanopore, facilitating the entry of a DNA molecule into the pore (Oxford Nanopore Technologies 2012d). As the DNA strand passes through the pore, the current fluctuations are translated into a DNA sequence by detecting the specific nucleotide bases passing through the pore (Oxford Nanopore Technologies 2012g). If a DNA template is prepared with a hairpin structure ligated to one end, both strands (sense and antisense) of the template would proceed through the pore as a single read (Oxford Nanopore Technologies 2012g).

Exonuclease sequencing utilizes a coupling of the protein nanopore with a processive enzyme that cleaves the individual nucleotides in sequential order from the DNA template. The nucleotides enter the nanopore and interact with cyclodextrin that has been covalently attached to the internal structure of the nanopore (Oxford Nanopore Technologies 2012j). The interaction of the nucleotide with cyclodextrin alters the electrical current, allowing for the sequential identification of the nucleotides (Oxford Nanopore Technologies 2012d). This approach also has the capability of detecting modified bases, such as methylated cytosine (Wallace et al. 2010).

Oxford Nanopore Technologies couples its patented sequencing methods with proprietary electronics enabling multiple nanopore-sensing processes to be performed in parallel, and allowing real time analysis and collection of data. The instrument that performs the data collection is known as a node, available in two different forms—the GridION and MinION. Nodes utilize a single-use cartridge. The cartridge contains an array chip that has multiple microwells. Within each microwell, a single nanopore is placed across a membrane (Oxford Nanopore Technologies 2012o). Each microwell has an electrode that can sense variations in current individually. The cartridge contains all the reagents required to perform sequencing analysis. Each cartridge is designed to have 2,000 nanopores, with expansion to approximately 8,000 nanopores expected in 2013 (Oxford Nanopore Technologies 2012k). Each cartridge of the GridION can be adapted to a 96-well-plate format, expanding the capability of cartridge to sequentially analyze 96 individual samples. When using the GridION with the 96-well plate adaptor, each sample can be run until the user-defined termination point has been reached, causing the sample to be expelled, cartridge flushed, and the next sample injected (Oxford Nanopore Technologies 2012m). The entire sequencing reaction takes place within this cartridge.

Each node is capable of tens of thousands of current recordings per second, collecting sequence data from the individual nanopores within the cartridge (Oxford Nanopore Technologies 2012f). Multiple nodes can be connected to a single desktop instrument with the capability to communicate with other nodes via a local network (Oxford Nanopore Technologies 2012h). The single desktop instrument can monitor and control multiple nodes individually or a cluster of nodes concurrently.

The MiniION is a single-use disposable DNA-sequencing platform that has a workflow similar to that of the GridION. The MiniION is set to be priced at less than $900 and will be the size of a standard Universal Serial Bus (USB) memory drive that can be plugged into any computer via the USB interface (Oxford Nanopore Technologies 2012k). The GridION and MiniION are capable of supporting multiple chemistries including the strand-sequencing method and the exonuclease-sequencing method. Just as multiple GridION nodes can be controlled simultaneously, multiple MiniION units may be connected by USB interface and be monitored and controlled concurrently.

Direct analysis of RNA strands is under development with the GridION platform. A processive enzyme is being developed, making it possible to analyze RNA directly, avoiding the need to build a cDNA library as is done conventionally (Oxford Nanopore Technologies 2012l). In addition to RNA analysis, the nanopore approach can be used to quantify microRNA (miRNA), which are noncoding RNAs that regulate gene expression and play a role in controlling several cellular processes such as proliferation, apoptosis, differentiation, and so forth (Wallace et al. 2010). Oxford Nanopore Technologies has developed probes complementary to the common miRNAs of interest (Oxford Nanopore Technologies 2012i). The binding of the probe to the miRNA causes the double-stranded complex to impede the advancement of the complex through the pore, causing a change in current that allows for the quantification and detection of miRNAs. At this writing, the nanopore approach for RNA applications was still under development and a release date had not been confirmed.

Oxford Nanopore Technologies will have many advantages that will truly set it apart from the current second-generation technologies. A major advantage of the real-time sequencing analysis is that it allows the user to adjust the parameters during the sequencing run. Variables such as temperature, salinity, the potential applied to the membrane, and movement of the sample can be adjusted while the sequencing run is under way. The capabilities of nanopore sequencing also allow sequence data collection to be terminated based on parameters set by the user. The data collection can be terminated after a certain time point (minutes or days), a specified coverage depth (10x or 1,000x), through a specific polymorphism, or when a designated sequence has bee sequenced. Therefore, the throughput of the nanopore-sequencing technology wi specified on the parameters set by the user. The technology has the potential to tens of gigabases per day (Oxford Nanopore Technologies 2012e). Oxford Nanopore Technologies report the ability to sequence a 48-kb genome of the Lambda phage as one complete fragment with both the sense and the antisense strand sequenced for a total of 100 kb (Oxford Nanopore Technologies 2012k). The Oxford Nanopore system does not require library preparation and dsDNA can be introduced directly (Oxford Nanopore Technologies 2012k). Further improvements will allow the current capacity of 2,000 nanopores to be quadrupled to 8,000 nanopores per cartridge. These cartridges will make it possible for a complete human genome to be sequenced in 15 minutes when utilizing 20 nodes simultaneously (Oxford Nanopore Technologies 2012k). Oxford Nanopore Technologies also suggests the possibility of dsDNA being sequenced directly from blood in some cases, with no sample preparation required (Oxford Nanopore Technologies 2012k).

14.7 Implications of Deep Sequencing in Forensic Investigations

PCR amplification with dye-labeled primers coupled with CE fragment analysis methods utilizing laser-excited fluorophores has been a standard approach to STR analysis in the scientific community for more than a decade. Levels of spectral overlap during fluorophore excitation and steric interactions of the fluorophores during DNA synthesis pose some interpretation and quantification limitations. A limitation of the CE method is the restricted "real estate" available in an electrophoretic run. For forensic purposes, assays must use predominantly small amplicon-sized products that tend to comigrate in the 100–300 bp region of the run. Sequencing methods utilizing the Sanger method with fluorescently labeled dideoxynucleotides and CE are more costly, are labor intensive, and are not as informative as deep-sequencing technologies when interrogating anything more than a few hundred nucleotides. Current forensic applications can incorporate the use of deep-sequencing technologies in the areas of degraded sample analysis, relationship testing, and missing persons identification.

14.7.1 Short Tandem Repeats

Genome studies have shown that more than 20,000 short tandem repeat (STR) loci are present within the human genome (Butler and Hill 2012). Currently, there are only 13 core loci searched by the Combined DNA Index System (CODIS) software in the United States. In April 2011, an expansion of the set of core STR loci was proposed by the FBI Laboratory aimed at reducing the likelihood of adventitious matches of evidentiary target profiles against the ever-increasing number of offender samples with the National DNA Index System (NDIS) and at improving discrimination power in casework comparisons (Butler 2012).

Limited studies have evaluated the analysis of STRs utilizing deep-sequencing technologies for forensic applications. The pyrosequencing approach has been successfully utilized to type 8 Y-STR loci (Edlund and Allen 2009), multiplex 13 STR markers (Scheible et al. 2011), and characterize 10 individuals at 5 STR loci (Fordyce et al. 2011). A challenge faced in several of these studies was the lack of expert software that allowed appropriate STR data interpretation. Van Neste et al. (2012) conclude that while it is feasible to use the GS FLX platform for STR analysis, the high cost combined with misinterpretation due to high error rate may limit its applications in STR analysis. Fordyce et al. (2011) developed an algorithm that facilitated the interpretation of STR sequencing data generated from the GS FLX platform, and Holland et al. (2011) utilized the NextGENe software (Softgenetics, LLC, College Station, Pennsylvania) to interpret the STR data they acquired utilizing the 454 GS Junior platform.

There are many advantages to using deep-sequencing technologies for STR typing. Fordyce et al. (2011) typed 5 STR loci for 10 individuals and found that 6 out of 10 samples had additional sequence polymorphisms not identified through capillary electrophoresis. These samples were found to have nucleotide substitutions, deletions, or insertions either in the flanking region or within the repeat structure not detectable by the conventional fragment analysis performed in forensic testing. Using deep-sequencing technologies, both the number of repeats and the position of sequence polymorphisms are captured simultaneously. For example, individuals who are typed as homozygous with conventional STR analysis methods may in actuality be same-length heterozygotes when nucleotide level variation is sequenced. Nucleotide variation among the repeat motifs has been reported for several STRs loci (Pemberton et al. 2009). These polymorphisms often arise due to a single nucleotide change within an individual repeat block of the locus motif and impact the allelic diversity and population distribution at the locus. Current CE methods define alleles based solely on the variation in repeat motif number, as inferred by amplicon length, and overlook the hidden variation present at the locus; for example, a heterozygous 11 allele (same length alleles with internal sequence polymorphisms) for a given loci will be defined as homozygous with CE since only fragment-length information is captured. By capturing sequence information of the STRs, one can detect the presence of sequence variants within the STRs. Capturing sequence polymorphisms of this nature makes it possible to increase allelic diversity, which then increases the discrimination power of the traditional forensic STR loci. Current methods, such as electrospray ionization mass spectrometry (ESI-MS) can detect the hidden variation present within a given STR locus but cannot identify the exact location of the variation within the locus (Planz and Hall 2012). The implementation of deep-sequencing technologies will allow the most complete characterization of the variation within microsatellite loci, as multiple polymorphisms within an STR amplicon may mask differentiated alleles both in the mass spectrometry and traditional CE assays.

An issue originally raised when using NGS for STR analysis is the low read lengths obtained for most systems. A strategy to overcome the read-length limitations may be to generate sequencing libraries from miniSTR (71–250 bp) amplicons. Read-length limitations may no longer be a concern as manufacturers are constantly improving protocols to achieve longer read lengths and more robust sequencing strategies. Ion Torrent systems are expected to soon achieve read lengths over 400 bp. Coupled with a recently introduced 96-barcode indexing adaptor system, the possibility of sequencing a suite of markers from 96 individuals for STRs and their entire mitochondrial genome in one sequencing run may be realized.

14.7.2 Mitochondrial DNA Analysis

Mitochondrial DNA plays a pivotal role in missing persons/unidentified remains cases, as well as cases where there is limited evidence, such as shed hairs. MtDNA is maternally inherited and each mitochondrion possesses 2–10 copies of its genome with a density of 1,000 organelles per cell (Budowle et al. 2003). The abundance of mtDNA within a cell makes it a viable resource in cases where the recovered DNA is degraded or quantity is low. The region of most interest in forensic applications is the control region that contains hypervariable region 1 (HV1) and hypervariable region 2(HV2).

In forensic applications, the mtDNA of an evidence sample is compared with either a reference sample for a direct comparison or multiple family-reference samples to assess a potential maternal relationship. There is considerable interest in the accurate detection of heteroplasmic sites and their role in interpretation in sample comparisons (Budowle et al. 2003). Heteroplasmy refers to the presence of multiple mtDNA sequence variants within a cell, tissue, or individual. Heteroplasmy can be length-based through the insertion of one or more nucleotides or site-specific with the presence of alternate nucleotides at a certain position. Detection of heteroplasmy was initially characterized using denaturing gradient gel electrophoresis (Tully et al. 2000). The common approaches used for mtDNA sequence analysis are very laborious and have discouraged many laboratories from implementing mtDNA testing. Analysis typically involves the amplification of sample DNA for the two hypervariable regions as one to eight different amplicons, followed by the standard Sanger sequencing process. The utilization of deep-sequencing technologies in mtDNA analysis has the potential to reduce the labor and cost associated with the traditional method. The analysis of an individual's entire mitochondrial genome can be accomplished with any of the current NGS platforms more labor- and cost-effectively than the two regions of the mitochondrial genome as currently assessed with Sanger sequencing.

Sanger methods can identify the presence of point heteroplasmy when the minor species is typically above 20%. Tang and Huang (2010) reported the detection of lower levels of heteroplasmy accurately down to 5% using the Illumina GA. They reported that mtDNA heteroplasmies greater than 5% were detected 100% of the time and also stated that the turnaround time from DNA sample preparation to a polymorphism report can be less than a week. He et al. (2010) reported that heteroplasmic variants were detected as low as 1 per 10,000 mitochondrial genome copies using the Illumina GA II with an average base coverage of 16,700x across the entire mitochondrial genome. This level of sensitivity enabled the authors to identify heteroplasmic variants between different tissues of the same individual. At this coverage depth it is possible to perform a quantitative analysis and identify individual misincorporated bases as a polymerase error. The GS Junior instrument by 454 Life Sciences identified low-level heteroplasmy in the HV1 region for 44% of the samples tested, a tremendous increase when compared to the 4% observed using the conventional Sanger dideoxynucleotide terminator method (Holland et al. 2011). These authors also used the multiplex identifier (MID) sequence approach that made it possible to sequence 44 different amplicons in nine instrument runs (Holland et al. 2011).

Due to the nature of mtDNA haplogroup structure in human populations, observing a haplotype in two samples with identical sequences in the limited control region assayed is not uncommon. Holland et al. (2011) stated that if two reference sources in a forensic case each provided the same control region sequence, the prospects for tapping into low-level heteroplasmy to differentiate the two would be high. Heteroplasmic variants in a given mixture were detected down to a ratio of 1:250 (Holland et al. 2011), which is an improvement from the 25% reported by the use of electrospray ionization mass spectrometry (PCR/ESI-MS) (Hares 2012). Even though the pyrosequencing approach has a potential to resolve mixtures at a lower ratio, a disadvantage observed was the poor resolution of homopolymeric sequences where sequencing artifacts were routinely observed (Holland et al. 2011). Sequencing of the entire mitochondrial genome from samples with identical control region sequences may also often differentiate samples by capturing private polymorphisms in the coding region of the molecule not interrogated with conventional typing. Due to the quantitative nature of deep-sequencing approaches, deconvolution of mixed samples may also be possible, especially when long read lengths are attainable.

In addition to its application in forensics, mitochondrial genome analysis has been used in evolutionary and disease association studies (Hazkani-Covo et al. 2010). DNA transfer has occurred throughout evolutionary history between the mitochondria and the nucleus resulting in regions of "Nuclear DNA of Mitochondrial Origin," termed NUMTs (Li et al. 2012). NUMTs are present in most eukaryotic genomes and the mechanism of integration has been thought to be during the process of double-stranded break repair (Lenglez et al. 2010). NUMTs are present throughout the human genome, the largest one being 14,654 bp, 90% of the size of the entire mitochondrial genome (Hazkani-Covo et al. 2010). Recently, it has been reported that there exist anywhere from 286 to 612 NUMTs depending on the search parameters used (Lenglez et al. 2010). Only about 33% of NUMTs in the human genome are due to direct insertion from mtDNA, the rest having been found to be duplications of preexisting NUMTs (Ramos et al. 2008). Nonselective amplification of NUMTs during mtDNA studies has led to multiple interpretation issues in mitochondrial disease studies where substitutions and heteroplasmic sites may be misrepresented to be actual mitochondrial polymorphisms (Lenglez et al. 2010). An example is the presence of an NUMT on chromosome 1 (5,842 bp) of humans that resulted in misdiagnosis of causes for hearing loss, cystic fibrosis, and low sperm motility (Yao et al. 2008).

Originally, it has been proposed that accidental amplification of NUMTs is unlikely to be observed due to a higher number of mitochondrial DNA molecules to be present in comparison with nuclear DNA (Goios et al. 2006). However, recently Petruzzella et al. (2012) used the 454 FLX system to demonstrate that heteroplasmic variants observed, linked to the Leber's Hereditary Optic Neuropathy (LHON) phenotype, were due to NUMTs. An NUMT has been identified that can misrepresent the mitochondrial control region, specifically within HV1. Using the GS Junior Instrument (454 Life Sciences Corporation), Bintz (2012) detected multiple heteroplasmic variants within the mitochondrial control region (from bases 16,089 to 59) that were traced to an NUMT present on chromosome 11. The increase in resolving power that deep-sequencing technologies offer today in comparison to dideoxynucleotide sequencing can increase the observation of low-level amplification of NUMTs. When utilizing deep-sequencing technologies, initial library preparation should consider the use of primers shown to exclude NUMTs, when analyzing mtDNA from a whole genome DNA preparation. The presence of NUMTs should be taken into consideration in applications that rely on mtDNA genome analysis, such as in forensic analyses, as even the slightest amplification of NUMTs can cause false representation of heteroplasmy and alter discrimination potential.

14.7.3 Single Nucleotide Polymorphisms

Single nucleotide polymorphisms are single-base substitutions or insertions/deletions within the genome that account for about 85% of the genetic variability in humans. Several hundred thousand SNPs have been characterized in different populations (Berglund et al. 2011) that may be analyzed for biogeographic ancestry and external visible characteristics (Kayser and Knijff 2011). Individual SNPs or short polymorphic DNA regions are often characterized as alleles when their frequency has an abundance of greater than 1% in a population (Brookes 1999). SNPs characterized for forensic analysis can be grouped into four groups: identity-testing SNPs, lineage-informative SNPs, ancestry-informative SNPs, and phenotype-informative SNPs (Budowle and Daal 2008).

SNPs have been discussed as potential replacements and/or augmentation for short tandem repeat analysis (Guha and Chakraborty 2007; Senge et al. 2011). STRs are currently the preferred markers for human identification due to their highly polymorphic nature and relative ease of analysis. Nonetheless, STRs have high mutation rates (10^{-3} to 10^{-5}), provide limited information (length but not sequence polymorphism determination), and involve challenging analysis while processing degraded samples due to long amplicons. Several characteristics of SNPs suggest a high potential for use in human identification. Millions of SNPs exist in the human genome that are mostly biallelic. SNPs have relatively low mutation rates (estimated at 10^{-8}), making them more stable when used in relationship testing or as historical biomarkers. The small amplicon size needed for SNP analysis enables processing of degraded samples. Major disadvantages of SNPs for human identification application include a poor ability to detect and deconvolute mixtures and a lower discrimination power than routinely used STR panels. A suite of 30–50 autosomal SNPs could provide match probabilities similar to those obtained with 10–15 STRs (Kayser and Knijff 2011). A panel of 52 SNPs was developed by the SNP*for*ID consortium for forensic and paternity analysis (Borsting et al. 2008; Musgrave-Brown et al. 2007) that achieves a random match probability of 5.0×10^{-19} (Shewale et al. 2012). A panel of 92 SNPs based on 44 worldwide populations has been developed for individual identification (Pakstis et al. 2010). These SNPs have high heterozygosity (>0.4) and low F_{st} (<0.06) values that would permit the use of a single allele-frequency database worldwide. Of these 92 SNPs, 45 demonstrate no genetic linkage and generate match probabilities of 10^{-17} in most individual populations and less than 10^{-15} in all populations studied (Kidd 2011).

Traditional SNP genotyping assays include allele-specific hybridization, primer extension, oligonucleotide ligation, and invasive cleavage (Sobrino et al. 2005). The most common method of SNP genotyping involves the use of Taqman probes. Custom panels can be assembled from a library of over 4.5 million predesigned SNP assays. Affymetrix offers a genome-wide SNP array that is capable of covering over 906,600 SNPs including autosomal, X and Y chromosomes, and the mitochondrial genome. This array has been used extensively in whole-genome association studies (Recent Publications on the Affymetrix Genome-Wide Human SNP Array 6.0 2012). Illumina has developed SNP arrays capable of interrogating over 4.3 million SNPs. However, these SNP arrays typically require microgram quantities of DNA for successful typing and this has been limited to population-level and clinical-sample testing.

Recently, phenotypic SNPs have received attention as sources of investigative leads. De Gruijter et al. (2011) sequenced DNA regions associated with human skin pigmentation from African, European, East Asian, and South Asian populations and identified 146 polymorphic loci. Other phenotypic traits, such as body height (Kayser and Knijff 2011), eye color (Spichenok et al. 2010), and hair color have been characterized (Berglund et al. 2011). Using deep-sequencing technologies, phenotypic and identity SNPs could be typed in parallel with STRs and mtDNA to better characterize an evidentiary sample. Recently it has been proposed that whole mitochondrial genomes for 25 to 50 individuals can be sequenced on a single Ion 316 chip (Ion Torrent), making it cost-effective to analyze SNP polymorphisms in the control region as well as haplogroup or private polymorphisms in the coding region of the molecule.

14.7.4 RNA Body Fluid Identification

The identification of body fluids can be done via mRNA profiling. Tissue-specific mRNA markers have been evaluated that permit differentiation between saliva, vaginal secretions, menstrual blood, peripheral blood, and semen (Hanson and Ballantyne 2012; Juusola and Ballantyne 2003). Traditional methods for the analysis of body fluids typically include extraction of RNA, then creation of a cDNA library, followed by PCR fragment analysis or sequencing (Richard et al. 2011). The cDNA libraries are generated using specific primers for genes expressed in different body fluids (Juusola and Ballantyne 2003; Setzer et al. 2008). Genes commonly used in profiling body fluids include: matrix metalloproteinase 7 (*MMP7*) for menstrual blood, *B*-spectrin (SPTB) and porphobilinogen deaminase (PBGD) for blood, histatin 3 (HTN3) and statherin (STATH) for saliva, transglutaminase 4 (*TGM4*) for semen protamine 1 (PRM1) and protamine 2 (PRM2) for semen, and mucin 4 (MUC4) for vaginal secretions (Richard et al. 2011). Setzer et al. (2008) have performed studies to evaluate the recovery and stability of RNA under multiple environmental conditions. SPTB, PBGD, HTN3, PRM1, PRM2, and MUC4 were detectable for more than 365 days under certain conditions. In some transcripts, the RNA detection decreased to 7 days when the sample was tested after being exposed to rain (Setzer et al. 2008). The avenue of using RNA profiling to identify body fluids has shown promising results and could play a key role in cases where a bloodstain has to be differentiated between peripheral blood and menstrual blood.

RNA sequencing has been performed using the Illumina GA, SOLiD, and 454 Life Sciences GS20 systems (Wang et al. 2009). Oxford Nanopore Technologies is developing methods of sequencing RNA strands directly, eliminating all downstream steps following RNA extraction (Oxford Nanopore Technologies 2012m).

14.8 Conclusions

For the past several decades, studies relying on DNA sequence data have utilized a single-sequencing technology. Laboratories and researchers were often limited by the capabilities of this method. With the introduction of multiple new sequencing technologies, laboratories have been able to expand the range of research and data obtained through sequencing. The recent availability of cost-effective, high-throughput sequencing platforms has increased the volume of genomic data available not only for the human genome, but for other animals, plants, and prokaryotes as well. At present, forensic DNA laboratories are limited to current methods for STR and mtDNA analysis and are unable to capture the statistical power of microvariants within the STR motifs, nor the cost and time savings associated with more effective sequencing strategies. The implementation of next-generation sequencing technologies would allow forensic DNA laboratories to efficiently sequence the STR loci for all samples being analyzed, thus increasing the resolution and power of discrimination. Deep-sequencing methods would also enhance the capabilities for mitochondrial DNA analysis, allowing both an expanded data range and quantitative interpretation of mixed samples or heteroplasmy. Technological advances are progressing rapidly as has been observed in recent years, and platforms appropriate for consideration in forensic applications have been and will continue to be adopted in the relevant scientific communities.

Implementation of deep-sequencing methods in the forensic lab environment will face several challenges. Forensic approaches have been platform-based for many years and the number of viable NGS alternatives continues to increase. Several platforms are becoming available that would be viable for forensic implementation, but validation hurdles may arise since each platform gathers its data using different technologies. Ultimately, the end result is DNA sequence, which in itself serves as a unifying constant across all platforms. A major challenge will not be faced in the laboratory benchwork and library preparation pathways, but by managing the large amount of data generated by these instruments. The development and validation of effective data pipelines that behave in a similar manner to present STR expert systems will be required to facilitate the rapid processing of sequence output into allelic comparisons. The realization of a paradigm shift in what is considered the "data" that are evaluated by forensic DNA analysts will be forthcoming as the interpretation of electropherograms of STR or mtDNA sequence will not exist. Analysts will have to rely on a larger suite of bioinformatics tools than they currently employ and will have to establish the same levels of confidence in their data-interpretation skills as they presently possess. Given the high level of statistical sampling and inherent redundancy in deep sequencing, this confidence should develop rapidly. However, integration of deep-sequencing data of designated forensic loci should be seamless for national and international databasing functions as the end-point data is generally the same as that presently employed. Deep-sequencing data generated on core STR panels are directly comparable with all the existing databases if the additional data from sequence polymorphisms is ignored and simple allele definitions incorporating finer levels of polymorphism detection can be easily applied (Planz et al. 2009). Expansion of mtDNA data to full mitochondrial genomes will foster the abandonment of current haplotype coding based on differences of sample sequences from the revised Cambridge Reference Sequence (rCRS) (Andrews et al. 1999). Utilizing the current technology for rapid comparison of sequence strings, as is routinely performed in Basic Local Alignment Search Tool (BLAST) searches (McEntyre and Ostell 2002), databasing of full mtDNA sequences becomes practical and eliminates nomenclatural issues frequently encountered in the present approach (Ingman and Gyllensten 2006; Stewart et al. 2001; Wilson et al. 2002).

The implementation of NGS approaches in forensic testing will undoubtedly draw attention, as the amount of genetic information capable of being evaluated in a single test will continue to increase. Longstanding debates regarding the "junk DNA" status of forensic markers (ENCODE Project Consortium 2004; Thurman et al. 2012) become moot, especially as interest and acceptance grows in phenotypic markers, RNA-based body fluid identification, and pharmacogenomics approaches to molecular autopsy. Even as the number of noncoding markers increases, possible functional genetic associations are probable (ENCODE Project Consortium 2004; Thurman et al. 2012). With the implementation of these newer technologies, the reality that greater ethical responsibilities accompany this enhanced discriminatory power must be realized and accepted as the next generation of forensic DNA testing is ushered in.

References

Alvarez-Cubero, M., M. Saiz, L. Martinez-Gonzalez et al. 2012. Genetic identification of missing persons: DNA analysis of human remains and compromised samples. *Pathobiology* 79:228–38.

Andrews, R. M., I. Kubacka, P. F. Chinnery, R. N. Lightowers, D. M. Turnbull, and N. Howell. 1999. Reanalysis and revision of the Cambridge reference sequence for human mitochondrial DNA. *Nat Genet* 23:147.

Applied Biosystems. 2007. Applied Biosystems announces early-access program for its next-generation sequencing platform. Press Release. June 4. http://marketing.appliedbiosystems.com/images/Product/Solid_Knowledge/Press_Release_Initial_Shipments_of_SOLiD_System.PDF (accessed September 2012).

Applied Biosystems. 2010. A theoretical understanding of 2 base color codes and its application to annotation, error detection, and error correction. Publication 139WP01-02 CO13982. Production Information Brochure.

Applied Biosystems. 2011a. 5500 Series Genetic Analysis Systems. Publication CO18235 0511. Production Information Brochure.

Applied Biosystems. 2011b. SOLiD System Accuracy with the Exact Call Chemistry module. Publication CO18253 0311. Production Information Brochure.

Babiker, H. M. A., C. M. Schlebusch, H. Y. Hassan, and M. Jakobsson. 2011. Genetic variation and population structure of Sudanese populations as indicated by 15 Identifiler sequence-tagged repeat (STR) loci. *Invest Genet* 2:1–13.

Bai, R. K. and L. J. Wong. 2004. Detection and quantification of heteroplasmic mutant mitochondrial DNA by real-time amplification refractory mutation system quantitative PCR analysis: A single-step approach. Clin Chem 50:6: 996–01.

Bainbridge, M. N., R. L. Warren, M. Hirst et al. 2006. Analysis of the prostate cancer cell line LNCaP transcriptome using a sequencing-by-synthesis approach. *BMC Genom* 7:246–56.

Barrot, C., C. Moreno, C. Sánchez, et al. 2011. Comparison of Identifiler, Identifiler Plus, and MiniFiler performance in an initial paternity testing study on old skeletal remains at the forensic and legal medicine area of the Government of Andorra (Pyrenees). *Forensic Sci Int Genet* Supp Ser 3:15–6.

Barrot, C., C. Sánchez, S. Jiménez, M. Ortega, E. Huguet, and M. Géne. 2011b. The Forensic Genetics Laboratory of the University of Barcelona: The evolution of paternity testing cases over the past 35 years. *Forensic Sci Int Genet* Supp Ser 3:17–8.

Bentley, D. R. 2006. Whole-genome re-sequencing. *Curr Opin Genet Dev* 16:545–52.

Berglund, E. C., A. Kiialainen, and A. C. Syvanen. 2011. Next-generation sequencing technologies and applications for human genetic history and forensics. *Invest Genet* 2:23–37.

Besaratinia, A., H. Li, J. I. Yoon, A. Zheng, H. Gao, and S. Tommasi. 2012. A high-throughput next-generation sequencing-based method for detecting the mutational fingerprint of carcinogens. *Nucleic Acids Res* 40:e116–28.

Bintz, B. 2012. Detection of low-level DNA sequences associated with nuclear mitochondrial pseudogenes (NUMTS) from human mitochondrial control region amplicons using massively parallel 454 pyrosequencing. http://ishinews.com/wpcontent/uploads/2012/10/Poster-Abstracts-text-list-FINAL.pdf (accessed November 2012).

Boers, S. A., W. A. van der Reijden, and R. Jansen. 2012. High-throughput multilocus sequence typing: Bringing molecular typing to the next level. *PLoS One* 7:e39630:1–8.

Borsting C., J. Sanchez, H. Hansen, A. Hansen, H. Brunn, and N. Morling. 2008. Performance of the SNPforID 52 SNP-plex assay in paternity testing. *Forensic Sci Int Genet* 4:292–300.

Bosch, E., F. Calafell, A. Perez-Lezaun et al. 2000. Genetic structure of north-west Africa revealed by STR analysis. *Eur J Hum Genet* 8:360–66.

Bras, J., A. Verloes, S. A. Schneider, S. E. Mole, and R. J. Guerreiro. 2012. Mutation of the parkinsonism gene ATP13A2 causes neuronal ceroid-lipofuscinosis. *Hum Mol Genet* 21:2646–650.

Brookes, A. J. 1999. The essence of SNPs. *Gene* 234:177–86.

Browning, S. R. and B. L. Browning. 2011. Haplotype phasing: Existing methods and new developments. *Nat Rev Genet* 12:703–14.

Budowle, B., M. W. Allard, M. R. Wilson, and R. Chakraborty. 2003. Forensics and mitochondrial DNA: Applications, debates, and foundations. *Annu Rev Genom Hum Genet* 4:119–41.

Budowle, B. and A. Daal. 2008. Forensically relevant SNP classes. *BioTechniques* 44(5):603–10.

Bueno, R., A. De Rienzo, L. Dong et al. 2010. Second-generation sequencing of the mesothelioma tumor genome. *PLoS One* 5:e10612:1–10.

Butler, J. M. 2007. Short tandem repeat typing technologies used in human identity testing. *BioTechniques* 43:ii–v.

Butler J. M. 2012. Houston DNA Training Workshop. http://www.cstl.nist.gov/strbase/pub_pres/2_STR_Artifacts.pdf (accessed September 2012).

Butler, J. M. and C. R. Hill. 2012. Biology and genetics of new autosomal STR loci useful for forensic DNA analysis. *Forensic Sci Rev* 24:15–26.

Carboni, I., S. Iozzi, A. L. Nutini, P. G. Macrì, F. Torricelli, and U. Ricci. 2011. 87 DNA markers for a paternity testing: Are they sufficient? *Forensic Sci Int Genet* Supp Ser 3:552–53.

Carneiro, M. O., C. Russ, M. G. Ross, S. Gabriel, C. Nusbaum, and M. A. Depristo. 2012. Pacific biosciences sequencing technology for genotyping and variation discovery in human data. *BMC Genom* 13:375–87.

Centers for Disease Control. 2012. Methicillin-resistant *Staphylococcus aureus* (MRSA) infections. http://www.cdc.gov/mrsa/ (accessed September 2012).

Chakraborty, R. and J. M. Opitz. 2005. Paternity testing with genetic markers: Are Y-linked genes more efficient than autosomal ones? *Am J Med Gen* 2:297–05.

Chen, S., J. Jiang, H. Li, and G. Liu. 2012. The salt-responsive transcriptome of *Populus simonii* x *Populus nigra* via DGE. *Gene* 504:203–12.

Chin, C. S., J. Sorenson, J. B. Harris et al. 2011. The origin of the Haitian cholera outbreak strain. *N Engl J Med* 364:33–42.

Clark, A. G. 1990. Inference of haplotypes from PCR-amplified samples of diploid populations. *Mol Biol Evol* 7:111–22.

Clarke, L., X. Zheng-Bradley, R. Smith et al. 2012. The 1000 Genomes Project: Data management and community access. *Nat Method* 9:459–62.

de Gruijter, M., O. Lao, M. Vermeulen et al. 2011. Contrasting signals of positive selection in genes involved in human skin-color variation from tests based on SNP scans and resequencing. *Invest Genet* 2:24–36.

Drobnic, K. 2001. PCR analysis of DNA from skeletal remains in crime investigation vases. *Probl Forensic Sci* 46:110–15.

Drossman, H., J. A. Luckey, A. J. Kostichka, J. D'Cunha, and L. M. Smith. 1990. High-speed separations of DNA sequencing reactions by capillary electrophoresis. *Anal Chem* 62:900–03.

Dudley, D. M., E. N. Chin, B. N. Bimber et al. 2012. Low-cost ultra-wide genotyping using Roche/454 Pyrosequencing for surveillance of HIV drug resistance. *PloS One* 7:e36494:1–11.

Dulik, M. C., A. C. Owings, J. B. Gaieski et al. 2012. Y-chromosome analysis reveals genetic divergence and new founding native lineages in Athapaskan- and Eskimoan-speaking populations. *Proc Natl Acad Sci* 109:8471–76.

Edlund, H. and M. Allen. 2009. Y chromosomal STR analysis using Pyrosequencing technology. *Forensic Sci Int Genet* 3:119–24.

Eid, J., A. Fehr, J. Gray et al. 2009. Real-time DNA sequencing from single polymerase molecules. *Science* 323:133–8.

Ellis, M. J., L. Ding, D. Shen et al. 2012. Whole-genome analysis informs breast cancer response to aromatase inhibition. *Nature* 486:353–60.

ENCODE Project Consortium. 2004. The ENCODE (ENCyclopedia Of DNA Elements) Project. *Science* 306:636–40.

Excoffier, L. and M. Slatkin. 1995. Maximum-likelihood estimation of molecular haplotype frequencies in a diploid population. *Mol Biol Evol* 12:921–7.

Eyre, D. W., T. Golubchik, N. C. Gordon et al. 2012. A pilot study of rapid benchtop sequencing of *Staphylococcus aureus* and *Clostridium difficile* for outbreak detection and surveillance. *BMJ Open* 2:e001124:1–9.

Farias-Hesson, E., J. Erikson, A. Atkins et al. 2010. Semi-automated library preparation for high-throughput DNA sequencing platforms. *J Biomed Biotechnol* 2010:617469:1–8.

Foquet, M., K. T. Samiee, X. Kong et al. 2008. Improved fabrication of zero-mode waveguides for single-molecule detection. *J Appl Phys* 103:034301:1–9.

Fordyce, S. L., M. C. Ávila-Arcos, E. Rockenbauer et al. 2011. High-throughput sequencing of core STR loci for forensic genetic investigations using the Roche Genome Sequencer FLX platform. *BioTechniques* 51:127–33.

454 Life Sciences Corporation. 2003. 454 Life Sciences obtains exclusive license from Pyrosequencing AB to [use] next generation sequencing technology for whole genome applications. Press Release. August 19. http://454.com/resources-support/news.asp?display=detail&id=32 (accessed August 2012).

454 Life Sciences Corporation. 2005. 454 Life Sciences installs first genome sequencing system at the Brand Institute. Press Release. April 25. http://454.com/resources-support/news.asp?display=detail&id=22 (accessed August 2012).

454 Life Sciences Corporation. 2007a. 454 Life Sciences and Baylor College of Medicine complete sequencing of DNA project. Press Release. May 31. http://454.com/resources-support/news.asp?display=detail&id=68 (accessed August 2012).

454 Life Sciences Corporation. 2007b. 454 Life Sciences announces the completion of its acquisition by Roche. Press Release. May 29. http://454.com/resources-support/news.asp?display=detail&id=67 (accessed August 2012).

454 Life Sciences Corporation. 2010. 454 Sequencing Systems: Get the complete picture. Publication 06272924001. Product Information Brochure.

Gharizadeh, B., Z. S. Herman, R. G. Eason, O. Jejelowo, and N. Pourmand. 2006. Large-scale Pyrosequencing of synthetic DNA: A comparison with results from Sanger dideoxy sequencing. *Electrophoresis* 27:3042–7.

Ghosh, W., A. George, A. Agarwal et al. 2011. Whole-genome shotgun sequencing of the sulfur-oxidizing chemoautotroph *Tetrathiobacter kashmirensis*. *J Bacteriol* 193:5553–4.

Gill, P., P. L. Ivanov, C. Kimpton et al. 1994. Identification of the remains of the Romanov family by DNA analysis. *Nat Genet* 6:130–5.

Goios, A., A. Amorim, and L. Pereira. 2006. Mitochondrial DNA pseudogenes in the nuclear genome as possible sources of contamination. *Int Congr Ser* 1288:697–9.

Grzybowski, T., B. A. Malyarchuk, J. Czarny, D. Miscicka-Sliwka, and R. Kotzbach. 2003. High levels of mitochondrial DNA heteroplasmy in single hair roots: Reanalysis and revision. *Electrophoresis* 24:1159–65.

Guha. S. and R. Chakraborty 2007. Genetic diversity of global human populations at STR, SNP, and Indel loci. *Am Soc Hum Genet* 2007. http://www.ashg.org/genetics/ashg07s/f10527.htm (accessed October 2012).

Hall, T., K. A. Sannes-Lowery, L. D. McCurdy et al. 2009. Base composition profiling of human mitochondrial DNA using polymerase chain reaction and direct automated electrospray ionization mass spectrometry. *Anal Chem* 81:7515–26.

Hanson, E. and J. Ballantyne. 2012. Highly specific mRNA biomarkers for the identification of vaginal secretions in sexual assault investigations. *Sci Justice.* DOI: 10.1016/j.scijus.2012.03.007:1–9.

Hares, D. R. 2012. Expanding the CODIS core loci in the United States. *Forensic Sci Int Genet* 6:52–4.

Hartman, D., O. Drummer, C. Eckhoff, J. Scheffer, and P. Stringer. 2011. The contribution of DNA to the disaster victim identification (DVI) effort. *Forensic Sci Int* 205:52–8.

Hazkani-Covo, E., R. M. Zeller, and W. Martin. 2010. Molecular poltergeists: Mitochondrial DNA copies (numts) in sequenced nuclear genomes. *PLoS Genet* 6:e10000834:1–11.

He, Y., J. Wu, D. C. Dressman et al. 2010. Heteroplasmic mitochondrial DNA mutations in normal and tumor cells. *Nature* 464:610–4.

Hillmer, A. M., F. Yao, K. Inaki et al. 2011. Comprehensive long-span paired-end-tag mapping reveals characteristic patterns of structural variations in epithelial cancer genomes. *Genome Res* 21:665–75.

Holland, M. M., M. R. McQuillan, and K. A. O'Hanlon. 2011. Second-generation sequencing allows for mtDNA mixture deconvolution and high resolution detection of heteroplasmy. *Croat Med J* 52:299–313.

Howden, B. P., C. R. McEvoy, D. L. Allen et al. 2011. Evolution of multidrug resistance during *Staphylococcus aureus* infection involves mutation of the essential two component regulator WalKR. *PLoS Pathog* 7:e1002359:1–15.

Howden, B. P., T. Seemann, P. F. Harrison et al. 2010. Complete genome sequence of *Staphylococcus aureus* strain JKD6008, an ST239 clone of methicillin-resistant *Staphylococcus aureus* with intermediate-level vancomycin resistance. *J Bacteriol* 192:5848–9.

Human Genome Project: About the HGP. 2012. http://www.ornl.gov/sci/techresources/Human_Genome/project/about.shtml (accessed August 2012).

Hyman, E. D. 1988. A new method of sequencing DNA. *Anal Biochem* 174:423–36.

Illumina. 2006. Illumina signs definitive agreement to acquire Solexa. Press Release. November 13. http://investor.illumina.com/phoenix.zhtml?c=121127&p=irol-newsArticle&ID=929959&highlight (accessed July 2012).

Illumina. 2010. Genome Analyzer. Publication 7702009034. Product Information Brochure.

Illumina. 2012a. HiSeq system applications. Publication 7702012003. Product Information Brochure.

Illumina. 2012b. MiSeq system applications. Publication 7702012023. Product Information Brochure.

Illumina. 2012c. One run, eight bacterial genomes: The potential of the MiSeq Personal Sequencer. Publication 7702012002. MiSeq interview: Tim Stinear.

Ingman, M. and U. Gyllensten. 2006. mtDB: Human Mitochondrial Genome Database, a resource for population genetics and medical sciences. *Nucleic Acids Res* 34, D749–51.

Jarne, P. and P. J. Lagoda. 1996. Microsatellites, from molecules to populations and back. *Trends Ecol Evol* 11:424–9.

Jiang, Y., J. Lu, E. Peatman et al. 2011. A pilot study for channel catfish whole genome sequencing and *de novo* assembly. *BMC Genom* 12:629–41.

Jones, S., X. Zhang, D. W. Parsons et al. 2008. Core signaling pathways in human pancreatic cancers revealed by global genomic analyses. *Science* 321:1801–6.

Juusola, J. and J. Ballantyne. 2003. Messenger RNA profiling: A prototype method to supplant conventional methods for body fluid identification. *Forensic Sci Int* 135:85–96.

Kalender Atak, Z., K. De Keersmaecker, V. Gianfelici et al. 2012. High accuracy mutation detection in leukemia on a selected panel of cancer genes. *PLoS One* 7:e38463:1–11.

Kayser, M. and P. Knijff. 2011. Improving human forensics through advances in genetics, genomics and molecular biology. *Nat Rev Genet* 12:179–92.

Kidd, K. 2011. Population Genetics of SNPs for Forensic Purposes. Department of Justice Final Report.

Koren, S., M. C. Schatz, B. P. Walenz et al. 2012. Hybrid error correction and de novo assembly of single-molecule sequencing reads. *Nat Biotechnol* 30:693–700.

Korlach, J., K. P. Bjornson, B. P. Chaudhuri et al. 2010. Real-time DNA sequencing from single polymerase molecules. *Method Enzymol* 472:431–55.

Köser, C.U., M. T. Holden, M. J. Ellington et al. 2012. Rapid whole-genome sequencing for investigation of a neonatal MRSA outbreak. *N Engl J Med* 366(24): 2267–75.

Koudijs, M. J., C. Klijn, L. van der Weyden et al. 2011. High-throughput semiquantitative analysis of insertional mutations in heterogeneous tumors. *Genome Res* 21:2181–9.

Leclair, B., C. J. Fregeau, K. L. Bowen, and R. M. Fourney. 2004. Enhanced kinship analysis and STR-based DNA typing for human identification in mass fatality incidents: The Swissair Flight 111 disaster. *J Forensic Sci* 49:939–53.

Lee, S. W., P. F. Markham, J. F. Markham et al. 2011. First complete genome sequence of infectious laryngotracheitis virus. *BMC Genom* 12:197–03.

Lenglez, S., D. Hermand, and A. Decottignies. 2010. Genome-wide mapping of nuclear mitochondrial DNA sequences links DNA replication origins to chromosomal double-strand break formation in *Schizosaccharomyces pombe*. *Genome Res* 20(9):1250–61.

Ley, T. J., E. R. Mardis, L. Ding et al. 2008. DNA sequencing of a cytogenetically normal acute myeloid leukaemia genome. *Nature* 456:66–2.

Li, H. X., D. Tong, H. Lu et al. 2011. Mutation analysis of 24 autosomal STR loci using in paternity testing. *Forensic Sci Int Genet* Supp Ser 3:159–60.

Li, M., R. Schroeder, A. Ko, and M. Stoneking. 2012. Fidelity of capture-enrichment for mtDNA genome sequencing: Influence of NUMTs. *Nucleic Acids Res* 40:e137:1–8.

Li, X., J. Cai, Y. Guo et al. 2011. Mitochondrial DNA and STR analyses for human DNA from maggots crop contents: A forensic entomology case from central-southern China. *Trop Biomed* 28:333–8.

Life Technologies. 2011. Product Information Brochure. Ion PGM. Publication C0315800711.

Life Technologies. 2012a. Ion Proton Systems installed and operational at the Baylor College of Medicine Human Genome Sequencing Center. Press Release. April 24. http://www.lifetechnologies.com/us/en/home/about-us/news-gallery/press-releases/2012/io-proto__8482_-systems-istalled-ad-operatioal-at-the-baylor-col.html (accessed September 2012).

Life Technologies. 2012b. Ion Torrent. Publication C0319110112. Product Information Brochure.

Lin, X., W. Tang, S. Ahmad et al. 2012. Applications of targeted gene capture and next-generation sequencing technologies in studies of human deafness and other genetic disabilities. *Hear Res* 288:67–6.

Liu, J., S. Li, J. Yin et al. 2011. Allele frequencies of 6 autosomal STR loci in the Xibo nationality with phylogenetic structure among Chinese populations. *Gene* 487:84–7.

Liu, P., T. S. Seo, N. Beyor, K. J. Shin, J. R. Scherer, and R. A. Mathies. 2007. Integrated portable polymerase chain reaction capillary electrophoresis microsystem for rapid forensic short tandem repeat typing. *Anal Chem* 79:1881–9.

Margulies, M., M. Egholm, W. E. Altman et al. 2005. Genome sequencing in microfabricated high-density picolitre reactors. *Nature* 437:376–80.

Marklund, S. and O. Carlborg. 2010. SNP detection and prediction of variability between chicken lines using genome resequencing of DNA pools. *BMC Genom* 11:665–72.

Maxam, A. M. and W. Gilbert. 1977. A new method for sequencing DNA. *Proc Natl Acad Sci USA* 74:560–64.

McEntyre, J. and J. Ostell. 2002. *The NCBI Handbook (Internet)*. National Center for Biotechnology Information: Bethesda, MD. http://www.ncbi.nlm.nih.gov/books/NBK21101/ (accessed November 2012).

Mellmann, A., D. Harmsen, C. A. Cummings et al. 2011. Prospective genomic characterization of the German enterohemorrhagic *Escherichia coli* O104:H4 outbreak by rapid next generation sequencing technology. *PLoS One* 6(7): e22751:1–9.

Menzies, B. R., M. B. Renfree, T. Heider, F. Mayer, and T. B. Hildebrandt. 2012. Limited genetic diversity preceded extinction of the Tasmanian tiger. *PLoS One* 7:e35433:1–7.

Metzker, M. L. 2010. Sequencing technologies—The next generation. *Nat Rev Genet* 11:31–46.

Meyer, B. S. and W. Salzburger. 2012. A novel primer set for multilocus phylogenetic inference in East African cichlid fishes. *Mol Ecol Resour* 12:1097–104.

Miller, W., V. M. Hayes, A. Ratan et al. 2011. Genetic diversity and population structure of the endangered marsupial Sarcophilus harrisii (Tasmanian devil). *Proc Natl Acad Sci USA* 108:12348–53.

Mitra, R. D., J. Shendure, J. Olejnik, O. Edyta-Krzymanska-Olejnik, and G. M. Church. 2003. Fluorescent *in situ* sequencing on polymerase colonies. *Anal Biochem* 320:55–65.

Musgrave-Brown, E., D. Ballard, K. Balogh et al. 2007. Forensic validation of the SNPforID 52-plex assay. *Forensic Sc Int Genet* 1:186–90.

Otto, T. D., M. Sanders, M. Berriman, and C. Newbold. 2010. Iterative Correction of Reference Nucleotides (iCORN) using second generation sequencing technology. *Bioinformatics* 26:1704–1707.

Oxford Nanopore Technologies. 2011. Oxford Nanopore congratulates collaborator Professor Mark Akeson on $3.6 million NHGRI grant. Press Release. August 23. http://www.nanoporetech.com/news/press-releases/view/23 (accessed September 2012).

Oxford Nanopore Technologies. 2012a. Biological nanopores. http://www.nanoporetech.com/technology/introduction-to-nanopore-sensing/biological-nanopores (accessed August 2012).

Oxford Nanopore Technologies. 2012b. DNA: An introduction to nanopore sequencing. http://www.nanoporetech.com/technology/analytes-and-applications-dna-rna-proteins/dna-exonuclease-sequencing- (accessed August 2012).

Oxford Nanopore Technologies. 2012c. DNA Exonuclease Sequencing. http://www.nanoporetech.com/technology/analytes-and-applications-dna-rna-proteins/dna-exonuclease-sequencing- (accessed August 2012).

Oxford Nanopore Technologies. 2012d. DNA: Strand sequencing. http://www.nanoporetech.com/technology/analytes-and-applications-dna-rna-proteins/dna-strand-sequencing (accessed August 2012).

Oxford Nanopore Technologies. 2012e. For customers. http://www.nanoporetech.com/about-us/for-customers (accessed August 2012).

Oxford Nanopore Technologies. 2012f. GridION node. http://www.nanoporetech.com/technology/the-gridion-system/gridion-node (accessed August 2012).

Oxford Nanopore Technologies. 2012g. Introduction to nanopore sensing. http://www.nanoporetech.com/technology/introduction-to-nanopore-sensing/introduction-to-nanopore-sensing (accessed August 2012).

Oxford Nanopore Technologies. 2012h. Management of nodes. http://www.nanoporetech.com/technology/the-gridion-system/management-of-nodes (accessed August 2012).

Oxford Nanopore Technologies. 2012i. microRNA. http://www.nanoporetech.com/technology/analytes-and-applications-dna-rna-proteins/microrna (accessed August 2012).

Oxford Nanopore Technologies. 2012j. Movie: An introduction to the GridION system. http://www.nanoporetech.com/technology/the-gridion-system/movie-an-introduction-to-the-gridion-system (accessed August 2012).

Oxford Nanopore Technologies. 2012k. Oxford Nanopore introduces DNA 'strand sequencing' on the high-throughput GridION platform and presents MinION, a sequencer the size of a USB memory stick. Press Release. February 17. http://www.nanoporetech.com/news/press-releases/view/39 (accessed August 2012).

Oxford Nanopore Technologies. 2012l. RNA. http://www.nanoporetech.com/technology/analytes-and-applications-dna-rna-proteins/rna (accessed August 2012).

Oxford Nanopore Technologies. 2012m. Single use cartridge. http://www.nanoporetech.com/technology/single-use-cartridge/single-use-cartridge (accessed August 2012).

Oxford Nanopore Technologies. 2012n. Solid state nanopores. http://www.nanoporetech.com/technology/introduction-to-nanopore-sensing/solid-state-nanopores (accessed August 2012).

Oxford Nanopore Technologies. 2012o. Summary. http://www.nanoporetech.com/about-us/summary (accessed August 2012).

Pacific Biosciences. 2011. Pacific Bioscience begins shipments of commercial PacBio RS Systems. Press Release. April 27. http://investor.pacificbiosciences.com/releasedetail.cfm?ReleaseID=572551 (accessed August 2012).

Pacific Biosciences. 2012. PacBio RS: Reveal the true biology. Product Information Brochure.

Pakstis, A. J., W. C. Speed, R. Fang et al. 2010. SNPs for a universal individual identification panel. *Hum Genet* 127(3):315–24.

Pemberton, T. J., C. I. Sandefur, M. Jakobsson, and N. A. Rosenberg. 2009. Sequence determinants of human microsatellite variability. *BMC Genom* 10:612–31.

Pena, S. D. and R. Chakraborty. 1994. Paternity testing in the DNA era. *Trends Genet* 10:204–9.

Petruzzella, V. R. Carrozzo, C. Calabrese et al. 2012. Deep sequencing unearths NUMTs under LHON-associated false heteroplasmic mitochondrial DNA variants. *Hum Mol Genet* 21(17):3753–64.

Pitterl, F., K. Schmidt, G. Huber et al. 2009. Increasing the discrimination power of forensic STR testing by employing high-performance mass spectrometry, as illustrated in indigenous South African and Central Asian populations. *Int J Legal Med* 124:551–8.

Planz, J. V., B. Budowle, T. A. Hall, A. J. Eisenberg, K. A. Lowery, and S. A. Hofstadler. 2009. Enhancing resolution and statistical power by utilizing mass spectrometry for detection of SNPs within the short tandem repeats. *Forensic Sci Int Genet Supp Ser* 2:529–31.

Planz, J. V. and T. A. Hall. 2012. Hidden variation in microsatellite loci: Utility and implications for forensic DNA analysis. *Forensic Sci Rev* 24: 27–42.

Planz, J. V., K. A. Sannes-Lowery, D. D. Duncan et al. 2012. Automated analysis of sequence polymorphism in STR alleles by PCR and direct electrospray ionization mass spectrometry. *Forensic Sci Int Genet* 6(5):594–606.

Poetsch, M., K. Bayer, Z. Ergin, M. Milbrath, T. Schwark, and N. von Wurmb-Schwark. 2011. First experiences using the new Powerplex ESX17 and ESI17 kits in casework analysis and allele frequencies for two different regions in Germany. *Int J Legal Med* 125:733–9.

Qiagen. 2012. Principle of Pyrosequencing Technology. http://www.pyrosequencing.com/DynPage.aspx?id=7454 (accessed August 2012).

Qiu, Q., G. Zhang, T. Ma et al. 2012. The yak genome and adaptation to life at high altitude. *Nat Genet* 44:946–9.

Quail, M., M. Smith, P. Coupland et al. 2012. A tale of three next generation sequencing platforms: Comparison of Ion Torrent, Pacific Biosciences and Illumina MiSeq Sequencers. *BMC Genom* 13:341–65.

Ramos, A., C. Santos, L. Alvarez, R. Nogués, and M. P. Aluja. 2008. Human mitochondrial DNA complete amplification and sequencing: A new validated primer set that prevents nuclear DNA sequences of mitochondrial origin co-amplification. *Electrophoresis* 30:1587–93.

Rasko, D. A., D. R. Webster, J. W. Sahl et al. 2011. Origins of the *E. coli* strain causing an outbreak of hemolytic–uremic syndrome in Germany. *N Engl J Med* 365:709–17.

Recent Publications on the Affymetrix Genome-Wide Human SNP Array 6.0. 2012. http://media.affymetrix.com/support/technical/other/snp6_array_publications.pdf (accessed November 2012).

Ricchetti, M., F. Tekaia, and B. Dujon. 2004. Continued colonization of the human genome by mitochondrial DNA. *PLoS Biology* 2: 1313–24.

Riccombeni, A., G. Vidanes, E. Proux-Wera, K. H. Wolfe, and G. Butler. 2012. Sequence and analysis of the genome of the pathogenic yeast Candida orthopsilosis. *PLoS One* 7:e35750:1–13.

Richard, M. L., K. A. Harper, R. L. Craig, A. J. Onorato, J. M. Robertson, and J. Donfack. 2011. Evaluation of mRNA marker specificity for the identification of five human body fluids by capillary electrophoresis. *Forensic Sci Int Genet* 6:452–60.

Robbins, C. M., W. A. Tembe, A. Baker et al. 2011. Copy number and targeted mutational analysis reveals novel somatic events in metastatic prostate tumors. *Genome Res* 21:47–5.

Roewer, L. 2009. Y chromosome STR typing in crime casework. *Forensic Sci Med Pathol* 5:77–84.

Ronaghi, M., M. Uhlen, and P. Nyren. 1998. A sequencing method based on real-time pyrophosphate. *Science* 281:363–65.

Rothberg, J. M., W. Hinz, T. M. Rearick et al. 2011. An integrated semiconductor device enabling non-optical genome sequencing. *Nature* 475:348–52.

Rucinski, C., A. L. Malaver, E. J. Yunis, and J. J. Yunis. 2011. Comparison of two methods for isolating DNA from human skeletal remains for STR analysis. *J Forensic Sci* 57:706–12.

Rusk, N. 2011. Torrents of sequence. *Nat Method* 8:44.

Sanger, F. and A. R. Coulson. 1975. A rapid method for determining sequences in DNA by primed synthesis with DNA polymerase. *J Mol Biol* 94:441–8.

Sanger, F., S. Nicklen, and A. R. Coulson. 1977. DNA sequencing with chain-terminating inhibitors. 1992. *Biotechnology* 24:104–8.

Schadt, E. E., S. Turner, and A. Kasarskis. 2010. A window into third-generation sequencing. *Hum Mol Genet* 19:R227–40.

Schatz, M. C., J. Witkowski, and W. R. McCombie. 2012. Current challenges in de novo plant genome sequencing and assembly. *Genome Biol* 13:1–7.

Scheible, M., O. Loreille, R. Just, and J. Irwin. 2011. Short tandem repeat sequencing on the 454 platform. *Forensic Sci Int Genet Supp Ser* 3:357–8.

Schneider, P. M. and P. D. Martin. 2001. Criminal DNA databases: The European situation. *Forensic Sci Int* 119:232–8.

Schwark, T., A. Heinrich, A. Preusse-Prange, and N. von Wurmb-Schwark. 2011. Reliable genetic identification of burnt human remains. *Forensic Sci Int Genet* 5:393–9.

Senge, T., B. Madea, A. Junge, M. A. Rothschild, and P. M. Schneider. 2011. STRs, mini STRs and SNPs—A comparative study for typing degraded DNA. *Legal Med (*Tokyo) 13(2):68–74.

Setzer, M., J. Juusola, and J. Ballantyne. 2008. Recovery and stability of RNA in vaginal swabs and blood, semen, and saliva stains. *J Forensic Sci* 53:296–305.

Shewale, J. G., L. Qi, and L. M. Calandro. 2012. Principles, practice, and evolution of capillary electrophoresis as a tool for forensic DNA analysis. *Forensic Sci Rev* 24:79–100.

Shulaev, V., D. J. Sargent, R. N. Crowhurst et al. 2011. The genome of woodland strawberry (*Fragaria vesca*). *Nat Genet* 43:109–16.

Smith, A. M., L. E. Heisler, R. P. St. Onge et al. 2010. Highly-multiplexed barcode sequencing: An efficient method for parallel analysis of pooled samples. *Nucleic Acids Res* 38:e142:1–7.

Smith, L. M., J. Z. Sanders, R. J. Kaiser et al. 1986. Fluorescence detection in automated DNA sequence analysis. *Nature* 321:674–79.

Sobrino, B., M. Brion, and A. Carracedo. 2005. SNPs in forensic genetics: A review on SNP typing methodologies. *Forensic Sci Int* 154:181–94.

Spichenok, O., Z. M. Budimlija, A. A. Mitchell et al. 2010. Prediction of eye and skin color in diverse populations using seven SNPs. *Forensic Sci Int Genet* 5:472–8.

Stewart, J. E., C. L. Fisher, P. J. Aagaard et al. 2001. Length variation in HV2 of the human mitochondrial DNA control region. *J Forensic Sci* 46(4):862–70.

Tang, S. and T. Huang. 2010. Characterization of mitochondrial DNA heteroplasmy using a parallel sequencing system. *BioTechniques* 48:287–96.

Thurman, R., E. Rynes, R. Humbert et al. 2012. The accessible chromatin landscape of the human genome. *Nature* 489:75–2.

Treangen, T. J. and S. L. Salzberg. 2011. Repetitive DNA and next-generation sequencing: Computational challenges and solutions. *Nat Rev Genet* 13:36–46.

Trovoada, M. J., C. Alves, L. Gusmao, A. Abade, A. Amorim, and M. J. Prata. 2001. Evidence for population sub-structuring in São Tomé e Príncipe as inferred from Y-chromosome STR analysis. *Ann Hum Genet* 65:271–83.

Tully, L. A., T. J. Parsons, R. J. Steighner, M. M. Holland, M. A. Marino, and V. L. Prenger. 2000. A sensitive denaturing gradient-gel electrophoresis assay reveals a high frequency of heteroplasmy in hypervariable region 1 of the human mtDNA control region. *Am J Hum Genet* 67:432–43.

Twine, N. A., K. Janitz, M. R. Wilkins, and M. Janitz. 2011. Whole transcriptome sequencing reveals gene expression and splicing differences in brain regions affected by Alzheimer's disease. *PLoS ONE* 6: e16266:1–13.

University of Munster. 2011. On Top of Outbreaks: Rapid Next Generation Sequencing for Diagnosis; Combating the superbug *Klebsiella Oxa-48* outbreak in a Dutch hospital. Press Release. August 15. http://www.uni-muenster.de/en/exec/upm.php?rubrik=Alle&neu=1&monat=201108&nummer=14629 (accessed October 2012).

Uthaipaisanwong, P., J. Chanprasert, J. R. Shearman et al. 2012. Characterization of the chloroplast genome sequence of oil palm (*Elaeis guineensis* Jacq.). *Gene* 500:172–80.

Valouev, A., J. Ichikawa, T. Tonthat et al. 2008. A high-resolution, nucleosome position map of *C. elegans* reveals a lack of universal sequence-dictated positioning. *Genome Res* 18:1051–63.

Van de Sande, W. W. 2012. Phylogenetic analysis of the complete mitochondrial genome of *Madurella mycetomatis* confirms its taxonomic position within the order Sordariales. *PLoS One* 7:e38654:1–10.

Van Neste, C., F. Van Nieuwerburgh, D. Van Hoofstat, and D. Deforce. 2012. Forensic STR analysis using massive parallel sequencing. *Forensic Sci Int Genet* 6:810–9.

Wallace, D. C., C. Stugard, and D. Murdock. 1997. Ancient mtDNA sequences in the human nuclear genome: A potential source of errors in identifying pathogenic mutations. *Proc Natl Acad Sci USA* 94:14900–905.

Wallace, E. V., D. Stroddart, A. J. Heron et al. 2010. Identification of epigenetic DNA modifications with a protein nanopore. *Chem Commun* 46:8195–97.

Wang, Y. and C. G. Lee. 2009. MicroRNA and cancer—Focus on apoptosis. *J Cell Mol Med* 13:12–23.

Wang, Z., M. Gerstein, and M. Snyder. 2009. RNA-Seq: A revolutionary tool for transcriptomics. *Nat Rev Genet* 10:57–63.

Wilson, M. R., M. W. Allard, K. Monson, K. W. Miller, and B. Budowle. 2002. Recommendations for consistent treatment of length variants in the human mitochondrial DNA control region. *Forensic Sci Int* 129:35–42.

Yao, Y. G., Q. P. Kong, A. Salas, and H. J. Bandelt. 2008. Pseudomitochondrial genome haunts disease studies. *J Med Genet* 45:769–72.

Yeung, S. H., S. A. Greenspoon, A. McGuckian et al. 2006. Rapid and high-throughput forensic short tandem repeat typing using a 96-lane microfabricated capillary array electrophoresis microdevice. *J Forensic Sci* 51:740–7.

Zech, W. D., N. Malik, and M. Thali. 2012. Applicability of DNA analysis on adhesive tape in forensic casework. *J Forensic Sci* 57:1036–41.

Zhang, T., X. Zhang, S. Hu, and J. Yu. 2011. An efficient procedure for plant organellar genome assembly, based on whole genome data from the 454 GS FLX sequencing platform. *Plant Method* 7:38–47.

Zhou, Y., F. Gao, R. Liu, J. Feng, and H. Li. 2012. *De novo* sequencing and analysis of root transcriptome using 454 pyrosequencing to discover putative genes associated with drought tolerance in *Ammopiptanthus mongolicus*. *BMC Genom* 13:266–79.

Sample-to-Result STR Genotyping Systems

Potential and Status

15

JENNY A. LOUNSBURY
JOAN M. BIENVENUE
JAMES P. LANDERS

Contents

15.1 Introduction 350
15.2 Sample Preparation and DNA Separation 353
  15.2.1 Sample Preparation 353
  15.2.2 DNA Extraction and PCR Amplification 354
  15.2.3 Microchip Electrophoresis of STR Fragments 357
15.3 Integrated Rapid DNA Analysis Platforms 358
  15.3.1 Substrate Choice and Device Size 358
  15.3.2 Surface Chemistry 360
  15.3.3 Flow Control 360
  15.3.4 Separation Speed 361
  15.3.5 Multiplexed STR Detection and Allele Calling 362
  15.3.6 Comparison of Next-Generation Prototypes 363
15.4 Implementation of Integrated Rapid DNA Analysis Platforms 368
15.5 Future Implications 370
15.6 Conclusions 370
References 371

Abstract: Forensic DNA analysis using short tandem repeats (STRs) has become the cornerstone for human identification, kinship analysis, paternity testing, and other applications. However, it is a lengthy, laborious process that requires specialized training and numerous instruments, and it is one of the factors that has contributed to the formation and expansion of a casework backlog in the United States of samples awaiting DNA processing. Although robotic platforms and advances in instrumentation have improved the throughput of samples, there still exists a significant potential to enhance sample-processing capabilities. The application of microfluidic technology to STR analysis for human identification offers numerous advantages, such as a completely closed system, reduced sample and reagent consumption, and portability, as well as the potential to reduce the processing time required for biological samples to less than 2 h. Development of microfluidic platforms not only for forensic use, but clinical and diagnostic use as well, has exponentially increased since the early 1990s. For a microfluidic system to be generally accepted in forensic laboratories, there are several factors that must be taken into consideration and the data generated with these systems must meet or exceed the same guidelines and standards that are applicable for the conventional methods. This review covers the current state of forensic microfluidic platforms starting with microchips for the individual DNA-processing steps of extraction, amplification, and electrophoresis. For fully integrated devices, challenges that come with microfluidic platforms are covered, including circumventing issues with surface chemistry, monitoring flow control, and proper allele calling. Finally, implementation and future implications of a microfluidic rapid DNA system are discussed.

15.1 Introduction

While looking ahead to the development of new, disruptive technologies that will change the face of forensic DNA analysis, it is interesting to look back at some of the critical developments that provided the underpinning for them. The early 1970s and 1980s were rampant with advances in biochemistry that provided the framework for modern molecular techniques. In 1975, O'Farrell described 2D gel electrophoresis and revolutionized our ability to interrogate the proteinaceous components of cells by exploiting a separation method that provided unprecedented resolution of proteins (O'Farrell 1975). The year 1985 brought new capabilities to the nucleic acids field in the form of two influential papers that seeded paradigm-shifting technologies that would change the face of human identification (HID) analysis. In March 1985, Professor (now Sir) Alec Jeffreys reported the discovery of "minisatellite" regions in the human genome that could be used as a tool to identify molecular differences between individuals (Jeffreys et al. 1985). He demonstrated that these minisatellites repeat throughout the genome and that the number of repeats varied from person to person, hence, variable number of tandem repeats (VNTR). Using restriction enzymes, Dr. Jeffreys was able to show the differences in the length of each fragment caused by the differences in the number of repeats; this technique is known as restriction fragment length polymorphism (RFLP). In December of the same year, Dr. Kary Mullis reported the amplification of target DNA regions using the polymerase chain reaction (PCR) (Saiki et al. 1985). By using primers with sequences specific to a target region and temperature-cycling a sample through three temperatures in the presence of a polymerase, as well as dNTPs, the target region can be exponentially amplified in copy number. Although it took half a decade before these two methods would be conjoined to yield an unparalleled force for human identification, the groundwork for modern DNA analysis had been laid. For reasons that are obvious, all three of these are citation classics (exceeding 400 citations).

As methods for interrogating the nucleic acid content of cells progressed and more information was acquired about the human genome, a class of minisatellite was reported, designated as short tandem repeats (STRs), which are typically two to eight nucleotide units that are repeated at various loci throughout the genome (Butler 2005). This allowed efforts by Al Edwards and Tom Caskey (Baylor College of Medicine) (Edwards et al. 1991, 1992) to develop the STR chemistry that formed the foundation of the HID kits that are now commercially available. Today's STR kits provide an average power of discrimination of approximately 10^{14} (the population of the earth is 10^9). The instrumentation required for performing STR analysis from biological samples includes centrifuges, liquid handlers, thermal cyclers, and capillary electrophoresis instrumentation, contributing to a large and expensive laboratory footprint. In addition, the time required to complete all the steps necessary is typically 7–10 h, depending on the sample type (see Figure 15.1). This, along with other factors, such as an increase in the demand for STR typing on a wider variety of samples, result interpretation (or data analysis), and insufficient numbers of trained personnel,

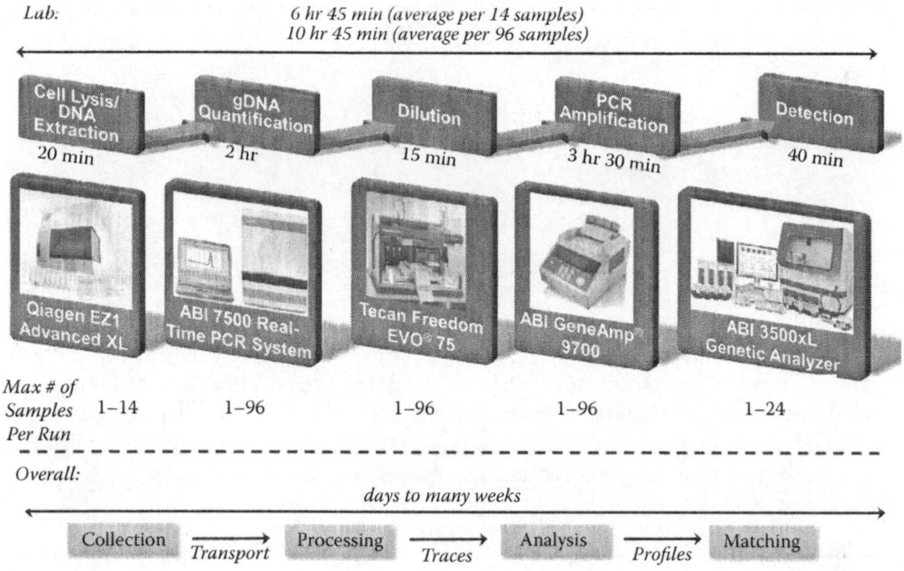

Figure 15.1 (See color insert.) Instrumentation and process times associated with conventional forensic DNA analysis. Note: The times associated with each step are per run only and do not include any preparation or transfer times. (Adapted from presentation of J. Bienvenue et al., Integrated microfluidics for forensic applications, *61st American Academy of Forensic Sciences Annual Meeting*, Denver, CO.)

has contributed to an increase in the casework backlog in the United States, which had risen to over 90,000 cases by December 2008 (National Institute of Justice 2011; Nelson 2011). However, this does not include the backlog of convicted offender and arrestee samples that are also awaiting processing (National Institute of Justice 2011). The development of an integrated microfluidic system for forensic DNA analysis has the potential to impact the way casework samples are processed. Such a system could be used for the processing of routine and/or reference samples, in much the same manner as robotic platforms are utilized in today's workflow, allowing the analyst to concentrate on more challenging samples including low copy number and degraded and/or mixed samples. In addition, an integrated system requires less specialized training (i.e., on the operation of just one instrument instead of five), reducing the cost of training analysts and potentially allowing for the hiring of more personnel to process samples.

The concept of a micro total analysis system (μTAS) was first described in 1990 as a total analysis system capable of "perform[ing] all sample-handling steps extremely close to the place of measurement" (Manz et al. 1990, p. 244). Research groups, startup companies, and major corporations worldwide focus on developing various aspects of μTAS devices with the hope of incorporating these devices into everyday workflows. The field of microfluidics offers numerous advantages that cannot only directly benefit forensic science, but also clinical and point-of-care applications as well, particularly those where a reduction in sample required and reagent volumes consumed can reduce the overall cost of analysis, potentially bringing what are currently expensive analytical techniques to resource- and/or funding-limited areas. In addition, the time required for sample analysis is expedited, which may benefit those fields where time-to-result matters. Furthermore, a completely closed microfluidic system nearly eliminates the need for sample transfer steps, minimizing the risk of sample

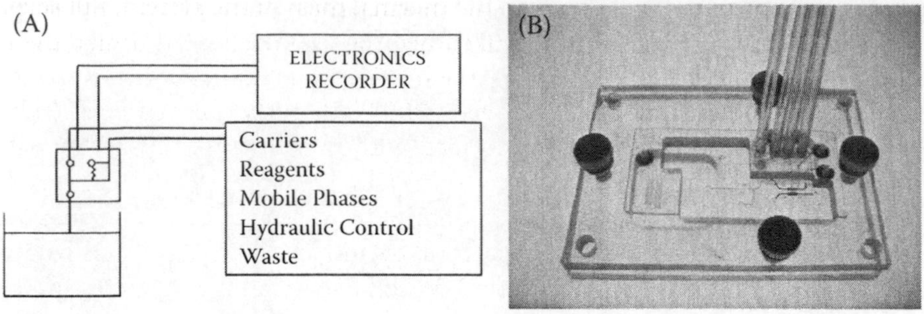

Figure 15.2 (A) Reproduction of a schematic of a micro total analysis system (μTAS) as proposed by Manz et al. (Manz, A. et al., 1990, Miniaturized total chemical analysis systems: A novel concept for chemical sensing, *Sensor Actuat B* 1: 244–8.) (B) Photograph of a sample-in-answer-out μTAS developed by Easley et al. (Easley, C. J. et al., 2006a, A fully integrated microfluidic genetic analysis system with sample-in-answer-out capability, *Proc Natl Acad Sci USA* 103: 19272–7.)

cross-contamination or mix-up. Ideally, the end goal for any of these miniaturized systems is a *sample-in–answer-out* instrument that performs all of the steps necessary for analysis, with minimal (if any) action required by the analyst (Figure 15.2). In the special case of forensic DNA analysis, these steps include sample preparation, DNA extraction/purification, DNA quantitation, STR PCR, and DNA separation and detection. This review will focus on the development of μTAS devices and/or instrumentation designed for one or more of the steps associated with forensic DNA analysis and, specifically, STR typing.

A distinct advantage associated with using a microfluidic system is the decrease in overall analysis time. If the time attributed to the overall analysis using conventional instrumentation where represented as the sum of the time for cell lysis/extraction (1.5 h), PCR (3.5 h), electrophoresis, detection and data conditioning (1–2 h), a total analysis time (excluding sample transfer and a quantification step) requires between 7 and >9 h, if a quantification step is needed. If microminiaturization delivers as is expected, the time for each of these processes should decrease dramatically.

It is difficult to predict the reduction in individual process times that would be necessary for microfluidic platforms to impact forensic DNA analysis, primarily because the sample-in–answer-out analysis time will largely be dependent on the application. Furthermore, there is much to be gained by microminiaturization of the processes associated with genetic analysis, as some of the individual processes in the conventional realm have been hampered in speed by physical (and sometimes biological) limitations. Consider PCR as an example for considering the limits on thermal cycling. The physical limits involve the hardware associated with the *Taq* polymerase-driven amplification, where the thermal cycling (between melt, anneal, and extend temperatures) in conventional systems is often carried out using block heaters. The thermal mass of the block limits the rate at which the temperature can be raised and lowered, therefore determining the minimal time for a cycle. The thermal mass of even large microscale volumes of PCR solution (25 mL) is small relative to the thermal mass of the block; hence, the thermal cycling of the block determines the rate (thin-walled polypropylene tubes minimize the time for thermal equilibrium between the PCR solution and the block).

Microscale-volume thermal cycling circumvents this in a number of ways. First, the thermal cycling of submicroliter volumes reduces the thermal mass of the solution even further, making it a smaller contributor to the thermal mass of the system. But second, and more important, is the fact that miniaturized heating approaches minimize the thermal mass of the surrounding substrate—that is, the heating and cooling of the substrate (glass, plastic) around the PCR chamber (Easley et al. 2007). This low thermal mass (substrate + solution) can be rapidly thermal-cycled; in fact, with noncontact heating approaches, 30 cycles can be completed in >5 min (Easley et al. 2006b). With unprecedentedly fast thermal cycling, individual cycles can be reduced from minutes to seconds, which brings *biological limits* into consideration. Assuming that thermal cycling could be accomplished in miniaturized systems with subsecond cycles (not unreasonable with ramp rates of 40–80°C/sec), the biological activity of *Taq* polymerase (i.e., the rate of base incorporation) becomes the limit. With literature values for base incorporation ranging from 60–150 bases per second (Grunenwald 2003; Innis et al. 1988), the upper end of this range would require dwelling at the extension temperature (72°C) for at least 3 s for a 450-base fragment. Consequently, breaking through the physical barriers for rapid thermal cycling places us in a regime where the activity of *Taq* polymerase becomes the rate-limiting factor for PCR. This enhances the importance of the "fast polymerases" that have become available and were exploited by Vallone et al. for fast STR amplification (Vallone et al. 2008).

15.2 Sample Preparation and DNA Separation

Prior to engaging the development of fully integrated microfluidic systems with sample-to-answer capabilities, the various processes associated with STR typing (DNA extraction, amplification, separation, and detection) needed to first be optimized for the microscale. While a fully integrated microdevice is the ultimate goal of most utility to the forensic community, standalone microfluidic systems for the individual processes could find utility replacing conventional methods in normal workflow, potentially providing modular microfluidic analysis systems and a reduction in processing time. This section will review the key literature on monotasking microfluidic devices that perform a single analytical step, and will point out the significant advantages associated with the microdevice.

15.2.1 Sample Preparation

To begin the process of forensic DNA analysis, it is necessary to remove the cellular material from the substrate it is on, such as cotton swabs or clothing. Sample preparation begins with elution of the cellular material, generally through the use of a detergent, and incubation for a set period of time, followed closely by a lysis step, where the cellular membranes are degraded, releasing the nucleic acids and other components. To realize a truly sample-in–answer-out microdevice, a sample-preparation domain must be included. More extensive information on microfluidic cell lysis, cell sorting, and sample-pretreatment microdevices is available in two excellent recent reviews (Chen and Cui 2009; Kim et al. 2009).

In the special case of sexual assault evidence, there exists an opportunity to separate the male cells from the female cells to produce two separate profiles. This is accomplished by exploiting the differences between sperm and vaginal epithelial cell membranes. A method called differential extraction uses gentle lysis conditions to lyse the epithelial cells, while the sperm cells are left intact. After the female DNA is removed, the sperm cells are then lysed under harsh conditions, usually using a reducing agent such as dithiothreitol

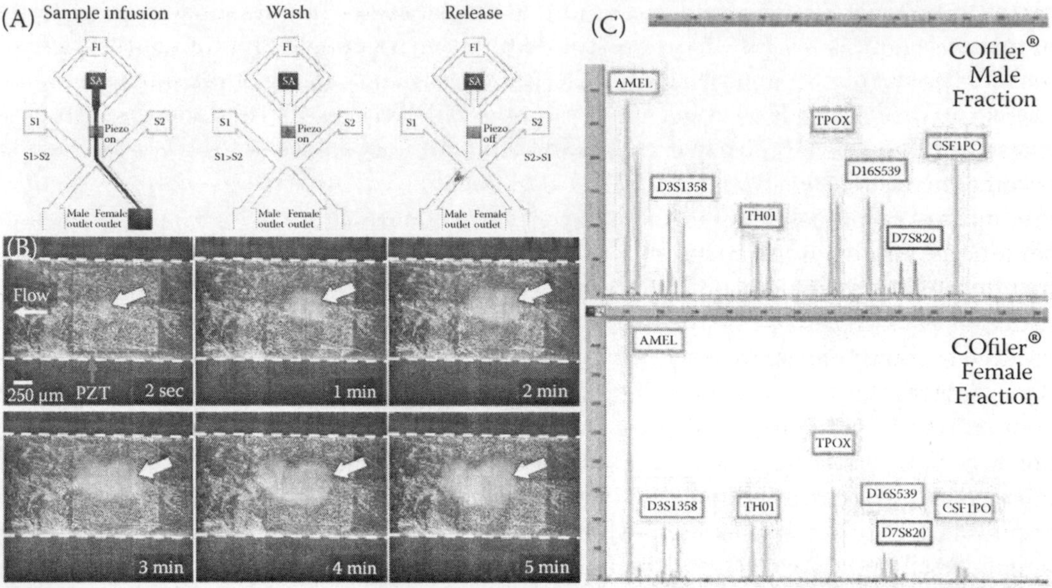

Figure 15.3 (See color insert.) (A) Process flow for an acoustic differential extraction (ADE) microdevice. Intact sperm cells are trapped by acoustic forces, while lysed female DNA flows to one outlet. After rinsing any remaining female DNA away, the flow is switched to the second outlet and the trapped sperm cells are released. (B) Acoustic trapping of 335 μL of dilute semen over the course of 5 min. The microchannel walls are indicated with white dashed lines, the PZT transducer is outlined in the first frame and the white aggregate (indicated by block arrows) consists of the trapped sperm cells. (C) Male (top) and female (bottom) STR profiles obtained after (ADE). (From Norris, J. V. et al., 2009a, Acoustic differential extraction for forensic analysis of sexual assault evidence, *Anal Chem* 81: 6089–95.)

(DTT) (Fujita and Kubo 2006; Gill et al. 1985; Sijen et al. 2010). Numerous groups have explored microfluidic cell-sorting methods, although the majority of these methods are still in their infancy and require further research. Horsman et al. used a microdevice that allowed the epithelial cells to settle to the bottom; after a certain interval, flow was introduced into the device, causing the sperm cells to migrate to the other side of the microchip (Horsman et al. 2005). Norris et al. (2009a) first lysed epithelial cells and then exploited acoustic forces to trap sperm cells within a microchannel, while the epithelial-cells lysate was directed toward the female outlet. Following removal of the cell lysate, the trapped sperm cells were released and the flow was switched so that the sperm cells were directed toward the male outlet (Figure 15.3). The trapping of sperm cells was also demonstrated using dielectrophoresis with a design that forces sperm cells away from the electrodes and toward the field minimum (Fuhr et al. 1998).

15.2.2 DNA Extraction and PCR Amplification

The standard DNA extraction technique used earlier for forensic analyses, the phenol-chloroform method (also known as organic extraction, PC, or PCIA), extracts nucleic acids from cellular debris through the partitioning of proteins into an organic layer and nucleic acids into an aqueous layer (Butler 2005). This method yields highly pure, intact DNA and is still in use at many forensic laboratories today, generally for more difficult cases (Brevnov et al. 2009; Steadman et al. 2008). However, the time-consuming nature of this method, the need for hazardous chemicals, and the numerous tube-opening steps make it less attractive and drive the search for alternative DNA extraction techniques. Presently, the most widely used method to extract DNA from a sample is solid phase extraction (SPE), generally using a silica-based solid phase, which can be nonmagnetic or paramagnetic, to reversibly bind the DNA under high-salt conditions. Impurities and PCR inhibitors are chromatographically removed, and the purified DNA eluted under low-salt conditions. Although this method is effective and certainly an improvement on PCIA, it requires iterative sample-handling steps, affording the opportunity for potential contamination of the sample.

The successful adaptation of SPE to a microdevice eliminates these sample-handling steps through the use of a completely closed system, reducing the incidence of sample contamination and providing purified, concentrated DNA in a small volume (~2 μL). The first report of SPE on a microchip utilized the surface chemistry of silicon dioxide posts within microchannels and demonstrated that a miniaturized method could provide a high surface area for DNA to bind and could increase concentration 10-fold as compared to the original sample (Christel et al. 1999). Similar devices with high surface area, but fabricated using polymeric substrates such as poly(methyl methacrylate) (PMMA) (Reedy et al. 2011b) or polycarbonate (PC) (Witek et al. 2006), have also been described.

Other groups have utilized both coated and uncoated silica-based beads, better representing current SPE methods, for capture of the DNA. Demonstration of DNA extraction from a packed bed of silica beads was first described by Tian et al. using a 500-nL capillary packed with silica beads and with recovery of approximately 70% of the DNA from the original sample (Tian et al. 2000). Using the information gained from this work, the methodology could be translated to a microchip platform, using a frit to hold the silica beads in place (Bienvenue et al. 2006, 2010; Breadmore et al. 2003). Alternately, magnetic silica-beads can be used in a packed-bed (Reedy et al. 2010a,b) or dynamic (Duarte et al. 2010, 2011; Karle et al. 2010) format and provide similar extraction efficiencies as the nonmagnetic beads. Although much work has been performed to optimize microchip SPE, it can be hindered by uneven packing of the bed or high back-pressure. Furthermore, there is a possibility of sample loss due to irreversible binding of DNA to the silica-based phase. That said, a number of groups and commercial entities have found success with SPE in other forms. Recently, a closed-tube, enzyme-based DNA preparation method has been developed that eliminates the need for a solid phase entirely (Moss et al. 2003). For more detailed information on these and other microfluidic DNA extraction techniques, recent reviews by Price et al. (2009) and Wen et al. (2008) provide a more in-depth analysis of current technologies.

Adaptation of PCR to a microfluidic device has been widely explored by numerous groups and generally falls into one of three categories: stationary, continuous-flow, or droplet. The advantages of microscale PCR are speed (faster thermal cycling of a smaller thermal mass), less time at 94°C and therefore longer *Taq* life, and finally, smaller volume leading to lower consumption of samples and reagents. The main disadvantage of microscale PCR is the surface area effects. While the volume change between a 1-μL PCR chamber relative to a 25-μL PCR tube is obviously 25X, the surface area to volume (SA/V) ratio, depending on the architecture of the microchip, can increase by as much as 100-fold (29-nL PCR chamber) (Koh et al. 2003). While this provides opportunity for rapid thermal cycling, it also provides increased surface area for the polymerase and other "master mix" components to adhere to, adversely affecting the reaction and potentially causing PCR to fail ("master mix" is the common nickname for a solution containing all the components necessary for PCR such as $MgCl_2$ and dNTPs, typically prepared in bulk and aliquoted so that each sample receives the same amount of each component). The majority of microdevices for stationary PCR generally have one or two PCR chambers where master mix is loaded and held in place while ceramic (Belgrader et al. 2001) or microfabricated (Lagally et al. 2001) heaters, Peltier elements (Manage et al. 2011), heated air (Lyon and Wittwer 2009; Wittwer et al. 1997) or microwave (Shaw et al. 2010) or infrared (Easley et al. 2007) radiation perform the necessary thermal cycling (Figure 15.4). This method is beneficial in that there are no pumps necessary; the sample is loaded, then not mobilized again until the PCR process is complete.

While the static-sample-chamber approach has been shown to yield forensically relevant data, continuous-flow PCR microdevices mobilize the sample from one temperature region on the microchip to another through the use of syringe pumps (Pjescic et al. 2010; Wu and Lee 2011) or ferrofluids (Sun et al. 2007, 2009). This method is advantageous because the heating elements do not need to be cycled through different temperatures. The time required for analysis is only limited by how fast the solution is flowing;

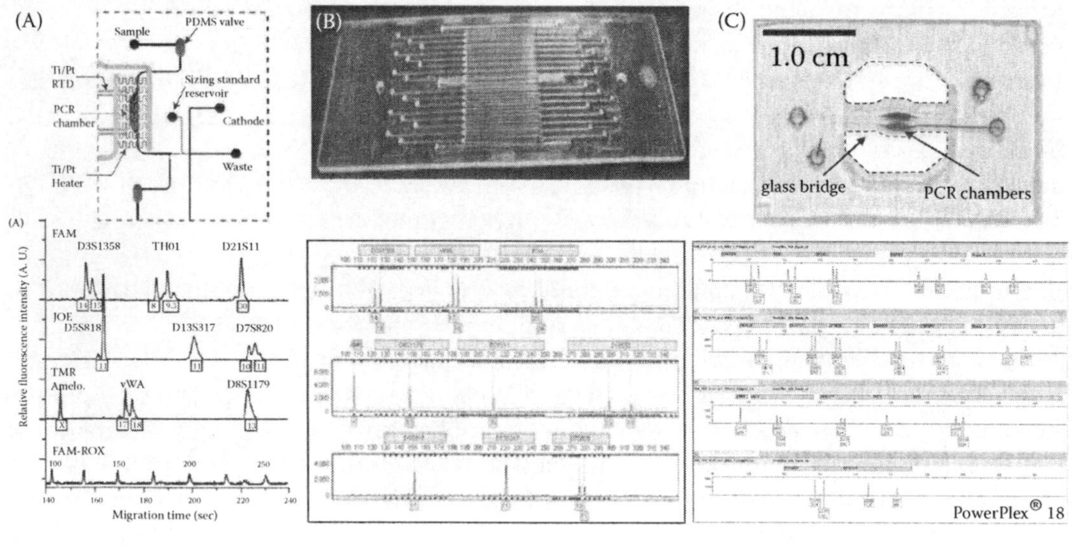

Figure 15.4 (See color insert.) Comparison of different types of stationary PCR. (A) Glass microchip with microfabricated heaters and resistance temperature detectors included in the microdevice. This device was able to produce Amelogenin peaks in less than 15 min. (From Lagally, E. T., C. A. Emrich, and R. A. Mathies, 2001, Fully integrated PCR-capillary electrophoresis microsystem for DNA analysis, *Lab Chip* 1: 102–7.) (B) Sixteen 7-μL PCR chambers are heated by a thermoelectric element and can yield full STR profiles (9 STR loci and Amelogenin) in 17.1 min. (From Giese, H., R. Lam, R. Selden, and E. Tan, 2009, Fast multiplexed polymerase chain reaction for conventional and microfluidic short tandem repeat analysis, *J Forensic Sci* 54: 1287–96.) (C) An IR-mediated PCR system uses 400-nL PCR chambers fabricated in glass and yields full STR profiles (17 STR loci and Amelogenin) in under 40 min. (From Easley, C. J., J. A. C. Humphrey, and J. P. Landers, 2007, Thermal isolation of microchip reaction chambers for rapid non-contact DNA amplification, *J Micromech Microeng* 17: 1758–66; Root, B. E., C. R. Reedy, K. A. Hagan et al., 2011, A Multichannel Microdevice for PCR Amplification and Electrophoretic Separation of DNA, *Presentation—15th Int Conf Miniaturized Systems for Chemistry and Life Sciences*. Seattle, WA.)

however, faster flow rates may not provide adequate dwell times at each temperature. The third approach, droplet PCR, allows for the amplification to be performed in pico- to nanoliter volume droplets, and is typically used in either a digital (Heyries et al. 2011; Pekin et al. 2011; Shen et al. 2010) or continuous-flow (Hua et al. 2010; Kiss et al. 2008; Schaerli et al. 2009) format. Droplet PCR is a very-high-throughput method, owing to the large number of pico- to nanoliter volume droplets that can be formed rapidly. While a number of successful applications have been developed using droplet and digital PCR, particularly for detecting low-copy-number targets with mutations (Markey et al. 2010; Vogelstein and Kinzler 1999), it remains open to question whether this will be applied to forensic DNA analysis and whether stochastic effects in the STR amplifications surface at these volumes.

A distinct advantage associated with using a microfluidic system is the decrease in overall analysis time. In the case of PCR, traditional polymerases, such as AmpliTaq Gold®, have lengthy initial denature times and do not provide fast enough extension rates to really benefit a microfluidic technique. Therefore, modified polymerases, such as SpeedSTAR™, have been used for tube-based PCR to reduce dwell times at the various temperatures and decrease the time required for PCR. When using SpeedSTAR in combination with another modified polymerase, PyroStart™, Vallone et al. (2008) were able to demonstrate the successful amplification of a 16-plex PCR (15 STR loci, plus Amelogenin) using a conventional thermal cycler in a 10-μL volume in less than 36 min. The use of these fast polymerases in combination with a rapid heating method has been used to successfully amplify a 10-plex STR reaction in a glass multichamber 7-μL microchip in 17.3 min (Giese et al. 2009). In addition, an 18-plex PCR reaction has recently been demonstrated on a polymeric substrate in less than 45 min, also using a combination of these modified polymerases (Root et al. 2011b). However, when the time for PCR is reduced (speed of thermal cycling increased), a decrease can be seen in the quality of the STR profile. Several issues arise, including allelic and/or locus dropout, which hinders the ability to call the profile. In addition, when using polymerases other than AmpliTaq Gold, there is an increase in the number of nonspecific artifacts and incomplete adenylation. To alleviate these issues, the PCR master mix must be augmented by the addition of more polymerase or other components, additives, and/or enhancers, and/or the STR primer concentrations must be modified to compensate for the changes in both the speed and volume of the reaction. More information about microfluidic PCR techniques can be found in several excellent recent reviews (Chen et al. 2007; Zhang and Xing 2007; Zhang and Ozdemir 2009).

15.2.3 Microchip Electrophoresis of STR Fragments

The separation of STR fragments requires single-base resolution since there are some common alleles that differ by only a single base pair, such as TH01 9.3/10. As a result of effort from the commercial sector, polymers for capillary-based separation of these amplified products are well-defined, as is the software that accepts the raw data and deconvolutes the multiple alleles from the multiple loci, detected in multiple colors into an interpretable profile of value to the forensic scientist. Perhaps one of the most challenging aspects of microminiaturization is the need to resolve the STR fragments in a separation length (L_{eff}) substantially shorter than the 37 cm presented in the capillary electrophoresis format—a distance for microchips, typically, no more than ~10 cm, but possibly longer if the channel is serpentine in design. Another important aspect of DNA separation is the detection of the resolved fragments postseparation. Commercially available STR kits have primers labeled with a fluorophore and as a result, four- or five-color laser-induced fluorescence detection is required.

Currently, most forensic laboratories use the Applied Biosystems Inc. (ABI) capillary electrophoresis series of instruments to separate and detect PCR-amplified STR fragments, and use a separation polymer, Performance Optimized Polymer (POP-4), as the sieving matrix. However, this polymer, when used in a microdevice with a shorter L_{eff}, does not provide the necessary resolution and, as a result, alternative polymers have been explored. These are typically acrylamide-based and provide the required resolution in a short separation channel. Evidence that use of optimizing novel separations polymers could provide resolution with shorter L_{eff} came early from a report by Bienvenue et al. (Bienvenue et al. 2005). They showed the resolution of the 9.3/10 TH01 locus on a commercial capillary instrument using a 3% w/v 600,000 Da poly(ethylene oxide) (PEO) solution, despite a slight reduction in the capillary L_{eff} from 36 cm to 30 cm and a reduction in time from 19.2 min to 13.5 min, illustrating the promise of these novel polymers.

Linear polyacrylamide (LPA) can be modified with the hydrophobic monomer N,N-dihexylacrylamide (DHA) to yield a copolymer that provides ~10% increase in DNA sequencing read length over LPA alone (Forster et al. 2008). In addition, this copolymer was capable of providing single-base resolution for the 9.3/10 alleles at the TH01 locus (Reedy et al. 2011b). Other groups have used Long Read LPA (Yeung et al. 2006) or LPA in a plastic (poly[methyl methacrylate]; PMMA) microchannel grafted with additional LPA (Llopis et al. 2007) to achieve single-base resolution. To increase the disposability of the microfluidic device, a nonreplaceable polymer, such as a photopolymerizable matrix, could be utilized and would eliminate any concern about cross-contamination between sample analyses (Epstein and Grates 2009; Tan et al. 2009; Ugaz et al. 2002). Additional information about advances in microchip electrophoresis technology (Figure 15.5) can be found in several recent reviews (Felhofer et al. 2010; Kailasa and Kang 2009; Sinville and Soper 2007).

15.3 Integrated Rapid DNA Analysis Platforms

The integration of all the processes involved in STR typing onto a single microdevice is not a trivial undertaking. En route to a fully integrated system, modular approaches have been explored using individual microfluidic devices for DNA analysis (extraction, PCR, and separation) (Reedy et al. 2011a) or integrated devices that fuse two of these processes (Hagan et al. 2011; Liu et al. 2007a). Each analytical process has inherent nuances when translated to a microdevice and these need to be carefully considered for each individual process as well as when interfacing two or more processes, in order for a fully integrated microdevice to be functional. This section of the review will focus on all of the aspects considered when designing, fabricating, and testing an integrated microdevice for STR analysis.

15.3.1 Substrate Choice and Device Size

For a significant portion of the evolutionary time of microdevice development, microchips have been fabricated in glass or silicon, substrates whose surface chemistry was well understood and could be modified to meet almost any application need. Moreover, with glass,

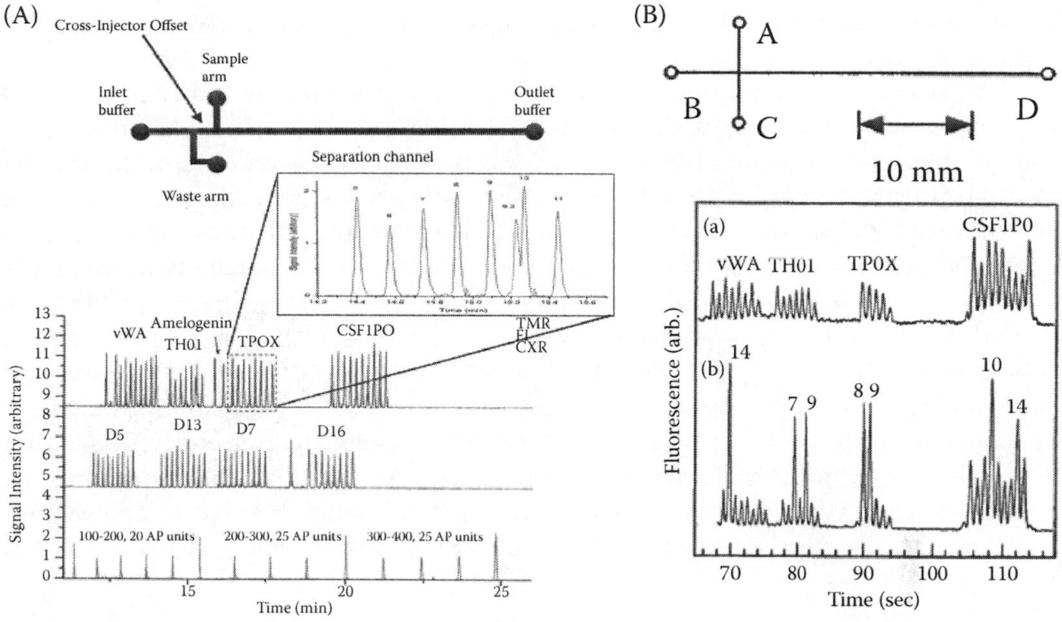

Figure 15.5 (A) Schematic of a cross-T microchip with a 11.5-cm separation channel. Resolution of the 9.3/10 alleles of the TH01 locus is achieved in 25 min. (From Mitnick, L. et al., 2002, High-speed analysis of multiplexed short tandem repeats with an electrophoretic microdevice, *Electrophoresis* 23: 719–26.) (B) Schematic of cross-T microchip with a 30-mm separation channel. Separation of the CTTv loci is achieved in under 2 min. (From Schmalzing, D. et al., 1997, DNA typing in thirty seconds with a microfabricated device, *Proc Natl Acad Sci USA* 94: 10273–8.)

its electrical and heat transfer properties, its chemical compatibility with acids and most solvents, and its transparency through the majority of the UV and visible spectrum are well understood. Easley et al. (2006a) reported a fully integrated microdevice fabricated in glass, capable of extracting DNA, performing PCR, and separating and detecting the product (*B. anthracis* in mouse blood or *B. pertussis* from a nasal swab) in sub-25 min. More recently, Liu et al. (2011) demonstrated a totally integrated glass device for the simultaneous analysis of two buccal swab samples for forensic STR typing.

However, the fabrication process for glass devices is time-consuming, expensive, and laborious, often requiring the use of a clean-room environment and harsh chemicals such as hydrofluoric acid (HF). The recent trend in the field of microfluidics is to use alternative substrates, with the goal being easier and cost-effective, inexpensive fabrication, ideally using high-throughput methods such as hot embossing, injection molding, or laser ablation. Alternative substrates that are not polymeric in nature have also been explored, including low-temperature co-fired ceramic (LTCC) (Henry et al. 1999). Polymeric substrates for microchip electrophoresis that have been widely explored include PMMA (Wang et al. 2002), poly(dimethylsiloxane) (PDMS) (Liu et al. 2000), polyester-toner (do Lago et al. 2003), and SU-8 photoresist (Sikanen et al. 2007). While the field seems to have converged on waxy polymers, it remains to be seen which polymeric substrate will find the best balance between cost and function.

If forensic DNA analysis (or any other application for that matter) is to fully exploit microfluidics, the footprint of the microdevice is an important characteristic. The microscale nature of the system allows for the seamless integration of sample preparation and analytical processes into a small footprint, one that for forensic applications will likely not be as small a compression as is possible. However, while this aids in the instrument development, it presents serious engineering challenges when attempting to unify the instrumentation for DNA extraction, amplification, electrophoretic separation, and detection all into one unit. A clear-cut goal is to have the integrated instrument be substantially smaller and more compact than the numerous pieces of large instrumentation, which each require specialized training, currently employed by the forensic community. Part of the motivation for the miniaturization of analytical techniques is to generate a platform capable of facilitating analysis outside of a laboratory setting, for example, in a doctor's office for clinical applications or at a crime scene for forensic DNA analysis. Therefore, not only are the processes themselves miniaturized, but the computer systems, fluorospectrometers, thermal cyclers, and other instrumentation must be reengineered (or reinvented) for smaller footprints (Lang 2004). For more details, a recent guidebook provides a thorough review of portable and miniaturized instrumentation (McMahon 2007).

15.3.2 Surface Chemistry

With miniaturization of biochemical processes, it is essential that the surfaces in the microfluidic architecture are adequately passivated so that minimal biofouling occurs. Although not problematic at the macroscale, this is particularly important in microfluidics where the SA/V ratios in microchambers and microchannels is high and, therefore, can adversely affect the analytical process. For example, in conventional PCR, adsorption to the polypropylene tube surface is minimal, and an adequate *Taq* polymerase concentration compensates for any loss due to wall adsorption. However, the same master-mix conditions used in a microfluidic device can have undesirable effects due to loss of polymerase to the walls. Furthermore, the chemistry of the untreated surface of the substrate being used can affect the quality of the separation in the electrophoresis channel, creating issues with electroosmotic flow (EOF; DNA separations generally require little to no EOF). To circumvent these issues, surface passivation (modification) can be exploited in one of two ways: through *static* coating, which is carried out prior to any analysis, or *dynamic* coating, which involves augmenting the content of the master mix so as to minimize wall loss of critical components applied during the analysis. Detailed discussion of substrate-specific surface modifications options can be found in two recent reviews (Belder and Ludwig 2003; Liu and Lee 2006).

15.3.3 Flow Control

Flow control is a vital aspect of microfluidic execution of analytical processes. This is particularly true for processes involving reagents that are toxic to downstream processes. For example, with SPE, the guanidine-HCl used to bind DNA to the silica beads and the isopropyl alcohol used to wash impurities away are potent inhibitors of PCR (Bessetti 2007); these can cause PCR failure even at low concentrations. Hence, a fluidic control that allows the "chemistries" for SPE and PCR to remain separated initially, but eventually fluidically connected, must be precise. In addition, fluidic movement control of sample or reagents through various functional domains of the chip is critical to successful analysis (e.g., if PCR solution is expelled from the PCR chamber during thermal cycling, amplification will fail). Several methods have been shown as effective for flow control in microfluidic chips.

Elastomeric membranes made of PDMS have been used extensively as active and passive microfluidic valves/pumps due to their ability to be deformed under pressure (positive or negative), with the extent of deformation governed by the Young's modulus. Two types of "active" valves have been evolved. Unger et al. (2000) exploited multiple layers of PDMS-based substrate on a glass base to fabricate "on-off" valves, switching valves, and pumps to demonstrate the control that can be achieved. Grover et al. (2003) expanded on this concept by developing glass-PDMS-glass valves that are suitable for use in arrays or large-scale integrated systems. Distinguishing these two approaches is the fact that the Unger valves are "normally open" valves, while the Grover valves are "normally closed"; the choice of valving is driven by the application and the function of the valve in the microfluidic system.

Other types of valve systems include magnetic or piezoelectric mechanical valves and electrochemical or phase-change nonmechanical valves (Oh and Ahn 2006), among others. Although these types of active valves have provided significant advances in fluid control on microchips (Iverson and Garimella 2008; Liu et al. 2007b), they are typically controlled by substantial external hardware, such as vacuum/pressure pumps, which increases the challenge in defining microfluidic systems that need to be portable. A novel method for flow control recently developed by Leslie et al. (2009) uses elastomeric membranes, but in a "passive" manner. They show that fluidic analogs to components found in electrical circuits (capacitors, resistors, and diodes) can be created and, arranged properly, a frequency-specific flow control could be obtained. Passive valves will eventually be a powerful, but elegantly simple, method for flow control without the need for bulky external hardware.

15.3.4 Separation Speed

The true value of adapting the processes associated with STR typing to a microchip is realized in terms of a reduction in analysis time. Reduction in sample, reagent, and chamber volumes in miniaturized systems can lead to a reduction in the amount of time needed for PCR, since heating and cooling occurs much more rapidly, as described above. The same advantages also benefit microchip electrophoresis, and can lead to reduced separation time as compared to separation using conventional capillary electrophoresis (i.e., the ABI series of instrument, which generally requires 30–35 min per electrophoresis run). Although separation of STR fragments has been demonstrated in less than 2 min using a 2.6-cm separation channel, single-base resolution has not been achieved (Schmalzing et al. 1997). This is exemplary of the compromise between ultrafast analysis and high-resolution separation; however, the standards set by the forensic community, which require single-base resolution (e.g., alleles 9.3 and 10 at the TH01 locus), are unforgiving in this respect, and while ultrafast separations are the goal, the quality of the separation cannot be sacrificed. By increasing the channel length to 11.5 cm and the separation time to 10 min, single-base resolution between 9.3 and 10 alleles at the TH01 locus was achieved (Schmalzing et al. 1999). Similar to STR typing, DNA sequencing also requires single-base resolution, and the demonstration of the separation of 600 bases in 6.5 min was recently reported (Fredlake et al. 2008).

Although ultrafast separation speeds have been achieved, conventional methods hold the advantage in terms of sample throughput. The ABI series of instruments have multicapillary arrays that have the capability to analyze up to 16 (ABI 3100) or 96 (ABI 3700) samples simultaneously. This same capability, combined with the fast separation times achieved in a microfluidic platform, would significantly increase the sample throughput of the device. Recently, a 16-channel microdevice was reported that uses LPA as the sieving matrix and is capable of single-base resolution (Goedecke et al. 2004). Furthermore, the Virginia Department of Forensic Science, in collaboration with Northwestern University and the Palm Beach County Sheriff's Office, recently tested a 96-channel microdevice developed at the University of California at Berkeley. The test demonstrated successful, simultaneous separation of 15 STR loci, plus Amelogenin, with single-base resolution, in 96 PCR samples in approximately 30 min (Greenspoon et al. 2008). These examples demonstrate that multisample separations of STR fragments in a reasonable amount of time are achievable and completely capable of the ever-important single-base resolution. Polymers similar to those reported in Fredlake et al. (2008) can be used to achieve ultrafast separations (sub-10 min) in unprecedently short separation distances (7 cm or less) (Norris et al. 2009a).

15.3.5 Multiplexed STR Detection and Allele Calling

Critically important to effectiveness of miniaturized forensic DNA analysis systems is the ability to detect four or five colors and successfully call alleles. The commercially available STR-typing kits used by forensic laboratories consist of fluorescently tagged primers, which allow for the detection of PCR products from multiple loci that may overlap in length and would otherwise be indistinguishable from one another (i.e., they can be spectrally resolved). ABI series instruments use an argon-ion laser to excite the fluorophores at 488 nm and a charge-couple device (CCD) camera for detection of the different emission colors (Butler et al. 2004). The use of multiple fluorescent tags has allowed for an increase in the number of STR loci that can be evaluated up to 26, greatly increasing the discriminatory power of the assay (Hill et al. 2009). Naturally, laser-induced fluorescence is the excitation method of choice, but the detection method for microchip electrophoresis can vary from the use of photo-multiplier tubes (PMTs), acousto-optic tunable filters (AOTFs), or CCD cameras.

Many electrophoretic microdevices for detection of STR fragments utilize multiple detectors (one for each color). For example, Shi and Anderson (2003) used four CCD cameras to detect each of the four colors. Contrary to this, Mitnick et al. (2002) utilized four PMT detectors, with corresponding dichroic mirrors, to detect each of the four wavelengths. Although these systems provide sensitive detection at each color, the systems become bulky due to the necessity of multiple detection components. To circumvent this, Karlinsey and Landers (2006) reported the development of an AOTF system for two-color detection of STR fragments and further expanded this to be capable of four- and five-color detection (Karlinsey and Landers 2008; Root et al. 2011a). This type of system is beneficial detection at higher numbers of colors (>5) because only the AOTF and a single photodetector are required, reducing the footprint of any instrument that would use this detection scheme.

With any of these detection schemes in use, the corresponding software must record the signal from the detector(s) and be capable of performing allele calls. The current, most widely utilized software, which is used with the ABI series of instruments, is called GeneMapper (most recent version is ID-X v3.2), which adjusts the baseline, deconvolutes the data, adds the allele bins, and makes allele calls with a simple click of the mouse, followed by a review of the data according to quality assurance standards (QAS) guidelines. For their integrated system, consisting of a plastic microfluidic cartridge for DNA extraction and PCR and a glass microchip for electrophoresis, Hopwood et al. used NanoIdentity® and GeneMarker® HID from SoftGenetics® to deconvolute the data and provide allele calls, respectively (Hopwood et al. 2010). Giddings et al. (1998) use a program called BaseFinder (University of North Carolina 1997, http://bioinfo.boisestate.edu/glabsoftware/BaseFinder/index.html) while other groups use in-house written programs (Goedecke et al. 2006; Karlinsey et al. 2005). In the development of integrated microfluidic systems for forensic DNA analysis, software (for both system control and data analysis) is an important component that cannot be overlooked or taken for granted, as the final success of the sample processing and analysis is dependent on the generation of a complete and accurately called genetic profile by an approved expert system. However, secondary conformation by a trained analyst will still be necessary for any integrated system to verify the quality of the profile, as required by the QAS issued by the FBI (Scientific Working Group on DNA Analysis Methods 2010).

15.3.6 Comparison of Next-Generation Prototypes

The preceding sections discussed numerous variables that must be considered and/or accounted for when designing and developing a truly integrated microdevice. As a result of the academic and commercial efforts of several groups, prototype microfluidic systems (and accompanying instrumentation) will soon be accessible to laboratories for human identification; some of these prototypes are currently being tested in forensic laboratory settings (Figure 15.6). The efforts of five academic or commercial entities and their resultant systems will be discussed here.

On the academic front, the Landers group at the University of Virginia has demonstrated the successful integration of SPE and PCR for STR analysis on a glass microdevice in ~1 h (Hagan et al. 2011). Epithelial cells were eluted and lysed from dried buccal swabs and flowed into the SPE domain of the microdevice. The purified DNA was eluted from the SPE bed, mixed with PCR master mix, and moved into a 500-nL PCR chamber. Thermal cycling was performed by an IR-PCR system and the STR products were separated and detected using a conventional ABI capillary electrophoresis instrument. This system was able to yield full profiles for both the MiniFiler (8 STR loci, plus Amelogenin) and Identifiler (15 STR loci, plus Amelogenin) STR kits from buccal swab samples in 1 h. This group has also developed a fully integrated glass microdevice (DNA extraction, PCR amplification, electrophoretic separation/detection) that could process submicroliter sample volumes in ~30 min, although not for STR analysis (Easley et al. 2006a). Whole blood was added to lysis buffer and DNA was extracted from the equivalent of 750 nL of whole blood in the SPE domain of the microchip. After extraction, the sample was mixed with PCR master mix and moved into a 550-nL PCR chamber. Thermal cycling was performed by an IR-PCR system in ~11 min and the sample was separated in a 4-cm microchip electrophoresis channel. Successful detection of *B. anthracis* in mouse blood and *B. pertussis* in

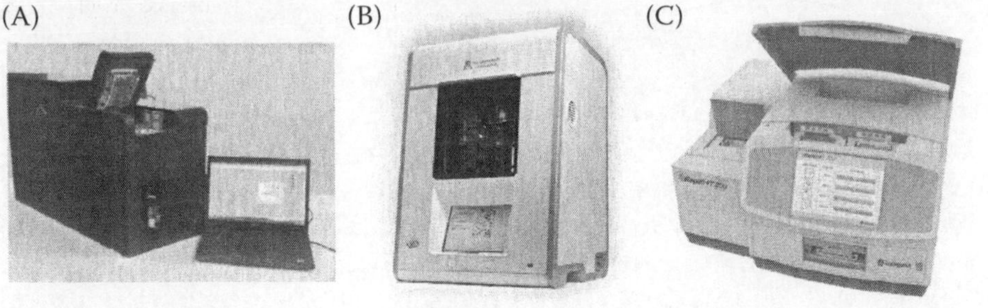

Figure 15.6 (See color insert.) Prototype microfluidic systems: (A) Photograph of the RapID™ system under development by MicroLab Diagnostics, a subsidiary of ZyGEM, Inc. (From Root, B. E. et al., 2011a, A Multichannel Microdevice for PCR Amplification and Electrophoretic Separation of DNA, *Presentation—15th Int Conf Miniaturized Systems for Chemistry and Life Sciences*. Seattle, WA.) (B) Photograph of the integrated, automated microfluidic system developed by the Forensic Science Service and the University of Arizona. (Center for ANBN [Applied Nanobioscience and Medicine], The University of Arizona: http://anbm.arizona.edu/research/current-projects-programs/rapid-dna-analysis.) (C) Photograph of the RapidHIT™ 200 Early Access Commercial System developed by IntegenX. (From El-Sissi, O. et al., 2010, RapidHIT 200 human DNA identification system, August 6, 2010, available at: http://integenx.com/applications-2/dna-identity/.)

human nasal aspirate was demonstrated in 24 min. Although this system was not used for STR analysis, it was one of the first demonstrations of a fully integrated microdevice that could provide *sample-in–answer-out* capability.

The Mathies group at the University of California at Berkley has developed a fully integrated glass microdevice that is capable of analyzing two samples simultaneously (Figure 15.7A) (Liu et al. 2011). The process begins by eluting and lysing epithelial cells from dried buccal swabs, then manually loading 20 μL of the lysed sample into the microdevice. Once loaded, DNA is extracted from the sample using silica-coated magnetic beads and the sample undergoes thermal cycling in a 250-nL PCR chamber. The thermal cycling is controlled by integrated microfabricated heaters and resistance temperature detectors. Following PCR, the product is mobilized into the electrophoresis domain where the DNA is separated and detected in a 14-cm separation channel. This system is unique in that it uses sequence-specific DNA purification prior to PCR to capture only those sequences used for the STR PCR; this improves the efficiency of the reaction and increases peak height. Two buccal swabs, provided by the Virginia Department of Forensic Science (Richmond, Virginia), each yielded full 9-loci profiles (8 STR loci, plus Amelogenin) in just under 3 h (Figure 15.7B). This system requires a minimum of 2.5 ng DNA in order to produce full profiles, but does not use disposable microchips (the glass microchips are washed with a piranha solution—7:3:H_2SO_4:H_2O_2) as seen with other prototypes.

The first generation of a platform developed by MicroLab Diagnostics (subsidiary of ZyGEM Corporation) and Lockheed Martin utilized a fully integrated polymeric microdevice for PCR and microchip electrophoresis. Prior to loading onto the microchip, epithelial cells from buccal swab samples were eluted and lysed using a microchip-optimized, enzyme-based DNA preparation process (*forensic*GEM) in <5 min. Once loaded onto the microchip, the instrument performed thermal cycling using infrared energy, followed by DNA separation/detection in <10 min using a 7-cm separation channel. Results show that this system yielded full profiles (17 STR loci, plus Amelogenin) in <90

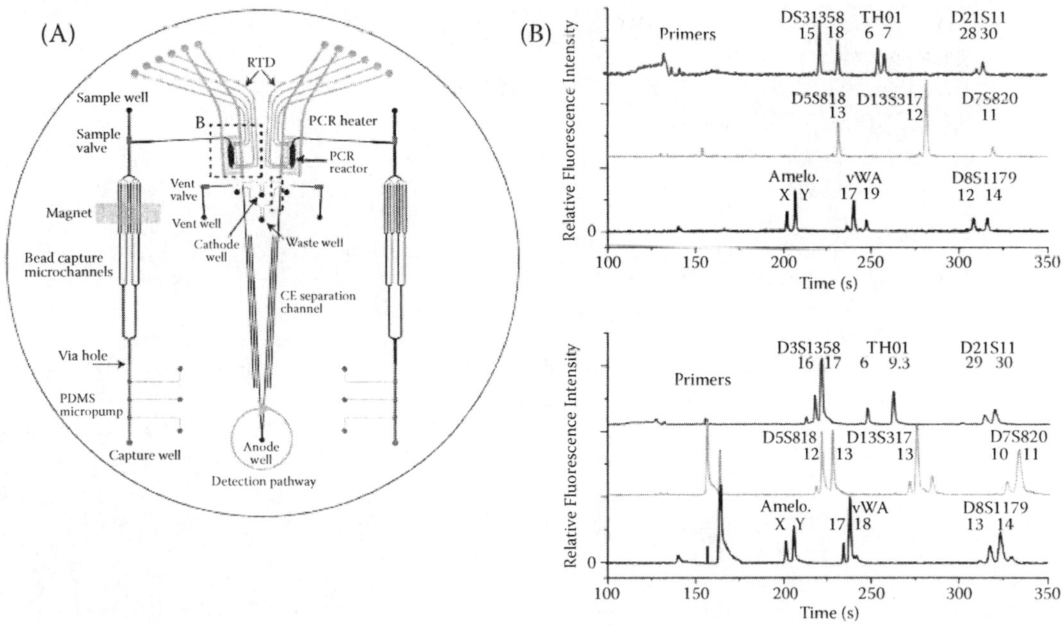

Figure 15.7 (See color insert.) A fully integrated device for DNA extraction, STR PCR, and microchip electrophoresis developed by the University of California at Berkley and the Virginia Department of Forensic Sciences. (A) A glass microchip, which has two domains for dual sample analysis, uses magnetic silica-based beads to capture DNA from lysed buccal cells, and contains microfabricated heaters and resistance temperature detectors for STR PCR. (B) Full STR profiles (9-plex, 4-color) from two dried buccal swabs obtained after a fully integrated run in under 3 h. (From Liu, P. et al., 2011, Integrated DNA purification, PCR, sample cleanup, and capillary electrophoresis microchip for forensic human identification, *Lab Chip* 11: 1041–8.)

min (Figure 15.8). The most recent second-generation instrument, the *IntrepID S2A90*™, has integrated the DNA preparation chemistry with amplification and separation/detection for *swab-in–profile-out*. Utilizing on-board reagents and a disposable, integrated microfluidic cartridge, which is the size of a 96-well plate and only a few millimeters thick, fully automated, simultaneous, end-to-end processing of four samples in <90 min is possible. With this system, the whole buccal swab is added to the disposable cartridge containing all of the reagents and fluidic architecture to process four samples. The cartridge is inserted into the instrument and analysis initiated by the user (Figure 15.9). The instrument processes the sample, without user intervention, from DNA liberation to profile, providing a full profile using either Applied Biosystems or Promega PCR chemistries, optimized for performance in the microscale amplification volumes exploited in this system (~1 μL). Similar to the prior prototype, this system relies on IR-mediated thermal cycling, with noncontact temperature sensing and short (7 cm) separation distances; however, it has the added feature of automated alignment for separate control of each sample during separation and detection to maximize sensitivity and enable the system to align on every new microchip for robust operation. In addition to providing full genetic analysis capability, with *sample-in–answer-out* technology, the system provides truly robust, portable, and ruggedized DNA analytics for field-forward operation, as well as lab-based processing, of buccal swab samples.

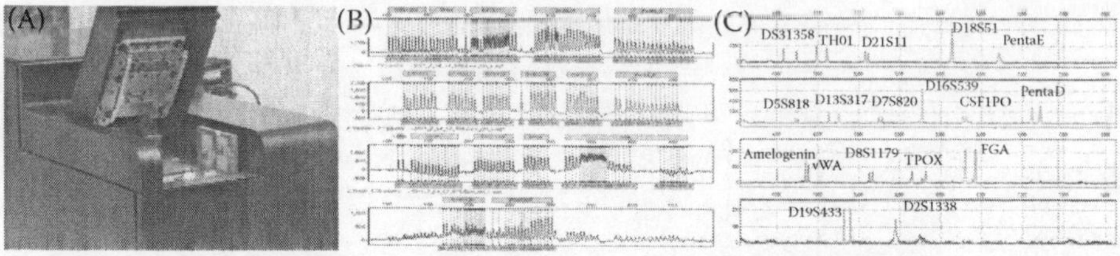

Figure 15.8 The first-generation RapID™ system by MicroLab Diagnostics, a subsidiary of ZyGEM, and Lockheed Martin. (A) Photograph of the instrument with the microchip insertion module open. This system was able to separate and detect the (A) PowerPlex® 18 allelic ladder and (B) PCR product after an integrated PCR-ME run using a 7-cm polymeric separation channel in less than 10 min. In addition, custom software is used to create the allelic bins, calls alleles, and give insight into overall sample quality. (From Root, B. E. et al., 2011a, A multichannel microdevice for PCR amplification and electrophoretic separation of DNA, *Presentation—15th Int Conf Miniaturized Systems for Chemistry and Life Sciences.* Seattle, WA.)

The Zenhausern Laboratory at the University of Arizona, in collaboration with the Forensic Science Service (Birmingham, United Kingdom), has described a two-microchip system; a polycarbonate cartridge for DNA extraction and PCR, and a glass microchip for electrophoresis (Hopwood et al. 2010; Hurth et al. 2010, http://anbm.arizona.edu/research/current-projects-programs/rapid-dna-analysis) (Figure 15.10A). The two components are assembled into a scaffold and placed in an instrument that automates the process. The process is initiated with elution and lysis of epithelial cells, followed by manual loading of 150 μL lysate onto the microchip. Magnetic ChargeSwitch beads are used for DNA extraction as the sample is moved throughout the cartridge and 10 μL of purified DNA is captured in the PCR chamber, which is preloaded with PCR master-mix reagents. After thermal cycling is carried out using a Peltier element for heating and cooling, the sample is moved

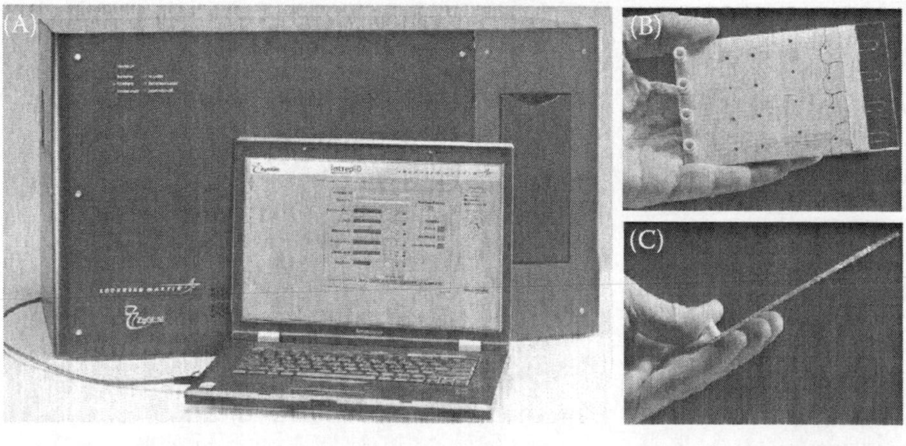

Figure 15.9 **(See color insert.)** The IntrepID S2A90™ system under development by MicroLab Diagnostics, a subsidiary of ZyGEM, and Lockheed Martin. (A) Photograph of the second generation prototype showing the size of the system relative to the laptop needed for instrument operation. Additional photographs show the (B) top and (C) side views of the disposable, integrated cartridge used, simultaneously analyzing up to four buccal swab samples. (From Root, B. E. et al., 2011a, A multichannel microdevice for PCR amplification and electrophoretic separation of DNA, *Presentation—15th Int Conf Miniaturized Systems for Chemistry and Life Sciences.* Seattle, WA.)

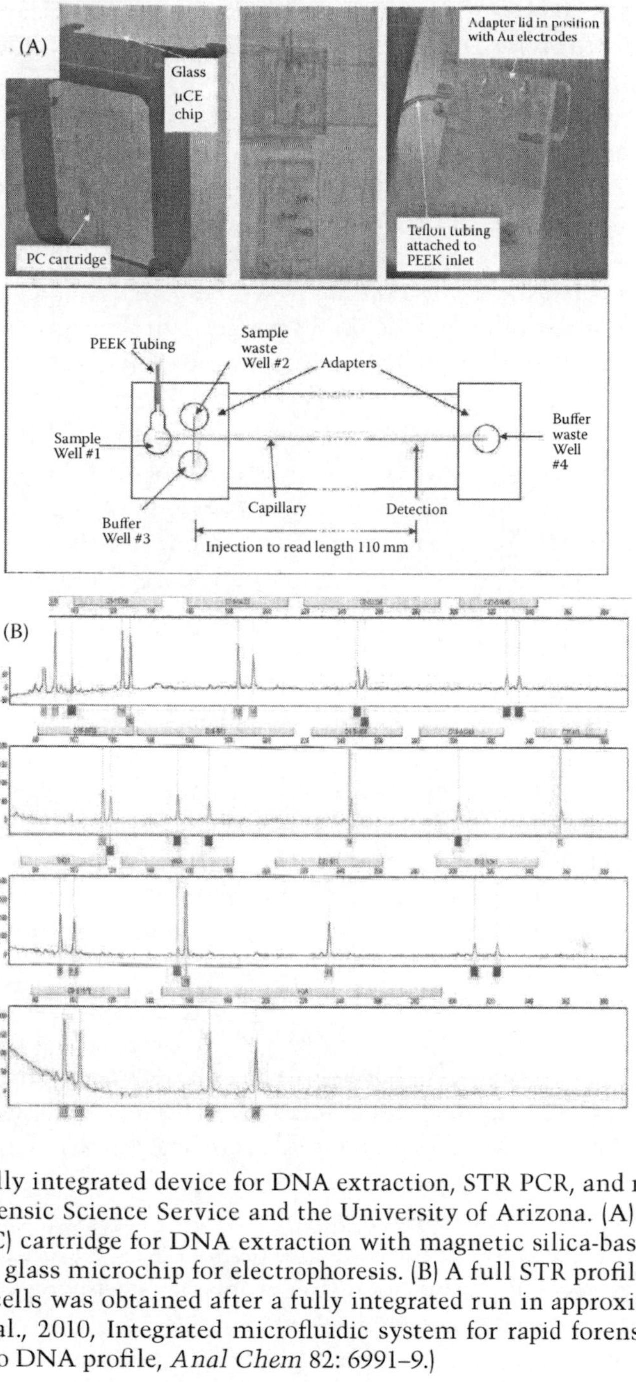

Figure 15.10 A fully integrated device for DNA extraction, STR PCR, and microchip electrophoresis by the Forensic Science Service and the University of Arizona. (A) This system uses a polycarbonate (PC) cartridge for DNA extraction with magnetic silica-based beads and STR amplification and a glass microchip for electrophoresis. (B) A full STR profile (16-plex, 5-color) from lysed buccal cells was obtained after a fully integrated run in approximately 4 h. (From Hopwood, A. J. et al., 2010, Integrated microfluidic system for rapid forensic DNA analysis: Sample collection to DNA profile, *Anal Chem* 82: 6991–9.)

through Teflon tubing from the polycarbonate cartridge to the glass microchip for electrophoresis in an 11-cm separation channel. This system was able to yield full profiles (15 STR loci, plus Amelogenin) in approximately 4 h (Figure 15.10B). However, in this system of two distinct microdevices, one is a nondisposable glass microchip with a pre-run preparation time of 75 min.

The RapidHIT™ 200 DNA Identification System (IntegenX, Pleasanton, California) is an automated platform for DNA analysis that can yield STR profiles from up to four dried buccal swab samples in ~2 h (Figure 15.11A) in a *swab-in–profile-out* manner. No elution of cellular material or other pretreatment steps are required for this system; the cotton swabs are added directly to the instrument (El-Sissi et al. 2010). Once initiated, the instrument first extracts DNA from the sample using magnetic beads (Promega DNA IQ kit) and a series of elastomeric valves to control flow. The purified DNA is mobilized into Teflon tubing that connects to the postamplification subsystem and thermal cycling is performed in the tubing by a custom thermal cycler using resistive heating and forced air cooling. Following PCR, the sample is mixed with formamide and size standard in the postamplification microchip, then injected into capillaries for DNA fragment separation. Software converts the raw data into a format compatible with GeneMapper ID-X software. This system was able to yield full profiles (15 STR loci, plus Amelogenin) from four buccal swabs, simultaneously, in ~2 h (Figure 15.11B). Although this system does not exploit a single, integrated microdevice and still uses capillary electrophoresis for separation and detection, it is fully automated and the DNA profiler cartridge is disposable, eliminating the possibility of cross-contamination between samples.

15.4 Implementation of Integrated Rapid DNA Analysis Platforms

While current DNA analysis requirements, technology, and policies limit the processing of DNA samples almost entirely to laboratory environments, the development of new integrated, portable, DNA analysis systems opens the possibility for analysis outside of the traditional laboratory setting. With this opportunity comes a number of practical implications and possible limitations for consideration; the community must take a close look at policies, validation requirements, quality assurance standards (QAS), and database requirements, to both control and enable this likely transition. This was already under way, with the creation of the SWGDAM Rapid DNA (R-DNA) Committee, in early 2011 (Callaghan 2011). This committee has been focusing on the issues salient to the successful acceptance of these platforms: first, into forensic laboratory workflows, and subsequently into other arenas. Areas for consideration include concordance with current analysis techniques, validation, quality assurance, pilot study programs, universal data file formats, appropriate system controls, performance evaluations, and developmental validations. By convening this panel prior to the full-scale production release of these instruments, the community is more ready for the introduction of this new technology and the successful implementation of these systems.

It should also be noted that, while the laboratory-based protocols and instrumentation required for DNA analysis currently limits the processing of samples to well-trained, skilled technicians or analysts, integrated rapid DNA platforms with *sample-in–answer-out* capabilities may render this requirement obsolete, particularly for convicted offender/arrestee samples. Integrated microfluidic technology will enable the minimally trained end user (e.g., a police officer, field analyst, or soldier) to perform what was once a complex scientific endeavor. Consequently, not only are the logistics and requirements for DNA analysis outside of the controlled laboratory environment to be considered, but

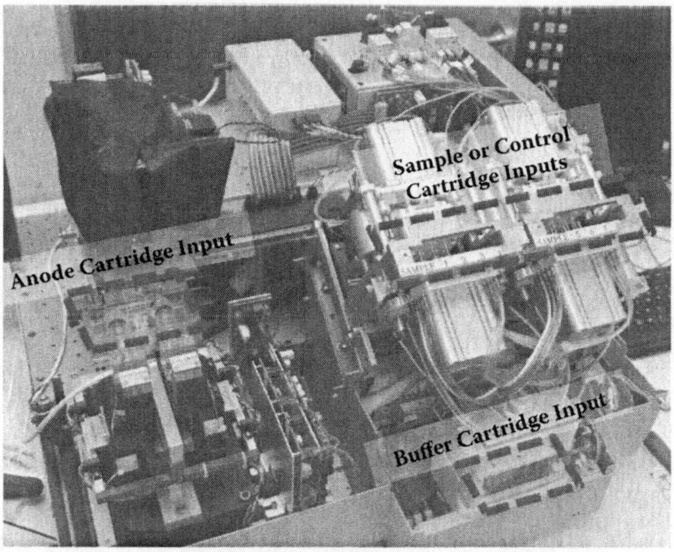

(A)

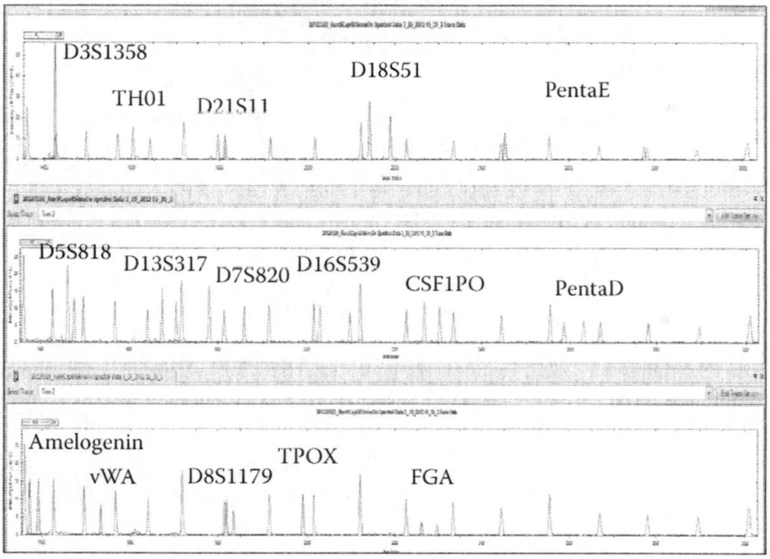

(B)

Figure 15.11 Photographs and results from the RapidHIT™ 200 DNA Identification System by IntegenX. Photographs of the assembled cartridge with the reaction tubing secured in the thermal cycler with four cotton-swab samples loaded (left) and DNA Profiler cartridge with two microchips connected by tubing (left-inset). Full STR profiles obtained from two of the four buccal swabs analyzed in a single fully integrated and automated run in approximately 2 h (center and right). (From El-Sissi, O. et al., 2010, RapidHIT 200 human DNA identification system, August 6, 2010, available online at: http://integenx.com/applications-2/dna-identity/.)

also the requirements and standards for system operators. What was once a domain of a select few has the potential to become a much broader and less controlled pool of individuals. Consequently, as the SWGDAM committee for R-DNA considers the full implications of this technology, the eventual end user becomes a critical component of this implementation.

15.5 Future Implications

The development of microfluidic systems for rapid genetic analysis has brought with it the promise of reduced volume requirements for both samples and reagents, less instrumentation, smaller instrument footprint, reduced cost, and throughput that brings a new dimension to sample processing. While it remains to be seen how these instruments impact the overall workflow in forensic laboratories, it is important to speculate on some of the more critical assertions that have been made, particularly those surrounding cost and the backlog. Although a reduction in reagent volume, necessary instrumentation, and hands-on analyst time would all point to a concomitant decrease in cost per sample, it is important to consider that reagent-licensing costs, manufacturing costs, and other fees may not decrease at all. Additionally, while projections may point to eventual cost-per-sample targets that are at or below current costs, those targets will be met over time as manufacturing methods are put through lean design challenges, production efforts are refined, and critical production volumes are met. Initial cost per sample may be significantly higher than current costs, until the volume of consumable sales required to sustain a lower price per sample is met.

When the backlog is considered, high-throughput robotic analysis methods currently in use may still provide a higher throughput (in spite of longer analysis times) than the initial, low- to midvolume production units provided. With sample throughput of four to eight samples for most initial platforms (including all controls and allelic ladders), the throughput may not initially compete with current robotic and other high-throughput methods for analysis (which may process batches in far greater numbers), although the other benefits of these platforms (portability, speed per batch, etc.) will still make them attractive and of increasing importance in the forensic laboratory workflow. Consequently, it is important that these two areas of promise for microfluidic systems—cost per analysis and the effect on sample backlogs—not be oversold when these systems are initially implemented. It will take time to realize these benefits and it is likely to require multiple iterations of the technology before the impact of these new methods and platforms are fully felt.

15.6 Conclusions

Microfluidic systems for forensic and clinical processes have been the subject of intense research and development since the concept was first conceived over two decades ago. Development has not been as expedient as expected for a variety of reasons. First, the electronic equivalents to fluidic microchips—the electronic microchip (i.e., Pentium processors)—were 20 years in evolution and supported by trillions of dollars in funding. Although microfluidics has been in development for more than two decades (since Manz et al. 1990), it has likely seen funding that is roughly four orders of magnitude lower at this point. Second, sample preparation was not seen as a priority in the first decade of development, where separations dominated the research and development. Finally, the development of integrated microfluidic systems was handcuffed early on by the absence of "valving" and/or fluidic control solutions that separated diverse chemistries and/or processes, yet allowed for fluidic connectivity when needed. Developments in the early 2000s (Grover et al. 2003; Unger et al. 2000) allowed for the unleashing of the potential power of microfluidics.

Microdevices for individual DNA analysis processes such as DNA extraction, STR amplification, and microchip electrophoresis have demonstrated the unique advantages offered by microfluidic systems. DNA extraction has been adapted to the microscale using numerous techniques, such as packed silica beads or high-surface-area polymeric posts, providing a concentrated DNA sample in a small volume, and has reduced the time required for extraction to 30 min or less. In addition, STR amplification has significantly benefited from translation to the microscale. After overcoming increased surface-area-to-volume ratio issues, 16-plex PCR has been performed in less than 30 min in a microchip, a reduction of approximately 2.5 h as compared to conventional methods. Finally, microchip electrophoresis has demonstrated that single-base resolution can be achieved in less than 10 min using a 10-cm or less separation channel by adjusting and optimizing the polymer used as the sieving matrix. Any of the single-process microdevices described above could be used as a stand-alone system, taking a modular approach to the incorporation of microfluidic systems into normal casework processing.

These advances have culminated and merged together in the development of fully integrated sample-in–answer-out microfluidic systems that are able to produce full STR profiles from lysed buccal cells in 2–4 h. Although these systems are only just now becoming commercially available in small numbers, they represent the tremendous strides that have been made in the translation of forensic DNA analysis techniques to the microscale. Future implementation of these systems in forensic laboratories requires that the quality of data produced using these systems meets or exceeds the very high standards and guidelines set forth by SWGDAM. Once this is demonstrated, the community will begin to see the positive impact of these systems in how, where, and by whom DNA analysis is accomplished in the future. Furthermore, as this technology improves to encompass higher sample throughput and as the cost per sample and per instrument decreases with increased adoption, one can envision not only the reduction of backlogs of certain sample types, but also the emergence of new applications.

References

Belder, D. and M. Ludwig. 2003. Surface modification in microchip electrophoresis. *Electrophoresis* 24: 3595-606.

Belgrader, P., S. Young, B. Yuan et al. 2001. A battery-powered notebook thermal cycler for rapid multiplex real-time PCR analysis. *Anal Chem* 73: 286–9.

Bessetti, J. 2007. An introduction to PCR inhibitors. *Profiles in DNA* 10: 9–10.

Bienvenue, J. M., N. Duncalf, D. Marchiarullo, J. P. Ferrance, and J. P. Landers. 2006. Microchip-based cell lysis and DNA extraction from sperm cells for application to forensic analysis. *J Forensic Sci* 51: 266–73.

Bienvenue, J. M., R. Giles, J. P. Landers, and S. Bell. 2009. Integrated microfluidics for forensic applications. *Presentation—61st American Academy of Forensic Sciences Annual Meeting*. Denver, CO.

Bienvenue, J. M., L. A. Legendre, J. P. Ferrance, and J. P. Landers. 2010. An integrated microfluidic device for DNA purification and PCR amplification of STR fragments. *Forensic Sci Int Genet* 4: 178–86.

Bienvenue, J. M., K. L. Wilson, J. P. Landers, and J. P. Ferrance. 2005. Evaluation of sieving polymers for fast, reproducible electrophoretic analysis of short tandem repeats (STR) in capillaries. *J Forensic Sci* 50: 842–8.

Breadmore, M. C., K. A. Wolfe, I. G. Arcibal et al. 2003. Microchip-based purification of DNA from biological samples. *Anal Chem* 75: 1880–6.

Brevnov, M. G., H. S. Pawar, J. Mundt, L. M. Calandro, M. R. Furtado, and J. G. Shewale. 2009. Developmental validation of the PrepFiler™ Forensic DNA Extraction Kit for extraction of genomic DNA from biological samples. *J Forensic Sci* 54: 599–607.

Butler, J. M. 2005. *Forensic DNA Typing: Biology, Technology, and Genetics of STR Markers*. Academic Press/Elsevier: Burlington, MA.

Butler, J. M., E. Buel, F. Crivellente, and B. R. McCord. 2004. Forensic DNA typing by capillary electrophoresis using the ABI Prism 310 and 3100 genetic analyzers for STR analysis. *Electrophoresis* 25: 1397–412.

Callaghan, T. 2011. Rapid DNA Analysis in the Police Booking Suite: "FBI Inititative for Reference Sample Point of Collection Analysis." *Presentation—22nd Int Symp Human Identification*. National Harbor, MD.

Chen, L., A. Manz, and P. J. R. Day. 2007. Total nucleic acid analysis integrated on microfluidic devices. *Lab Chip* 7: 1413–23.

Chen, X. and D. F. Cui. 2009. Microfluidic devices for sample pretreatment and applications. *Microsystem Technologies-, Micro-, and Nanosystems-Information Storage and Processing Systems* 15: 667–76.

Christel, L. A., K. Petersen, W. McMillan, and M. A. Northrup. 1999. Rapid, automated nucleic acid probe assays using silicon microstructures for nucleic acid concentration. *J. Biomech Eng: T ASME* 121: 22–7.

do Lago, C. L., H. D. T. da Silva, C. A. Neves, J. G. A. Brito-Neto, and J. A. F. da Silva. 2003. A dry process for production of microfluidic devices based on the lamination of laser-printed polyester films. *Anal Chem* 75: 3853–8.

Duarte, G. R. M., C. W. Price, B. H. Augustine, E. Carrilho, and J. P. Landers. 2011. Dynamic solid phase DNA extraction and PCR amplification in polyester-toner based microchip. *Anal Chem* 83: 5182–9.

Duarte, G. R. M., C. W. Price, J. L. Littlewood et al. 2010. Characterization of dynamic solid phase DNA extraction from blood with magnetically controlled silica beads. *Analyst* 135: 531–7.

Easley, C. J., J. A. C. Humphrey, and J. P. Landers. 2007. Thermal isolation of microchip reaction chambers for rapid non-contact DNA amplification. *J Micromech Microeng* 17: 1758–66.

Easley, C. J., J. M. Karlinsey, J. M. Bienvenue et al. 2006a. A fully integrated microfluidic genetic analysis system with sample-in-answer-out capability. *Proc Natl Acad Sci USA* 103: 19272–7.

Easley, C. J., J. M. Karlinsey, and J. P. Landers. 2006b. On-chip pressure injection for integration of infrared-mediated DNA amplification with electrophoretic separation. *Lab Chip* 6: 601–10.

Edwards, A., A. Civitello, H. A. Hammond, and C. T. Caskey. 1991. DNA typing and genetic mapping with trimeric and tetrameric tandem repeats. *Am J Hum Genet* 49: 746–56.

Edwards, A., H. A. Hammond, L. Jin, C. T. Caskey, and R. Chakraborty. 1992. Genetic-variation at 5 trimeric and tetrameric tandem repeat loci in 4 human-population groups. *Genomics* 12: 241–53.

El-Sissi, O., H. Franklin, B. Nielsen, S. Pagano, R. Belcinski, G. Bogdan, and S. Jovanovich. 2010. RapidHIT 200 human DNA identification system; August 6, 2010. Available online at: http://integenx.com/applications-2/dna-identity/.

Epstein, D. and K. Grates. 2009. Evaluating Rugged and Portable Chemical Testing Platforms. *Presentation—25th Natl Envir Monitor Conf.* San Antonio, TX.

Felhofer, J. L., L. Blanes, and C. D. Garcia. 2010. Recent developments in instrumentation for capillary electrophoresis and microchip-capillary electrophoresis. *Electrophoresis* 31: 2469–86.

Forster, R. E., T. N. Chiesl, C. P. Fredlake, C. V. White, and A. E. Barron. 2008. Hydrophobically modified polyacrylamide block copolymers for fast, high-resolution DNA sequencing in microfluidic chips. *Electrophoresis* 29: 4669–76.

Fredlake, C. P., D. G. Hert, C. W. Kan et al. 2008. Ultrafast DNA sequencing on a microchip by a hybrid separation mechanism that gives 600 bases in 6.5 minutes. *Proc Natl Acad Sci USA* 105: 476–81.

Fuhr, G., T. Muller, V. Baukloh, and K. Lucas. 1998. High-frequency electric field trapping of individual human spermatozoa. *Hum Reprod* 13: 136–41.

Fujita, Y. and Kubo, S. 2006. Application of FTA technology to extraction of sperm DNA from mixed body fluids containing semen. *J Legal Med* 8: 43–7.

Giddings, M. C., J. Severin, M. Westphall, J. Wu, and L. M. Smith. 1998. A software system for data analysis in automated DNA sequencing. *Genome Res* 8: 644–65.

Giese, H., R. Lam, R. Selden, and E. Tan. 2009. Fast multiplexed polymerase chain reaction for conventional and microfluidic short tandem repeat analysis. *J Forensic Sci* 54: 1287–96.

Gill, P., A. J. Jeffreys, and D. J. Werrett. 1985. Forensic application of DNA fingerprints. *Nature* 318: 577–9.

Goedecke, N., B. McKenna, S. El-Difrawy, L. Carey, P. Matsudaira, and D. Ehrlich. 2004. A high-performance multilane microdevice system designed for the DNA forensic laboratory. *Electrophoresis* 25: 1678–86.

Goedecke, N., B. McKenna, S. El-Difrawy et al. 2006. Microfluidic DNA forensics by the simple tandem repeat method. *J Chromatogr A* 1111: 206–13.

Greenspoon, S. A., S. H. I. Yeung, K. R. Johnson et al. 2008. A forensic laboratory tests the Berkeley Microfabricated Capillary Array Electrophoresis Device. *J Forensic Sci* 53: 828–37.

Grover, W. H., A. M. Skelley, C. N. Liu, E. T. Lagally, and R. A. Mathies. 2003. Monolithic membrane valves and diaphragm pumps for practical large-scale integration into glass microfluidic devices. *Sensor Actuat B-Chem* 89: 315–23.

Grunenwald, H. 2003. Optimization of Polymerase Chain Reactions. PCR Protocols, 2nd ed. *Method Mol Cell Biol* 226: 89–99.

Hagan, K. A., C. R. Reedy, J. M. Bienvenue, A. H. Dewald, and J. P. Landers. 2011. A valveless microfluidic device for integrated solid phase extraction and polymerase chain reaction for short tandem repeat (STR) analysis. *Analyst* 136: 1928–37.

Henry, C. S., M. Zhong, S. M. Lunte, M. Kim, H. Bau, and J. J. Santiago. 1999. Ceramic microchips for capillary electrophoresis-electrochemistry. *Anal Commun* 36: 305–7.

Heyries, K. A., C. Tropini, M. VanInsberghe et al. 2011. Megapixel digital PCR. *Nat Method* 8: 649–51.

Hill, C. R., J. M. Butler, and P. M. Vallone. 2009. A 26plex autosomal STR assay to aid human identity testing. *J Forensic Sci* 54: 1008–15.

Hopwood, A. J., C. Hurth, J. Yang et al. 2010. Integrated microfluidic system for rapid forensic DNA analysis: Sample collection to DNA profile. *Anal Chem* 82: 6991–9.

Horsman, K. M., S. L. R. Barker, J. P. Ferrance, K. A. Forrest, K. A. Koen, and J. P. Landers. 2005. Separation of sperm and epithelial cells in a microfabricated device: Potential application to forensic analysis of sexual assault evidence. *Anal Chem* 77: 742–9.

Hua, Z. S., J. L. Rouse, A. E. Eckhardt et al. 2010. Multiplexed real-time polymerase chain reaction on a digital microfluidic platform. *Anal Chem* 82: 2310–6.

Hurth, C., S. D. Smith, A. R. Nordquist et al. 2010. An automated instrument for human STR identification: Design, characterization, and experimental validation. *Electrophoresis* 31: 3510–7.

Innis, M. A., K. B. Myambo, D. H. Gelfand, and M. A. D. Brow. 1988. DNA Sequencing with Thermus-Aquaticus DNA-Polymerase and direct sequencing of polymerase chain reaction-amplified DNA. *Proc Natl Acad Sci USA* 85: 9436–40.

IntegenX. RapidHIT 200 human DNA identification system. Available online at: http://integenx.com/applications-2/dna-identity/ (accessed August 6, 2010).

Iverson, B. D. and S. V. Garimella. 2008. Recent advances in microscale pumping technologies: A review and evaluation. *Microfluid Nanofluid* 5: 145–74.

Jeffreys, A. J., V. Wilson, and S. L. Thein. 1985. Hypervariable minisatellite regions in human DNA. *Nature* 314: 67–73.

Kailasa, S. K. and S. H. Kang. 2009. Microchip-based capillary electrophoresis for DNA analysis in modern biotechnology: A review. *Sep Purif Rev* 38: 242–88.

Karle, M., J. Miwa, G. Czilwik, V. Auwarter, G. Roth, R. Zengerle, and F. von Stetten. 2010. Continuous microfluidic DNA extraction using phase-transfer magnetophoresis. *Lab Chip* 10: 3284–90.

Karlinsey, J. M. and J. P. Landers. 2006. Multicolor fluorescence detection on an electrophoretic microdevice using an acoustooptic tunable filter. *Anal Chem* 78: 5590–6.

Karlinsey, J. M. and J. P. Landers. 2008. AOTF-based multicolor fluorescence detection for short tandem repeat (STR) analysis in an electrophoretic microdevice. *Lab Chip* 8: 1285–91.

Karlinsey, J. M., J. Monahan, D. J. Marchiarullo, J. P. Ferrance, and J. P. Landers. 2005. Pressure injection on a valved microdevice for electrophoretic analysis of submicroliter samples. *Anal Chem* 77: 3637–43.

Kim, J., M. Johnson, P. Hill, and B. K. Gale. 2009. Microfluidic sample preparation: Cell lysis and nucleic acid purification. *Integr Biol* 1: 574–86.

Kiss, M. M., L. Ortoleva-Donnelly, N. R. Beer et al. 2008. High-throughput quantitative polymerase chain reaction in picoliter droplets. *Anal Chem* 80: 8975–81.

Koh, C. G., W. Tan, M. Q. Zhao, Z. H. Fan, and A. J. Ricco. 2003. Integrating polymerase chain reaction, valving, and electrophoresis in a plastic device for bacterial detection. *Anal Chem* 75: 4591–8.

Lagally, E. T., C. A. Emrich, and R. A. Mathies. 2001. Fully integrated PCR-capillary electrophoresis microsystem for DNA analysis. *Lab Chip* 1: 102–7.

Lang, M. 2004. The Pocket Laboratory. *Siemens' Pictures of the Future* 74–6.

Leslie, D. C., C. J. Easley, E. Seker et al. 2009. Frequency-specific flow control in microfluidic circuits with passive elastomeric features. *Nat Phys* 5: 231–5.

Liu, J. K. and M. L. Lee. 2006. Permanent surface modification of polymeric capillary electrophoresis microchips for protein and peptide analysis. *Electrophoresis* 27: 3533–46.

Liu, P., X. Li, S. A. Greenspoon, J. R. Scherer, and R. A. Mathies. 2011. Integrated DNA purification, PCR, sample cleanup, and capillary electrophoresis microchip for forensic human identification. *Lab Chip* 11: 1041–8.

Liu, P., T. S. Seo, N. Beyor, K. J. Shin, J. R. Scherer, and R. A. Mathies. 2007a. Integrated portable polymerase chain reaction-capillary electrophoresis microsystem for rapid forensic short tandem repeat typing. *Anal. Chem.* 79: 1881–9.

Liu, R. H., P. Grodzinski, J. Yang, and R. Lenigk. 2007b. Self-Contained, Fully Integrated Biochips for Sample Preparation, PCR Amplification and DNA Microarray Analysis. *Integrated Biochips for DNA Analysis*. Austin, TX, USA.

Liu, Y., J. C. Fanguy, J. M. Bledsoe, and C. S. Henry. 2000. Dynamic coating using polyelectrolyte multilayers for chemical control of electroosmotic flow in capillary electrophoresis microchips. *Anal Chem* 72: 5939–44.

Llopis, S. L., J. Osiri, and S. A. Soper. 2007. Surface modification of poly(methyl methacrylate) microfluidic devices for high-resolution separations of single-stranded DNA. *Electrophoresis* 28: 984–93.

Lyon, E. and C. T. Wittwer. 2009. LightCycler technology in molecular diagnostics. *J Mol Diag* 11: 93–101.

Manage, D. P., Y. C. Morrissey, A. J. Stickel et al. 2011. On-chip PCR amplification of genomic and viral templates in unprocessed whole blood. *Microfluid Nanofluid* 10: 697–702.

Manz, A., N. Graber, and H. M. Widmer. 1990. Miniaturized total chemical analysis systems: A novel concept for chemical sensing. *Sensor Actuat B* 1: 244–8.

Markey, A. L., S. Mohr, and P. J. R. Day. 2010. High-throughput droplet PCR. *Methods* 50: 277–81.
McMahon, G. 2007. *Analytical Instrumentation: A Guide to Laboratory, Portable and Miniaturized Instruments*. Wiley: West Sussex, England.
Mitnick, L., L. Carey, R. Burger et al. 2002. High-speed analysis of multiplexed short tandem repeats with an electrophoretic microdevice. *Electrophoresis* 23: 719–26.
Moss, D., S. A. Harbison, and D. J. Saul. 2003. An easily automated, closed-tube forensic DNA extraction procedure using a thermostable proteinase. *Int J Legal Med* 117: 340–9.
National Institute of Justice. 2011. Defining, Counting and Reducing the Casework Backlog. http://www.dna.gov (accessed August 12, 2011).
Nelson, M. 2011. Making sense of DNA backlogs—Myths vs. reality. *National Institute of Justice Special Report;* https://www.ncjrs.gov/pdffiles1/nij/232197.pdf.
Norris, J. V., M. Evander, K. M. Horsman-Hall, J. Nilsson, T. Laurell, and J. P. Landers. 2009a. Acoustic differential extraction for forensic analysis of sexual assault evidence. *Anal Chem* 81: 6089–95.
Norris, J. V., B. E. Root, O. N. Scott et al. 2009b. Fully Integrated, Multiplexed STR-Based Human Identification Using a Single Microfluidic Chip and Automated Instrument. *Presentation—20th Int Symp Human Identification*. Las Vegas, NV.
O'Farrell, P. H. 1975. High resolution two-dimensional electrophoresis of proteins. *J Biol Chem* 250: 4007–21.
Oh, K. W. and C. H. Ahn. 2006. A review of microvalves. *J Micromech Microeng* 16: R13–R39.
Pekin, D., Y. Skhiri, J. C. Baret et al. 2011. Quantitative and sensitive detection of rare mutations using droplet-based microfluidics. *Lab Chip* 11: 2156–66.
Pjescic, I., C. Tranter, P. L. Hindmarsh, and N. D. Crews. 2010. Glass-composite prototyping for flow PCR with *in situ* DNA analysis. *Biomed Microdevices* 12: 333–43.
Price, C. W., D. C. Leslie, and J. P. Landers. 2009. Nucleic acid extraction techniques and application to the microchip. *Lab Chip* 9: 2484–94.
Reedy, C. R., J. M. Bienvenue, L. Coletta et al. 2010a. Volume reduction solid phase extraction of DNA from dilute, large-volume biological samples. *Forensic Sci Int Genet* 4: 206–12.
Reedy, C. R., K. A. Hagan, B. C. Strachan et al. 2010b. Dual-domain microchip-based process for volume reduction solid phase extraction of nucleic acids from dilute, large volume biological samples. *Anal Chem* 82: 5669–78.
Reedy, C. R., K. A. Hagan, D. J. Marchiarullo et al. 2011a. A modular microfluidic system for deoxyribonucleic acid identification by short tandem repeat analysis. *AnalChim Acta* 687: 150–8.
Reedy, C. R., C. W. Price, J. Sniegowski, J. P. Ferrance, M. Begley, and J. P. Landers. 2011b. Solid phase extraction of DNA from biological samples in a post-based, high surface area poly(methyl methacrylate) (PMMA) microdevice. *Lab Chip* 11: 1603–11.
Root, B. E., K. A. Hagan, C. R. Reedy et al. 2011a. A multisample platform for automated genetic analysis for human ID. *Presentation—22nd Int Symp Human Identification*. National Harbor, MD.
Root, B. E., C. R. Reedy, K. A. Hagan et al. 2011b. A multichannel microdevice for PCR amplification and electrophoretic separation of DNA. *Presentation—15th Int Conf Miniaturized Systems for Chemistry and Life Sciences*. Seattle, WA.
Saiki, R. K., S. Scharf, F. Faloona et al. 1985. Enzymatic amplification of beta-globin genomic sequences and restriction site analysis for diagnosis of sickle-cell anemia. *Science* 230: 1350–54.
Schaerli, Y., R. C. Wootton, T. Robinson et al. 2009. Continuous-flow polymerase chain reaction of single-copy DNA in microfluidic microdroplets. *Anal Chem* 81: 302–6.
Schmalzing, D., L. Koutny, A. Adoourian, P. Belgrader, P. Matsudaira, and D. Ehrlich. 1997. DNA typing in thirty seconds with a microfabricated device. *Proc Natl Acad Sci USA* 94: 10273–8.
Schmalzing, D., L. Koutny, D. Chisholm, A. Adoourian, P. Matsudaira, and D. Ehrlich. 1999. Two-color multiplexed analysis of eight short tandem repeat loci with an electrophoretic microdevice. *Anal Biochem* 270: 148–52.

Scientific Working Group on DNA Analysis Methods (SWGDAM). 2010. SWGDAM Interpretation Guidelines for Autosomal STR Typing by Forensic DNA Testing Laboratories. http://www.fbi.gov/about-us/lab/codis/swgdam-interpretation-guidelines (accessed February 24, 2012).

Shaw, K. J., P. T. Docker, J. V. Yelland et al. 2010. Rapid PCR amplification using a microfluidic device with integrated microwave heating and air impingement cooling. *Lab Chip* 10: 1725–8.

Shen, F., W. B. Du, J. E. Kreutz, A. Fok, and R. F. Ismagilov. 2010. Digital PCR on a SlipChip. *Lab Chip* 10: 2666–72.

Shi, Y. and R. C. Anderson. 2003. High-resolution single-stranded DNA analysis on 4.5 cm plastic electrophoretic microchannels. *Electrophoresis* 24: 3371–7.

Sijen, T., C. C. G. Benschop, D. C. Wiebosch, and A. D. Kloosterman. 2010. Post-coital vaginal sampling with nylon flocked swabs improves DNA typing. *Forensic Sci Int Genet* 4: 115–21.

Sikanen, T., L. Heikkila, S. Tuornikoski et al. 2007. Performance of SU-8 microchips as separation devices and comparison with glass microchips. *Anal Chem* 79: 6255–63.

Sinville, R. and S. A. Soper. 2007. High resolution DNA separations using microchip electrophoresis. *J Sep Sci* 30: 1714–28.

Steadman, S. A., J. D. McDonald, J. S. Andrews, and N. D. Watson. 2008. Recovery and STR amplification of DNA from RFLP membranes. *J Forensic Sci* 53: 349–58.

Sun, Y., Y. C. Kwok, and N. T. Nguyen. 2007. A circular ferrofluid driven microchip for rapid polymerase chain reaction. *Lab Chip* 7: 1012–7.

Sun, Y., P. F. P. Lee, N. T. Nguyen, and Y. C. Kwok. 2009. Rapid amplification of genetically modified organisms using a circular ferrofluid-driven PCR microchip. *Anal Bioanal Chem* 394: 1505–8.

Tan, E., H. Giese, and D. M. Hartmann. 2009. Microfluidic DNA Extraction and Purification from Forensic Samples: Towards Rapid, Fully Integrated STR Analysis. *Final Report to the National Institute of Justice.* Available online at https://www.ncjrs.gov/pdffiles1/nij/grants/226810.pdf.

The University of Arizona. MiDAS Rapid DNA Analysis. http://anbm.arizona.edu/research/current-projects-programs/rapid-dna-analysis (accessed on December 2, 2011).

Tian, H. J., A. F. R. Huhmer, and J. P. Landers. 2000. Evaluation of silica resins for direct and efficient extraction of DNA from complex biological matrices in a miniaturized format. *Anal Biochem* 283: 175–91.

Ugaz, V. M., S. N. Brahmasandra, D. T. Burke, and M. A. Burns. 2002. Cross-linked polyacrylamine gel electrophoresis of single-stranded DNA for microfabricated genomic analysis systems. *Electrophoresis* 23: 1450–9.

Unger, M. A., H. Chou, T. Thorsen, A. Scherer, and S. R. Quake. 2000. Monolithic microfabricated valves and pumps by multilayer soft lithography. *Science* 288: 113–6.

University of North Carolina. 1997. BaseFinder6.
http://bioinfo.boisestate.edu/glabsoftware/BaseFinder/index.html (accessed January 03, 2013).

Vallone, P. M., C. R. Hill, and J. M. Butler. 2008. Demonstration of rapid multiplex PCR amplification involving 16 genetic loci. *Forensic Sci Int Genet* 3: 42–5.

Vogelstein, B. and K. W. Kinzler. 1999. Digital PCR. *Proc Natl Acad Sci USA* 96: 9236–41.

Wang, J., M. Pumera, M. P. Chatrathi et al. 2002. Towards disposable lab-on-a-chip: Poly(methylmethacrylate) microchip electrophoresis device with electrochemical detection. *Electrophoresis* 23: 596–601.

Wen, J., L. A. Legendre, J. M. Bienvenue, and J. P. Landers. 2008. Purification of nucleic acids in microfluidic devices. *Anal Chem* 80: 6472–9.

Witek, M. A., S. D. Llopis, A. Wheatley, R. L. McCarley, and S. A. Soper. 2006. Purification and preconcentration of genomic DNA from whole cell lysates using photoactivated polycarbonate (PPC) microfluidic chips. *Nucleic Acids Res* 34: e74.

Wittwer, C. T., K. M. Ririe, R. V. Andrew, D. A. David, R. A. Gundry, and U. J. Balis. 1997. The LightCycler™: a microvolume multisample fluorimeter with rapid temperature control. *BioTechniques* 22: 176–81.

Wu, W. and N. Y. Lee. 2011. Three-dimensional on-chip continuous-flow polymerase chain reaction employing a single heater. *Anal Bioanal Chem* 400: 2053–60.

Yeung, S. H. I., S. A. Greenspoon, A. McGuckian et al. 2006. Rapid and high-throughput forensic short tandem repeat typing using a 96-lane microfabricated capillary array electrophoresis microdevice. *J Forensic Sci* 51: 740–7.

Zhang, C. S. and D. Xing. 2007. Miniaturized PCR chips for nucleic acid amplification and analysis: Latest advances and future trends. *Nucleic Acids Res* 35: 4223–37.

Zhang, Y. H. and P. Ozdemir. 2009. Microfluidic DNA amplification: A review. *Anal Chim Acta* 638: 115–125.

Training

V

Training of Forensic DNA Scientists—A Commentary

16

MEREDITH A. TURNBOUGH
ARTHUR J. EISENBERG
LISA SCHADE
JAIPRAKASH G. SHEWALE

Contents

16.1 Introduction 381
16.1.1 Worldwide Expansion of DNA Analysis 381
16.1.2 Backlog—A Challenge to the Forensic DNA Laboratory 382
16.2 Single-Source and Casework Samples 383
16.3 Training as a Tool to Overcome Laboratory Challenges 384
16.3.1 Quality Assurance Standards 384
16.3.2 Training Sources 386
16.3.3 Examples of Training Entities 386
16.3.4 Center for Forensic Excellence—A Model for Training 387
Acknowledgments 388
References 388

Abstract: For the past two decades, forensic DNA analysis has rapidly expanded in both utility and value to criminal investigations. As the number of crime-scene and convict/arrestee samples has continued to grow, many forensic DNA laboratories find themselves struggling to test samples in a timely fashion, and database and casework sample backlogs continue to present a major challenge. One issue many forensic laboratories face is limited availability of resources for training new analysts. High-quality training enables analysts to effectively perform various aspects of DNA profiling and is essential to ensuring consistent, high-quality results. This is well documented in the guidelines established in the FBI's Quality Assurance Standards for Forensic DNA Testing Laboratories in the United States as well as internationally by agencies like INTERPOL. A facility dedicated to training analysts on both theoretical and practical aspects of automated sample processing accelerates the establishment and expansion of high-throughput forensic DNA laboratories. This chapter discusses various aspects of sample-processing training and the entities that provide such training programs.

16.1 Introduction

16.1.1 Worldwide Expansion of DNA Analysis

Genotyping of biological samples is now routinely performed in human identification laboratories worldwide for various applications like paternity testing, forensic casework,

DNA databasing, the search for missing persons, family lineage studies, identification of human remains, mass disasters, relationship testing, and so forth. The predominant genotyping method currently practiced is profiling for short tandem repeat (STR) loci. STRs are microsatellite loci that reside as repeat units of nucleotide sequence typically ranging from two to six nucleotides (Butler 2005; Edwards et al. 1991; Moretti et al. 2001). Polymorphic STR loci—that is, loci exhibiting variation in the number of repeat units among individuals—are useful in distinguishing between two individuals. Forensic DNA analysis is probably the fastest-growing area of crime investigation in the past two decades since the first report of the ability to detect many highly variable genetic loci simultaneously by Jeffreys et al. (1985). The success of DNA profiling in resolving crime can be gauged from the survey by the International Criminal Police Organization (INTERPOL) in 2008 comprising 172 member countries (INTERPOL 2009). Based on this survey, the number of countries utilizing DNA profiling for forensic casework increased from 53 in 1999 to 120 in 2009, and the number of countries adopting a national DNA database increased from 16 to 54. DNA databases in the United Kingdom, United States, and China, established in 1995, 1998, and 2001, respectively, are presently the largest DNA databases. The demand for DNA profiling is increasing for a variety of reasons, including acceptance of DNA results by court systems, legislation passed by governing bodies, increased funding, the generation of DNA databases, success in resolving cases worldwide, increasing awareness, expanding the scope to resolving property crimes, inflow of cold cases, postconviction testing, and advancements in DNA analysis technologies. The number of samples from convicted offenders and arrestees combined increased substantially, which is evident from the data published by different agencies (Federal Bureau of Investigation 2012; National Policing Improvement Agency: Statistics 2012; National Policing Improvement Agency: The National DNA Database 2012; Nelson 2011). The end result is increased backlogs in spite of the expansion of sample-processing capabilities of the forensic laboratories.

16.1.2 Backlog—A Challenge to the Forensic DNA Laboratory

The increased demand for DNA testing is, in part, a result of the increase in samples submitted for DNA testing from crime scenes that would not have been considered viable sources of DNA or that would have previously been classified as lower-priority crimes (i.e., property crimes). Expanded legislation to collect DNA samples from convicted felons and arrested persons has contributed significantly to increased demand. Thus, the number of samples collected can rapidly exceed the planned capacity of forensic laboratories. It is not surprising that backlogs of database and casework samples in forensic laboratories in the United States have been a continuous issue for the past decade (U.S. Department of Justice Audit Report 2009).

Forensic laboratories in developed as well as developing countries face similar challenges to meet the sudden increase in the number of samples for DNA profiling. Funding is an ever-present concern that weighs particularly heavily on developing nations, but has become increasingly problematic for developed nations in light of recent global economic events. Even if funding was not an issue, the challenges include creating physical space; assessment, procurement, and installation of instruments; selection, optimization, and validation of appropriate chemistries; development of appropriate standard operating procedures and quality assurance programs; automation and integration of the workflow; accreditation; and training of staff members. Of these, training of staff members is of particular relevance to the current discussion.

Training of staff members can become a complex exercise due to factors such as people coming on board with diverse backgrounds and levels of experience, trainees having varying learning skills, trainer fatigue during periods of high turnover, training being conducted by the most recently trained person, and the lack of a standardized training program. Thus, a dedicated facility providing customized training on both theoretical aspects and/or hands-on training for automated workflows helps to accelerate the establishment or expansion of high-throughput forensic DNA laboratories.

16.2 Single-Source and Casework Samples

A typical STR profiling exercise is a lengthy procedure employing multiple steps such as sample collection, sample preservation, evidence examination, extraction of DNA, assessment of DNA recovered, amplification of target loci, separation of amplified products by capillary electrophoresis, data analysis, results interpretation, and report generation. Automation and workflow integration streamlines the entire process. The different samples processed in a forensic laboratory can be grouped into reference samples and evidence samples. Reference samples are single-source samples that are collected using standardized devices, available in relatively abundant quantities, stored under controlled conditions, and provide greater yield of high-quality DNA. Casework evidence samples, on the other hand, may be mixtures of two or more contributors, deposited on highly variable substrates, often present in limited quantities, potentially exposed to extreme environmental conditions, and provide limited yield of lower quality of DNA. It is obvious that the workflows for these two groups of samples are different, though core principles are the same.

Continuous efforts have been made to improve the DNA-profiling methodology for both evidence samples and reference samples. Adoption of strategies like automation, workflow integration, and use of laboratory information management systems (LIMS) has increased the throughput of both these types of samples. Major improvements in the processing of single-source samples include direct amplification technology (eliminating the need for extraction and quantification of DNA) and expert subsystems for data analysis. AmpFℓSTR Identifiler Direct (Wang et al. 2011) and PowerPlex 18D (Promega Corporation, Madison, Wisconsin; Oostdik et al. 2011) are examples of direct amplification genotyping multiplex kits. Expert systems such as GeneMapper *ID-X* (Applied Biosystems 2009, 2010) and FSS-i3 software (Roby and Christen 2007) are capable of analyzing data with stringent requirements, thereby eliminating the need for analysis by a second analyst, which is a major improvement in enhancing the throughput of the laboratory. The typical workflow for single-source samples validated at our laboratory is presented in Figure 16.1.

As mentioned earlier, processing of evidentiary samples presents a different set of challenges such as removal of polymerase chain reaction (PCR) inhibitors during extraction of DNA and amplification of challenged samples that may be degraded or inhibited, or that contain limited quantities of DNA. With the advent of improved, automatable extraction chemistries (Liu et al. 2012; Stray et al. 2009) and improved STR typing kits such as the AmpFℓSTR MiniFiler PCR amplification kit (Mulero et al. 2008), which is designed for use with degraded samples, as well as robust STR amplification kits like Identifiler Plus

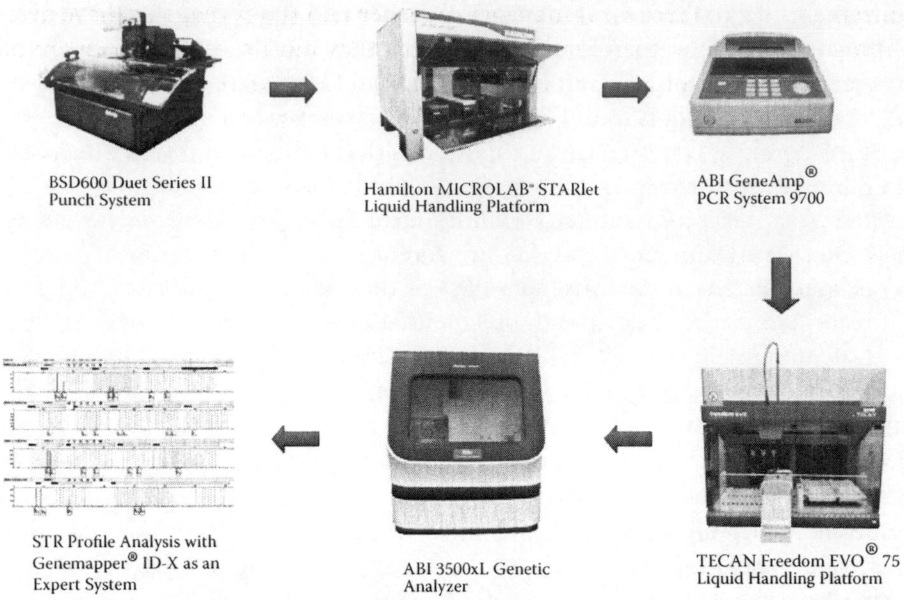

Figure 16.1 **(See color insert.)** Optimized and validated workflow for high-throughput processing of reference samples for generating STR profile.

(Wang et al. 2011), NGM SElect (Applied Biosystems 2011), and PowerPlex 16 HS (Promega Corporation, Madison, Wisconsin; Ensenberger and Fulmer 2009), which are examples of multiplex STR kits with improved sensitivity that enable amplification of samples containing relatively high levels of PCR inhibition. Thus, it is now possible to obtain a complete genotype from a wider variety of biological samples. It is important to note that this expanding array of sample types may widen the gap of technical expertise between the analysts processing single-source samples and evidence samples in a forensic laboratory. Training of laboratory staff for the processing of single-source samples can be standardized and completed in a few weeks, and although this would certainly contribute to the foundational knowledge of a forensic casework analyst, training of staff for the full range of samples and analysis techniques, as well as expert testimony, for full forensic casework analysis would be modular and can vary from months to years.

16.3 Training as a Tool to Overcome Laboratory Challenges

16.3.1 Quality Assurance Standards

In the guidelines established by the DNA Advisory Board's Quality Assurance Standards for Forensic DNA Testing Laboratories that took effect October 1, 1998, training was mentioned only nine times. It is found once under Standard 3, Quality Assurance Program, stating that the quality manual should address personnel qualifications and training; four times under Standard 5.1, which states that laboratory personnel shall have the education, training, and experience commensurate with the examination and testimony provided as well as requiring a documented training program, continuing education program, and recordkeeping on all of the above; and four times under Standard 5.2, which addresses the requirements for the technical manager or leader and states that they have oversight of the training of other laboratory personnel and that they must themselves possess adequate training commensurate with their job descriptions and in statistics/population genetics. In contrast, the term training is found *34 times* in the latest version of the Quality Assurance Standards for Forensic DNA Testing Laboratories that took effect July 1, 2009, *81 times* in the FBI Quality Assurance Standards Audit Document that accompanies those standards, and *36 times* in the Quality Assurance Standards for DNA Databasing Laboratories. In the FBI's new Quality Assurance Standards for Forensic and DNA Databasing Laboratories, training is again briefly mentioned as a part of the Quality Assurance Program in that training records must be maintained, but the real difference comes under the personnel section of the document. Under Standard 5.1, the training program *must* include a training manual, teach and assess skills and knowledge, require an individual to demonstrate competency, assess any previous training for new hires, document continuing education and training (internal or external), and maintain records of all of the above. Standards 5.2–5.4 state the many ways in which the technical leader should review and assess training of laboratory personnel, as well as training requirements for Combined DNA Index System (CODIS) administrators, analysts, and technicians. Training is also mentioned as one of the criteria that must be evaluated and approved for each analyst, CODIS administrator, and technical leader during the laboratory's annual audit. The Audit Document asks for proof of training records, training manuals, documented training program, adequate assessment and direction of personnel training by the technical leader, and many other instances that further demonstrate the vital role that training plays in every forensic laboratory.

In addition to the above guidelines for the United States, INTERPOL has provided guidelines for a typical training program (INTERPOL 2009):

- Training must be based on documented program with clearly identified learning outcomes.
- Training programs should be written in accordance with international accreditation guideline formats.
- Training programs should be competency-based, with a formal assessment resulting in a formal authorization, thus allowing the individual to perform the work of which he/she has successfully demonstrated a command.
- These programs should be delivered on a command basis allowing for differences in learning rates, but with upper time limits to ensure efficient and cost-effective training schedules.
- Alongside all relevant technical aspects of the methods employed, the programs should also cover QA, proficiency testing, and audits.
- Training on statistical approaches to DNA evidence is essential.
- Training must also include comprehensive understanding of the relevant legislation.
- Training programs must include expert evidence training, preferably utilizing local Public Prosecutor/Defender office personnel, and subjects such as video techniques.
- Training programs should also include awareness information regarding privacy and civil liberty issues.
- These programs are to be delivered within forensic laboratory training systems, via associated tertiary education institutes, or in formalized programs with legal offices responsible for prosecution/defense and court services.

It is important to note that the need for proper training is addressed by almost all agencies' guidelines that are involved in accreditation of forensic laboratories, directly or indirectly. Documents ILAC-G19 and NABL 113 published by the International Laboratory Accreditation Corporation (ILAC 2012) and the National Accreditation Board for Testing and Calibration of Laboratories (NABL 2012), respectively, exemplify this.

16.3.2 Training Sources

Training is clearly of utmost importance in the forensic DNA laboratory, but what is the best way to obtain the initial and ongoing training required for laboratory personnel? A certain level of theoretical knowledge is assumed to have been gained during specific undergraduate and graduate coursework required for many positions, but while education and training overlap in some areas, they are distinct entities. Education in a forensic context can of course be obtained from traditional sources such as colleges, universities, and scientific meetings, but where can training be obtained? The most common answer is to have existing personnel train all new employees in-house as they are hired. This solution works well for new hires in laboratories with fairly low turnover rates or that are large enough to have staff positions dedicated to training, but it cannot always address the need for continuing education or training on subject matter that is new to the entire staff and it can be difficult to maintain a consistent training program if the training duties rotate among staff members. Internal training carries the added burden of taking up valuable sample-processing time in the operational laboratory in terms of facilities, equipment, and personnel that would otherwise be processing cases. Sources for continuing education and training related to new equipment, software, or techniques are often workshops at scientific meetings, private companies, government (or government-funded) entities, and some academic institutions. Regardless of the source, hands-on experiential training, as opposed to passive listening, is often the most difficult type of training to obtain, but also the most valuable.

16.3.3 Examples of Training Entities

There are some good examples of entities that offer forensic training services. Bode Technology Group, Forensic Training Network, and Sorenson Forensics are all private companies/groups based in the United States that offer training in some form. Bode offers workshops, Forensic Training Network offers online courses (funded by the National Institute of Justice [NIJ]) as well as workshops/seminars, and Sorenson Forensics offers per-request training programs. Forensic Science Services was an example of a government-owned organization that provided training, consultancy, and other support to agencies in the United Kingdom and internationally before the unfortunate shuttering of the agency in 2012. U.S. academic institutions that provide training include the Marshall University Forensic Science Center (MUFSC), State University of New York at Albany's Northeast Regional Forensic Institute (NERFI), and the University of North Texas Health Science Center (UNTHSC) Center for Forensic Excellence. MUFSC offers 3- to 5-day NIJ-funded training programs (not for scholastic credit) on a variety of DNA-related topics. NERFI offers an NIJ-funded, 16-week, 12 scholastic credit–hour program with modules on molecular biology, benchwork, forensic statistics, and moot court exercises. They also offer 3- to 5-day workshops in a variety of DNA-related topics. The UNTHSC Center for Forensic

Excellence offers a 4-week, 8 scholastic credit–hour certificate program in DNA databasing as well as 1-week, 2 scholastic credit–hour certificate programs, all of which are discussed in greater detail below.

16.3.4 Center for Forensic Excellence—A Model for Training

The ideal situation for training would be to have an exact replica of the trainee's home laboratory where analysts and technicians could use the same equipment to train as they will for reference samples or databasing without taking up time or space in the operational laboratory. This is, of course, impossible for the vast majority of laboratories. The Center for Forensic Excellence (CFE) was established to fill that niche. It exists as a kind of core facility for the forensic community; a space that is outfitted exactly as a databasing laboratory should be, but with the sole purpose of training forensic scientists on state-of-the-art equipment (Figure 16.1). Yet another objective of this international program is to provide training to forensic scientists from many nations. Due to the increasing global volume of legislation for development of DNA databases, the first (and still the most prominent) curriculum offered by the CFE is a 4-week, 8 scholastic credit–hour program in high-throughput DNA databasing. The three-part curriculum combines education and training with 45 classroom hours covering the fundamentals of DNA databasing including sample collection, DNA extraction, PCR, real-time PCR (DNA quantification), STR biology and genetics, capillary electrophoresis methodology and instrumentation, STR profile interpretation (including relevant software), evidence handling and chain of custody, information tracking and LIMS systems, population statistics, and statistical analysis (statistical analyses are shown to demonstrate the practical utility of a database). There are 15 classroom hours on quality assurance/quality control (QA/QC), validation studies, validation reports, fundamentals of sound result interpretation policies, laboratory audit preparation, and accreditation. The students spend 120 hours in the laboratory where they perform sample collection, generation of appropriately formatted sample sheets and input files for all instrumentation, semiautomated paper punching, use of two different liquid-handling robotic systems (one for PCR amplification setup and one for capillary electrophoresis setup), PCR amplification, capillary instrument spatial and spectral calibration, capillary electrophoresis, data-collection software manipulation, STR profile analysis (including software analysis methods and settings), and troubleshooting at every step along the way. This process is repeated until each student can perform all tasks confidently and without instructor intervention. The students learn to set up, operate, maintain, and troubleshoot each of the following instruments and software programs: the BSD600-Duet Series Punching System, the Hamilton MICROLAB® STARlet Liquid Handling Platform, the TECAN Freedom EVO 75 Liquid Handling Platform, the ABI 3500xL Genetic Analyzer, and the GeneMapper ID-X Software. Lecture sessions are taught by faculty and staff in the Department of Forensic and Investigative Genetics at the UNTHSC. For those who are already processing samples in their home laboratory, but want or need to upgrade their skills to include automation, the CFE offers a 1-week, 2 scholastic credit–hour, instrumentation-only program that will allow students to spend every minute in the lab mastering the fundamental operations of the robotic systems as well as the unique validation and contamination controls required when using automated systems.

All CFE programs feature continuous student teach-back, whereby the instructors can assess and correct deficiencies in knowledge almost immediately, and are capped off by exams testing the students' theoretical knowledge of the topics covered in lecture as well as laboratory practical exams where students are required to demonstrate proficiency starting from sample materials and proceeding through all steps to generating and interpreting results without instructor intervention. Longer programs feature frequent quizzes to ensure that knowledge gaps are detected in time to take corrective action in the form of extended tutoring sessions, extra time on problematic instrumentation, and verbal quizzing to determine comprehension.

The CFE also has the flexibility to design custom curricula for laboratories that are hiring several people at once or that would like to send a group of analysts and/or technicians to upgrade their skills in accordance with the needs of a changing laboratory environment. All of the CFE's educational/training opportunities are offered through the UNTHSC Graduate School of Biomedical Sciences, so in addition to earning a certificate for completion of the program, students also earn graduate credit hours from an accredited university.

Acknowledgments

We thank Leonard Klevan, Lisa Calandro, and Lindy Kauffman from Life Technologies and Jamboor K. Vishwanatha, Bruce Budowle, Ranajit Chakraborty, Jianye Ge, and Carla Lee from the University of North Texas Health Sciences Center for their support and encouragement.

References

Applied Biosystems. 2011. AmpF*l*STR® NGM SElect™ PCR amplification kit. User Guide. Part #4458841 Rev. D.

Applied Biosystems. 2009. GeneMapper® *ID-X* Software v1.1.1 Update. User Bulletin. Part #4444446 Rev. A.

Applied Biosystems. 2010. GeneMapper® *ID-X* Software v1.2. User Bulletin. Part #4462639 Rev. A.

Butler, J. M. 2005. *Forensic DNA Typing*, 2nd ed. Elsevier Academic Press: Burlington, MA.

Edwards A., A. Civitello, H. A. Hammond, and C. T. Caskey. 1991. DNA typing and genetic mapping with trimeric and tetrameric tandem repeats. *Am J Hum Genet* 49:746–56.

Ensenberger, M. G. and P. M. Fulmer. 2009. The PowerPlex® 16 HS system. *Profiles in DNA* 12:9–11.

Federal Bureau of Investigation: CODIS-NDIS Statistics, http://www.fbi.gov/about-us/lab/codis/ndis-statistics (accessed March 13, 2012).

International Laboratory Accreditation Corporation: Guidelines for Forensic Science Laboratories, ILAC-G19, http://www.ilac.org/documents/g19_2002.pdf (accessed May 22, 2012).

INTERPOL. 2009. *INTERPOL Handbook on DNA Data Exchange and Practice: Recommendations from the INTERPOL DNA Monitoring Expert Group*, 2nd ed. International Criminal Police Organization: Lyon, France.

Jeffreys, A. J., V. Wilson, and S. L. Thein. 1985. Hypervariable "minisatellite" regions in human DNA. *Nature* 314:67–73.

Liu, J. Y., C. Zhong, A. Holt et al. 2012. Automate Express™ forensic DNA extraction system for extraction of genomic DNA from biological samples. *J Forensic Sci* 57:1022–30.

Moretti, T. R., A. L. Baumstark, D. A. Defenbaugh, K. M. Keys, J. B. Smerick, and B. Budowle. 2001. Validation of short tandem repeats (STRs) for forensic usage: Performance testing of fluorescent multiplex STR systems and analysis of authentic and simulated forensic samples. *J Forensic Sci* 46:647–60.

Mulero, J. J., C. W. Chang, R. E. Lagacé et al. 2008. Development and validation of the AmpF*l*STR® MiniFiler™ PCR amplification kit: A MiniSTR multiplex for the analysis of degraded and/or PCR inhibited DNA. *J Forensic Sci* 53:838–52.

National Accreditation Board for Testing and Calibration of Laboratories: Specific Guidelines for Accreditation of Forensic Science Laboratories & Checklist for Assessors, NABL 113, http://www.nabl-india.org/nabl/asp/users/documentMgmt.asp?cp=2&docType=both (accessed May 22, 2012).

National Policing Improvement Agency: Statistics, http://www.npia.police.uk/en/13338.htm (accessed March 13, 2012).

National Policing Improvement Agency: The National DNA Database, http://www.npia.police.uk/en/8934.htm (accessed March 13, 2012).

Nelson, M. 2011. Making Sense of DNA Backlogs, 2010—Myths vs. Reality. National Institute of Justice Special Report #NIJ 232197.

Oostdik, K., M. Ensenberger, B. Krenke, C. Sprecher, and D. Storts. 2011. The PowerPlex® 18D System: A Direct Amplification STR System with Reduced Thermal Cycling Time, http://www.promega.com/resources/articles/profiles-in-dna/2011/the-powerplex-18d-system-a-direct-amplification-str-system-with-reduced-thermal-cycling-time/ (accessed March 13, 2012).

Roby, R. K. and A. D. Christen. 2007. Validating expert systems: Examples with the FSS-i3™ Expert Systems Software. *Profiles in DNA* 10(2):13–15.

Stray, J., A. Holt, M. Brevnov, L. M. Calandro, M. R. Furtado, and J. G. Shewale. 2009. Extraction of high quality DNA from biological materials and calcified tissues. *Forensic Sci Int Genet* Suppl Series 2:159–60.

U.S. Department of Justice Audit of the Convicted Offender DNA Backlog Reduction Program, Audit Report 09-23, March 2009. http://www.usdoj.gov/oig/reports/OJP/a0923/final.pdf (accessed March 13, 2012).

Wang, D. Y., C. W. Chang, R. E. Lagacé, N. J. Oldroyd, and L. K. Hennessy. 2011. Development and validation of the AmpFℓSTR® Identifiler® Direct PCR Amplification Kit: A multiplex assay for the direct amplification of single-source samples. *J Forensic Sci* 56:835–45.

Index

A

Acousto-optic tunable filters (AOTF), 262
Additional Y-STRs, 221–245
availability of additional Y-STRs for forensic applications, 233–240
haplotype resolution, 233
most commonly used markers, 233, 234–238
most useful, noncore complex Y-STRs, 239
multicopy markers, 240
mutation rate, 239
simple, single-copy Y-STRs, 239
forensic value of additional Y-STRs, 226–233
false exclusions, 226
global markers, 229
gold standard Yfiler set, 230
haplotype resolution, 230
HGDP-CEPH dataset, 232
identical by state, 227
incorrect nonexclusions, 232
male-lineage differentiation, 227–232
markers with high genetic diversity, 229
paternity/familial testing, 232–233
Rapidly Mutating Y-STRs, 227
single-step mutations, 232
ultra-high discrimination set, 230
universal strategy, 229
next-generation commercial Y STR kits, 241
commercial kits, 241
male-lineage differentiation, 241
practical implications, 240–241
nonexclusion, 240
paternity/familial testing, 241
reference databases, 240
Y-STRs in forensics and the need for additional markers, 222–226
crime stain, 226
currently applied Y-STRs in forensics, 223–224
databases, 223
differential DNA extraction, 222
disaster victim identification, 222
frequency surveying method, 223
genetic diversity, 224
genotyping, 222
haplotype frequencies, 22
identical haplotypes, 226
markers tested, 224
minimal haplotype, 224
mutations, 226
need for additional Y-STRs in forensics, 224–226
paternal lineage resolution, 224
PCR amplification, 222
sexual assault cases, 222
time discrepancy, 223
X-Y homologues, 222
Y-chromosome in forensics, 222–223
Adenosine diphosphate (ADP), 105
Adenosine triphosphate (ATP), 105, 314
ADP; *See* Adenosine diphosphate
AFDIL; *See* Armed Forces DNA Identification Laboratory
AIMs; *See* Ancestry informative markers
ALlele FREquency Database (ALFRED), 283
ALS; *See* Alternative light source
Alternative light source (ALS), 8
AmpFℓSTR Identifiler PCR Amplification Kit (Life Technologies), 165
AmpFℓSTR NGM Kit (Life Technologies), 169
AmpliTaq Gold®, 357
Ancestry Informative Marker Indel panels, 287
Ancestry informative markers (AIMs), 297–303
Ancient DNA, 258
Anhydrobiosis, 25
AOTF; *See* Acousto-optic tunable filters
Armed Forces DNA Identification Laboratory (AFDIL), 22
ATP; *See* Adenosine triphosphate
Automated extraction of DNA, 49–51
benchtop systems, 50–51
high-throughput systems, 51
laboratory information management system, 49
major challenge, 49
Automate Express™ (Applied Biosystems), 50
Autosomal SNPs and indels, applications of, 279–310
ancestry-informative marker SNPs, 297–303
admixture ratio, 302
allele-frequency data, 302
Bayesian classifier, 302
forensic services company, 298
Kosoy AIMs, 302
reference populations, 302
whole genome scan, 298
discrimination power of STRs using SNPs, enhancement of, 287–291

adding SNPs to extend polymorphisms available for complex relationship testing, 289–291
distant relationships, 290
kit discordancy, 287
nomenclature coding, 290
repeat regions, 289
SNPs in or around STRs, 287–289
typing systems, 289
uncommon occurrence, 289
indels, 286–287
amplicon generation, 286
Ancestry Informative Marker Indel panels, 287
heterozygote peak height ratio, 286
major advantage, 286
paraffin-embedded tissues, 286
skeletal materials, 286
original rationale of forensic SNP analysis, 282–286
DNA database searches, 282
ideal SNP candidates, 283
identification applications, 283
missing persons identification, 282
nucleosome binding, 283
random match probabilities, 283
thought experiment, 283
overview of SNP definition, 280
genome variant catalog, 270
SNP structure, 280
SNP typing applied to predictive forensic tests for common pigmentation variation, 304–306
differentiation of phenotypes, 305
forensic eye-color analysis, 304
Irisplex, 305
phenotyping ambiguity, 305
SNaPshot assay, 305
tri-allelic SNPs, 303–304
utility of SNP databases, 291–297
chromosome location query function, 297
data mining, 292
divergence, 298
example query, 293
fine-scale genetic maps for forensic markers using HapMap high-density SNP data, 295–297
hair-color variation, 293
HapMap project, 292, 295
hotspots, 296
online SNP data and how to access it for forensic purposes, 291–295
primer details, 291
Principal Component Analysis, 292
viable forensic SNP genotyping systems in 2012, 280–281
MALDI-TOF, 281
sample-tracking test, 281
Autosomal STR loci (new), biology and genetics of, 181–198
additional autosomal STR loci, 191
autosomal STR loci in current commercial kits, 185–191
CSF1PO, 185
D10S1248, 190
D12S391, 190
D13S317, 188
D16S539, 188
D18S51, 189
D19S433, 189
D1S1656, 189
D21S11, 198
D22S1045, 190
D2S1338, 189
D2S441, 189–190
D3S1358, 187
D5S818, 188
D6S1043, 190
D7S820, 188
D8S1179, 188
F13A01, 190
F13B, 190
FESFPS, 191
FGA, 185
HUMLIPOL, 191
LPL, 191
Penta C, 191
Penta D, 189
Penta E, 189
SE33, 190
TH01, 185
VWA, 185–187
commercial kits, 183–184
manufacturers, 183
multicolor fluorescence detection, 183
core loci, 182–183
Human Genome Project, 182
law enforcement data-sharing efforts, 183
tetranucleotide STR loci, 182
expansion of U.S. core loci, 194–195
DYS391, 194
European Standard set, 194
INTERPOL, 195
kinship associations, 194
TPOX, 194
microsatellites, 182
microvariants, 182
polymerase chain reaction, 182
relative variability of autosomal STR loci, 191–194
genotype combinations, 192
locus performance, 191
observed heterozygosity, 191
polymorphism information content, 191

repeated DNA sequences, 181
simple sequence repeats, 182
tetranucleotide STRs, 182

B

Baecchi staining, 88
BaseFinder, 363
Basic Local Alignment Search Tool (BLAST), 338
Beckman, 51
Becton Dickinson, 51
Benchtop systems, 50–51
casework samples, 50
LySep column, 50
BioRobot EZ1 (Qiagen), 50
Bioterrorism monitoring, 107
BLAST; *See* Basic Local Alignment Search Tool
BLUESTAR®, 8
Bode PunchPrep™, 49
Bode Technology Group, 250
Body-fluid–specific markers, 89, 90

C

Capillary electrophoresis (CE), 131–162, 312
advantages, 132–133
deep-sequencing technologies, 312
DNA sequencing, 136–137
preferred samples, 136
sequence comparison, 137
typical workflow, 137
electrophoretic microdevices and integrated systems, 155
evolution of CE systems and utility in forensic DNA analysis, 138–147
CCD camera, 146
data-collection software, 146
detection, 145–147
development of polymer and capillaries, 140–142
electroosmotic flow, 141
electropherogram, 146
electrophoresis, 144
emitted fluorescence, 146
entangled polymer buffers, 140
formic acid, 144
free zone electrophoresis, 138
hydroxyethylcellulose sieving matrices, 139
injection and run parameters, 142–145
intercalating dye, 139
laser-induced fluorescence, 138
linear polyacrylamide, 141
matrix generation, 146
multicolor fluorescence detection, 139
PDMA polymer, 141
polyacrylamide gel-filled capillaries, 140
polymer viscosity, 145
sample stacking, 144
separation of DNA fragments, 142
single-stranded DNA, 145
spectral calibration, 146
virtual filters, 146
forensic DNA analysis, 133–137
DNA sequencing, 136–137
fragment analysis and STR typing, 133–136
SNP typing, 137
fragment analysis and STR typing, 133–136
allelic ladder, 135
calibration curves, 134
electrophoresis, 135
fluorescence, threshold for detection of, 135
human identification applications, 133
size standards, 134
sizing precision, 135, 136
variable-bin method, 135
gas chromatography, 132
high-performance liquid chromatography, 132
human identification, 131
instruments for forensic DNA analysis, 147–155
CCD camera, 152
hardware improvements, 151
key features of 3500 genetic analyzers, 149–152
normalization and consistency of 3500 genetic analyzers, 152–155
radio frequency identification technology, 152
sizing precision, 149
–laser-induced fluorescence (CE–LIF), 87
microsatellite loci, 200
mRNA marker analysis, 133
other applications, 155
requirements of CE for STR analysis, 137–138
short tandem repeat, 131, 132
single-base resolution capability, 131
SNP typing, 137
base substitutions, 137
identification of missing persons, 137
oligo ligation assay, 137
variable number of tandem repeats, 132
Car bomb, 7
CCD camera; *See* Charge-couple device camera
cDNA; *See* Complementary DNA
CE; *See* Capillary electrophoresis
CE–LIF; *See* Capillary electrophoresis–laser-induced fluorescence
Cellulose-digesting enzymes, 56
Center for Forensic Excellence (CFE), 387–388
Centricon® filter, 48
CFE; *See* Center for Forensic Excellence
Charge-couple device (CCD) camera, 146, 152, 362
ChargeSwitch (Invitrogen), 48, 51
Chelating agents, 44
Chelex® DNA extraction method, 43

Codon wobble positions, 266
Combined DNA Index System (CODIS), 164, 200
  administrators, 385
  core loci, 21
  database, 262
Complementary DNA (cDNA), 86
Contact DNA, 6
Corneodesmosin, 86
Crime scene, forensic DNA evidence collection at, 3–17
  forensic technical plan and documentation of crime scene, 6
    photograph the item in place, 6
    sketch the location, 6
    take measurements, 6
    take written notes, 6
  personal protective equipment, 5–6
  recovery of biological DNA evidence, 6–17
    alternative light source, 8
    buccal swabs, 15
    car bomb, 7
    chain of custody, 16
    chemical enhancements, 8
    cigarette butts, 14
    clothing, 14
    contact DNA, 6
    crime scene examination, 13
    degradation, 16
    DNA recovery supplies, 9–10
    fingerprints and DNA, 9
    known samples, 15
    magnifying glass, 12
    packaging DNA evidence, 15–16
    paper items, 13
    probative and nonprobative evidence, 8
    recovery of biological DNA evidence from various substrates, 11–14
    reference samples, 15
    sources of DNA evidence, 7
    swabbing techniques for touch DNA, 11
    touch DNA, 6, 7
    transfer DNA, 6
    transporting and storing DNA evidence, 16–17
    types of DNA samples, 15
    vehicle-borne improvised explosive device, 7
  theory of evidence transference, 4
Cyclodextrin, 330
Cytochrome P450, 85

D

Databases, SNP, 291–297
  chromosome location query function, 297
  data mining, 292
  divergence, 298
  example query, 293
  fine-scale genetic maps for forensic markers using HapMap high-density SNP data, 295–297
  hair-color variation, 293
  HapMap project, 292, 295
  hotspots, 296
  online SNP data and how to access it for forensic purposes, 291–295
  primer details, 291
  Principal Component Analysis, 292
Deep-sequencing technologies, 311–347
  applications of second-generation sequencing technologies, 319–322
    conservation genetics applications, 320
    Human Genome Project, 319
    immunity, 320
    laryngotracheitis virus, 320
    messenger RNA, 322
    methicillin-resistant *Staphylococcus aureus*, 320
    whole-genome sequencing data, 321
  deoxyribonucleotide triphosphates, 312
  dideoxynucleotide terminator sequencing, 313
  fragment separation and analysis, 313
  gold standard for DNA sequencing, 313
  implications, 332–337
    allele-specific hybridization, 336
    capillary electrophoresis, 333
    disadvantages, 336
    electrospray ionization mass spectrometry, 333
    genotyping assays, 336
    invasive cleavage, 336
    Leber's Hereditary Optic Neuropathy phenotype, 335
    missing persons identification, 332
    mitochondrial DNA analysis, 333–335
    multiplex identifier sequence, 334
    NUMTs, 335
    oligonucleotide ligation, 336
    primer extension, 336
    repeat motif number, 333
    RNA body fluid identification, 337
    short tandem repeats, 332–333
    single nucleotide polymorphisms, 335–336
  junk DNA status, forensic markers, 338
  length heteroplasmy, 314
  Maxam-Gilbert method, 313
  minus reaction, 312
  mitochondrial DNA, 312
  next-generation sequencing technologies, 329–332
    array chip, 330
    cyclodextrin, 330
    exonuclease sequencing, 330
    lipid bilayer membrane, 329
    nanopore sequencing technology, 331

node, 330
patented sequencing methods, 330
protein pores, 329
paternity testing, 312
performance comparison, 324–329
Ion Proton, 327
"lower throughput" benchtop MiSeq system, 327
pyrosequencing chemistry, 326
run time estimation, 327
sequencing cycles, number of, 326
Plus-Minus Sequencing, 312
polymerase chain reaction, 312
resolution at nucleotide level, 312
Sanger dideoxynucleotide terminator method, 313
second-generation sequencing technologies, 314–319
DNA library formation, 317
elongation, 135
Exact Call Chemistry, 317
FlowChip, 316
fluorescence, color of, 316
ligation reactions, 316
pyrosequencing technology, 314
sequencing primer, 315
twofold coverage, 317
third-generation sequencing technologies, 322–324
bacteria, 323
DNA library preparation, 322
incorporated vs. unincorporated dNTP, 323
polyerase error, 322
SMRT Cell, 323
zero-mode waveguides, 323
Denaturing gradient gel electrophoresis (DGGE), 267
Denaturing HPLC (dHPLC), 257
Deoxyribonucleotide triphosphates (dNTPs), 312
Deparaffinization, 45
Dideoxynucleotide terminator sequencing, 313
Differential extraction of DNA, 55–58
cell elution, 56–57
cellulose-digesting enzymes, 56
comparison of DNA contents, 56
epithelial DNA, 55
fluorescence *in situ* hybridization, 58
fractions, 55
gravitational filtration, 55
laser microdissection, 57–58
phosphate-buffered saline, 55
robotic system, 57
sample collection, 56
sperm DNA, 55
time since intercourse, 56
utility of DNAse in removal of contaminating DNA, 57
Differex™ Kit (Promega), 55
Disaster victim identification, 222; *See also* Mass disaster recovery
Distant relationships, 290
Dithiothreitol (DTT), 44, 58, 353–354
Divergence, 298
DNA IQ system, 51
DNAprint™ Genomics Inc., 298
DNeasy® Kit (Qiagen), 44
dNTPs; *See* Deoxyribonucleotide triphosphates
Dried stains, tissue origin of; *See* RNA profiling, identification of the tissue origin of dried stains using
Dry matrix storage, 24
DTT; *See* Dithiothreitol
Duplex real-time PCR, 113
Dye-based measurements, 104
Dynabeads® DNA Direct (Invitrogen), 48

E

ECC; *See* Exact Call Chemistry
EDNAP; *See* European DNA Profiling Group
EDTA; *See* Ethylenediamine tetra acetic acid
Electroosmotic flow (EOF), 141
Electropherogram, 146
Electrospray ionization (ESI), 201
Electrospray ionization mass spectrometry (ESI-MS), 207, 333
Elute cards, 43
End-point PCR, 105–107
duplex assay, 106
fluorescent end-point assay, 106
NADH dehydrogenase subunit 1, 106
qualitative method for identification, 106
quantitative template amplification technology, 106
ENFSI; *See* European Network of Forensic Science Institutes
Enzymatic digestion, cell elution, 56
EOF; *See* Electroosmotic flow
Epidermal cornified cell envelope, 86
ESI; *See* Electrospray ionization
ESI-MS; *See* Electrospray ionization mass spectrometry
ESSplex kit (Qiagen), 190
Estrogen receptor 1, 83
Ethylenediamine tetra acetic acid (EDTA), 68
European DNA Profiling Group (EDNAP), 92, 140
European Network of Forensic Science Institutes (ENFSI), 183
Evidence collection; *See* Crime scene, forensic DNA evidence collection at
Evidence transference, theory of, 4
Exact Call Chemistry (ECC), 317
Exonuclease sequencing, 330
EZ1 DNA Tissue Kit (Qiagen), 71

F

Federal Bureau of Investigation (FBI), 102, 250
Fingerprint DNA Finder Kit, 47
Fingerprints, 9
FlowChip, 316
Fluorescence
  bleed-through of, 147
  color of, 316
  evolution of CE systems, 146
  measurement of, 106
  multicolor, 139, 183
  reporter dye, 112
  signal, real-time PCR, 107
  threshold for detection of, 135
Fluorescence *in situ* hybridization (FISH), 58
forensicGEM kit, 48
Forensic Quality Services (FQS), 262
Formalin, 45
Fragment analysis (capillary electrophoresis), 133–136
  allelic ladder, 135
  calibration curves, 134
  electrophoresis, 135
  fluorescence, threshold for detection of, 135
  human identification applications, 133
  size standards, 134
  sizing precision, 135, 136
  variable-bin method, 135
Free zone electrophoresis, 138
FTA cards, 43, 49, 166
Fucosyltransferase 6, 85

G

Gas chromatography, 132
Genotyping, assessment of DNA extracted from forensic samples prior to, 101–128
  assessment of the quantity and quality of DNA, 102
  end-point PCR, 105–107
    duplex assay, 106
    fluorescent end-point assay, 106
    NADH dehydrogenase subunit 1, 106
    qualitative method for identification, 106
    quantitative template amplification technology, 106
  hybridization, 105
    adenosine diphosphate, 105
    adenosine triphosphate, 105
    quantum dot-based hybridization assay, 105
    READase™ polymerase, 105
  methods for DNA assessment, 104–123
    end-point PCR, 105–107
    hybridization, 105
    real-time PCR, 107–123
    spectrophotometry and fluorescence spectroscopy, 104–105
  need for assessment, 102–104
    extent of DNA degradation, 104
    method selection, 103
    mixture ratio for male and female nuclear DNA, 103
    presence of PCR inhibitors, 103
    quantity of human male nuclear DNA, 103
    quantity of human nuclear DNA, 103
    quantity of mtDNA, 104
    undetected quantification results, 104
  real-time PCR, 107–123
    advantages, 107
    assays, 112–116
    chemistries, 111–112
    correlation between assays, 122
    critical design factors, 107
    detection of PCR inhibitors, 113, 115, 117
    DNA degradation, 122
    downstream genotyping systems, 113
    duplex real-time PCR, 113
    fluorescence signal, 107
    fluorescent dyes, 107
    homologous region of human sex chromosomes, 115
    human DNA quantification, 112, 113
    human male DNA quantification, 113
    human telomerase reverse transcriptase target gene, 113
    important concepts in assay design, 116–123
    intercalating dyes, 111
    internal PCR control, 113
    limit of quantification and detection, 116
    mechanisms of inhibition, 118
    minor groove binder, 111
    mixtures of human male and female DNA, 119
    molecular beacons, 111
    mtDNA quantification, 113
    NADH dehydrogenase subunit 1 target gene, 113
    normalized inhibitor factor, 117
    no template controls, 116
    nuclear and mtDNA quantification, 114
    physicochemical properties of the plastics, 117
    Plexor® chemistry, 112
    quadruplex real-time PCR, 116
    quantification standard, 107
    quantifier assay, 111
    retinoblastoma susceptibility gene, 114
    Scorpions chemistry, 112
    singleplex real-time PCR assays, 112
    size of amplicons, 116
    slot-blot, 122
    species specificity, 116

standard DNA, 107
TaqMan chemistry, 111
triplex real-time PCR, 114
restriction fragment length polymorphism, 102
single nucleotide polymorphism genotyping, 102
spectrophotometry and fluorescence spectroscopy, 104–105
dye-based measurements, 104
fluorescent dyes, 104
UV absorbance, 104
standard reference material 2372, 123
Genotyping, extraction of DNA from forensic biological samples for, 39–64
automated extraction of DNA, 49–51
benchtop systems, 50–51
high-throughput systems, 51
laboratory information management system, 49
major challenge, 49
benchtop systems, 50–51
casework samples, 50
LySep column, 50
Chelex kit, 51, 52
comparison studies, 51–54
differential extraction of DNA, 55–58
cell elution, 56–57
cellulose-digesting enzymes, 56
comparison of DNA contents, 56
epithelial DNA, 55
fluorescence *in situ* hybridization, 58
fractions, 55
gravitational filtration, 55
laser microdissection, 57–58
phosphate-buffered saline, 55
robotic system, 57
sample collection, 56
sperm DNA, 55
time since intercourse, 56
utility of DNAse in removal of contaminating DNA, 57
DNA extraction methods, 42–49
basic steps, 42
off-the-shelf kits for isolation of DNA from forensic samples, 48
one-tube DNA extraction protocols, 43–44
PCR compatible reagents, 48–49
two-step DNA extraction protocols, 44–48
high-throughput systems, 51
liquid-handling robots, 51
open systems, 51
importance of DNA extraction, 40–42
diploid genomes, 41
extraction methodology, 41
human identification laboratories, 40
improper storage conditions, 40
mass disaster recovery, samples from, 41
missing persons testing, 40
paternity applications, 41
polymerase chain reaction, 40
restriction fragment length polymorphism, 41
Y-chromosome STRs, 41
need for extraction chemistries, 48
one-tube DNA extraction protocols, 43–44
chelating agents, 44
Chelex® method, 43
dithiothreitol, 44
elute cards, 43
FTA paper, 43
one-tube extraction kits, 44
sodium dodecyl sulfate, 44
thermostable proteinases, 44
reference samples, 48
two-step DNA extraction protocols, 44–48
centrifugal filter units, 48
deparaffinization, 45
DNA separation and isolation, 46–48
ethanol precipitation, 46
formalin, 45
gold standard, 46
hair, 44
lysis, 44–46
non-DNA contaminants, 46
nucleic acids, 47
organic extraction, 46
paraffin-embedded tissues, 45
phenol-chloroform, 46
phosphate-buffered saline, 45
saliva deposition, 45
solid phase, 46
synchronous coefficient of drag alteration, 47
telogen hairs, 45
GlobalFiler™ Express PCR Amplification kits (Life Technologies), 170
Global markers (Y-STRs), 229
Gold standard
DNA and RNA quantitation, 111
DNA separation and isolation, 46
DNA sequencing, 313
Yfiler set, 230
GridION (Oxford Nanopore Technologies), 329

H

Hamilton, 51
Handling of DNA extracts; *See* Storage and handling of DNA extracts, optimizing
HapMap project, 292, 295
Hemastix® reagent strip, 47
Hemoglobin beta (HBB), 92
Heteroplasmy
mtDNA analysis, 255, 271
pyrosequencing and deep sequencing, 271
HF; *See* Hydrofluoric acid

HGP; *See* Human Genome Project
HID; *See* Human identification
HID EVOlution™ Extraction System (Applied Biosystems), 52
High-performance liquid chromatography (HPLC), 132
High-throughput systems, 51
HMBS; *See* Hydroxymethylbilane synthase
Hoechst-33258 dyes, 104
Housekeeping genes, 86, 88
HPLC; *See* High-performance liquid chromatography
hTERT target gene; *See* Human telomerase reverse transcriptase target gene
Human beta defensin 1, 83
Human evolution, DNA banks established for studying, 22
Human Genome Project (HGP), 182, 319
Human identification (HID), 131
  analysis, 350
  applications, capillary electrophoresis, 133
  kits, 350
  laboratories, 40, 381
Human remains, extraction of DNA from, 65–77
  bone and tooth as sources of DNA, 65–67
    example studies, 66–67
    missing persons, profiling of human remains in resolving, 66
  extraction of DNA, 67
    deamination of bases, 67
    oxidative damage, 67
    PCR inhibitors, 67
  isolation of DNA, 70
    commercial isolation systems, 70
    demineralization procedure, 70
    organic extraction, 70
    phenol-chloroform, 70
  kits and protocols, 71–72
  lysis, 68–69
    chelating agent, 68
    demineralization protocol, 69
    Pleistocene bones, 69
    proteinase K lysis, 69
    sodium dodecyl sulfate, 68
  precautions, 67–68
    clean environment, importance of, 67
    PCR amplification, 68
  preparation of samples, 68
    cleaning, 68
    common samples from tooth, 68
    selection of bone, 68
  success rate, 73, 74
Human telomerase reverse transcriptase (hTERT) target gene, 113
Human thyroid peroxidase gene, 185
Hydrofluoric acid (HF), 359
Hydroxyethylcellulose sieving matrices, 139
Hydroxymethylbilane synthase (HMBS), 83
Hypoxanthine, 257

I

ICMP; *See* International Commission on Missing Persons
Identical by descent (IBD), 203
Identical by state (IBS), 227
Identifiler® Direct STR kit (Applied Biosystems), 42
Indel-plex (Qiagen), 286
Indels; *See* Autosomal SNPs and indels, applications of
Intercalating dyes, 111
Internal PCR control (IPC), 113
International Commission on Missing Persons (ICMP), 66
INTERPOL, 195, 382, 385
Investigator Argus X-12 Kit (Qiagen), 170, 176
Investigator ESSplex SE Kit (Qiagen), 174
Investigator® Quantiplex system (Qiagen), 113
Ion Personal Genome Machine™ (Life Technologies), 317
Ion Proton, 327
Ion Torrent systems, 317
IPC; *See* Internal PCR control
iPrep™ (Invitrogen), 50
Irisplex, 305

J

Junk DNA status, forensic markers, 338

K

Keratin, 83
Kits; *See also* STR genotyping kits, next-generation
  discordancy, 287
  manufacturers, competition among, 177
Kosoy AIMs, 302

L

Laboratory information management systems (LIMS), 49, 383
Laryngotracheitis virus, 320
Laser capture microdissection (LCM), 57
Laser-induced fluorescence (LIF), 138, 362
Laser microdissection (LM), 57–58
Law enforcement data-sharing efforts, 183
LCN; *See* Low copy number
Leber's Hereditary Optic Neuropathy (LHON) phenotype, 335
Leuco crystal violet, 8
LHON phenotype; *See* Leber's Hereditary Optic Neuropathy phenotype
LIF; *See* Laser-induced fluorescence

Life Technologies, 170
LIMS; *See* Laboratory information management systems
Linear polyacrylamide (LPA), 141, 358
Liquid-handling robots, 51
LM; *See* Laser microdissection
Loricrin, 86
Low copy number (LCN), 20
Low-temperature co-fired ceramic (LTCC), 359
LPA; *See* Linear polyacrylamide
Luciferase, 269
Luminol, 8
Lyse and Go PCR Reagent, 49

M

MagAttract® DNA Mini M48 Kit (Qiagen), 71
MALDI-TOF, 281
Mass disaster recovery
- degradation of samples, 20
- ESI-MS, 273
- evidence samples, 41
- genotyping method, 103
- only source for human identity, 65–66
- STR profiling, 133
- worldwide analysis, 381–382

Matrix generation, 146
Maxam-Gilbert method, 313
Messenger RNA (mRNA), 82–93
- analytical methodologies, 86–87
  - capillary electrophoresis–laser-induced fluorescence, 87
  - complementary DNA, 86
  - primers, 86
  - quantitative (real-time) PCR, 87
  - reverse transcriptase–polymerase chain reaction, 86
- biology, 82–83
- profiling, 83–93
- terminal differentiation of cells, 82
- tissue mRNA markers, 83–86
  - blood-specific markers, 83
  - corneodesmosin, 86
  - cytochrome P450, 85
  - epidermal cornified cell envelope, 86
  - estrogen receptor 1, 83
  - fucosyltransferase 6, 85
  - housekeeping genes, 86
  - human beta defensin 1, 83
  - hydroxymethylbilane synthase, 83
  - keratin, 83
  - loricrin, 86
  - myeloid cell nuclear differentiation antigen, 83
  - saliva-specific markers, 83
  - semen-specific markers, 83
  - transcription elongation factor, 86
  - ubiquitin conjugating enzyme, 86
  - vaginal-specific markers, 83
  - whole transciptome profiling, 85
- validation for casework, 87–93
  - Baecchi staining, 88
  - body-fluid–specific markers, 89, 90
  - hemoglobin beta, 92
  - housekeeping genes, 88
  - multiplex development, 88
  - stability in environmentally compromised samples, 87

Methicillin-resistant *Staphylococcus aureus* (MRSA), 320
MGB; *See* Minor groove binder
Microcon filters, 48
MicroRNA (miRNA), 93–96
- biology, 93
  - amplicon size, 93
  - body-fluid–specific biomarkers, 93
  - miRNA-induced silencing complex, 93
  - premature ribosome dissociation, 93
  - regulation of developmental processes, 93
- profiling, 94–96
  - body fluid identification, 96
  - normalized expression data, 94
  - studies, 94

Microsatellite loci, hidden variation in, 199–219
- allozyme electrophoresis, 215
- capillary electrophoresis, 200
- degenerate primer cocktails, 200
- electrospray ionization, 201
- molecular biology of interrupted repeats, 201–203
  - exons, 201
  - identical by descent, 203
  - polymorphism, 201
  - population geneticists, 202
  - stutter alleles, 203
  - template, 202
- mutation modeling for STRs, 203–207
  - identity testing, 203
  - interrupted motif blocks, 205
  - microsatellite alleles, 207
  - microsatellite markers, 203
  - motif categories, 206
  - parentage evaluation, 203
- population observations, 207–212
  - allele family variations, 210
  - allelic diversity, 209
  - deep-sequencing applications, 207
  - electrospray ionization mass spectrometry, 207
  - mutation rates, 210
  - population substructure, 209
  - reference sequence, 210
- relationship testing, 199
- repeat motif pattern, disrupted, 200

STR loci, 199
value to forensic investigations, 212–215
allele sharing, 213
hidden polymorphisms, 213
low-complexity regions, sequencing of, 212
relationship testing, 213
sequencing errors, 212
stutter product percentages, 212
MID sequences; *See* Multiplex identifier sequences
MinilON (Oxford Nanopore Technologies), 329
MiniSTR kits, 170–171
Minor groove binder (MGB), 111
miRNA; *See* MicroRNA
miRNA-induced silencing complex (miRISC), 93
Missing persons
databases, 22, 253, 262
deep sequencing, 332
false exclusions, 226
genotyping method, 103
identification of, 133, 137
implementation of SNP assays, 213
kinship associations, 194
linear array method, 265
mtDNA, 333
next-generation international STR kits, 170
profiling of human remains in resolving, 66
samples from, 41
search for, 381–382
testing, 40
use of SNPs to identify, 282
Y-chromosome, 222
Mitochondrial DNA (mtDNA), 312
Mitochondrial DNA (mtDNA) analysis, 249–278
alternative methods and future potential, 263–272
coding region SNPs, 266
codon wobble positions, 266
common profiles, 265
expanded sequence analysis, 265–266
hair shafts, 264
heteroplasmy, 271
kit-based system, 266
linear array method, 265
low-level heteroplasmy, 271
luciferase, 269
mass spectrometry, 268–269
mixtures and heteroplasmy investigation (DGGE and dHPLC), 267–268
multiplex identifier sequences, 271
mutational hotspots, 271
population study, 265
potential obstacles, 263
private polymorphisms, 263
pyrosequencing and deep sequencing, 269–272
screening methods, 263–265
sequencing error rates, 272
sulfurylase, 269
current practice, 251–263
admissibility hearings, 262
ancient DNA, 258
cases, 260
courtroom experiences, 262
degraded sample, 258
demand, customers, and sample type, 251–252
denaturing HPLC, 257
evolution of protocols, 252–254
frequent crime scene sample, 254
heteroplasmy, 255
hypoxanthine, 257
interpretation guidelines, 255
mitochondrial DNA profile, 253
mixed nucleotide bases, 256
mixture interpretation, 255–261
"naturally occurring" mixture, 255
overlapping amplification products, 253
PCR amplification, 252
pet hairs, 254
phylogenetic knowledge, 257
regulation and accreditation, 262–263
Sanger sequencing, 256
skeletal samples, 258, 259
SNP assays, 254
statistical approach, 257
statistics and databases, 261–262
yield gel, 252
golden age of mtDNA testing, 250
hypervariable regions, 250
STR analysis, 250
Mitotyping Technologies, 250
MNDA; *See* Myeloid cell nuclear differentiation antigen
Molecular beacons, 111
mRNA; *See* Messenger RNA
MRSA; *See* Methicillin-resistant *Staphylococcus aureus*
mtDNA; *See* Mitochondrial DNA
Multiplex identifier (MID) sequences, 271, 334
Myeloid cell nuclear differentiation antigen (MNDA), 83

N

NADH dehydrogenase subunit 1, 106
National Center for Forensic Science (NCFS), 175
National Institute of Standards and Technology (NIST), 24, 183, 261
Next-generation sequencing (NGS), 314
Next-generation sequencing technologies, 329–332
array chip, 330
cyclodextrin, 330
exonuclease sequencing, 330
lipid bilayer membrane, 329

nanopore sequencing technology, 331
node, 330
patented sequencing methods, 330
protein pores, 329
Next-generation STR genotyping kits, 163–180
Combined DNA Index System, 164
competition among kit manufacturers, 177
contents of commercial STR kit, 164–165
allelic ladder, 165
amplification step, 164–165
primer mix, 164
validated kit, 165
genotypic concordance, 176
kits that amplify SE33 marker, 171–174
German market, 171
microvariants, 171
modified PCR cycling conditions, 172
tetranucleotide STR, 171
miniSTR kits, 170–171
next-generation international STR kits, 170
nonstandard STR markers in forensics, 174
PCR cycling times, 163
second-generation multiplex, 164
sex-chromosome STR kits, 174–176
database, 175
Investigator Argus X-12 Kit, 176
marker linkage groups, 176
"minimal haplotype" loci, 174
rapid mutating Y-STRs, 176
X-chromosomal STR markers, 176
STR kits in Europe, 169–170
data-sharing initiatives, 169
ENFSI loci, 169
range of sample types, 170
STR kits in the United States, 165–169
direct amplification strategies, 168
enhanced buffer system, 165
FTA cards, 166
PCR buffer formulations, 168
PCR inhibitors, 165
NGS; *See* Next-generation sequencing
NIST; *See* National Institute of Standards and Technology
Normalized inhibitor factor (NIF), 17
No template controls (NTC), 116
Nucleic acids, elution of, 47
NucleoSpin XS column, 57
NUMTs, 335

O

Oligo ligation assay (SNP typing), 137
OliGreen® dyes, 104
On-ladder alleles, 188
Optical density (OD), 104
Orchid Cellmark, 250
Organic extraction, 46, 70, 354
Oxford Nanopore Technologies, 329, 330, 331

P

Pacific Biosciences™, 323
Paraffin-embedded tissues, 45
Paternity applications
CE systems, 147
deep-sequencing technologies, 312
DNA extraction, 41
genotyping, 103
next-generation Y-STR kits, 241
nonstandard STR markers, 174
SNP typing, 137
STR kits, 170
STR profiling, 133
worldwide laboratories, 381
Y-STRs, 222, 223, 232
PBS; *See* Phosphate-buffered saline
PCA; *See* Principal Component Analysis
PCR; *See* Polymerase chain reaction
Performance Optimized Polymer, 358
PerkinElmer, 51
Personal protective equipment (PPE), 5–6
Phosphate-buffered saline (PBS), 45, 55
Photo-multiplier tubes (PMT), 362
Phylogenetic knowledge, 257
PIC; *See* Polymorphism information content
PicoGreen® dyes, 104
Pleistocene bones, 69, 70
Plexor® chemistry, 112
Plus-Minus Sequencing, 312
PMT; *See* Photo-multiplier tubes
Polymerase chain reaction (PCR), 350; *See also* Real-time PCR
amplification, 68
mtDNA analysis, 252
Y-STRs, 222
autosomal STR loci, 182
bias, 322
cycling times, 163
deep-sequencing technologies, 312
end-point, 105–107
duplex assay, 106
fluorescent end-point assay, 106
NADH dehydrogenase subunit 1, 106
qualitative method for identification, 106
quantitative template amplification technology, 106
enhancement reagents, 21
inhibitor, 49, 67, 103
quantitative real-time, 107
reverse transcriptase–, 86
Polymorphism information content (PIC), 191
PowerPlex 16 System, 165
PowerPlex 18D (Promega), 42
PowerPlex ESX-16 (Promega), 169

PowerPlex Fusion System (Promega), 170
PPE; *See* Personal protective equipment
PrepFiler Automated Forensic DNA Extraction Kit, 52
PrepFiler Filter Plate, 50
PrepFiler Forensic DNA Extraction kit (Applied Biosystems), 47, 71
PrepFiler LySep™ columns, 50
PrepFiler system, 51
Prep-N-Go™, 49
Principal Component Analysis (PCA), 292
Promega, 42, 183
Proteinase K, 46

Q

QAS; *See* Quality assurance standards
Qdot-based hybridization assay; *See* Quantum dot-based hybridization assay
QIAamp Blood Kit (Qiagen), 48
QIAamp DNA Micro Kit (Qiagen), 71
QIAcube® robotic platform (Qiagen), 69
Qiagen, 51
QiaSafe, 25
qPCR; *See* Quantitative (real-time) PCR
Q-TAT; *See* Quantitative template amplification technology
Quadruplex real-time PCR, 116
Quality assurance standards, 363, 368, 384–386
Quantification standard, 107
Quantifiler Duo DNA Quantification kit, 117, 118
Quantifiler® Human DNA Quantification Kit (Applied Biosystems), 26
Quantitative (real-time) PCR (qPCR), 87, 107
Quantitative template amplification technology (Q-TAT), 106
Quantum dot (Qdot)-based hybridization assay, 105

R

Radio frequency identification (RFID) technology, 152
RAMP; *See* Random match probabilities
Random match probabilities (RMP), 283
Rapidly Mutating Y-STRs, 176, 227
READase™ polymerase, 105
Ready Amp™ Genomic DNA Purification System, 48
Real-time PCR, 107–123
  advantages, 107
  assays, 112–116
    detection of PCR inhibitors, 113, 115
    downstream genotyping systems, 113
    duplex real-time PCR, 113
    homologous region of human sex chromosomes, 115
    human DNA quantification, 112, 113
    human male DNA quantification, 113
    human telomerase reverse transcriptase target gene, 113
    internal PCR control, 113
    mtDNA quantification, 113
    NADH dehydrogenase subunit 1 target gene, 113
    nuclear and mtDNA quantification, 114
    quadruplex real-time PCR, 116
    retinoblastoma susceptibility gene, 114
    singleplex real-time PCR assays, 112
    triplex real-time PCR, 114
  chemistries, 111–112
    intercalating dyes, 111
    minor groove binder, 111
    molecular beacons, 111
    Plexor® chemistry, 112
    quantifier assay, 111
    Scorpions chemistry, 112
    TaqMan chemistry, 111
  critical design factors, 107
  fluorescence signal, 107
  fluorescent dyes, 107
  important concepts in assay design, 116–123
    correlation between assays, 122
    detection of PCR inhibitors, 117
    DNA degradation, 122
    limit of quantification and detection, 116
    mechanisms of inhibition, 118
    mixtures of human male and female DNA, 119
    normalized inhibitor factor, 117
    no template controls, 116
    physicochemical properties of the plastics, 117
    size of amplicons, 116
    slot-blot, 122
    species specificity, 116
  quantification standard, 107
  standard DNA, 107
Relative fluorescence unit (RFU), 145
Restriction fragment length polymorphism (RFLP), 41, 102
  DNA storage, 32
  sensitivity range, 101
  STR genotyping systems, 350
  transition from, 163, 164, 215
Retinoblastoma susceptibility gene, 114
Reverse transcriptase–polymerase chain reaction (RT-PCR), 86
RFID technology; *See* Radio frequency identification technology
RFLP; *See* Restriction fragment length polymorphism
RFU; *See* Relative fluorescence unit
RNA profiling, identification of the tissue origin of dried stains using, 81–99

conventional methods for body fluid stain analysis, 82
infrequent use of body fluid identification methods, 82
messenger RNA, 82–93
analytical methodologies, 86–87
Baecchi staining, 88
biology, 82–83
blood-specific markers, 83
body-fluid–specific markers, 89, 90
capillary electrophoresis–laser-induced fluorescence, 87
complementary DNA, 86
corneodesmosin, 86
cytochrome P450, 85
epidermal cornified cell envelope, 86
estrogen receptor 1, 83
fucosyltransferase 6, 85
hemoglobin beta, 92
housekeeping genes, 86, 88
human beta defensin 1, 83
hydroxymethylbilane synthase, 83
keratin, 83
loricrin, 86
messenger RNA profiling, 83–93
multiplex development, 88
myeloid cell nuclear differentiation antigen, 83
primers, 86
quantitative (real-time) PCR, 87
reverse transcriptase–polymerase chain reaction, 86
saliva-specific markers, 83
semen-specific markers, 83
stability in environmentally compromised samples, 87
terminal differentiation of cells, 82
tissue mRNA markers, 83–86
transcription elongation factor, 86
ubiquitin conjugating enzyme, 86
vaginal-specific markers, 83
validation for casework, 87–93
whole transciptome profiling, 85
microRNA, 93–96
amplicon size, 93
biology, 93
body fluid identification, 96
body-fluid–specific biomarkers, 93
miRNA-induced silencing complex, 93
normalized expression data, 94
premature ribosome dissociation, 93
profiling, 94–96
regulation of developmental processes, 93
studies, 94
molecular-genetics-based approach, 82
Roche Applied Science LINEAR ARRAY Mitochondrial DNA HVI/HVII Region-Sequence Typing Kit, 264
Romanov family (Tsar Nicholas II), 66
RT-PCR; *See* Reverse transcriptase–polymerase chain reaction

S

Sample-in–answer-out technology, 352, 365
SampleMatrix, 25, 29
Sanger sequencing, 256, 313
SCODA; *See* Synchronous coefficient of drag alteration
Scorpions chemistry, 112
SDS; *See* Sodium dodecyl sulfate
Second-generation multiplex (SGM), 164
Sex-chromosome STR kits, 174–176
database, 175
Investigator Argus X-12 Kit, 176
marker linkage groups, 176
"minimal haplotype" loci, 174
rapid mutating Y-STRs, 176
X-chromosomal STR markers, 176
SGM; *See* Second-generation multiplex
Short tandem repeat (STR), 131, 132, 199; *See also* Additional Y-STRs; STR genotyping kits, next-generation; STR genotyping systems, sample-to-result
Simple sequence repeats (SSRs), 182
Single Molecule, Real Time (SMRT®) Technology (Pacific Biosciences™), 323
Single nucleotide polymorphism (SNP), 102, 137
Singleplex real-time PCR assays, 112
Single-stranded DNA (ssDNA), 145
Sizing precision, 135
Slicprep™ 96 Device (Promega), 50
SMRT Cell, 323
Snipper open-access Bayesian classification portal, 302
SNP; *See* Single nucleotide polymorphism
Sodium dodecyl sulfate (SDS), 44, 68
Spectral calibration, 146
ssDNA; *See* Single-stranded DNA
SSRs; *See* Simple sequence repeats
Standard DNA, 107
Storage and handling of DNA extracts, 19–37
DNA storage and handling strategies, 23–33
additional DNA storage modalities, 32–33
anhydrobiosis, 25
cold storage, 24–25
DNA stability, 31
dry matrix storage, 24
dry storage comparisons, 25–32
FTA technology, 25
QiaSafe, 25

restriction fragment length polymorphism, 32
SampleMatrix, 25, 29
shipping stress, 32
shipping study, 29
trehalose, 25
tube characteristics, 23–24
yield of recovery, 31
factors influencing DNA stability, 20–21
forensic DNA storage issues, 20
extrinsic differences, 21
low copy number, 20
touch samples, 20
importance of sample storage, 22–23
biodiversity DNA databanks, 22
dry-storage DNA damage, 23
forensic DNA databanks and casework samples, 22
mechanisms of DNA loss, 22–23
missing persons databases, 22
nonforensic DNA databanks and biobanks, 22
typing strategies of nonoptimal samples, 21
CODIS core loci, 21
PCR enhancement reagents, 21
STR; *See* Short tandem repeat
STR genotyping kits, next-generation, 163–180
Combined DNA Index System, 164
competition among kit manufacturers, 177
contents of commercial STR kit, 164–165
allelic ladder, 165
amplification step, 164–165
primer mix, 164
validated kit, 165
genotypic concordance, 176
kits that amplify SE33 marker, 171–174
German market, 171
microvariants, 171
modified PCR cycling conditions, 172
tetranucleotide STR, 171
miniSTR kits, 170–171
next-generation international STR kits, 170
nonstandard STR markers in forensics, 174
PCR cycling times, 163
second-generation multiplex, 164
sex-chromosome STR kits, 174–176
database, 175
Investigator Argus X-12 Kit, 176
marker linkage groups, 176
"minimal haplotype" loci, 174
rapid mutating Y-STRs, 176
X-chromosomal STR markers, 176
STR kits in Europe, 169–170
data-sharing initiatives, 169
ENFSI loci, 169
range of sample types, 170
STR kits in the United States, 165–169
direct amplification strategies, 168
enhanced buffer system, 165
FTA cards, 166
PCR buffer formulations, 168
PCR inhibitors, 165
STR genotyping systems, sample-to-result, 349–377
biological limits, 353
future implications, 370
human identification analysis, 350
implementation of integrated rapid DNA analysis platforms, 368–370
quality assurance standards, 368
sample-in–answer-out capabilities, 368
SWGDAM committee for R-DNA, 370
integrated rapid DNA analysis platforms, 358–368
acousto-optic tunable filters, 362
BaseFinder, 363
charge-couple device camera, 362
comparison of next-generation prototypes, 363–368
elastomeric membranes, 361
flow control, 360–361
genetic profile, 363
hydrofluoric acid,. 359
laser-induced fluorescence, 362
low-temperature co-fired ceramic, 359
multiplexed STR detection and allele calling, 362–363
photo-multiplier tubes, 362
quality assurance standards guidelines, 363
sample-in–answer-out technology, 365
separation speed, 361–362
substrate choice and device size, 358–360
surface chemistry, 360
swab-in–profile-out, 365
microfluidic platforms, 349
micro total analysis system 351
minisatellite regions, 350
paradigm-shifting technologies, 350
polymerase chain reaction, 350
restriction enzymes, 350
restriction fragment length polymorphism, 350
sample-in–answer-out instrument, 352
sample preparation and DNA separation, 353–358
differential extraction, 353
dithiothreitol, 353–354
DNA extraction and PCR amplification, 354–357
linear polyacrylamide, 358
microchip electrophoresis of STR fragments, 357–358
nucleic acids, 353
organic extraction, 354
Performance Optimized Polymer, 358

sample preparation, 353–354
thermal mass, 352
variable number of tandem repeats, 350
STR typing (capillary electrophoresis), 133–136
allelic ladder, 135
calibration curves, 134
electrophoresis, 135
fluorescence, threshold for detection of, 135
human identification applications, 133
size standards, 134
sizing precision, 135, 136
variable-bin method, 135
Stutter alleles, 203
Sulfurylase, 269
Swab-in–profile-out, 365
Synchronous coefficient of drag alteration (SCODA), 47

T

TaqMan chemistry, 111
Tecan Freedom EVO® 150 robot, 52
Time since intercourse (TSI), 56
Tissue origin of dried stains; *See* RNA profiling, identification of the tissue origin of dried stains using
Touch DNA, 6
Touch samples, 20
Training of forensic DNA scientists, 381–389
backlog (challenge to forensic DNA laboratory), 382–383
developing countries, 382
INTERPOL, 382
single-source and casework samples, 383–384
challenges of sample processing, 383
laboratory information management systems, 383
multiplex STR kits, 384
PCR inhibitors, 383
training as a tool to overcome laboratory challenges, 384–388
Center for Forensic Excellence, 387–388
CODIS administrators, 385
DNA databasing, 387
examples of training entities, 386–387
internal training, 386
INTERPOL guidelines, 385
quality assurance standards, 384–386
training sources, 386
worldwide expansion of DNA analysis, 381–382
Transcription elongation factor, 86
Transfer DNA, 6
Trehalose, 25
Triplex real-time PCR, 114
TSI; *See* Time since intercourse

U

Ubiquitin conjugating enzyme (UCE), 86
UHD set of Y-STRs; *See* Ultra-high discrimination set of Y-STRs
UltraClean® Forensic DNA Isolation Kit (MoBio Laboratories), 46
Ultra-high discrimination (UHD) set of Y-STRs, 230
U.S. Scientific Working Group on DNA Analysis Methods, 174
UV absorbance, 104

V

Variable number of tandem repeats (VNTR), 132, 350
Vehicle-borne improvised explosive device (VBIED), 7
Virtual filters, 146
von Willebrand Factor gene, 185

W

Whole genome scan (WGS), 298
Whole transciptome profiling, 85
Wizard® Genomic DNA Purification Kit (Promega), 48
World Trade Center disaster, 66

X

X-chromosomal STR markers, 176

Y

Y-chromosome Haplotype Reference Database (YHRD), 224
Y-chromosome short tandem repeat (Y-STR), 41, 221; *See also* Additional Y-STRs
YOPRO-1 intercalating dye, 139
Young's modulus, 361

Z

Zero-mode waveguides (ZMW), 323

£ 63.99

21